BIOLOGY OF SKINKS OF THE GENUS *LAMPROPHOLIS*

Shelley Burgin and Harold Heatwole

CRC Press
Taylor & Francis Group
Boca Raton London New York

CRC Press is an imprint of the
Taylor & Francis Group, an **informa** business

Designed Cover Image: (top) *L. couperi* from Mt. Nebo, Qld.; (bottom, left to right) *L. couperi* from Mt. Nebo, Qld.; *L. elongata* from Ngulin Nature Reserve, Nowendoc, NSW; *L. adonis* from Bulburin National Park, Qld. (photos by Stephen Mahony); (bottom, background) Mt. Lewis habitat (photo by Stephen Zozaya); Back cover: (left to right) *L. robertsi* from Mt. Lewis, Qld.; *L. caligula* from Barrington Tops, NSW; *L. mirabilis* from Magnetic Island, northern Qld. (photos by Stephen Mahony)

This edition first published 2027
by CRC Press
2385 NW Executive Center Drive, Suite 320, Boca Raton FL 33431

and by CRC Press
4 Park Square, Milton Park, Abingdon, Oxon, OX14 4RN

CRC Press is an imprint of Informa UK Limited

First published by CSIRO Publishing. This edition published by arrangement with the Commonwealth Scientific and Industrial Research Organisation (CSIRO), Australia.

ISBN: 978-1-041-01445-4 (hbk)
ISBN: 978-1-041-01446-1 (pbk)

Typeset in Leo
by Envisage Information Technology

Contents

Dedicated to our students whose theses (BAppSc, Hons, MSc, and PhD) or later published works are cited herein. In alphabetical order they are Lynette Anderson, Caroline Belmont, Bronwyn Crome, Bruce Firth, Daniel Giffney, Anne Jardine, Jane Jelbart, Purani Keyur, Elaina McCann, Richard Shine, Fiona Warner, and Danny Wotherspoon, and especially to the late Malumo Philip Simbotwe and the late Kenneth Zimmerman, whose lives and promising careers were prematurely cut short by the ravage of disease. We are reminded of how fortunate we are to have been privileged to pursue our unabated fascination with herpetology over an extended and rewarding lifetime.

Preface

Our journey began when the present junior author (HH) served as an external examiner of the senior author's (SB) PhD on the genus *Lampropholis* in 1989. Over the years, we both have supervised various theses, dedicated in part or in whole to the biology of this taxon, and have even dabbled in the topic ourselves. Much of the available information on this genus previously was buried in unpublished theses. Being unpublished does not mean that the research was of poor quality but often that the authors, upon completion of their degrees, moved on to other topics or fields, such as medicine or teaching, rather than continuing in research. With the permission of these authors, we have mined their data to make it more accessible to the scientific public. Indeed, the motivation for producing this book arose from the realisation that the effort and public funds invested in the research on which these theses were based would be wasted if not reviewed and published in a more accessible form. Because they are not readily available otherwise, we have presented the data in these theses as tables and figures in a more detailed way than we have for work already published in journals to which readers have easier access. We have tried to cite all the relevant original literature available on this genus from the inception of the first descriptions of the species. We have not cited secondary sources of literature, such as field guides to Australian species, unless they contained some original material; rather, we sought the original sources on which they relied and have cited those.

Our goal was to assemble the diffuse body of literature on what is known of the biology of *Lampropholis* and, more importantly, on what is *not* known about it. We hope our book will serve as a guide for fruitful lines of future research and stimulate the next generation of herpetologists to provide a more complete understanding of this fascinating taxon and assist conservationists in dealing with their charges more effectively. By the time we had assembled the scattered information and begun to write, the number of publications on this genus began to increase and the number of investigators involved in ongoing projects also escalated and expanded into topics not previously covered. In corresponding with some of these researchers, we realised that many of them had long-term projects planned and that the number of papers would soon be appearing at an exponential rate. This trend was gratifying as it fulfilled our hope that *Lampropholis* would become a model taxon. However, we realised that we needed to set a cut-off date because otherwise we would simply be expanding the book indefinitely. We set that time to be 28 February 2024. One paper published after that date, however, has been cited. We felt that the discovery of a new species, *Lampropholis isla*, and its formal description and naming by Hoskin in 2025 warranted making an exception and, where warranted, we have discussed that important paper in various chapters.

The wide audience intended for this book ranges from professional herpetologists to the general public unfamiliar with many of the technical terms peculiar to the field and we found it difficult to gauge the best level at which to describe the life of these reptiles. On the one hand, substituting technical terms by lengthy interpretations of their meanings whenever the concept they represented arose would be redundant and tedious for biologists. On the other hand, not doing so would render many of the important statements unintelligible and boring for some of the naturalists and most of the general public. Consequently, some readers may merely wish to skip over the more technical sections and concentrate on the parts of greater interest to them. When possible, however, if merely a few words would suffice to define a technical term, we have done so.

The scientific names of these species (see Cogger 2018 and The Reptile Database [nd]) may not be familiar to all readers but are used in this book. With a little attention on the part of the reader, however, it may be more comprehensible in the long-term than the myriad of other names assigned to various species. It is essential for persons working for national parks or in governmental offices dealing with protecting wildlife, or who otherwise need a fundamental understanding of this genus, to be familiar not only with basic scientific terminology but also with the scientific names of the species with which they deal; otherwise they would be unable to effectively avail themselves of the essential information required for competent execution of their duties.

The common names of the species of *Lampropholis* have had a complicated history and we pondered at length as to which ones we should use in this book.

Common names are those that arise from the general public as a means of identifying and communicating about the local species with which they are familiar. Many widespread species, however, transcend local regions and extend beyond linguistic borders, with the result that a given species may have multiple common names, even in the same language, or the differences between closely related species may not be recognised and several species may share a common name. For example, initially, the two most common and most widespread species (*Lampropholis delicata* and *Lampropholis guichenoti*) were so similar in appearance that often they were not easily distinguished by the public, so collectively both were called 'penny lizards'.

The International Commission on Zoological Nomenclature (ICZN) regulates the naming of animals according to the International Code of Zoological Nomenclature (see Chapter 14) and the scientific name is regarded as the only official one. Thus, when one wishes to leave no doubt as to the species intended, that is the name that should be used.

That having been said, however, it should be noted that even scientific names do change sometimes. For example: (1) as evolutionary relationships become better understood, a species may be moved to a different genus so as to be grouped with those species to which it is genetically most closely related; (2) when a species is described and named unknowingly and independently by two different authors, the later name has to be discarded and priority given to the earlier one; or (3) when two apparently distinct 'species' prove merely to be variants of the same species. These situations cause confusion enough without the invention of contrived 'common' names adding further unnecessary complexity.

Note that the complete official designation for a species includes the name of the genus to which it belongs, followed by its individual name, then the name of the person(s) who originally described it, and, finally, the date of publication of the original description. If, for any of the reasons described above, the scientific name is altered, the name(s) of the describer(s) and the date of description are enclosed in parentheses; otherwise, they are not. The generic and specific names are always italicised or underlined unless embedded in italicised text.

The complete designations for the 14 species officially recognised as belonging to the genus *Lampropholis* are listed in Chapter 1. In subsequent chapters we dispense with the name(s) of the describer(s) and the date of description except when that information is specifically required.

We struggled for a long time deciding on a title for this book. The potential readership hopefully spans the spectrum of hobbyists; naturalists; conservationists; the public at large; governmental officials, such as park rangers, legislators who formulate laws protective of reptiles, and officers charged with enforcing those laws; and professional biologists, especially herpetologists and those otherwise studying in the field. Should we use the official name required were the species to write a will before becoming extinct, or to apply for a passport to disperse to a new geographic region? In such a case the book should be dubbed 'The Biology of *Lampropholis*' and the name of one of the species applying for a passport should be '*Lampropholis delicata*'. We felt that this title might eliminate a large segment of the readership who were not acquainted with scientific nomenclature and who would not realise that the book treated the personal lives of the tiny lizards adorning the gardens around their home. Using that title would be much like writing a biography of 'Marion Robert Morrison' and expect the public to even pick up the book to see who the subject really was. If, instead, the person perusing the shelves of the bookstore saw the title *The Life and Times of John Wayne* (the screen name of Marion Robert Morrison), it might spark more interest. Many people recognise the screen names (for movie stars), *noms de plume* (for writers), and pseudonyms or aliases (in general) rather than the official (real) names of celebrities. Such aliases are contrived artificial names mandated by an organisation, or by the persons themselves. Their real names are the official ones listed on their birth certificate and passport. In addition, a person may have a nickname that is applied by family and/or friends because of some personal attribute or to shorten the real name. Examples are 'Bob' for Robert or, strangely, 'Bluey' for someone with red hair. To relate these categories to the names of lizards, the scientific name is the official, real name of a species; the contrived so-called 'common' name mandated by an agency is equivalent to an alias; and the name used by family, friends, and the public, who recognise that it is neither the official name nor an alias, but rather merely a nickname they choose to use.

Because contrived names are mandated for use by some governmental institutions and other organisations, we have acceded to this varied nomenclature by providing the readership with all available names, common, contrived, and scientific (see Appendix 1). The name 'sunskinks' may, in time, become a 'common' name because it was mandated to accompany the scientific names in *Reptiles and Amphibians of Australia*, updated 7th edition (Cogger 2018). That monumental work serves as the authority for identifying Australian lizards; hence 'sunskinks' may, in time, replace the truly common names in general use and become generally adopted by people in all walks of life and in all parts of the country (but likely not overseas). That time has not yet come, however, and by the time it does, our present book probably will be out of print and in need of revision.

The names 'penny lizard' and 'garden skink' are truly common names, recognisable over a wide area in Australia and in popular books. 'Grass skink' and 'litter skink' were also apparently spontaneously generated by the public but less commonly used, and perhaps only in fewer localities. The only common names used for *Lampropholis* outside of Australia are for *L. delicata* in places where it has been introduced: 'plague skink' in New Zealand and 'rainbow skink' in Hawaii (see Chapter 3). '*Lampropholis*' is featured in the main title because it is the official name and because all other names would be meaningless to international readers. The explanatory material in Appendix 1 is designed to allow the wider readership targeted by this book to relate whichever names are familiar to them to the official scientific nomenclature that otherwise is used throughout the book. When we use the name '*Lampropholis*' alone, it refers collectively to all species in the genus.

The scientists associated with the Australian Society of Herpetologists (ASH) Taxonomic Committee maintain an 'official' list of 'common and scientific names' for Australia that is updated regularly. Some of the contrived names are more complex than the scientific ones. For example, it would seem easier to remember the scientific name (*guichenoti*) rather than the contrived 'pale-flecked garden sunskink', without confusing it with the 'dark-flecked garden sunskink' (*delicata*). These two contrived names are grammatically flawed as they suggest that *L. guichenoti* lives in pale-flecked gardens and *L. delicata* in dark-flecked ones, not the meanings these names were designed to convey.

If there must be a set of contrived names, linking them to the scientific name would allow anyone delving into either the popular or scientific literature to readily recognise which species was intended. Occasionally this has occurred; for example, one of the least-used contrived names for *L. couperi* is 'Couper's sunskink' and one for *L. guichenoti* is 'Guichenote's skink' (see Appendix 1). Others that could be treated similarly are 'Cogger's skink'[1] for *L. coggeri* and 'Roberts' skink' for *L. robertsi*.

Even those not so obviously linked as the above could have more useful derivative names; is not 'Mirabilis skink' (for *L. mirabilis*) as easy to remember as 'saxacoline sunskink'? We were tempted to present a new set of simpler contrived names to replace those currently listed, but that would merely have added yet another layer to the confusion; after all, some species already have a number of contrived and common names (see Appendix 1).

The suffix '*-ensis*' in Latin signifies 'from the locality indicated by the prefix'. A more sensible contrived name for *L. bellendenkerensis* is thus Bellenden Ker skink rather than 'southern montane sunskink', an alternative one also proposed (see Appendix 1). Although cumbersome, another such usage has been designated: 'Mount Elliot rainforest sunskink' for *L. elliotensis* (but why not just 'Mount Elliot skink'?).

We conducted a brief survey of people in regions where *Lampropholis* occurs and asked them what the tiny dark-coloured lizards in their gardens, around their houses, or seen during walks in the bush are called. 'Penny lizard' was the most frequent name given (see Appendix 2) and was the only one given by lay persons other than the indefinite responses of 'small lizard' and 'I don't know'. The second-most frequent name was 'garden skink', again a truly common name, used well before the invention of contrived names (see Cogger 1989). In the survey, this name was offered only by professional herpetologists. Perhaps its persistence in that group was because it was a common name used both for *L. guichenoti* and *L. delicata* (Appendix 1). Only one respondent (a professional herpetologist) listed a contrived name that relates directly to the scientific name, and that was 'delicate skink' (referring to *L. delicata*).

No exhaustive search was made for the use of popular names in papers published by professional herpetologists in scientific journals or books, but a quick perusal of some papers, titles, and abstracts showed that when contrived names accompanied the scientific ones, a wide

1 Note, however, that the similar contrived name 'Cogger's island skink' already is used for *Geomyersia coggeri* from the Admiralty Islands (Greer 1982).

range similar to that shown in Appendix 1 was represented. As well, additional truly common names, not represented in the survey, were seen in the older literature that preceded the invention of contrived names. Two such were 'ground skink' and 'litter skink' used for *L. delicata* in Tasmania, south-eastern mainland Australia, and the Bass Strait islands (Rawlinson 1974).

Our small sample of the nomenclature is not adequate to give precise estimates of the proportions of people using the various names, or how that may vary geographically. However, it does illustrate the confusion that adding contrived common names to the list has caused. It also shows that the public at large has not universally adopted the contrived names for *Lampropholis,* and that professional herpetologists' usage is inconsistent.

Discussions with many people have contributed over the years to our ideas about lizard ecology, some of which also are mentioned in the Dedication. Notable among them, in alphabetical order, are Ingrid Belan, Caroline Belmont, David Briscoe, Elizabeth Cameron, Harold Cogger, Bronwyn Crome, Christopher Daniels, the late John de Bavay, Barbara Done, Steve Donnellan, Bruce Firth, Sharon Fraser, Lori Freer, David Horton, the late Clifford Johnson, Menna Jones, Jean Joss, Susan King, the late John Legler, Harvey Lillywhite, Dan Lunney, Michael Mahony, Jeffrey Miller, Harvey Miller, Joaquin Molinari, John Parmenter, the late John Pernetta, Adrian Renshaw, Barry Richardson, Pauline Ross, the late Owen Sexton, the late Malumo Simbotwe, Rick Shine, Bryan Stuart, Vilma Suarez, Janet Taylor, Frank Torres, John ('Charlie') Veron, Grahame Webb, Sue Wilson, and Geoffrey Witten. Sue King improved the book in many ways by contributing her expertise in the ever-advancing technology of computing and in the preparation of visual aids. Glen Shea assisted by providing literature we had been unable to access otherwise. Harold Cogger generously updated his key to include the species of *Lampropholis* described since publication of the latest edition of his manual (see Box 14.1). He and Caroline Belmont, David Chapple, Ondi Crino, Bruce Firth, Sue King, Ko-Huan Lee, and Rick Shine each reviewed some sections of our early manuscript and provided valuable criticism and advice; any residual errors, misinterpretations, or faults, however, are ours alone.

Shelley Burgin,
Varsity Lakes, Queensland
27 May 2024

Harold Heatwole,
Guyra, NSW
27 May 2024

REFERENCES

Cogger HG (1989) *Australian Reptiles in Colour.* Treasure Press, Melbourne.

Cogger HG (2018) *Reptiles and Amphibians of Australia.* Updated 7th edn. CSIRO Publishing, Melbourne.

Greer AE (1982) A new species of *Geomyersia* (Scincidae) from the Admiralty Islands, with a summary of the genus. *Journal of Herpetology* **16**, 61–66. doi:10.2307/1563905

Hoskin CJ (2025) A new species of *Lampropholis* skink (Lacertilia: Scincidae) from Scawfell Island, mid-east Queensland, Australia. *Australian Journal of Taxonomy* **85**, 1–13. doi:10.54102/ajt.6uihp

Rawlinson PA (1974) Biogeography and ecology of the reptiles of Tasmania and the Bass Strait area. In *Biogeography and Ecology of Tasmania*. Monographiae Biologicae 25. (Ed. WD Williams) pp. 291–338. Dr. W. Junk, The Hague.

The Reptile Database (nd) <http://www.reptile-database.org> (accessed May 2024).

1 Introduction

Lampropholis delicata and *Lampropholis guichenoti* are arguably two of the commonest species of lizards in eastern Australia. Almost everyone has them in their gardens and is likely to encounter them when going for a walk in the bush. These lizards are small, readily available, have short generation times, are easy to maintain in captivity and do not have restrictive diets. Multiple studies of their attributes have been conducted at different sites, encompassing a range of habitats, vegetation, topography, climate, use of land, and history of natural and anthropogenic disturbances. Not surprisingly, the taxonomic composition of biotic communities varies from place to place over such a range of conditions, and species of *Lampropholis* not only face challenges from varying physical environments in different areas, but they also find themselves embedded within different assemblages of lizards with whom they interact in diverse ways (see Chapter 11). The entire gamut of potential competitors, predators and prey changes across the spectrum of environmental conditions. The fact that these two species are so eurytopic means that they are important functional components of different natural ecosystems, as well as of those modified by humans, and they can provide insights into the dynamics of biotic communities.

Conversely, the genus *Lampropholis* also contains some species that occupy fewer habitats and have limited distributions peripheral to, or embedded in, the ranges of the two widespread species. Such stenotopic species pose intriguing questions about evolution and local adaptation, provide fruitful material for the study of biochemical taxonomy, and satisfy the urge of biologists to seek novel species. This book examines vignettes of the biology of the species of *Lampropholis* in time and space as a way of gaining an entree into an understanding of these organisms, and we hope that it will stimulate comparative work.

The attributes described above are precisely those that make for a successful model genus, and *Lampropholis* has reached that status in recent years (see Chapter 15).

There are now 15 species within the genus *Lampropholis* that are recognised as valid by The Reptile Database (nd) including a new species formally described in 2025. In alphabetical order, they are: *Lampropholis adonis* Ingram 1991; *Lampropholis amicula* Ingram and Rawlinson 1981; *Lampropholis bellendenkerensis* Singhal, Hoskin, Couper, Potter, and Moritz 2018; *Lampropholis caligula* Ingram and Rawlinson 1981; *Lampropholis coggeri* Ingram 1991; *Lampropholis colossus* Ingram 1991; *Lampropholis couperi* Ingram 1990; *Lampropholis delicata* (De Vis 1888); *Lampropholis elliotensis* Singhal, Hoskin, Couper, Potter, and Moritz 2018; *Lampropholis elongata* Greer 1997; *Lampropholis guichenoti* (Duméril and Bibron 1839); *Lampropholis isla* Hoskin 2025; *Lampropholis mirabilis* Ingram and Rawlinson 1981; *Lampropholis robertsi* Ingram 1991; and *Lampropholis similis* Singhal, Hoskin, Couper, Potter, and Moritz 2018 (Plate 1).

Their hierarchical classification is Domain Eukaryota; Kingdom Animalia; Phylum Chordata; Class Reptilia; Superorder Lepidosauria; Order Squamata (lizards and snakes); Family Scincidae (skinks); Subfamily Eugongylinae[1]; genus *Lampropholis*.

REFERENCES

De Vis CW (1888) A contribution to the herpetology of Queensland. *Proceedings of the Linnean Society of New South Wales* **2**, 811–826. doi:10.5962/bhl.part.29213

Duméril AMC, Bibron G (1839) *Erpétologie Générale, ou Histoire Naturelle Complète des Reptiles*. Roret, Paris.

Greer AE (1997) A new species of *Lampropholis* (Squamata: Scincidae) with a restricted, high altitude distribution in eastern Australia. *Australian Zoologist* **30**(3), 360–368. doi:10.7882/AZ.1997.014

Hoskin CJ (2025) A new species of *Lampropholis* skink (Lacertilia: Scincidae) from Scawfell Island, mid-east Queensland, Australia. *Australian Journal of Taxonomy* **85**, 1–13. doi:10.54102/ajt.6uihp

Ingram G (1990) Five new skinks from Queensland rainforests. *Memoirs of The Queensland Museum* **30**(3), 443–453.

Ingram G, Rawlinson P (1981) Five new species of skinks (genus *Lampropholis*) from Queensland and New South Wales. *Memoirs of The Queensland Museum* **20**(2), 311–317.

Shea GM (2021) Nomenclature of supra-generic units within the family Scincidae (Squamata). *Zootaxa* **5067**(3), 301–351. doi:10.11646/zootaxa.5067.3.1

Singhal S, Hoskin CJ, Couper P, Potter S, Moritz C (2018) A framework for resolving cryptic species: a case study from the lizards of the Australian Wet Tropics. *Systematic Biology* **67**(6), 1061–1075. doi:10.1093/sysbio/syy026

The Reptile Database (nd) <http://www.reptile-database.org> (accessed May 2024).

1 Alternatively interpreted as either in the Tribe Eugongylini within the Subfamily Lygosominae *sensu lato* or with its status raised to that of a subfamily (Eugongylinae) closely allied with the Subfamily Lygosominae *sensu stricto* (see Chapter 14) or separated into its own tribe of Lamprophoiini (for a history of the nomenclature above the generic level, see Shea 2021).

2
Distribution

INTRODUCTION

The distribution of animals can be viewed from at least four different perspectives: geographic location, elevation above sea level, the kind of habitat (including type of vegetation) occupied, and the composition of assemblages of interacting species. All of these are considered in the present chapter, although habitat is treated in more detail and distribution in relation to the type of biotic assemblage is addressed mainly in Chapter 11.

The environment impinging upon *Lampropholis* has been changing over the past 10 million years (Chapple *et al.* 2011b). The rainforests that dominated the landscapes of eastern Australia gave way to drier woodlands and open forests in the latter part of the Miocene. Accompanying these drier conditions was a lowering of sea level that uncovered previously marine-inundated regions in south-eastern Australia. During the early Pliocene there was a brief return of warmer and wetter conditions, but by the end of that epoch, open woodlands, sclerophyllous vegetation, and grasslands again dominated eastern Australia. Some of these changes were cyclic, or at least occurred more than once. For example, in north Queensland, a thin corridor of rainforest, called the Black Mountain Corridor, at present connects the northern and southern blocks of rainforest, but during dry periods in the past this corridor was replaced by dry forests that separated the rainforest into two sections (see Chapple *et al.* 2011b). Similarly, in the Pleistocene contrasting cool-dry and warm-wet conditions intensified and there were repeated cycles of contractions of mesic habitats alternating with expansions of drier ones, thereby forming barriers that periodically isolated populations of *Lampropholis* and led to divergences between them. In addition, intermittent lowering of sea level allowed expansion of available terrestrial habitats, and hence expansion of the ranges of *Lampropholis* that subsequently were isolated during periods of high water. Populations also diverged by adapting to elevationally stratified habitats.

It is within the dynamic context of these changing conditions that the adaptive interaction between the genetic constitution of the species of *Lampropholis* and their environments must be understood. During these kaleidoscopic events, various barriers isolated populations leading to different degrees of genetic and morphological divergence via genetic drift and/or adaptation to different conditions. In some cases the level of divergence reached allopatric speciation, sometimes into cryptic species; in other situations it was deemed merely to have resulted only in inhibiting or preventing gene flow between localities to the extent that only intraspecific differentiation among populations occurred to form clades and subclades (see Chapters 11 and 14). Eventually, barriers of aridity limited westward expansion.

GEOGRAPHIC BARRIERS TO DISPERSAL AND GENE FLOW

Dry barriers inhibited dispersal and isolated populations; in some cases dry habitats surrounded elevated peaks containing remnant patches of moist forest (e.g. Kroombit Tops, Blackdown Tableland), leading to the genetic divergence of the disjunct populations isolated on mountain tops. In other cases, populations in mesic areas diverged when separated by drier barriers at equivalent elevations. The first such divergence isolated clade 1 from the rest of the species somewhere in the region of present-day Gladstone (southern Queensland) in the Late Miocene (Fig. 2.1), followed at various times in the Pliocene by divergence between subclades 1a and 1b across the Burdekin Gap; between subclades 1b and 1c and subclades 1a and 1b across the St Lawrence Gap; between subclade 9a and clades 4–5 across the relatively dry Hunter Valley; and between coastal clade 9 and inland subclade 7a by the McPherson/Main ranges (southern NSW barrier). The impact of the last-mentioned barrier is great because the divergence over the short distance across it is greater than between the population on the inland side of it and the distant populations in South Australia. There is an extensive geographical hiatus between subclades a and b of clade 7.

The East Gippsland barrier, although the site of important phylogeographic breaks in some other taxa (Chapple *et al.* 2011b), did not prove to be an obstruction to the dispersal of *L. delicata* from extreme south-eastern NSW into southern Victoria.

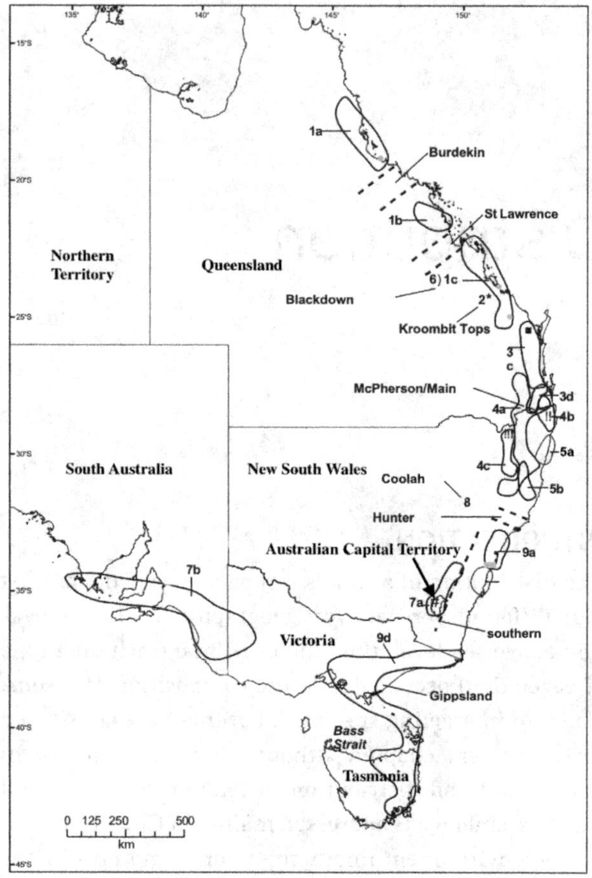

Fig. 2.1. Distribution of the clades and subclades (numbered polygons) of *Lampropholis delicata* in eastern Australia and the barriers (dashed lines) that led to their divergence. Note the land bridge that connected Tasmania and mainland Australia during low sea levels in the Pleistocene glacial periods. Modified from Chapple *et al.* (2011b).

GEOGRAPHIC RANGES

Maps of the geographic distribution of the species of *Lampropholis* within Australia are available from various sources, including Merrick *et al.* (2006, their Figure 1), Bell *et al.* (2010), Cogger (2018), Singhal *et al.* (2018), Chapple *et al.* (2019), and are shown in Figs 2.1 and 2.2, as well as in Plate 2. In brief, the native distribution of the genus is eastern and south-eastern mainland Australia (eastern Queensland south of the Cape York Peninsula, eastern NSW, Victoria, southern South Australia) and Tasmania.

The two widespread eurytopic species

Lampropholis delicata and *L. guichenoti* are widespread throughout much of eastern Australia (Cogger 2018; Chapple *et al.* 2019) and occupy a variety of relatively mesic environments and urbanised habitats with different types of vegetation, physical environments, reptilian assemblages, and other fauna. For example, these species are abundant in many of the suburbs of Sydney (Shea 2010). They overlap greatly throughout their respective geographic ranges; both occupy a broad band of coastal NSW, Victoria, and South Australia. Rawlinson (1971) noted that in Victoria both species inhabited the Bassian zone and the warm temperate and cool temperate zones, but neither occurred in the Eyrean zone or in the Murray River Corridor.

Rawlinson (1974) hypothesised *L. delicata* to be a relic with a long history in southern Australia and, if so, there would have been ample time for this species to cross into Tasmania via the land bridge connecting the island to mainland Australia. However, the genetic study by Chapple *et al.* (2011b) indicates that this species only colonised Victoria and Tasmania from coastal southern NSW no earlier than the Late Pleistocene; so, if *L. delicata* did travel overland to Tasmania, the migration would have taken place before the land connection between mainland Australia and Tasmania was severed following the last glaciation. Given the known history of human-assisted dispersal of *L. delicata* to various Pacific islands (see Chapter 3), the possibility that this species may have reached Tasmania via inadvertent transport by humans, rather than via a land bridge, remains a possibility. Neither species occurs on either Rodondo Island or Curtis Island in Bass Strait between mainland Australia and Tasmania (Rawlinson 1973).

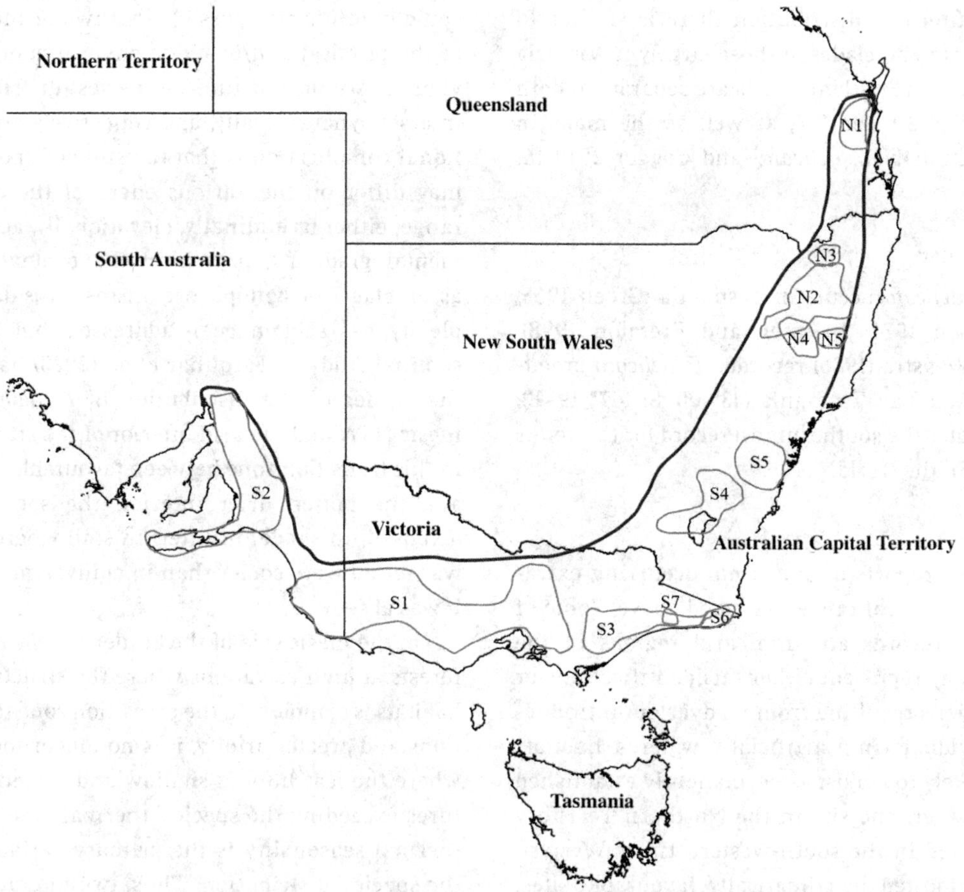

Fig. 2.2. Distribution of the clades and subclades (numbered polygons) of *Lampropholis guichenoti* in eastern Australia. N1–N5, northern clades; S1–S7, southern clades. The clades and subclades are circumscribed in thin dark lines. Modified from Figure 2 in Chapple *et al.* (2011a). Note that had all these clades and subclades been included collectively in a single area, as is usually done to denote the geographic range of a species (see heavy dark line), it would have significantly exaggerated the size of the area constituting this species' distribution and underestimated the extent of the vacant area that potentially could be occupied by other species without overlapping with *L. guichenoti*. For further discussion of this error in demarcating the species' distributional ranges, see Chapters 4 and 16. Printed with permission.

Latitudinal range of distributions

The native latitudinal range of *L. delicata* in Australia (Fig. 2.1) is more extensive than that of *L. guichenoti* (Fig. 2.2). Coastally they overlap as far north as Rainbow Beach, south-eastern Queensland (Latitude 25.9043° S), but beyond that locality only the range of *L. delicata* extends farther north to the base of Cape York Peninsula, to approximately 17.5° S, a difference of ~8.4° in latitude. This means that the Burdekin Gap and the St Lawrence Gap are outside the range of *L. guichenoti* and are therefore not relevant to the differentiation of its populations. The gaps that do occur within its range and that potentially could influence genetic divergence among its populations are the McPherson Range, the Great Dividing Range (including the part in southern NSW), the Hunter Valley, and the Gippsland Basin.

Width of east-coastal range of distributions

Within the central parts of their ranges these two species overlap over a broad band of coastal NSW, Victoria, and South Australia, but with *L. guichenoti* occurring farther inland into drier areas of NSW than the limits of *L. delicata*. This pattern suggests that perhaps at least some populations of *L. guichenoti* are tolerant of slightly drier conditions in the interior than is *L. delicata*, an hypothesis that requires further testing.

Southern coastal range of distributions

On the south-eastern coast of Australia, *L. delicata* had expanded its distribution westward into South Australia as far as the Eyre Peninsula prior to 12,750 years ago, but that extension subsequently became fragmented (Rawlinson 1974). The same may be true for *L. guichenoti* because its pattern of distribution there is similar to that of *L. delicata* and clades of those largely in Victoria and those mostly in South Australia are separate in both species (see Figs 2.1 and 2.2, as well as the maps in Chapple *et al.* 2011b [*L. delicata*] and Cogger 2018 [*L. guichenoti*]).

Tasmanian range of distributions

Lampropholis delicata occurs in Tasmania (Green 1965, 1981; Rawlinson 1974; Driessen and Brereton 1998). Wapstra and Wapstra (1986) recorded *L. delicata* breeding at Blackmans Bay, Tasmania (43° 00′ S; 147° 19′ E), which constitutes the southernmost record for the genus *Lampropholis* in the world.

Outliers

There are some reports of *L. delicata* occurring extralimitally to the natural range described above. Some of these putative records are from arid regions of the country and may represent either misidentifications or small populations resulting from inadvertent introductions of individuals into artificially watered habitats and thus unlikely to persist as permanently established colonies. However, one site in the Northern Territory and six clustered in the south-western tip of Western Australia are located in climatically favourable sites. Similarly, 17 sites for *L. guichenoti* were reported for south-western Australia (http://www.ala.org.au). If valid, these records may represent established populations originating as introductions from eastern Australia. Given the volume of commercial traffic between eastern Australia and the south-western corner of Western Australia, and especially the propensity of *L. delicata* for transportation in such cargo, inadvertent introduction of these species to that location would not be surprising. However, our attempts to verify some purported records were unsuccessful and, until there is convincing evidence for their veracity, they should be viewed with scepticism.

Edge effects at the boundaries of geographic distributions

As a species approaches the edges of its geographic range there is not always a sharp demarcation between favourable and unfavourable habitat, but sometimes a gradual spatial gradient in the factor or factors limiting farther extension of its range. Conceivably at such a border there could be a favourable mosaic of patches of habitat, some of which could still support the species and others that could not. It is also possible that two or more attributes of the physical and/or biotic environment, which individually would not limit a species' distribution, could interact synergistically, and collectively do so. An additional complication is that the suite of operative factors may differ on the various edges of the distributional range, either latitudinally, elevationally, across environmental gradients, or within the context of different assemblages of syntopic organisms. This daunting complexity has seldom been addressed, but fortunately a seminal study by Farquhar *et al.* (2023) has done so for the border of the distribution of *L. delicata* between mesic NSW and the arid interior of the state. In general, in this transition zone between favourable microhabitats and the hotter, drier interior, the specific locations (exposed, in shade, in litter, in soil) where this species was found were cooler than in equivalent habitat where it was absent.

On the mesic side of the border *L. delicata* is found in forests at high elevations where the structure of microhabitats is complex. In the transition zone to lower elevations and greater aridity, it is no longer found in places where the leaf litter is shallow and subject to temperatures exceeding the species' thermal preference. There, thermal seasonality is the primary variable predicting the species' distribution. Thus, two interacting limiting factors affect the suitability of habitats: the structure (thickness) of the leaf litter and favourable thermal microclimates. This finding, however, may not be the entire story. For example, the kind of vegetation may affect the type, thickness, and quality of the leaf litter; shade could allow the use of thinner layers of leaf litter; the presence of a new, more arid-adapted species could increase competition for the diminishing number of patches of the thick litter; and so on.

Clades of *Lampropholis delicata*

Chapple *et al.* (2011b) identified the barriers that separated populations of *L. delicata* as well as a land bridge that temporarily connected previously isolated populations, during the climatic and other environmental changes from the Miocene to the present. These barriers were the Burdekin Gap (dry woodland and savanna separating northern rainforests from mid-eastern Queensland forests), St Lawrence Gap (a dry corridor separating mid-eastern Queensland forests from south-eastern Queensland forests), Kroombit Tops (a high-elevation moist habitat surrounded by drier eucalypt woodland – an isolated cool, wet refuge for species adapted to rainforests and wet forests), the McPherson Range (a spur of wet forest along the Great Dividing Range near the Queensland/New South Wales [NSW] border and a barrier for species adapted to lowland dry forests), the Hunter Valley (a dry, open, lowland river valley between the eastern biogeographic region and the northern limit of the South East Forest region), southern NSW (a transition from the lowland coastal region and the elevated Southern Highlands region of the Great Dividing Range in NSW), the Murray Basin (a low-lying area of repeated marine incursion bordered on the west by the Mt Lofty Ranges), East Gippsland (a low-lying coastal region subject to repeated marine incursions and bordered on the north by the Great Dividing Range), and Bass Strait (a shallow strait now separating Tasmania from mainland Australia but periodically connecting these landmasses when sea levels were lower during Pleistocene glacial periods (Figs 2.1 and 2.2).

Clades of *L. guichenoti*

The gaps that occur within the range of *L. guichenoti* and that could potentially influence genetic divergence among its populations are the McPherson Range, the Great Dividing Range (including the part in southern NSW), the Hunter Valley, and the Gippsland Basin (Fig. 2.1). Isolation of populations resulted in two lineages: a northern one that diversified into five clades and a southern one that diversified into seven (Fig. 2.2).

Chapple *et al.* (2011a) obtained sequence data from mitochondrial genes from 123 individuals from 64 populations across the entire range of *L. guichenoti* and used these skinks as a basis for phylogeographic analysis (see Chapter 13). Chapple *et al.* (2011a) found a deep break of ~14% that they estimated to have occurred across the Hunter Valley in northern NSW during the Middle to Late Miocene that separated populations north of the Hunter Valley (northern lineage: clades N1–N5) from those south of it (southern lineage: clades S1–S7; Fig. 2.2).

Stenotopic species with restricted ranges

There are 13 species that, unlike *L. delicata* and *L. guichenoti*, have much smaller distributional ranges and occupy a more restricted range of habitat. Maps of their geographic ranges have been portrayed by Cogger (2018) and Chapple *et al.* (2019), as well as by various websites, such as The Reptile Database (http://reptile-database.org/); maps of the ranges of individual species also appear in the various publications cited below. Plate 2 contains a map illustrating the geographic ranges of all of the stenotopic species, colour-coded and showing the overlap of ranges between species.

The lower left map in Plate 2 shows the ranges of *L. delicata* and *L. guichenoti* (light blue and pink shading respectively; also see Figs 2.1 and 2.2 for details of the distribution of these two species) overlaid by the species of skinks in the genus *Lampropholis*. Inset maps (blue, red, and green outlines) do not include the ranges of *L. delicata* and *L. guichenoti* so as to highlight species with ranges that are more restricted.

Lampropholis adonis

Cogger (2018) portrayed and mapped the distribution of *Lampropholis adonis* as two apparently isolated coastal populations (one in mid-eastern Queensland and the other in south-eastern Queensland) with a third one isolated inland to the west of the Great Dividing Range. A map from the Department of Environment, Tourism, Science and Innovation, Queensland (2013) suggested that the location of the inland population would be in the Carnarvon National Park. More recently, Shea *et al.* (2018a) presented a map with more-detailed geographic information indicating that the inland population is not a restricted one, but rather a segment of a more widely, but sparsely, distributed taxon within the Fitzroy Basin.

East of the Great Dividing Range, the distribution of *L. adonis* stretches from Cape Gloucester in the north to the Great Sandy National Park in the south. Although the map of Shea *et al.* (2018a) does not provide place names, it is sufficiently detailed to indicate that the two coastal populations are separated by the Fitzroy River.

Although the distribution of *L. adonis* needs more-detailed study, as currently known it is a species of the Fitzroy Basin west of the Great Dividing Range with two distinct subpopulations north and south of the Fitzroy River and east of the Great Dividing Range. More intensive sampling in the Fitzroy Basin and a molecular genetic analysis are required to determine the relationships between subpopulations and to ascertain whether they are sufficiently related genetically to be considered a single species.

Lampropholis amicula

This species forages in open forest and heathlands in the eastern area of northern NSW and southern Queensland (Cogger 2018). In the eastern segment of its range, it has been subject to selective logging and silvicultural management for nearly 100 years, and to intermittent controlled burnings by humans and wildfire. It was in this area that Partridge *et al.* (2003) studied *L. amicula*. Within the three sites that were sampled in subtropical eucalypt forests, the interval between fires averaged 7.2 ± 1.6 SEM years. Of the seven most common species of reptiles encountered, two were in the genus *Lampropholis*. Of these, *L. delicata* was the species observed most frequently. *Lampropholis amicula* was common at two of the study sites but absent from the third, a pattern that may have resulted from land management because this species favoured areas of 'top disposal burns' (a silvicultural burn immediately following selective logging). Hannah *et al.* (1998) undertook research in south-eastern Queensland into the impact on reptiles of prescribed burning every 2–5 years since 1973, compared with lands unburnt for ~80 years. Two of the three species of *Lampropholis* (*L. amicula* and *L. couperi*) were not encountered in burnt areas but were present in areas where fire had been excluded. The third species, *L. delicata*, was present in much reduced numbers in burnt areas.

Garden *et al.* (2007) observed that, at least in urban settings, *L. amicula*, in addition to *L. delicata*, occurred in areas with a moderate (20–50%) cover of weediness, provided such cover facilitated thermoregulation or gave refuge from potential predators. Hodgkison *et al.* (2007) compared the faunal diversity of skinks in 20 'eucalypt-based golf courses' (10 residential areas and 10 eucalypt fragments) across rapidly urbanising areas of Brisbane and the Gold Coast in south-eastern Queensland. These sites were on relatively flat lowlands that had remained unburnt for at least 10 years. Five species of reptiles, including *L. amicula*, had significantly higher relative abundance in eucalypt remnants than they did on golf courses.

Lampropholis bellendenkerensis

At the molecular level, *L. bellendenkerensis* may be reliably distinguished from *L. robertsi*, its closest congener, by 13 nucleotide differences in the NADH dehydrogenase mitochondrial gene. Between species, four of these nine genes result in nine differences in amino acids (Singhal *et al.* 2018).

Lampropholis bellendenkerensis is restricted to the uplands (>900 m asl) of Mt Bellenden Ker and Mt Bartle Frere, together with the highest areas of the southern Atherton Tablelands, which includes Mt Baldy and Mt Fisher, Longlands Gap, and the area associated with Tully Falls (Plate 2).

Lampropholis caligula

There are limited published data on *L. caligula*, and even its original description was brief (Ingram and Rawlinson 1981). Like most other *Lampropholis* species, *L. caligula* is a ground-dwelling species. Its range is restricted to upland areas of north-eastern NSW. Cogger (2018) recorded its distribution as the 'Barrington Tops region', although he questioned whether its range might not be wider. More recently, Eldridge *et al.* (2020) recorded its presence at Coolah Tops National Park farther inland from Barrington Tops. The map provided by Shea *et al.* (2018b) suggests that *L. caligula* may have a continuous distribution between those two protected areas. Cogger (2018) reported that the species foraged in litter or in open snowgrass areas in upland wet sclerophyll forests and woodlands, and Eldridge *et al.* (2020) commented that, in May, the activity of reptiles (including *L. caligula*) was low due to cooler, dry conditions.

Lampropholis coggeri, *L. robertsi*, and *L. similis*: Three related species from rainforests in Queensland

Lampropholis coggeri is endemic to the montane tropical rainforests of north-eastern Queensland. Its current range extends from Bluewater Range Track, adjacent to Townsville, northwards to near Cooktown and Palm Island (Ingram and Rawlinson 1981; Cogger 2018). Using palaeomodelling, Bell *et al.* (2010) predicted the palaeodistribution of *L. coggeri*. Their model and the species' current range suggested a stable, widespread, and connected distribution. These data were consistent with the species having historically low dispersal and high persistence within multiple isolated montane regions. Morphologically, the only character of *L. coggeri* that differed between the northern and southern major genetic lineages was the body size of males.

Singhal and Moritz (2012) investigated the outcomes of secondary contact following the retreat of glacial cycles that resulted in the retraction of rainforest and associated fauna during the late Quaternary. During this period, genetic divergence occurred and, in some instances, cryptic phylogeographic lineages evolved. Palynological data and ecoclimatic modelling revealed that the natural forested habitat of *L. coggeri* contracted, leaving two major refugia during the glacial cycles. Subsequent warming of the climate over the past 3,000–8,000 years resulted in the rainforest, together with the associated specialist species, including *L. coggeri*, expanding rapidly. Genetic divergence may have varied widely, and thus cryptic, morphologically similar pairs of lineages with varying extents of genetic divergence may have occurred. Singhal and Moritz (2012) found that there was strong selection against morphologically cryptic hybrids at contact zones and that hybrid zones were extremely narrow (averaging ~400 m wide). In addition to these narrow hybrid zones, the authors also noted that there was strong and ubiquitous genetic disequilibrium. For example, compared with previous studies and other systems, reproductive isolation evolved without apparent morphological differentiation; the clines are very narrow, and the lineages were likely to have been in contact during interglacial periods when there was the potential for the taxa to merge. Despite such opportunities, all loci assayed by Singhal and Moritz (2012) indicated that *L. coggeri* remained a distinct, viable species. It was noted, however, that there is an ever-growing list of taxa, including lizards, spiders, and moss, that have been shown to have deep phylogeographic structure with little or no overt morphological or ecological divergence.

Lampropholis robertsi, a montane specialist, is endemic to upland tropical rainforests in the Wet Tropics of north-eastern Queensland (Ingram and Rawlinson 1981; Bell *et al.* 2010) at between 1,000 and 1,600 m asl, habitat that typically harbours species of extraordinary local endemism due to the influences of past geological and climatic conditions (Bell *et al.* 2010). Palaeomodelling generally accurately predicted the present distribution of *L. robertsi*, although there was some overprediction. For example, two isolated areas at high elevations, Lamb Range and Windsor Uplands, were predicted to be appropriate habitat for *L. robertsi*; however, despite repeated surveys in these areas, Bell *et al.* (2010) reported that no *L. robertsi* were located. Based on spatial modelling, Bell *et al.* (2010) predicted that the range of *L. robertsi* would have been fragmented during the warm wet periods of the mid-Holocene and thus its distribution would have been restricted to the highest peaks.

Lampropholis robertsi mainly forages along the edge of clearings in the forest and along streams and tracks, where it preys on small insects, including ants (Cogger 2018). With the caveat that the data should be considered preliminary due to the small sample size and lack of information for males in their sample, Bell *et al.* (2010) observed that divergence in the shape of the body occurred and was more pronounced than in *L. coggeri*.

Lampropholis similis is known from the Bellenden Ker Range (excluding the Malbon Thompson Range to the east) south to the Hervey Range and The Pinnacles, near Townsville (Singhal *et al.* 2018).

Lampropholis colossus

Lampropholis colossus is a rainforest species, restricted to the Bunya Mountains in south-eastern Queensland (Cogger 2018). It is not a grassland specialist, but rather basks in sunlit patches at the edge of dry rainforest adjacent to grassy balds (i.e. areas with a dense layer of tussock grasses with few trees or shrubs and limited patches of exposed substrate or leaf litter) in savanna woodland and dry rainforest at higher elevations (Smith *et al.* 2015). *Lampropholis colossus* has also been observed in forest clearings along tracks and beside streams (Cogger 2018). There was a dramatic reduction in montane grassy bald areas in the Bunya Mountains (Moravek *et al.* 2013) that between 1951 and 1991 decreased in area by 26% (Fensham and Fairfax 1996). If this trend has continued since then, it is likely that the habitat of *L. colossus* has continued to shrink, although other changes in the management of the balds may have, at least indirectly, supported their maintenance. For example, Peeters and Butler (2014) reported that such areas have appeared to become more substantial under grazing by livestock.

Lampropholis couperi

This species occurs in rainforest in south-eastern Queensland. Its northern boundary ends at the southern bank of the Fitzroy River south of Rockhampton (see map in Cogger 2018) and its most southerly reported extent is Mt Glorious, north-west of Brisbane. In such habitat, *L. couperi* typically basks in sunny patches where there are breaks in the canopy (e.g. along tracks and on the edges of streams; Cogger 2018). Hannah *et al.* (1998) described the habitat of *L. couperi* in the Bauple State Forest as within the humid subtropics (annual rainfall 1,000–1,200 mm), but within predominantly dry sclerophyll forest growing on low ridges (160 m elevation). This species and *L. amicula* (see above) were not encountered in areas that had been burnt annually for 42 years, although *L. delicata* was present there, albeit in very low numbers compared with its density in unburnt habitat.

Kanowski *et al.* (2006) investigated factors that affected reptiles' use of reforested sites in previously cleared rainforest in both tropical and subtropical locations. They observed that *L. couperi* typically was associated with leaf litter.

Shea *et al.* (2018c) suggested that *L. couperi* likely is more widespread than previously recorded, and thus may not be as vulnerable as its limited distribution would suggest.

Lampropholis elliotensis

This species is a rainforest species that is small with dark sides. It has pentadactyl limbs that usually are separated by several rows of scales when adpressed. The lower eyelid is moveable and contains a transparent disc. *Lampropholis elliotensis* may be distinguished from its sibling species *L. similis* and *L. coggeri* by 17 nucleotide differences in the mitochondrial gene NADH dehydrogenase, four of which may result in 15 differences in amino acids among species (Singhal *et al.* 2018).

Lampropholis elongata

This species is an inhabitant of uplands on the Great Dividing Range in an area near Walcha, mid-northern NSW, on the New England Tableland. This is an area of cool, upland eucalypt forest and grassland where *L. elongata* forages in the litter (Cogger 2018). Graham *et al.* (2023) reported that this species had not been seen for almost a decade before they undertook their fieldwork. Using a combination of pitfall traps and active search within the known range and habitat, they found that it was relatively easy to observe these skinks, especially when they were in the bases of tussocks of the grass *Poa sieberiana*. For further discussion of the preferred habitat of this species and its relationship with *L. delicata*, see section *Morphological and historical correlates of the ecological niche* in Chapter 11.

Lampropholis isla

This species is restricted to areas of rainforest on Scawfell Island (Hoskin 2025), a small islet of 21.8 km^2 lying 60 km off the coast of Queensland, north-east of Mackay. This island is part of the South Cumberland Islands National Park and rises to a peak of 397 m.

Lampropholis mirabilis

Described by Ingram and Rawlinson (1981), *L. mirabilis* occurs on Mt Elliot and Cape Cleveland in Queensland and on Magnetic Island just offshore (Czechura and Covacevich 1985; Burgin 1989; Ehmann 1992). More recently, Valentic (1997) reported it from Jourama Falls, Paluma Range National Park, more than 100 km from the nearest other known locality.

Lampropholis mirabilis inhabits the edges of, and clearings in, rainforests and occurs in monsoon forest, dense woodlands, and vine forests where there are creeks and gullies (Ingram and Rawlinson 1981; Burgin 1989; Cogger 2018). It is a conspicuous climber (Greer and Wadsworth 2003) and is typically associated with rocky areas (Greer 2007), such as large granite boulders (Ingram and Rawlinson 1981). Burgin (1989) reported that *L. mirabilis* was not observed to forage in areas of full sunlight. Although considered rare because of its restricted distribution, it may be locally abundant in its preferred habitat (Covacevich *et al.* 1982; Burgin 1989). This species diverged from *L. coggeri* in the shape of its body.

ELEVATIONAL RANGES

Lampropholis delicata and *L. guichenoti* occupy extensive spans of elevation as well as having wide areal geographic ranges (Fig. 2.3). Pengilley (1972) reported an individual of *L. guichenoti* at Snowy Flats in the Brindabella Range, NSW, but indicated that neither this species nor *L. delicata* were common above elevations of 1,220 m at that location. *Lampropholis delicata* introduced into Hawaii have established populations at elevations up to 1,760 m, i.e. higher than recorded for Australia (see Chapter 3).

By contrast, most of the stenotopic species with small geographic ranges (*L. bellendenkerensis*, *L. caligula*, *L. colossus*, *L. elongata*, and *L. robertsi*) occupy a more limited elevational span and are montane species restricted to the cooler habits at higher elevations; of such species, only *L. coggeri* extends over an elevational extent rivalling the spans *of L. delicata* and *L. guichenoti*.

CLIMATE AND PHYSIOLOGY AS DETERMINANTS OF GEOGRAPHIC DISTRIBUTION

In recognition that climate has a key impact on an animal's physiology, which, in turn, can have a profound effect on geographic distributions, Anderson *et al.* (2023) tested the hypothesis that species with distributions over vast areas (*L. delicata* and *L. guichenoti*) have broader physiological tolerances than those with restricted geographic ranges (*L. adonis*, *L. amicula*, *L. caligula*, *L. couperi*, *L. isla*, *L. mirabilis*, and *L. similis*). The alternative hypothesis is that individual populations (e.g. clades or subclades) of broadly distributed species do not have wide tolerances, but rather that each infraspecific population adapts to the specific range of local conditions they encounter and do not have a wider span of tolerances than those of species with restricted geographic ranges. To test their hypothesis, Anderson *et al.* (2023) compared the two widespread species *L. guichenoti* and *L. delicata* (but with widely separated populations of the latter treated as though they were different 'species') with eight species with small geographic ranges. The physiological categories they used were metabolism, water balance, thermal physiology, thermoregulatory behaviour, and ecological performance. They tested whether models that incorporated physiology defined species' distributions better than those relying solely on climatic data. Their results confirmed that widespread species can tolerate and perform better over broader ranges of temperature than species that are restricted to smaller geographic ranges (Anderson *et al.* 2023). They also found differences in the body temperatures experienced in the field, but not in thermal preferences between species with large ranges and those with small ranges. Metabolism and water balance were not correlated with the size of the geographic range. Clearly, physiological responses to climate are central in establishing the boundaries of the geographic ranges of these skinks, and the incorporation of behavioural processes occurring at local scales can refine models of distribution. The maps of the distributional ranges of species and the lines enclosing the real ranges of the species (Figs 2.1, 2.2, and Plate 2) indicate that all continuous areas of suitable habitat currently are occupied, and that it is unlikely that expansions of range will occur for any species. In fact, given human-induced ongoing changes in the physical environment, ranges likely will shrink in the future. Unfortunately, Anderson *et al.* (2023) only published the parameters of the statistical tests and a description of the models, so actual data on such features as the thermal tolerances and the metabolic rates cannot be presented here.

Although *L. delicata* and *L. guichenoti* are very similar in almost every aspect of their biology, and frequently occur syntopically, there must be nuances of difference between them. Faunal surveys of specific areas have been conducted (see Appendix 3). However, the intention of most of these surveys was merely to prepare a faunal list, so often only rough indications of abundance were provided.

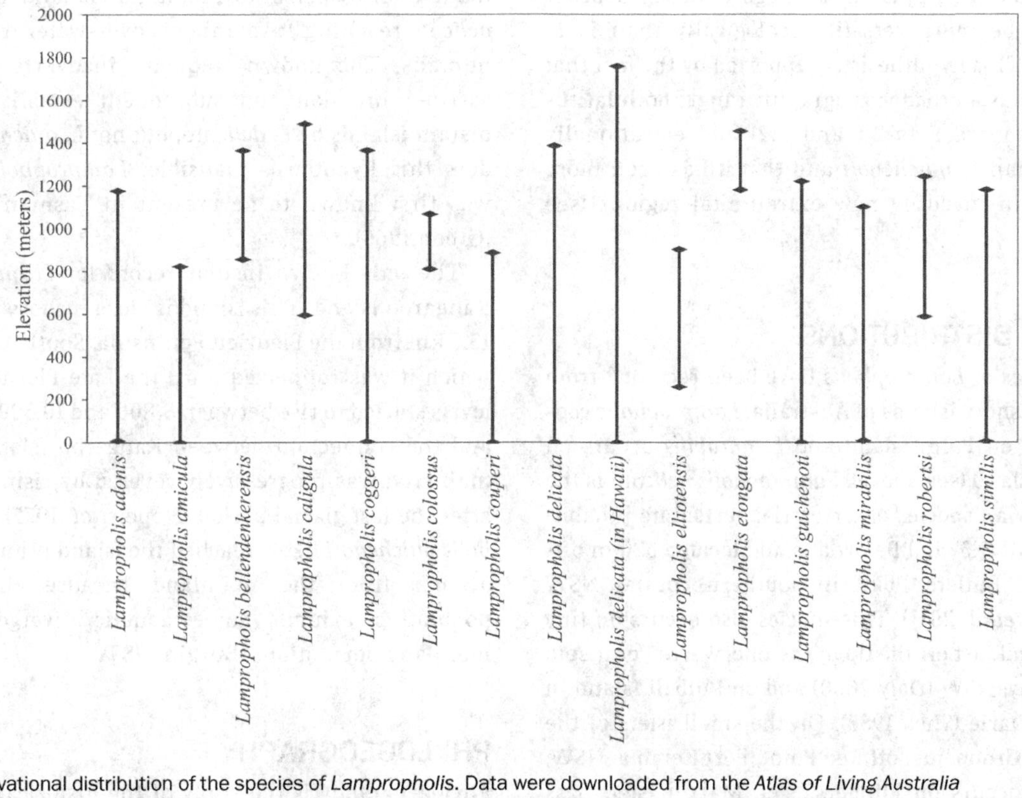

Fig. 2.3. Elevational distribution of the species of *Lampropholis*. Data were downloaded from the *Atlas of Living Australia* (doi:10.26197/5d7f2348165a0; accessed 22 March 2024), trimmed to exclude outliers requiring verification. Image prepared by Susan King, supplemented by information from Pengilley (1972), Baker (1979), and Greer (1997). Note that the recently described *L. isla* is not included.

These faunal surveys found that, at most sites, these two species co-occur; however, the surveys also indicate that within the geographic range encompassing both species, they differ in their relative abundances from site to site and, at some places, one or the other may occur alone (Appendix 3). At 17 areas *L. guichenoti* either was present when *L. delicata* was absent or was more abundant than *L. delicata*; the reverse was true, with *L. delicata* being the sole or dominant species, at 22 areas. These results are very approximate because in many studies assessments of abundance were not made for each individual site, but rather whether the species was present or absent from the entire region. In such cases, one species could have been dominant in some types of vegetation, say, and the other species dominant in others. Also, there could be ecological segregation into different microhabitats within a given habitat that would not have been evident from the broad scope of most of the studies listed in Appendix 3 (see also Chapter 11). More quantitative, fine-grained studies are needed across the full geographic range occupied by these two species jointly. Nevertheless, despite their ecological similarities and joint occupation of many sites, the data in Appendix 3 suggest that *L. delicata* often may be more versatile ecologically than is *L. guichenoti*. This hypothesis is supported by the fact that *L. delicata* has a broader geographic range, both latitudinally (compare Figs 2.1 and 2.2) and elevationally (Fig. 2.3) than *L. guichenoti* and that it has been more successful in invading new extralimital regions (see Chapter 3).

INSULAR DISTRIBUTIONS

Some species of *Lampropholis* have been recorded from small near-shore islands of Australia. *Lampropholis coggeri* occurs on Palm Island and *L. mirabilis* occurs on Magnetic Island (see above). *Lampropholis delicata* is the only non-avian species of terrestrial vertebrate inhabiting the small (1.57 ha) Belowla Island located 520 m offshore from Butlers Point in southern coastal NSW (Blackmore *et al.* 2011). This species also occurs on tiny Bowen Island just off the tip of the Bherwerre Peninsula in Jervis Bay, NSW (Daly 2000) and on Pulbah Island in Lake Macquarie (Shea 1985). On the small islets of the Five Island Group, just off Red Point, Port Kembla, NSW, *L. delicata* occurs on Flinders Islet, Martin Islet, Bass Islet, and Big Islet, but *L. guichenoti* does not. The morphological characteristics of *L. delicata* on these islands accord with the data for this species shown in Table 14.3 in Chapter 14). *Lampropholis guichenoti and L. delicata* both occur on Montague Island, offshore from Narooma, NSW. *Lampropholis isla* is restricted to Scawfell Island in the South Cumberland group (Plate 2).

Lampropholis delicata has been introduced inadvertently by humans into New Zealand, Hawaii, and Lord Howe Island, and has become established in all those places (see Chapter 3). It also occurs in Tasmania.

Contrary to the conclusion of Rawlinson (1974) that *L. delicata* was a relic with a long history in southern Australia, a genetic study by Chapple *et al.* (2011b) indicated that this species did not colonise Victoria from eastern NSW until the Late Pleistocene, so the land bridge between Victoria and Tasmania and intervening islands would have begun to be submerged by the rising sea level about that time. For this reason, one hypothesis for the difference in distribution between these two species could be that *L. delicata* reached Victoria just before the land bridge closed, with *L. guichenoti* only arriving in Victoria somewhat later, after the disappearance of the land bridge. Alternatively, perhaps both species may have reached Victoria too late for dispersal to Tasmania via land, but with *L. delicata* reaching Tasmania via over-water transport by humans. The known frequent, inadvertent human-assisted invasion and subsequent establishment on distant islands by *L. delicata*, but not *L. guichenoti*, renders this hypothesis plausible. *Lampropholis delicata* was first known to be present in Tasmania in 1964 (Green 1965).

The only known insular record for *L. guichenoti* is Kangaroo Island. This island is close inshore, lying only 13.5 km from the Fleurieu Peninsula, South Australia, to which it was connected until the Late Pleistocene. Sea levels started to rise between 8,800 and 10,500 years ago, and the connection between Kangaroo Island and the mainland was progressively severed by rising sea level after the last glacial period (Hope *et al.* 1977). *Lampropholis guichenoti* likely reached the island prior to its separation from the mainland because the insular population exhibits some genetic divergence from mainland populations (Burgin 1989).

PHYLOGEOGRAPHY

Phylogeography is the study of the historical processes responsible for the transition from past to present geographic distributions of genealogical lineages and combines genetics (particularly population genetics), palaeogeography (geological changes in topography and geomorphology), climatic oscillations, and changes in vegetation. These ecological and geological vagaries form barriers that isolate populations, prevent gene flow between them, and promote genetic divergence. An important item in the toolkit of this discipline, especially at the infraspecific level, is the analysis of mitochondrial DNA (mtDNA) (Avise *et al.* 1987), which allows calculation of the times when clades were formed.

Chapple *et al.* (2011a, 2011b) conducted a detailed and comprehensive analysis of the mtDNA both of *L. delicata* and *L. guichenoti*. Nine geographically defined, genetically divergent clades were identified for *L. delicata* (see Figs 2.1 and 2.2). These diverged during the Late Miocene to Pliocene times. Interbreeding of populations introduced into New Zealand from four different isolated sources in Australia supports maintaining those entities within *L. delicata*, rather than considering them separate species (Chaplin *et al.* 2022).

REFERENCES

Anderson RO, Tingley R, Hoskin CJ, White CR, Chapple DG (2023) Linking physiology and climate to infer species distributions in Australian skinks. *Journal of Animal Ecology* **92**(10), 2094–2108. doi:10.1111/1365-2656.14000

Avise JC, Arnold J, Ball RM, Bermingham E, Lamb T, Neigel JE, Reeb CA, Saunders NC (1987) Intraspecific phylogeography: The mitrochondrial DNA bridge between population genetics and systematics. *Annual Review of Ecology and Systematics* **18**, 489–522. doi:10.1146/annurev.es.18.110187.002421

Baker JK (1979) The rainbow skink, *Lampropholis delicata*, in Hawaii. *Pacific Science* **33**(2), 207–212.

Bell RC, Parra JL, Tonione M, Hoskin CJ, Mackenzie JB, Williams SE, Moritz C (2010) Patterns of persistence and isolation indicate resilience to climate change in montane rainforest lizards. *Molecular Ecology* **19**(12), 2531–2544. doi:10.1111/j.1365-294X.2010.04676.x

Blackmore CJ, Jarman MR, Craven P, Carlile N (2011) Seabird Islands No. 3/1. Belowla Island, New South Wales. *Corella* **35**, 82–85.

Burgin S (1989) The taxonomy and phylogenetic relationships of Australian scincid lizards (Scincidae: Lygosominae). PhD thesis. Macquarie University, Sydney.

Chaplin K, Smith Date K, Bray RD, Miller KA, Lutz ML, Razeng E, Thompson MB, Chapple DG (2022) Intraspecific hybridisation of an invasive lizard on Lord Howe Island. *Australian Journal of Zoology* **69**(5), 184–196. doi:10.1071/ZO21045

Chapple DG, Chapple SNJ, Thompson MB (2011a) Biogeographic barriers in south-eastern Australia drive phylogeographic divergence in the garden skink, *Lampropholis guichenoti*. *Journal of Biogeography* **38**(9), 1761–1775. doi:10.1111/j.1365-2699.2011.02531.x

Chapple DG, Hoskin CJ, Chapple SNJ, Thompson MB (2011b) Phylogenetic divergence in the widespread delicate skink (*Lampropholis delicata*) corresponds to dry habitat barriers in eastern Australia. *BMC Evolutionary Biology* **11**(1), 191. doi:10.1186/1471-2148-11-191

Chapple DG, Tingley R, Mitchell NJ, Macdonald SL, Keogh JS, Shea GM, Bowles P, Cox NA, Woinarski JCZ (2019) *The Action Plan for Australian Lizards and Snakes 2017.* CSIRO Publishing, Clayton South, Vic.

Cogger HG (2018) *Reptiles and Amphibians of Australia.* Updated 7th edn. CSIRO Publishing, Clayton South, Vic.

Covacevich J, Ingram GJ, Czechura GV (1982) Rare frogs and reptiles of Cape York Peninsula, Australia. *Biological Conservation* **22**(4), 283–294. doi:10.1016/0006-3207(82)90023-4

Czechura GV, Covacevich J (1985) Poorly known reptiles in Queensland. In *Biology of Australasian Frogs and Reptiles.* (Eds G Grigg, R Shine and H Ehmann) pp. 471–476. Surrey Beatty & Sons and the Royal Zoological Society of New South Wales, Sydney.

Daly G (2000) Island populations: Reptiles and amphibians of the Jervis Bay region on the South Coast of New South Wales. *Herpetofauna* **30**(1), 11–17.

Department of Environment, Tourism, Science and Innovation, Queensland (2013) Carnarvon National Park – facts and maps. In Wetland*Info*. <https://wetlandinfo.des.qld.gov.au/wetlands/facts-maps/national-park-carnarvon/> (accessed 20 February 2024).

Driessen MM, Brereton R (1998) A range extension for the delicate skink, *Lampropholis delicata*, in Tasmania. *Herpetofauna* **28**(1), 44–45.

Ehmann H (1992) *Encyclopaedia of Australian Animals: Reptiles*. Angus and Robertson, Pymble, NSW.

Eldridge MDB, Ingleby S, King AG, Mahony SV, Parnaby HE, Beatson CA, Divljan A, Frankham GJ, Hay AC, Major RE, Reader SE, Sadlier RA, Tsang LR (2020) Australian Museum surveys of the vertebrate fauna of Coolah Tops National Park, NSW. *Technical Reports of the Australian Museum Online* **30**, 1-26. doi:10.3853/j.1835-4211.30.2020.1757

Farquhar JE, Russell W, Chapple DG (2023) Identifying the abiotic factors that determine the inland range limits of a mesic-adapted lizard species. *Integrative and Comparative Biology* **64**(1), 55–66. doi:10.1093/icb/icad124

Fensham RJ, Fairfax RJ (1996) The disappearing grassy balds of the Bunya Mountains, south-eastern Queensland. *Australian Journal of Botany* **44**(5), 543–558. doi:10.1071/BT9960543

Garden JG, McApline CA, Possingham HP, Jones DN (2007) Habitat structure is more important than vegetation composition for local-level management of native terrestrial reptile and small mammal species living in urban remnants: a case study from Brisbane, Australia. *Austral Ecology* **32**, 669–685. doi:10.1111/j.1442-9993.2007.01750.x

Graham KA, Mahony SV, Chapple DG, Farquhar JE (2023) The long unknown: rediscovery of the long sunskink, *Lampropholis elongata* (Squaata: Scincidae) – after almost a decade, and after 50 years of data deficiency. *Austral Ecology* **48**(5), 877–884. doi:10.1111/aec.13314

Green RH (1965) Two skink lizards newly recorded from Tasmania. *Records of the Queen Victoria Museum* **New Series no. 19**, 1–4.

Green RH (1981) Distribution of the delicate skink. *Tasmanian Naturalist* **68**, 8.

Greer A (1997) A new species of *Lampropholis* (Squamata: Scincidae) with a restricted, high altitude distribution in eastern Australia. *Australian Zoologist* **30**(3), 360–368. doi:10.7882/AZ.1997.014

Greer A (2007) *The Biology and Evolution of Scincid Lizards*. Published online only. <www.academia.edu/35305801/The_Biology_and_Evolution_of_Scincid_Lizards_doc>

Greer AE, Wadsworth L (2003) Body shape in skinks: the relationship between relative hind limb length and relative snout-vent length. *Journal of Herpetology* **37**(3), 554–559. doi:10.1670/138-02N

Hannah DS, Smith GC, Agnew G (1998) Reptile and amphibian composition in prescribed burnt dry sclerophyll forest, southern Queensland. *Australian Forestry* **61**(1), 34–39. doi: 10.1080/00049158.1998.10674714

Hodgkison S, Hero J-M, Warnken J (2007) The efficacy of skall-scale conservation efforts, as assessed on Australian golf courses. *Biological Conservation* **135**, 576–586.

Hope JH, Lampert RJ, Edmondson E, Smith MJ, Van Tets GF (1977) Late Pleistocene faunal remains from Seton Rock Shelter, Kangaroo Island, South Australia. *Journal of Biogeography* **4**(4), 363–385. doi:10.2307/3038194

Hoskin CJ (2025) A new species of *Lampropholis* skink (Lacertilia: Scincidae) from Scawfell Island, mid-east Queensland, Australia. *Australian Journal of Taxonomy* **85**, 1–13. doi:10.54102/ajt.6uihp

Ingram G, Rawlinson P (1981) Five new species of skinks (genus *Lampropholis*) from Queensland and New South Wales. *Memoirs of the Queensland Museum* **20**(2), 311–317.

Kanowski JJ, Reis TM, Catterall CP, Piper SD (2006) Factors affecting the use of reforested sites by reptiles in cleared rainforest landscapes in tropical and subtropical Australia. *Restoration Ecology* **14**(1), 67–76. doi:10.1111/j.1526-100X.2006.00106.x

Merrick JR, Archer M, Hickey GM, Lee MSY (2006) *Evolution and Biogeography of Australasian Vertebrates*. Auscipub, Oatlands, NSW.

Moravek S, Luly J, Grindrod J, Fairfax R (2013) The origin of grassy balds in the Bunya Mountains, southeastern Queensland, Australia. *The Holocene* **23**, 305–315.

Partridge DA, Lewis T, Tran CT, Castley JG (2003) Fire and habitat variables explain reptile community abundance and richness in subtropical open eucalypt forests. *International Journal of Wildland Fire* **32**, 1089–1108. doi:10.1071/WF22001

Peeters PJ, Butler DW (2014) *Eucalypt Open-forests: Regrowth Benefits Management Guideline*. Department of Science, Information Technology, Innovation and the Arts, Brisbane.

Pengilley R (1972) Systematic relationships and ecology of some lygosomine lizards from southeastern Australia (2 Vols). PhD thesis. Australian National University, Canberra.

Rawlinson PA (1971) Reptiles of Victoria. In *Victorian Amphibians and Reptiles*. (Ed. VH Arnold) pp. 11–36. *Victorian Yearbook* 85, Melbourne.

Rawlinson PA (1973) Natural history of Curtis Island, Bass Strait – 4. The reptiles of Curtis and Rodondo islands. *Papers and Proceedings of the Royal Society of Tasmania* **107**, 153–170. doi:10.26749/rstpp.107.153

Rawlinson PA (1974) Biogeography and ecology of the reptiles of Tasmania and the Bass Strait area. In *Biogeography and Ecology of Tasmania*. Monographiae Biologicae 25. (Ed. WD Williams) pp. 291–338. Dr. W. Junk, The Hague.

Shea GM (1985) Island herpetofaunas in New South Wales: A review. *Herpetofauna* **16**(2), 30–38.

Shea GM (2010) The suburban terrestrial reptile fauna of Sydney – winners and losers. In *The Natural History of Sydney*. (Eds D Lunney, P Hutchings and D Hochuli) pp. 154–197. Royal Zoological Society of NSW, Sydney. doi:10.7882/FS.2010.015

Shea G, Wilson S, Venz M, Hobson R (2018a) *Lampropholis adonis* Ingram. The IUCN Red List of Threatened Species 2018: e.T10947289A109472952. <https://www.iucnredlist.org/species/109472898/109472952>

Shea G, Cogger H, Greenlees M (2018b) *Lampropholis caligula*. The IUCN Red List of Threatened Species 2018: e.T109473000A109473024. <https://www.iucnredlist.org/species/109473000/109473024>

Shea G, Venz M, Sanderson C, Hobson R, Wilson S (2018c) *Lampropholis couperi*. The IUCN Red List of Threatened Species 2018:e.T109473083A109473088. <https://www.iucnredlist.org/species/109473083/109473088>

Singhal S, Moritz C (2012) Strong selection against hybrids maintains a narrow contact zone between morphologically cryptic lineages in a rainforest lizard. *Evolution* **66**, 1474–1489.

Singhal S, Hoskin CJ, Couper P, Potter S, Moritz C (2018) A framework for resolving cryptic species: a case study from the lizards of the Australian Wet Tropics. *Systematic Biology* **67**, 1061–1075. doi:10.1093/sysbio/syy026

Smith G, Fensham R, Ferguson D, Hogan L, Mathieson M (2015) Fauna of the grassland–forest landscape mosaics of the Bunya Mountains, eastern Australia. *Australia Zoologist* **37**(3), 302–310. doi:10.7882/AZ.2014.035

Valentic R (1997) A range extension for the skink *Lampropholis mirabilis*. *Monitor (Victorian Herpetological Society)* **8**(3), 154.

Wapstra M, Wapstra E (1986) Breeding record of the delicate skink *Leiolopisma delicata* in southern Tasmania. *Tasmanian Naturalist* **87**, 5–6.

3
Lampropholis delicata: An invasive stowaway

INTRODUCTION

Lee (2002) posited the term 'invasive species' as typically referring to 'introduced species that cause negative impacts on the environment, human activities, or human health'. Species with unknown effects were excluded, although the definition did include those taxa that posed 'major threats to biodiversity ... [and] ecosystem integrity'. Other authors (e.g. Simberloff *et al.* 1997; Holway and Suarez 1999; Mack *et al.* 2000; O'Dowd *et al.* 2003; Whitney and Gabler 2008; Capinha *et al.* 2017) focused on species that caused environmental damage and had impacts on the structure and composition of communities and/or disrupted ecosystem-level services and processes. Often, such discussions linked invasive species with the economic costs associated with pests (e.g. see Dukes and Mooney 2004; Whitney and Gabler 2008; Marbuah *et al.* 2014; Hanley and Roberts 2019), although, compared with the large number of human-assisted dispersals (either deliberate or accidental), a relatively small number of species have been demonstrated to have significant economic impacts (Dukes and Mooney 2004; Holmes *et al.* 2009). Invasive species, however, may contribute to the impairment of ecosystems more generally (Pejchar and Mooney 2009; McGeoch *et al.* 2010; Pyšek and Richardson 2010). Such impairment may be exacerbated in the future because the number of invasions by species worldwide has increased substantially in past decades to the point where invasive species now constitute a substantial element of the change that is occurring in global ecosystems (Dukes and Mooney 1999; Pyšek and Richardson 2010; Pyšek *et al.* 2020; Cuthbert *et al.* 2021) and climatic change is likely to exacerbate the establishment and survival of even more species in the future (McGeoch and Latombe 2016).

Lampropholis delicata has been accidently introduced beyond its natural range and is the only Australian endemic *Lampropholis* known to have established alien populations overseas (Lever 2003; Kraus 2009; Chapple *et al.* 2013a, 2014). It has been introduced into Hawaii (Baker 1979; Kraus 2002; Lever 2003; Hutchinson *et al.* 2005; Zug 2013), New Zealand (West 1979; Gill and Whitaker 2001; Gill *et al.* 2001; Lever 2003; Peace 2004; Hutchinson *et al.* 2005; Harris *et al.* 2021), and Lord Howe Island (Lever 2003; Hutchinson *et al.* 2005; Brown *et al.* 2007; Brown and Baker 2009; Chapple and Thompson 2009; Chapple *et al.* 2014, 2016a, b; Tingley *et al.* 2016) and is now considered one of the most successful Australian invasive reptiles of the Pacific region (Chapple and Thompson 2009). In this chapter, we review the status of *L. delicata* outside of its native range, together with its potential to become more widely dispersed as an invasive species.

THE PROCESS OF INVASION AND ESTABLISHMENT

A species can reach a new home overseas by various means, such as being blown there by the wind (Dijkstra 2007), stuck in the mud on a migratory bird's legs (Viana *et al.* 2016), transported in or on flotsam by ocean currents (Heatwole and Levins 1972), as an inadvertent stowaway in goods or plants being transported by humans (White and Shine 2009; Hulme 2021), or deliberately introduced by humans (Michaelides *et al.* 2015). Only the last two means are open to small lizards, and transport as stowaways is probably the most common. Even that is fraught with difficulty and demands special attributes on the part of the lizards.

Chapple *et al.* (2012) and Chapple and Wong (2016) divided the process of invasion into four steps: (1) transport (uptake from the native habitat and carriage across the water barrier; (2) introduction (disembarkation at the new site); (3) establishment (survival under the new conditions and reproducing there); and (4) spread (dispersal from the point of entry into new habitats). Each of these phases poses new challenges that require special attributes of the invader. For example, a tendency to explore new sites may assist survival at the first stage of invasion and the later one of leaving cover to find a suitable environment. Chapple and Wong (2016) emphasised that suites of behaviour can be integrated into an invasion syndrome that could determine the success, or otherwise, of a potential invader. They further indicated that such syndromes, in addition to behaviour, may include life history, morphology, and physiology, all of which can mediate the response of potential invaders to new environments. The sexes may differ in their structure of such syndromes (see Chapter 4).

During spread from the point of entry and establishment at a new location, an invading species is confronted by a novel physical environment and meets a different assemblage than the one in its former native habitat – either a new set of competitors and predators or, alternatively, a vacant niche into which it can expand. Consequently, compensatory responses may be expected. Miller *et al.* (2017) addressed this issue and found that phenotypic changes occurred over a relatively short period of time. Even in their native Australian localities of origin, there was sexual dimorphism in both the body size and shape of *L. delicata*. During establishment and expansion into new habitats in their new geographic location, there were phenotypic shifts in the shape of the body, but not in its size. In two of the three invasive populations, males and females of *L. delicata* responded differently. In Hawaii, there were progressive changes in the shape of the bodies of males, but not of females, whereas in New Zealand it was the reverse. Similarly, following colonisation of Hawaii and Lord Howe Island, there were changes in the frequency of a colour morph with a mid-lateral stripe (increased over time in females; decreased in males) that was known to have sex-specific effects on fitness (Miller *et al.* 2017) (see also Chapter 5).

Chapple *et al.* (2022) also addressed the dynamics of the establishment of invasive species in a new land, using the introduction of *L. delicata* into the various Pacific islands as an example of the selective filtering out of individuals with certain attributes in favour of others that responded differently. There were differences in repeatability both within and among individuals in attributes such as general level of activity and the degree of exploratory behaviour and boldness between populations in Australia and the islands to which they were introduced, as well as among populations from different groups of islands. Collectively, the results of Chapple *et al.* (2022) suggest that biological invasions favour risk-prone individuals that progressively become more adept at altering their behavioural responses to change. As the relative numbers of these more aggressive individuals increase, they could become more challenging to the native species in the invaded communities.

Smith *et al.* (2017) found that in New Zealand the effect of *Rattus rattus* (black rat) on introduced *L. delicata* was less that of other alien predators, presumably because in its native land *L. delicata* had co-occurred with *Rattus fuscipes* (Australian bush rat) for many thousands of years and therefore probably had developed strategies that pre-adapted it to countering the impact of the ecologically similar *R. rattus* in its new land.

Young *et al.* (2022) investigated the interplay of genetics and physiology in the Australian population from which the invasive species originated and in populations arising from introductions to Lord Howe Island. Although they found some significant correlations between variables, such as metabolic rate varying significantly with temperature and body mass, and some differences in Critical Thermal Maxima (CTMax) and Critical Thermal Minima (CTMin) between locations on Lord Howe, or between locations on Lord Howe and Australia, the preferred temperatures did not differ significantly among populations. In general, the differences between localities on the mainland and Lord Howe Island, or between locations on Lord Howe Island, were fewer than expected and Young *et al.* (2022) summarised their findings as follows:

> *Overall we showed that ecophysiological traits of ectotherms can change over invasion processes. The changes of sprint speeds and cold tolerance may respond to the novel environmental conditions of the islands, but also the genetic background of the populations may play an important role in shaping physiological traits of the invasive populations. Overall, our study shows that genetic variation may not translate to phenotypic variation, indicating that the environmental*

conditions may overwhelm the genetic differences between populations. This lack of phenotypic variation between populations may suggest equal adaptive capacity under new thermal environments.

Bezzina *et al.* (2014) reasoned that a species' intelligence may influence its invasive potential and compared the ability of *L. delicata* (a species that is a notorious invader) to that of the closely related species *L. guichenoti* (which is not) to learn to select the correct arm of a Y-maze to glean food as a reward. Bezzina *et al.* (2014) discovered that over repeated tests both species shortened the time it took them to find the food, but, paradoxically, *L. guichenoti* found the food sooner than did *L. delicata*, even though at the end of repeated trials both species were still just as likely to turn the wrong rather than correct way (i.e. no learning had occurred). On further analysis, it was discovered that the difference between the two species was that *L. delicata* stopped more often for periods of immobility during the search than did *L. guichenoti*.

INVASION OF THE HAWAIIAN ISLANDS

All 27 species of extant reptiles in the Hawaiian Islands are believed to have been introduced by humans (McKeown 1996; Fisher 1997; Austin 1999; Eldredge 1999; Fisher and Ineich 2012). The most recent human-mediated importation was *L. delicata.* The first known specimen of this species within Hawaii was found in 1909 in the Moanalua Valley on Oahu. Initially identified as the endemic moth skink *Lipinia noctua* (now *Ornithuroscincus noctua*), it is in the collection of the Bernice Pauahi Bishop Museum in Honolulu (Baker 1979). It was described as a new species, namely *Lygosoma hawaiiensis*, by Loveridge (1939), and subsequently as *Lygosoma metallicum* by Brongersma (1942). In a review of the Hawaiian collection, Oliver and Shaw (1953) determined that these two names were synonyms. Hunsaker and Breese (1967) and Baker (1976) initially continued to use the name *L. metallicum*, but later the species was confirmed to be of Australian origin and not natively Hawaiian at all. Baker (1979) published correspondence from Alan Greer (from the Australian Museum, Sydney) confirming that the species was the Australian endemic *L. delicata.* Its origin was subsequently identified as Brisbane, south-eastern Queensland (Chapple *et al.* 2013a).

Lampropholis delicata from Australia made its way to the Hawaiian island of Oahu sometime about 1900 (Oliver and Shaw 1953), where it became established (Greer 1997a; Ehmann 1992; Smith 2016). Despite its lengthy residency on Oahu, Hunsaker and Breese (1967) found no evidence that it had been transferred from that island to other islands of Hawaii before the late 1960s. However, by the 1970s, it was known to have spread rapidly across the Hawaiian archipelago (Baker 1976). For example, specimens were collected from Kauai in 1963 (Hunsaker and Breeze 1967), the island of Hawaii by 1972 (Baker 1979), and Maui and Moloka'i by 1967 (S. McKeown, pers. obs.), as recorded in the addendum to Baker (1979). Indeed, Baker (1979) reported that there were already established populations on all Hawaiian Islands that had substantial populations of humans (Fig. 3.1). By 1999, *L. delicata* was established in urban areas of Lanai (Kraus 2002). Climatic condition modelling (Peace 2004) has been widely used to predict the potential distribution of species (Hutchinson and Bischof 1983; Nix 1986; Busby 1991), and the actual spread of *L. delicata* to all major inhabited islands in Hawaii conformed to expectations.

Unlike many of the more recent deliberate introductions of exotic reptiles smuggled into the Hawaiian Islands for pets (Kraus 2002, 2003; Witmer *et al.* 2007), the expansion of the range of *L. delicata* into New Zealand has been assumed to have been due to the accidental translocation of individuals to new localities in nursery plants and building materials from established populations associated with human settlements; it is also likely that vegetable produce and personal effects sometimes concealed individuals and served as a vehicle for their transport (Chapple *et al.* 2016a, 2016b).

Within much of its natural range, *L. delicata* is one of the most common species of urban reptile (e.g. Burgin 1989; White and Burgin 2004; Cogger 2010, 2018; Shea 2010). At sites such as suburban gardens (Burgin 1993; Cogger 2018), they are often present in great abundance (Mather 1986; Forsman and Shine 1995a, 1995b; Cogger 2018), sometimes to the extent of constituting the dominant species in the assemblage (Mather 1986; Taylor and Fox 2001a, 2001b). Baker (1979) estimated that at some Hawaiian sites *L. delicata* reached densities similar to those occurring in some of their native Australian habitat (300–400 individuals/100 m^2; see Chapter 11). Others also have commented on their high numbers in Hawaii as being 'remarkable' (Oliver and Shaw 1953; Baker 1979), 'abundant', or 'dense' (Quay 1973), leading Witmer *et al.* (2007) to dub this species the 'plague skink'.

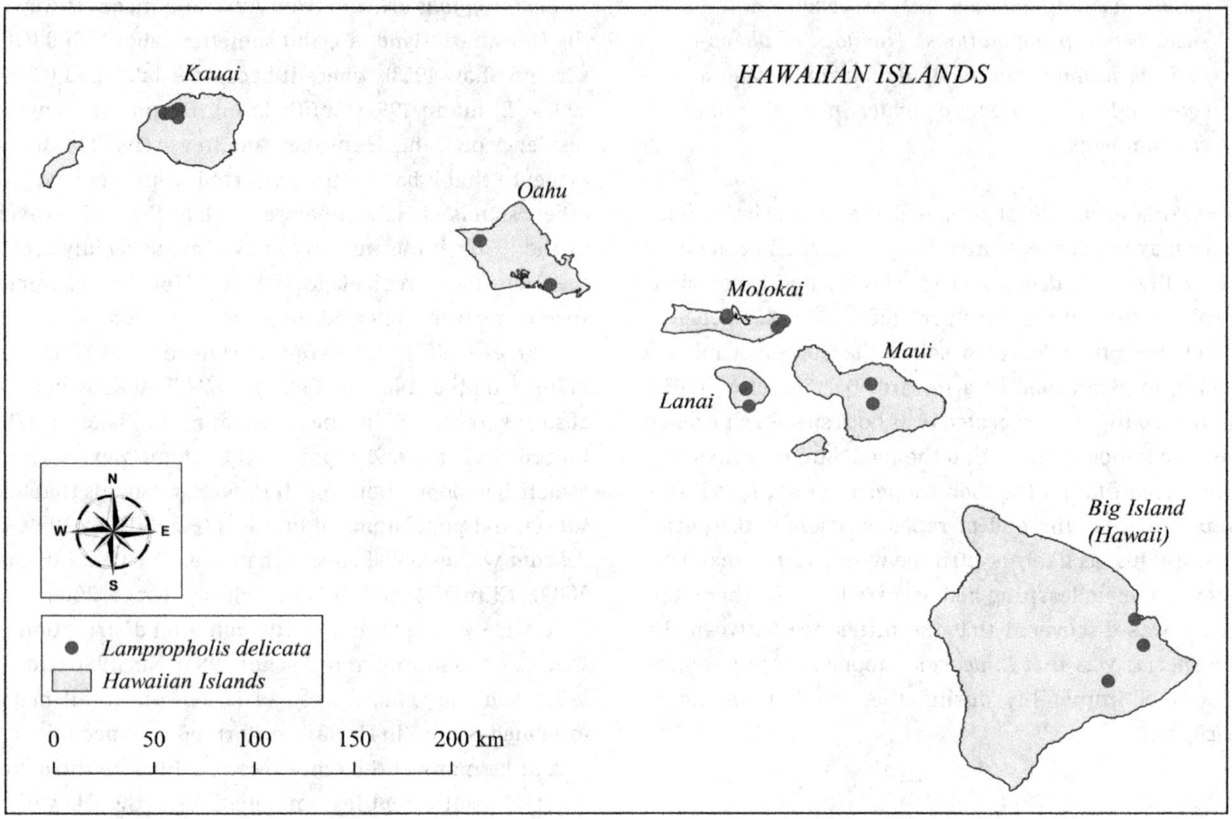

Fig. 3.1. Distribution of *Lampropholis delicata* in the Hawaiian Islands. Figure prepared by Susan King using QGIS 2.18.3 Open Source Geospatial Foundation Project http:/qgis.org with data from Chapple *et al.* (2013a), iNaturalist (2024), and the *Atlas of Living Australia* (doi:10.2619/5d712348165a0; accessed 22 March 2024), trimmed to exclude outliers requiring verification.

Although common in urban gardens along much of the Australian eastern seaboard and some areas beyond, *L. delicata* inhabits a wide diversity of vegetated habitats from low shrubby heath to the edges of rainforests (Cogger 2018; also see Chapter 11). In the Hawaiian Islands, its centre of establishment typically has been urban, due to human-mediated transport, rather than by natural spread of a species from its point of entry. Thus, sites initially occupied in Hawaii did not necessarily reflect the species' preferred habitat (Forsman and Shine 1995a, 1995b; Cogger 2018). In some areas this species has now expanded into a wide range of Hawaiian habitats, including, in addition to urban areas, the edges of rainforest, wet and dry sclerophyll habitats, and adjacent upland areas (Baker 1979).

At its location of origin in south-eastern Queensland, *L. delicata* ranges from 'dry lowland through wet upland' (Cogger 2018; also see Chapter 11), equivalent to the elevational range that Baker (1979) identified in Hawaii where the species has now become well established at higher elevations, as well as at lowland sites. For example, it has been observed at 1,220 m on Kauai and at 1,760 m on the island of Hawaii (Baker 1979). On Kauai, dense populations were recorded in association with bogs at elevations as high as 1,345 m (Kraus 2002). All other reptilian species in Hawaii inhabit the lowlands and, typically, have not been recorded above 600 m elevation (Baker 1979).

The widespread establishment of this species across the islands of Hawaii, its movement into higher elevations than those occupied by any other species of lizard there (Baker 1979), and its presence in plague numbers, at least in some areas (Witmer *et al.* 2007), demonstrates that it is, indeed, an invasive species in Hawaii.

It has been suggested that *L. delicata* may have had a negative effect on the Pacific cosmopolitan lizard *Ornithuroscincus noctua* (Peace 2004), thought to have been introduced into Hawaii previously by early Polynesian colonists (McKeown 1996; Austin 1999). *Lampropholis delicata* increased in abundance and spread within New Zealand, whereas *O. noctua* declined. Various causes for these events can be assessed.

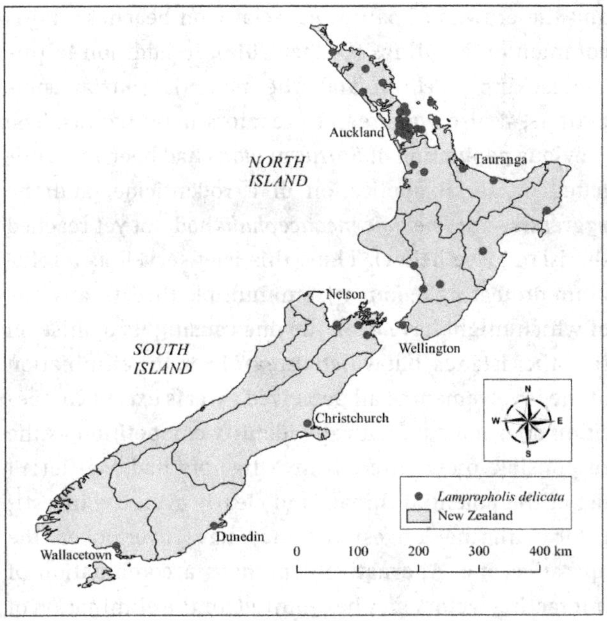

Fig. 3.2. Distribution of *Lampropholis delicata* in New Zealand. Map created by Susan King using QGIS 2.18.3, using data downloaded from the *Atlas of Living Australia* (doi:10.2619/5d712348165a0; accessed 22 March 2024) and trimmed to exclude outliers requiring verification, and supplemented by data from Chapple *et al.* (2013a, 2013b), Department of Conservation (2023), New Zealand Herpetological Society (2023), and Kelly (2023).

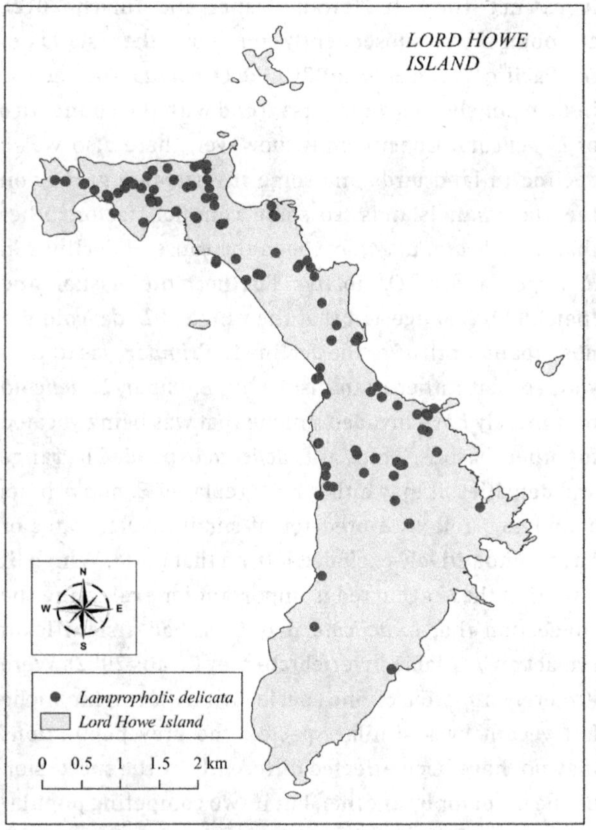

Fig. 3.3. Distribution of *Lampropholis delicata* in Lord Howe Island. Map created by Susan King using QGIS 2.18.3, using data downloaded from the *Atlas of Living Australia* (doi:10.2619/5d712348165a0; accessed 22 March 2024), trimmed to exclude outliers requiring verification, and supplemented by data from Chapple *et al.* (2013a).

Direct interspecific aggression by *L. delicata* is not a likely cause of the decline in *O. noctua* because no combat of *L. delicata* with any other species of lizard has ever been recorded in the literature, nor has it been observed by the senior author despite decades of observation of this species. Likewise, neither is there any report of *L. delicata* preying on lizards, nor has any reptile ever been found in its stomach contents (see Chapter 8).

Both *O. noctua* (McKeown 1996, 1978; Chapple *et al.* 2013b) and *L. delicata* (Belmont 1977; Crome 1981; Webb 1983; Simbotwe 1986; Lunney *et al.* 1989; Warner 1995) are insectivorous (see Chapter 8), forage in equivalent habitats (Table 3.1), and are similar in size and appearance. Direct competition could, therefore, be expected to occur between these two species and may have been a factor in the decline of *O. noctua*, especially because *L. delicata* increased in abundance and distribution and ultimately became the most abundant and widespread species of lizard in Hawaii (Oliver and Shaw 1953; Hunsaker and Breese 1967; McKoy and Busack 1970; Quay 1973; Baker 1979; Chapple *et al.* 2013a). McKeown (1978) proposed destruction of habitat and/or predation by mongooses as alternative causes of the decline of *O. noctua*. However, both these species have a high tolerance of destruction of habitat and are successful even in suburban gardens.

Nearly all species of lizards introduced onto islands come from predator-rich continental areas and therefore possess protective adaptations against predators (Dumont 2015). For example, *L. delicata* has more adaptive behavioural responses to mammalian predators than does a native New Zealand skink (Case and Bolger 1991).

Another Hawaiian species of lizard, namely *Emoia impar*, was contemporaneous with *L. delicata* and *O. noctua*. This species, once common, was last reported in Hawaii in the 1960s, and extensive searches in 1988 and 2008 failed to encounter any individuals, with the result that the United States Geological Survey declared it 'extinct' (*sic*[1]) in Hawaii (Fisher and Ineich 2012), although it was subsequently found on other islands of the Pacific (Wood *et al.* 2013). Like *O. noctua*, the decline in *E. impar* showed an inverse trend with the abundance of *L. delicata*. Concurrently, however, there also was a decline in land birds and some invertebrate groups on the Hawaiian Islands, so some common factor, other than *L. delicata*, may have been the cause of declines in *E. impar* and/or *O. noctua*. Furthermore, Fisher and Ineich (2012) suggested that the range of *L. delicata* did not expand until *after* the decline in *E. impar* was underway, so that rather than displacing *E. impar, L. delicata* may merely have invaded a niche that was being vacated for other reasons. Thus, as *L. delicata* expanded its range and densities, it may either have replaced *E. impar* in its abandoned role as a predator of small invertebrates or have competitively excluded it from that guild. Which (if either) of these occurred is important for evaluating the suggestion that *L. delicata* may have had a significant impact on available invertebrate prey (Kraus 2002). Were it merely a matter of one species filling a trophic niche left vacant by a similar species, the prey populations may not have been affected differently by the succession of one predator by another, but if two competing populations were drawing down the density of prey to the extent that one competitor was eliminated, the effect on the prey species may have been great. Ironically, however, rather than the lizards having an impact on the arthropod prey, it may have been the reverse; Fisher and Ineich (2012) suggested that a historical widespread invasion by ants may have altered the distributional patterns of some reptiles.

The above considerations may be valid for most of the islands in Hawaii. However, after *E. impar* had disappeared from the other islands in the Hawaiian chain and was considered to have been extirpated[1] throughout Hawaii, it was found to have persisted on one tiny islet, Mokapu, just off Moloka'i. On this islet, the only other species of lizards present were the skink *Cryptoblepharus poecilopleurus* and the gecko *Lepidodactylus lugubris* (Wood *et al.* 2013). Neither of these are likely competitors with *E. impar* because the former occupies trees, usually within 50 m of the beach, and may forage intertidally on invertebrates such as small crustaceans and polychaetes (Blom 2015), whereas the latter inhabits low vegetation and the crowns of palms, especially on beaches, and is common on buildings (Cogger 2018). In addition to this islet lacking *L. delicata* and other potential saurian competitors, it also was free of predators upon lizards. The previous population of *Rattus exulans* had been exterminated by aerial application of a rodenticide, and the aggressive ant *Pheidole megacephala* had not yet reached the island (see below). Thus, this islet served as a refugium protecting *E. impar* from multiple threats, any one of which might have been the one causing its demise on the other islands, but which threat? Even the elimination of the involvement of all perceived threats except competition does not automatically identify competition as the responsible one. Different investigators had a different list of the potential threats and clearly even the investigator with the longest list may have overlooked the operative one. Alternatively, perhaps a combination of interacting factors may be required for the elimination of some species. The cause of the disappearance of *E. impar* remains a mystery; default does not constitute proof and the characterisation of *E. impar*'s extirpation from most of the islands of Hawaii as 'cryptic' is an apt conclusion. Indeed, Fisher and Ineich (2012) concluded that:

> *A review of the plausible causal factors indicates that the spread of the introduced big-headed ant* Pheidole megacephala *is the most likely factor in this lizard's* [Emoia impar's] *decline. The introduction and spread of a similar skink* Lampropholis delicata *across the islands appear to temporally follow the decline of* E. impar, *although there is no evidence of competition between these species ... This study highlights the cryptic nature of this early species extinction* [sic] *as evidence that current biogeographical patterns of non-charismatic or enigmatic reptiles across the Pacific may be the historical result of early widespread invasion by ants. Conservation and restoration activities for reptiles in the tropical Pacific should consider this possibility and evaluate all evidence prior to any implementation.*

Some confusion may have arisen from a difficulty in distinguishing between *Emoia cyanura* and *E. impar*. Both are similar in size, shape, and polymorphic colour

1 A species is extinct only if it no longer exists anywhere in the world. The correct term for a species that still exists somewhere else in the world but no longer exists in a particular area where it once occurred is 'extirpated'.

pattern (Bruna *et al.* 1996a; Fisher and Ineich 2012), so that some records purported to be of one of these species could have been of the other, or both may have been co-existing. In any event, it seems unlikely that interaction of these two species would have led to the complete elimination of either in Hawaii, because on various other Pacific islands, such as Rarotonga (Bruna *et al.* 1996b; Schwaner and Ineich 1998), Henderson (Havery *et al.* 2018), and Mo'orea (McElroy 2014), they are sympatric although not completely syntopic. For example, they appear to choose different kinds of places for oviposition and *E. cyanura* exhibits preferences for, and tolerances of, higher temperatures and tends to selectively occupy the warmer, more open habitats at lower elevations. In addition, *E. impar* exhibits numerical dominance in the shadier, cooler forests at higher elevations (Table 3.1), and the relative abundance of the two species differs in response to disturbance (Schwaner and Ineich 1998). Such a pattern could have arisen in various ways: (1) by the two species simply responding solely to variation in the thermal environment to which each independently had become adapted, and accordingly sorting themselves out along elevational gradients in temperature and degree of canopy cover (a more vagile generalist, such as *E. cyanura*, may be able to handle physiological stress better and thus dominate in the more harsh environment of the lowlands); (2) by evolutionarily diverging from similar, or identical, ecological niches into somewhat divergent ones (character displacement), thereby reducing the intensity of competition between them; and/or (3) by the two species actively competing with each other, with each having the competitive advantage at its own end of an environmental gradient and each intruding upon the other from their positions of strength at opposite ends of the gradient.

The interaction between members of a given pair of species may not be identical on every island on which they chance to meet, and it may not be valid to extrapolate from one island or archipelago to another. Theoretically, it would seem possible that, depending on the kinds of habitats available, two species could share an island, each exclusively occupying the habitat to which it is best adapted but excluding the opposite member of the pair from that favoured habitat. However, under the same circumstances except with one member of the species-pair being absent, its counterpart could expand into both habitats. Different islands may have slightly different environments, or mix of habitats, and a species' competitive abilities, or physiological tolerances to the local physical conditions, may reflect the selective pressures exerted on it by the biotic and physical milieu prevailing in its region of origin. Bruna *et al.* (1996b) indicated that *E. cyanura* and *E. impar* occur on several different islands, but in each case the matrix of the other species in the assemblage differed in number and taxonomic composition, with varying degrees of similarities in ecological niches. The comparative study of these two species was conducted by Bruna *et al.* (1996b) on Rarotonga where there are three other species of lizards (*Emoia trossula, Ornithuroscincus noctua*, and *Cryptoblepharus* sp.). These species were considered to be ecologically segregated from *E. cyanura* and *E. impar* by virtue of being arboreal. However, in Hawaii, *O. noctua* forages on tree trunks and on the ground (Table 3.1). *Emoia cyanura* and *E. impar* may therefore interact – at least peripherally. Zug (1991), however, did not consider it likely (although not impossible) that they interacted sufficiently to competitively exclude either of the species of *Emoia* and went on to comment that, in Fiji, there was an even greater number of potential competitors that could further complicate interactions within the structure of the assemblage.

Clarifying the way the different biotic milieus affect the relationships of these two *Emoia* species could contribute to a better understanding of the community dynamics of insular populations and help elucidate how the nuances of the structure and composition of insular lizard assemblages influence their vulnerability to invasive species (see Chapter 11). For example, although the literature does not address the topic of the interaction of *O. noctua* and *E. impar*, if these two species were ecologically sufficiently similar that both competed with *L. delicata*, either they competed along different axes with the invader or these two species also competed. To elucidate the differing effects of *L. delicata* on the two *Emoia* species would require a study of how each species fitted into their three-way interaction. This would best be achieved by comparative studies in localities where: (1) all three species were still present together; (2) only *L. delicata* and *O. noctua* coexisted; (3) only *L. delicata* and *E. impar* coexisted; or (4) only one of these three species was present. Without such a series of studies, conclusions about the impact of the invasion of *L. delicata* remain speculative. Unfortunately, the extirpation of *E. impar* (except from one small islet on which neither of the other relevant species occurs) renders a more complete understanding of the dynamics of this particular invasion impossible. Perhaps it was competition with the invader, but also perhaps not.

Table 3.1. Comparison of some characteristics of *Lampropholis delicata* with those of the native species of lizards from Lord Howe Island and some of the introduced species in Hawaii.

Species	Length and weight (mean ± sd)	Diet and foraging	Activity period and habitat
Ornithuroscincus noctua[A], Hawaii	SVL: 41.7 ± 3.6 mm (range 34.0–48.0 mm) in females (*n* = 50); 38.8 ± 3.6 mm (range 33–45 mm) in males (*n* = 31)[1]	Forages for: invertebrates in leaf litter around roots of large trees, and on rock walls in suburban gardens and orchards;[2] and 'tree trunks' (32%), 'under loose bark' (32%), 'forest floor' (13%), and 'palm-leaf axils' (13%; *n* = 31)[3]	Reproductively active year-around[1] Predominantly diurnal, shelters under surface objects/loose topsoil at night[2] May forage nocturnally; arboreal and secretive[3,4]
Emoia cyanura, Hawaii	SVL: 45 ± 0.6 mm (range 33–35 mm; *n* = 70)[5] SVL: 46.6 ± 0.6 mm (range 33.0–65.0 mm)[6]	Primarily terrestrial, occasionally forages low on base/lower branches of trees or axils of palms/bananas[6] Insectivorous diet[7]	Diurnal and basks in patches of sun; preference for disturbed sunny areas[4, 5, 8–10] Tends to inhabit warmer, disturbed habitats and forests with open canopies at lower elevations than does *E. impar*[4]
Emoia impar, Hawaii	SVL: 43.5 ± 0.4 mm (range 29.0–50.0 mm). Body 71.2 mm[4]	Forages on forest floor (68%), tree trunks (31%)[3] 'Leaf litter skinks'[5]	Diurnal[8] and basks in patches of sun[3, 4] Tends to occupy cooler forests with moderate to closed canopies at higher elevations than does *E. cyanura*[4,10,11] Only observed on forest floor[3]
Christinus guentheri[B], Lord Howe Island	SVL: (maximum 91 mm)[12] Weight: ≤12 g[12]	Forages on the ground and in trees for insects and other invertebrates; also forages for nectar of *Lagunaria patersonia* (sally wood) and exudates from fruits[12]	Nocturnal; uncommon on Lord Howe Island but abundant on the small satellite islands, also on satellite islands of Norfolk Island; scansorial, particularly on slopes of boulders and rock faces; arboreal[11]
Oligosoma lichenigerum[C] Lord Howe Island	SVL: 78 mm (maximum 96 mm) SVL: 'about 89 mm'[13] Weight: 12 g[13]	Consumes invertebrates (beetles, spiders, moths, and ants) in the leaf litter; also forages in littoral/supra-littoral zone for invertebrates[12,14] In winter, forages in rocky areas where it breaks and consumes seabirds' eggs[12,14]	Nocturnal in summer but largely diurnal in winter; rests beneath stones, fallen timber, boulders, beach boulders, upturned boats, holes in volcanic rocks, and splits in trees[12,14]
Lampropholis delicata, introduced to Lord Howe Island	SVL: 35–51 mm Weight: 1.1 g[6]	Forages among leaf litter/ground debris[4] Eats a wide range of small invertebrates, mostly insects[7]	–

[A]Formerly *Lipinia noctua*; [B]Formerly *Phyllodactylus guentheri*; [C]Formerly *Leiolopisma licenigera*.
[1]Goldberg and Kraus (2012); [2]McKeown (1978, 1996); [3]Buden (1998, 2013); [4]Bruna *et al.* (1996b); [5]McElroy (2014); [6]Fitzsimons and Thomas (2016); [7]Wilkinson and Priddel (2011); [8]Ineich (1987); [9]Zug (1991); [10]Schwaner and Ineich (1998); [11]Department of Environment and Climate Change (2007); [12]Cogger (1983); [13]OE&H (2019a); [14]Chapple *et al.* (2011).

Summary

Lampropholis delicata reached the island of Oahu in the Hawaiian Islands from south-eastern Queensland at the latest by 1909, probably via inadvertent stowage in cargo. It was progressively transported, probably via the same means, to the other islands in the chain during the remainder of the 20th century. From establishment in its initial urban points of entry, it dispersed into diverse non-urban habitats like those occupied in Australia and where it reached similarly high population densities as those in its native land.

Competition with *L. delicata* is just one candidate for the cause of declines in *O. noctua* and *E. impar*, and the role of *L. delicata*, if any, in the decline of native species of lizards is not clear.

INVASION OF NEW ZEALAND

New Zealand has ~136 known taxa of reptiles, of which 76 species are skinks, all in the endemic genus *Oligosoma* (Hitchmough *et al.* 2021). West (1979) reported four species of Australian lizards intercepted in cargos of vegetables and fruits shipped from Australia to New Zealand. Two of these were then considered as belonging in the genus *Lampropholis* (*L. mustelina* and *L. challenger*), but now both have been reassigned to the closely related genus *Saproscincus*. Later, Chapple *et al.* (2016b) reported that 99 foreign species of lizards (in 13 families) had been successfully intercepted entering New Zealand, 36 species of which were of Australian origin. Despite this heavy influx of potential immigrants, for a long time the only alien species of reptile known to have become established in New Zealand was *L. delicata* (Robb 1974; West, 1979; Gill *et al.* 2001; Chapple *et al.* 2008, 2016a, 2016b; Chapple 2016; Harris *et al.* 2021; Hitchmough *et al.* 2021). In addition to reptiles, Pili *et al.* (2023) reported three species of frogs (*Litoria ewingii*, *Ranoidea aurea*, and *Ranoidea raniformis*) to have established populations or become invasive.

The transport of *L. delicata* from Australia to New Zealand occurred multiple times via accidental transport in cargo (Chapple *et al.* 2013b, 2016a, b). Gill *et al.* (2001) found that 75% of the intercepted herpetofauna entering New Zealand (including *L. delicata*) was found in shipped cargo and that ~50% of these consignments entered through the urban entry point of Auckland.

Over the more than 60 years since the accidental introduction of *L. delicata* into New Zealand (Chapple *et al.* 2016a), the species has become naturalised (Robb 1974, 1986; West 1979; Gill *et al.* 2001; Hitchmough *et al.* 2012) and is now considered an invasive species with an expanding range. Based on climate modelling, it has been estimated that the range of *L. delicata* may ultimately encompass most of the North Island (except for some inland areas), and some regions of the South Island (Lever 2003; Peace 2004; Chapple *et al.* 2013b; see also Fig. 1 in Peace 2004). Despite the widespread distribution of *L. delicata* in New Zealand, its ecology there remained virtually unknown at the beginning of the present century (Peace 2004), although more recently there has been increasing interest in this species (Killick *et al.* 2014; Chapple *et al.* 2016a, 2016b; Tingley *et al.* 2016; Michelangeli *et al.* 2017; Harris *et al.* 2021).

The expansion of the range of *L. delicata* within New Zealand from its current established populations may well be enhanced by additional accidental introductions into new localities. This prediction is supported by the intermittent intercepts of accidental arrivals in personal effects, freight, and cargo (Gill *et al.* 2001; Chapple *et al.* 2016a, b). Although *L. delicata* is the only alien reptilian species known to have become established in New Zealand so far (Robb 1974; West 1979; Gill *et al.* 2001), there is a risk that other species could also become established from such accidental imports (Blackburn *et al.* 2011; Chapple *et al.* 2012, 2013a). For example, *L. guichenoti* has been intercepted from freight entering New Zealand from Australia at various ports of entry, including Auckland, Wellington, and Rotorua (Chapple *et al.* 2013b).

Lampropholis delicata was initially observed in New Zealand in the mid-1960s in railway yards in South Auckland, North Island (Gill and Whitaker 2001; Lever 2003; Chapple *et al.* 2013b). It was assumed that individuals had arrived in a shipment of wooden railway sleepers (Gill and Whitaker 2001; Lever 2003). This founding population apparently originated from a single genetic lineage of *L. delicata* (Chapple *et al.* 2013a). Although not able to dismiss multiple introductions from the same origin, the first introduction originated from a forestry centre in inland northern New South Wales (NSW).

The observations that *L. delicata* lay their eggs communally (see Chapter 4) and often shelter under or in cracks in logs (Greer 1989, 1997b) provide a realistic mechanism for such a mode of movement between inland Australia and Auckland. However, few eggs of reptiles have been intercepted by New Zealand's biosecurity, and none of these belonged to *Lampropholis* (Gill *et al.* 2001). Perhaps immigrating propagules consist of juveniles or adults, either leaving Australia in that stage or hatching en route.

At the point of accidental entry in Auckland (Chapple *et al.* 2013b), *L. delicata* first became well established in the greater Auckland area, Coromandel Peninsula, Tauranga, and Te Puke (SRARNZ 2003; Peace 2004). The species remained localised in these areas for ~15 years from the mid-1960s onward. Although the numbers of *L. delicata* in Auckland and Waikato have remained substantial, the species has also expanded its range rapidly (Peace 2004) and at least 22 human-mediated translocations have occurred from sites within the area of the initial establishment to extralimital sites across the North Island and to Nelson, Christchurch, and Dunedin on the South Island (Fig. 3.2). Effectively, all sources for these populations were consistent with the founders being derived from introductions into Auckland. However, a further 13 'new' introductions from Australia have been confirmed and additional ones may also have been recent introductions. These identifiable new haplotypes of *L. delicata* in New Zealand originated from six sites within the Australian range of the species – all from populations endemic to areas, both coastal and inland, between Brisbane and Sydney. In New Zealand, they have been collected from both the North Island and South Island, with more than half the detections within New Zealand having been from outside the previous known range within the country (Chapple *et al.* 2013a; Fig. 3.2).

Due to both natural movement and human-mediated translocation, *L. delicata* has the potential to spread from its current established range both inland and south on the South Island of New Zealand. This suggestion was supported by the climatic modelling of Peace (2004), a technique that was used previously to predict the potential distribution of species (Hutchinson and Bischof 1983; Nix 1986; Busby 1991; Mitchell *et al.* 2016; Petford and Alexander 2021), and the observed spread of *L. delicata* to all major inhabited islands of Hawaii (Baker 1979).

Climate-condition modelling also predicted that *L. delicata* had the potential to expand its range much more widely on the North Island. For example, this species was predicted to be able to establish in appropriate habitats on effectively all the area north of Rotorua and south of Palmerston North, together with a broad swathe of lands along the eastern coast, and into additional isolated patches of the North Island. Compared with the predicted ultimate distribution of the species on the North Island, its predicted distribution on the South Island was much more restricted, particularly around Nelson, with the potential for a disjunct distribution as far south as Christchurch. However, with an admixture of genetic material from different source areas of Australia, presumably with different climatic tolerance, *L. delicata* was predicted to be able to expand its range farther than predicted by previous modelling (Kolbe *et al.* 2004; Peace 2004; Chapple *et al.* 2013a), and indeed may have already done so. The potential for further introductions is evident from the observation that some 65% of new introductions of *L. delicata* from Australia to New Zealand were intercepted after evading detection at the border. Many of these individuals also have been observed from areas beyond those where the species had become established (Chapple *et al.* 2013b). Based on these observations, human-mediated 'jump dispersal' has apparently been of greatest consequence in the rapid spread of *L. delicata* across New Zealand due to stowaways in freight originating from the Auckland area (Chapple *et al.* 2013a). Now the distribution of *L. delicata* in New Zealand extends from the northern tip of the North Island to the southern tip of the South Island, but this does not mean that the species has continuous distribution across all of New Zealand. Using the common practice of describing the geographic range of a species by enclosing all known locality records in a common polygon distorts the reality of its distribution much in the same way enclosing locations of individuals into a common polygon distorts the configuration of home ranges (see Chapter 16 and Fig. 2.2 in Chapter 2).

The occurrences of *L. delicata* in the South Island consist mainly of scattered records near cities or settlements of humans. Does this mean that natural habitats over much of the South Island are unsuitable for this species and it can only persist in urbanised areas? Some of the records are only of a few individuals and perhaps not indicative of established populations. Excerpts from the southernmost record of this species in New Zealand (Wallacetown) read:

> *'A Biosecurity and Biodiversity Operations report included in the chief executive's report in the agenda for the Environment Southland meeting on August 23 said there had been a report of plague skinks (*Lampropholis delicate [sic]*) in Wallacetown.... A Department of Conservation spokesperson said two plague skink incursions have been reported in Southland in separate incidents in recent months' and 'They're tiny, they're unwanted, and no-one seems to know how they got there.' (Kelly 2023)*

Thus, only two individuals were reported and so it is not known whether a permanent population was established.

Interaction of *Lampropholis delicata* with native species

There was some concern about the effect *L. delicata* might have on native New Zealand species because, in captivity, it was found to successfully compete for food with one of the endemic species of lizard in New Zealand, namely *Cyclodina aenea* (West 1979), the smallest of the New Zealand skinks (Gill and Whitaker 2001). However, *L. delicata* was not mentioned as a threat to any of the species of that genus in the recovery plan for the genus *Cyclodina* (Towns 1999), despite a substantial reduction in the historic range of *C. aenea*.

Wells *et al.* (2023) observed quantified agonistic encounters between *L. delicata* and the native skink *Oligosoma kakerakau* at sites used for basking in New Zealand. They found that 72% of the encounters between these two species resulted in aggression and that the native skink was more aggressive towards *L. delicata* than it was towards individuals of its own species. Juveniles of *O. kakerakau*, however, were more vulnerable than conspecific adults to encounters with *L. delicata* and were more likely to exhibit submissive behaviours and flee (Wells *et al.* 2023). The authors considered the native species (especially its juveniles) to be threatened by interspecific interactions and that it might be subject to competitive exclusion from sites suitable for basking.

Summary

Lampropholis delicata originally reached New Zealand from Australia, almost certainly as stowaways in cargo arriving in the Auckland area. Subsequently, it spread naturally as well as by human-assisted jump dispersal, supplemented by new invasions, including those of different haplotypes from various localities in NSW, so that now *L. delicata* is widespread in the North Island and occurs at scattered localities in the South Island, even to the southernmost tip, although some of the records may not represent permanently established populations. *Lampropholis guichenoti* has reached New Zealand in freight, but so far has been intercepted successfully and no populations have become established.

INVASION OF LORD HOWE ISLAND

In 1971, the only species of lizards occurring on Lord Howe Island were the gecko *Phyllodactylus* (now *Christinus*) *guentheri* and the skink *Leiolopisma* (now *Oligosoma*) *lichenigerum* (Cogger 1971), but at least by the 1990s *L. delicata* had been introduced there and become established (Lever 2003; Hutton *et al.* 2007; Hara *et al.* 2010; Chapple *et al.* 2013a, 2014). Subsequently, *L. delicata* expanded its range to encompass nearly the whole island (Schulz 2009; Chapple *et al.* 2013a; Fig. 3.3). Its populations on the island represent lineages from seven separate introductions from four genetically different geographic areas from their natural range between Brisbane and Sydney (Chapple *et al.* 2013a). Mitochondrial DNA and microsatellite analyses indicate intraspecific hybridisation among individuals from all these four sources (Chaplin *et al.* 2022; see Chapter 13).

Accidental introductions of *L. delicata* to Lord Howe Island were probably associated with the import of products by air and/or sea (Hara *et al.* 2010), for example in cargo (Hutton 2003; Chapple *et al.* 2013a, 2014; Moule *et al.* 2015) of horticultural materials (Gill *et al.* 2001; Lever 2003; Hara *et al.* 2010) or personal effects (Lever 2003; Chapple *et al.* 2011, 2013a). The multiple introductions identified by Chapple *et al.* (2013a) first reached an urban destination. Brown *et al.* (2007), for example, found that *L. delicata* initially became established in the urban settlement on the island and subsequently expanded outwards to Intermediate Hill and the northern hills. More recently, it became widespread across the island in all vegetation at low elevations (Chapple *et al.* 2014). *Lampropholis delicata* is commonly observed in gardens and other human-modified habitats (Hara *et al.* 2010).

Relationship of *Lampropholis delicata* with other resident species of lizards

Both *C. guentheri* and *O. lichenigerum* are listed as 'Vulnerable' under the NSW *Threatened Species Conservation Act 1995 No. 101* (Legislation.nsw.gov.au). Both are larger (snout-to-vent length 80 mm) than the introduced *L. delicata* (snout-to-vent length 45 mm; Cogger 2018).

Christinus guentheri is a nocturnal, arboreal species occupying rocks and trees; its favoured tree is the Norfolk Island hibiscus, *Lagunaria patersonia* (Cogger 2018). It obtains nectar from this plant's flowers and, additionally, licks exudate from the fruit (Cogger *et al.* 2006) – not a recorded dietary component of any Australian *Lampropholis*. However, its diet resembles that of *Lampropholis* in that it also eats beetles, spiders, moths, and ants that are foraged both at ground level and in trees.

The ecological overlap between this species and *L. delicata* would seem to be minimal because of: (1) the scarcity of plant-derived food in the diet of *L. delicata*, (2) the difference in time of daily activity between the two species, (3) the lack of arboreal foraging in *L. delicata*, and (4) the difference in the size of the body between these two species, which in turn would affect their potential competition for food. The size of the body is related to width of gape that, in turn, influences the size of prey that can be managed. Adult *L. delicata* therefore likely would compete for food only with juveniles of *C. guentheri*.

Oligosoma lichenigerum forages across a wide range of habitats from lowland to montane rainforests. It occurs on the main island of Lord Howe and also is found on the small offshore islets where there is extensive cover of native grass (OE&H 2019b). It is effectively nocturnal in winter and feeds on amphipods and insects in the littoral and supra-littoral zones (Cogger 2018); in summer it is diurnal and consumes the contents of seabirds' eggs which the skinks break by rolling them across the rocks until they break. Thus, in part of the year

L. delicata shares its time of activity and one of its kinds of habitat with *O. lichenigerum* but overlap is minimal and these species largely differ in type of food consumed and the microhabitats where they forage for it. Although not explicitly demonstrated, it may be assumed that there is limited overlap in diet between the introduced *L. delicata* and the native *C. guentheri*.

In summary it seems unlikely that *L. delicata* competes significantly with either of the native species and probably would neither affect their populations substantially, nor be affected by them.

At the time of release of the most recent *Lord Howe Island Biodiversity Management Plan* (Brown *et al.* 2007), the major threats to the two native lizards of the Island were likely invasive weeds and the destruction of habitat by clearing and trampling (Cogger *et al.* 2006; Cogger 2018). *Lampropholis delicata* is common in urban and other disturbed habitats and thus, unless the weeds become so dense as to impede movement, they can serve as habitat for this introduced skink. In fact, removal of dense weeds, such as asparagus fern (*Asparagus aethiopicus*) in forested areas (Brown *et al.* 2007), may initially reduce habitat for *L. delicata*. Accordingly, weediness on Lord Howe Island is likely to give an advantage to the introduced species over the two native lizards.

INTRASPECIFIC FLEXIBILITY IN ADAPTING TO NEW ENVIRONMENTS

Species introduced into a novel environment may encounter a different array of niches than experienced in their land of origin. There likely are assemblages of native species in the new land that occupy niches in different ways than those the invaders previously encountered, and there may be vacant niches similar to those they experienced before being translocated that are suitable, at least in part, for their unopposed occupancy. The versatility of the immigrant in successfully locking into the structure of the new assemblages and new environments is paramount in determining whether it persists in its new location. A species with rigid requirements would be less likely to become established than one with more malleable prerequisites.

Tingley *et al.* (2016) examined the intraspecific flexibility of *L. delicata* in adjusting to the structure of the niches encountered in the novel physical and biotic environments it found upon being introduced into New Zealand, Lord Howe Island, and Hawaii. They recognised three outcomes of responses to the new conditions: (1) 'niche stasis' (the finding and filling of niches familiar to those experienced in their native Australia), (2) 'niche unfilling', a euphemism for failing to colonise a similar niche in their new location as the one they occupied in their native range, i.e. the proportion of the occupied native niche that did not overlap the occupied invaded niche, and (3) 'niche expansion' (extending their breadth of niche beyond that occupied in their native range).

They showed that *L. delicata* exhibited considerable intraspecific variation in the extent it executed such shifts in niche and that had they relied solely upon data from Australia and New Zealand they would have underestimated the flexibility of this species and would have falsely concluded that it was not capable of persisting in climates beyond the limits of the niche it occupied in its country of origin.

POTENTIAL FOR INVASION OF *LAMPROPHOLIS* INTO OTHER COUNTRIES

Despite their ecological similarities, between 2001 and 2008, *L. delicata* was intercepted entering New Zealand 13 times and *L. guichenoti* was intercepted four times. In addition to the differences in numbers of intercepts, the origin of the species was also different. The sources of *L. delicata* were from locations between Brisbane and Sydney, with the majority from south-eastern Queensland/northern New South Wales. By contrast, the *L. guichenoti* that were intercepted originated from more southerly climes (Melbourne, Adelaide, and between Bega and Eden in southern New South Wales). *Lampropholis delicata* was also more likely to have arrived in New Zealand by air, whereas *L. guichenoti* was most likely to have arrived by sea. All except a single individual of the two species intercepted were alive, and most individuals of both species arrived in shipping containers (Chapple *et al.* 2013b). Due to their near identical biology and opportunity to stowaway, there is no obvious reason why *L. delicata* and not *L. guichenoti* should succeed in becoming established multiple times as an alien species in New Zealand, or why their origin should be from different latitudes. However, with the continued entry of these two species, at least into New Zealand, together with substantial jump-dispersal within the country (Chapple *et al.* 2013b), it is likely that more alien populations and/or genetic lineages will become established outside of the species' current feral range, including on Lord Howe Island.

The more individuals that successfully evade biosecurity, the more likely it is that these individuals will become established in a new area (Lockwood *et al.* 2007; Chapple *et al.* 2011). Despite New Zealand's high-quality biosecurity practices, *L. delicata* has clearly been successful in avoiding detection at the border and becoming established in the country (Chapple *et al.* 2013a). Although *L. delicata* and *L. guichenoti* have very similar biology, the latter is apparently a less frequent stowaway than the former. One possible explanation arose from a laboratory-based study by Chapple *et al.* (2011) who found that *L. delicata* has a greater tendency to explore retreats and hide, and thus it has a greater chance of becoming trapped in cargo than does *L. guichenoti*. *Lampropholis*, other than the above two species, have limited ranges, relatively lower numbers, and occupy areas away from dense human populations and transport hubs (e.g. national parks; forestry areas). These attributes render them less subject to accidental translocation by humans.

Movement of passengers and cargo between countries has occurred over centuries (Lockwood *et al.* 2007; Kraus 2009; Gill *et al.* 2001), and such movements have provided ample opportunities for the accidental transport of small animals such as lizards. For example, while there is no recorded evidence of travel by humans directly between Australia and Hawaii, New Zealand, or Lord Howe Island in pre-history, there has been human traffic between the Australian mainland and islands to the north of Australia (Indonesia, New Guinea) for centuries. The distance between Cape York, the northern extreme of Australia, and the southern coast of Papua New Guinea is only 150 km and the area contains over 100 islands, islets, reefs, and cays.

The senior author (SB) never experienced any scrutiny of personal possessions when entering Papua New Guinea, even as late as the early 1980s onward,[2] but did hear occasional comment about lizards and/or eggs being found in shipments of personal possessions arriving in Papua New Guinea from Australia. There also is substantial movement of boats between Indonesia and Australia, for example, associated with illegal fishing (Vince 2007). Despite such movement between these countries, there is no documented evidence of any *Lampropholis* species becoming established in Papua New Guinea, Indonesia, or the islands/islets in between.

Based on current knowledge of the successful invasion of *L. delicata* to Pacific islands, it would be the most likely candidate for northward translocation. Apart from being the only species of *Lampropholis* to become established as an invasive population outside of Australia, it's distribution stretches into north-eastern Australia (see Chapter 2).

Several other species, including *L. coggeri*, *L. mirabilis*, *L. robertsi*, also occur in north-eastern Australia. However, since none of these species reach the coastal limits of far-northern Australia, movement to the islands of the north before the advent of Europeans in Australia would have required movement of cargo over great distances across the country on foot, followed by a sea voyage. Modern commerce between Australia and its northern neighbours would seem to increase the probably of northward translocation of *Lampropholis* although there is no evidence that this has occurred. Also, these species are represented by smaller numbers than the two widespread ones and that may limit their opportunities for accidental dispersal by humans. Williamson and Fitter (1996) suggested that a species' abundance probably is the key characteristic influencing the successful introduction of invasive species.

Within Australia *L. mirabilis* occurs on Magnetic Island (see Chapter 2) and it is possible that it was translocated there from the mainland by Aboriginal peoples. However, it is also possible that its isolation on the island may have resulted from a once-continuous distribution being interrupted when the island was separated from the mainland by a rise in sea level ~6,000–10,000 years ago.

As mentioned above, despite their ecological similarities, *L. delicata* and *L. guichenoti* have been stowaways from multiple[3] locations on the eastern coast of Australia, although only *L. delicata* have become established in New Zealand, Lord Howe Island, and Hawaii (see above). Despite their similar biology and opportunity to stowaway, *L. delicata* but not *L. guichenoti* has succeeded in becoming established as an alien on distant islands; why their origin should be from different latitudes when the species broadly overlap over a substantial portion of their ranges (see Cogger 2018) is unknown. If it is assumed that the number of intercepts reflects the same proportion of individuals that successfully evade detection,

2 She resided in Papua and travelled between there and Australia various times.
3 For an account of the four different locations in Australia that were sources for *L. delicata* introduced into Lord Howe Island, see Chapter 14.

L. delicata would have a much higher probability of establishment than would *L. guichenoti*. With the continued entry of these two species, at least into New Zealand, together with substantial jump-dispersal within the country (Chapple *et al.* 2013b), it is likely that more alien populations and/or genetic lineages will become established outside the species' native range.

Chapple *et al.* (2011, p. 278) succinctly described an invasive as one that had 'transitioned through each stage of the introduction process (transportation, establishment, and spread)'. This description would, however, describe any species that had become established outside of its natural range and does not provide a workable basis on which to consider if a species will become detrimental within its new environment. Lee (2002) provided three criteria to determine whether a species was detrimental as well as invasive. He considered that an invasive species was one that had negative impacts on: (1) the environment, (2) human activities, or (3) human health. There is no indication that the presence of *L. delicata* has an impact on human activities or human health, and because of the species' predilection to inhabit urban gardens, it is expected that any issues in this regard would already have become obvious, both within its native range and beyond. The use of the word 'plague' to describe *L. delicata* populations (Witmer *et al.* 2007; Chapple and Thompson 2009), indicates that this skink had already been viewed as an undesirable alien species in Hawaii (Witmer *et al.* 2007), and in New Zealand (Chapple and Thompson 2009). In addition, the species fits the strongest criteria (E) of invasive species of Blackburn *et al.* (2011, p. 337) – 'fully invasive species, with individuals dispersing, surviving and reproducing at multiple sites across a greater or lesser spectrum of habitats and extent of occurrence'.

Oceanic islands, such as those where *L. delicata* has successfully become established, have assemblages of invertebrates, including insects, that often are particularly vulnerable to extinction. For example, in New Zealand, endemism of arthropods, may occur at the generic level (Gillespie and Roderick 2002). The islands that make up Hawaii also have low insect diversity with high endemism of evolutionary significance on a global scale (New 2008). Many of these taxa have narrow geographical and/or ecological limits. As such they are unusually vulnerable to selection pressure from non-native species (Gagne and Christensen 1985). Particularly in areas such as the upland bogs of Kaual where *L. delicata* has become established in 'plague' numbers, and where no other reptilian species has occurred historically (Kraus 2002), the impact of introduced predators such as *Lampropholis* might have a deleterious effect on native invertebrates and lead to a loss of genetic biodiversity or even to the extinction of some invertebrate prey and a consequent disruption to the dynamics of the local ecosystem.

While Lee (2002, p. 5) used the definition of an invasive species as one that had 'negative impacts on the environment, human activities or human health', he specifically excluded from the definition species with unknown harmful effects that were major threats to biodiversity and to the integrity of the ecosystem. A major issue with excluding the unknown is that factors that could have the greatest impact may be ignored. For example, large populations of *L. delicata* undoubtedly could be a major threat to the integrity of native ecosystems. However, even when such threats are identified they often do not illicit immediate remedial action. An apparently innocuous small skink, such as *L. delicata,* is likely to continue its effect as an invasive species without recognition of the impact it may be having on native ones. It is highly unlikely that it will become known as one of the 'world's 100 worst species' but it will certainly take a heavy toll on its invertebrate prey as it continues to expand its range on the islands it has already invaded and very likely on some that it will invade in the future. It may also be joined by its congener *L. guichenoti*.

REFERENCES

Austin CC (1999) Lizards took express train to Polynesia. *Nature* **397**(6715), 113–114. doi:10.1038/16365

Baker JK (1976) The occurrence and ecological significance of metallic skinks on the islands of Hawai'i and Kaua'i. In *Proceedings of the First Conference in Natural Sciences, Hawaii Volcanoes National Park*. 19–20 August, 1976. (Ed. CW Smith) pp. 11–17. Cooperative National Park Studies Unit, Department of Botany, University of Hawaii, Manoa.

Baker JK (1979) The rainbow skink, *Lampropholis delicata*, in Hawaii. *Pacific Science* **33**(2), 207–212.

Belmont CM (1977) Resource partitioning and habitat selection in the ground skinks. BSc (Hons) thesis. University of New England, Armidale, NSW.

Bezzina CA, Amiel JJ, Shine R (2014) Does invasion success reflect superior cognitive ability? A case study of two congeneric lizards species (*Lampropholis*, Scincidae). *PLoS One* **9**(1), e86271. doi:10.1371/journal.pone.0086271

Blackburn TM, Pyšek P, Bacher S, Carlton JT, Duncan RP, Jarošik V, Wilson JRU, Richardson DM (2011) A proposed unified framework for biological invasions. *Trends in Ecology and Evolution* **26**(7), 333–339. doi:10.1016/j.tree.2011.03.023

Blom MPK (2015) Habitat use and new locality records for *Cryptoblepharus poecilopleurus* (Squamata: Scincidae) from French Polynesia. *Herpetology Notes* **8**, 579–582.

Brongersma LD (1942) Notes on scincid lizards. *Zoologische Mededelingen* **24**, 125–152.

Brown D, Baker L (2009) The Lord Howe Island Biodiversity Management Plan: An integrated approach to recovery planning. *Environmental Management and Restoration* **10**(S1), S70–S78. doi:10.1111/j.1442-8903.2009.00449.x

Brown D, Baker L, McKay K, Murphy M (2007) *Lord Howe Island Biodiversity Management Plan*. Department of Environment and Climate Change, Sydney.

Bruna EN, Fisher RN, Case TJ (1996a) Morphological and genetic evolution appear decoupled in Pacific skinks (Squamata: Scincidae: *Emoia*). *Proceedings of the Royal Society B: Biological Sciences* **263**(1371), 681–688. doi:10.1098/rspb.1996.0102

Bruna EN, Fisher RN, Case TJ (1996b) New evidence of habitat segregation between two cryptic species of Pacific skinks (*Emoia cyanura* and *E. impar*). *Copeia* **1996**(4), 998–1005. doi:10.2307/1447665

Buden DW (1998) *The Reptiles of Kapingamarangi Atoll, Micronesia*. National Museum of Natural History, Smithsonian Institution, Washington, DC.

Buden DW (2013) Reptiles of Sorol Atoll, Yap, Federated States of Micronesia. *Pacific Science* **67**(1), 119–128. doi:10.2984/67.1.9

Burgin S (1989) The taxonomy and phylogenetic relationships of Australian scincid lizards (Scincidae: Lygosominae). PhD thesis. Macquarie University, Sydney.

Burgin S (1993) *Lampropholis*: the new 'laboratory' animals. In *Herpetology in Australia: A Diverse Discipline*. (Eds D Lunney and D Ayres) pp. 279–281. Royal Zoological Society of New South Wales, Sydney. doi:10.7882/HIA.1993

Busby JR (1991) BIOCLIM: A bioclimate analysis and prediction system. *Plant Protection Quarterly* **6**(1), 8–9.

Capinha C, Seebens H, Cassey P, García-Díaz P, Lenzner B, Mange T, Moser D, Sysek P, Rödder S, Scalera R, Winter M, Dullinger S (2017) Diversity, biogeography and the global flow of alien amphibians and reptiles. *Diversity and Distributions* **23**(1), 1313–1322. doi:10.1111/ddi.12617

Case TJ, Bolger DT (1991) The role of introduced species in shaping the distribution and abundance of island reptiles. *Evolutionary Ecology* **5**, 272–290.

Chaplin K, Smith Date K, Bray R, Miller K, Lutz M, Razeng E, Thompson M, Chapple D (2022) Intraspecific hybridisation of an invasive lizard on Lord Howe Island. *Australian Journal of Zoology* **69**(5), 184–196. doi:10.1071/ZO2145

Chapple DG (2016) Synthesising our current knowledge of New Zealand lizards. In *New Zealand Lizards*. (Ed. DG Chapple) pp. 1–12. Springer, Cham. doi:10.1007/978-3-319-41674-8_1

Chapple DG, Thompson MB (2009) Isolation and characterization of microsatellite loci from the invasive delicate skink (*Lampropholis delicata*), with cross-amplification in other Australian *Eugongylus* group species. *Conservation Genetics Resources* **1**, 55–58. doi:10.1007/s12686-009-9013-z

Chapple DG, Wong BBM (2016) 2 – The role of behavioural variation across different stages of the introduction process. In *Biological Invasions and Animal Behaviour*. (Eds JS Weis and D Sol) pp. 7–25. Cambridge University Press, Cambridge. doi:10.1017/CBO9781139939492.003

Chapple DG, Patterson GB, Bell T, Daugherty CH (2008) Taxonomic revision of the New Zealand copper skink (*Cyclodina aenea*: Squamata: Scincidae) species complex, with descriptions of two new species. *Journal of Herpetology* **42**(3), 437–452. doi:10.1670/07-110.1

Chapple DG, Simmonds SM, Wong BBM (2011) Know when to run, know when to hide: can behavioural differences explain the divergent invasion success of two sympatric lizards? *Ecology and Evolution* **1**(3), 278–289. doi:10.1002/ece3.22

Chapple DG, Simmonds SM, Wong BBM (2012) Can behavioural and personality traits influence success of unintentional species introductions? *Trends in Ecology and Evolution* **27**(1), 57–64. doi:10.1016/j.tree2011.09.010

Chapple DG, Miller KA, Kraus F, Thompson MB (2013a) Divergent introduction histories among invasive populations of the delicate skink (*Lampropholis delicata*): has the importance of genetic admixture in the success of biological invasions been overemphasized? *Diversity and Distributions* **19**(2), 134–146. doi:10.1111/j.1472-4642.2012.00919.x

Chapple DG, Whitaker AH, Chapple SNJ, Miller KA, Thompson MB (2013b) Biosecurity interceptions of an invasive lizard: Origin of stowaways and human-assisted spread within New Zealand. *Evolutionary Applications* **6**(2), 324–339. doi:10.1111/eva/12002

Chapple DG, Miller KA, Chaplin K, Barnett L, Thompson MB, Bray RD (2014) Biology of the invasive delicate skink (*Lampropholis delicata*) on Lord Howe Island. *Australian Journal of Zoology* **62**(6), 498–506. doi:10.1071/ZO14098

Chapple DG, Reardon JT, Peace IE (2016a) Origin, spread and biology of the invasive plague skink (*Lampropholis delicata*) in New Zealand. In *New Zealand Lizards*. (Ed. D Chapple) pp. 341–359. Springer, Cham. doi:10.1007/978.3.319.41674.8.13

Chapple DG, Knegtmans J, Kikillus H, van Winkel D (2016b) Biosecurity of exotic reptiles and amphibians in New Zealand: Building upon Tony Whitaker's legacy. *Journal of the Royal Society of New Zealand* **46**(1), 66–84. doi:10.1080/03036758.2015.1108344

Chapple DG, Naimo AC, Brand JA, Michelangeli N, Martin JN, Goulet CT, Brunton GH, Sih A, Wong BBM (2022) Biological invasions as a selective filter driving behavioral divergence. *Nature Communications* **13**(1), 5996. doi:10.1038/s41467-022-33755-2

Cogger HG (1971) The reptiles of Lord Howe Island. *Proceedings of the Linnean Society of New South Wales* **96**(1), 23–40.

Cogger HG (1983) *Reptiles and Amphibians of Australia*. Rev. edn. AH and AW Reed, Sydney.

Cogger HG (2010) Sydney's herpetofauna: Some observations on its history of discovery, biogeographic significance and contribution to the debate on effective reserve size. In *The Natural History of Sydney*. (Eds D Lunney, P Hutchings and D Hochuli) pp. 198–218. Royal Zoological Society of New South Wales, Sydney.

Cogger HG (2018) *Reptiles and Amphibians of Australia*. Updated 7th edn. CSIRO Publishing, Clayton South, Vic.

Cogger HG, Muir G, Shea G (2006) *A Survey of the Terrestrial Reptiles of Norfolk Island March 2005: 2. A Qualitative Assessment of the Relative Abundance of the Norfolk and Lord Howe Island Reptiles* Christinus guentheri *and* Oligosoma lichenigera *and Comparison with Previous Survey*. Department of Environment and Heritage, Canberra.

Crome B (1981) The diet of some ground-layer lizards in three woodlands of the New England Tableland of Australia. *Herpetofauna* **13**(1), 4–11.

Cuthbert RN, Diagne C, Haubrock PJ, Turbelin AJ, Courchamp F (2021) Are the '100 of the world's worst' invasive species also the costliest? *Biological Invasions* **24**(2021) doi:10.1007/s10530-021-02568-7

Department of Conservation (2023) Plague skinks. <https://www.doc.govt.nz/nature/pests-and-threats/animal-pests-and-threats/plague-skinks/> (accessed 18 June 2023).

Department of Environment and Climate Change (DECC) (2007) *Appendices Lord Howe Island Biodiversity Management Plan*. DECC, Sydney.

Dijkkstra K-D (2007) Gone with the wind: Westward dispersal across the Indian Ocean and island speciation in *Hermicordulia* dragonflies (Odonata: Corduliidae). *Zootaxa* **1438**, 27–48.

Dukes JS, Mooney HA (1999) Does global change increase the success of biological invaders? *Trends in Ecology and Evolution* **14**(4), 135–139. doi:10.1016/S0169-5347(98)01554-7

Dukes JS, Mooney HA (2004) Disruption of ecosystem processes in western North America by invasive species. *Revista Chilena de Historia Natural* **77**(3), 411–437.

Dumont CT (2015) *An Investigation into Declining Skink Populations and their Behavioural Responses to Introduced Mammalian Predators*. MSc thesis. University of Canterbury, Canterbury, New Zealand.

Ehmann H (1992) *Encyclopaedia of Australian Animals: Reptiles*. Angus & Robertson, Sydney.

Eldredge LG (1999) Numbers of Hawaiian species: Supplement 4. *Bishop Museum Occasional Papers* **58**, 72–76.

Fisher RN (1997) Dispersal and evolution of the Pacific Basin gekkonid lizards *Gehyra oceanica* and *Gehyra mutilata*. *Evolution* **51**(3), 906–921. doi:10.1111/j.1558-5646.1997.tb03672.x

Fisher R, Ineich I (2012) Cryptic extinction of a common Pacific lizard *Emoia impar* (Squamata, Scincidae) from the Hawaiian Islands. *Oryx* **46**(2), 187–195. doi:10.1017/S0030605310001778

Fitzsimons JA, Thomas JL (2016) *Emoia cyanura* (Brown-tailed copper-striped skink). Nocturnal foraging and foraging on buildings. *Herpetological Review* **47**, 669.

Forsman A, Shine R (1995a) Parallel geographic variation in body shape and reproductive life history within the Australian scincid lizard *Lampropholis delicata*. *Functional Ecology* **9**(6), 818–828. doi:10.2307/2389979

Forsman A, Shine R (1995b) The adaptive significance of colour pattern polymorphism in the Australian scincid lizard *Lampropholis delicata*. *Biological Journal of the Linnean Society* **55**(4), 273–291. doi:10/1111/j.1095-8312.1995-8312.1995.tb01066.x

Gagne WC, Christensen CC (1985) Conservation status of native terrestrial invertebrates in Hawai'i. In *Hawai'i's Terrestrial Ecosystems: Preservation and Management*. (Eds CP Stone and JM Scott) pp. 105–126. Cooperative National Park Resources Studies Unit, University of Hawaii, Manoa.

Gill BJ, Whitaker AH (2001) *New Zealand Frogs and Reptiles*. David Bateman, Auckland.

Gill BJ, Bejakovich T, Whitaker AH (2001) Records of foreign reptiles and amphibians accidentally imported to New Zealand. *New Zealand Journal of Zoology* **28**(3), 351–359. doi:10.1080/03014223.2001.9518274

Gillespie RG, Roderick GK (2002) Arthropods on islands: Colonisation, speciation, and conservation. *Annual Review of Entomology* **47**, 595–632. doi:0066-4170/02/0101-0595$14.00

Goldberg SR, Kraus F (2012) Note on reproduction of the moth skink, *Lipinia noctua* (Squamata: Scincidae) from the Pacific Islands. *Current Herpetology* **31**(1), 58–60. doi:10.5358/hsj.31.58

Greer A (1989) *The Biology and Evolution of Australian Lizards*. Surrey Beatty and Sons, Chipping Norton, NSW.

Greer AE (1997a) *The Biology and Evolution of Australian Snakes*. Surrey Beatty & Sons, Chipping Norton, NSW.

Greer A (1997b) A new species of *Lampropholis* (Squamata: Scincidae) with a restricted, high altitude distribution in eastern Australia. *Australian Zoologist* **30**, 360–368. doi:10.7882/AZ.1997.014

Hanley N, Roberts M (2019) The economic benefits of invasive species management. *People and Nature* **1**(2), 124–137. doi:10.1002/pan3.31

Hara AH, Jacobsen CM, Marr SR, Niino-DuPonte RY (2010) Hot water as a potential disinfection treatment for an invasive anuran amphibian, the coqui frog, *Eleutherodactylus coqui* Thomas (Leptodactylidae), on potted plants. *International Journal of Pest Control* **56**(3), 255–263. doi:10.1080/09670870903560114

Harris J, Smith CR, van Winkel D, Brunton DH, Goulet CT, Chapple DG (2021) Does the invasive plague species (*Lampropholis delicata*) compete with native skinks species in New Zealand? *Austral Ecology* **46**(3), 463–474. doi:10.1111/aec.12984

Havery S, Oppel S, Cole N, Dunnfield N (2018) Density of three skink species on a sub-tropical Pacific Island estimated with hierarchical distance sampling. *Herpetological Conservation and Biology* **13**(3), 507–516.

Heatwole H, Levins R (1972) Biogeography of the Puerto Rican Bank: Flotsam transport of terrestrial animals. *Ecology* **53**(1), 112–117. doi:10.2307/1935715

Hitchmough R, Anderson P, Barr B, Monks J, Lettink M, Reardon J, Tocher M, Whitaker T (2012) *Conservation Status of New Zealand Reptiles*. New Zealand Threat

Classification Series 2, Department of Conservation, Wellington.

Hitchmough R, Barr B, Knox C, Lettink M, Monks JM, Patterson GB, Reardon JT, van Winkel D, Rolfe J, Michek P (2021) *Conservation Status of New Zealand Reptiles*. New Zealand Classification Threat Series 35, Department of Conservation, Wellington.

Holmes TP, Aukema JE, Von Holle B, Liebhold A, Sills E (2009) Economic impacts on invasive species in forests. *Annals of the New York Academy of Sciences* **1162**(1), 18–38. doi:10.1111/j.1749-6632.2009.04446.x

Holway DA, Suarez, AV (1999) Animal behaviour: An essential component of invasion biology. *Trends in Ecology and Evolution* **14**(8), 328–330. doi:10.1016/S0169-5347(99)01636-5

Hulme PE (2021) Unwelcome exchange: International trade as a direct and indirect driver of biological invasions worldwide. *One Earth* **4**(5), 666–679. doi:10.1016/j.oneear.2021.04.015.

Hunsaker DH, Breese P (1967) Herpetofauna of the Hawaiian Islands. *Pacific Science* **21**, 423–428.

Hutchinson MF, Bischof RJ (1983) A new method for estimating the spatial distribution of mean seasonal and annual rainfall applied to the Hunter Valley, New South Wales. *Australian Meteorological Magazine* **31**(3), 179–184.

Hutchinson MN, Thompson MB, Stewart JR (2005) *Lampropholis delicata* (delicate skink, rainbow skink) introduction. *Herpetological Review* **36**, 450–451.

Hutton I (2003) *The Australian Geographic Book of Lord Howe Island*. Australian Geographic, Sydney.

Hutton I, Parkes JP, Sinclair ARE (2007) Reassembling island ecosystems: the case of Lord Howe Island. *Animal Conservation* **10**(1), 22–29. doi:10.1111/j.1469-1795.2006.00077.x

iNaturalist (2024) Dark-flecked garden sunskink (*Lampropholis delicata*). <https://www.inaturalist.org/taxa/38293-Lampropholis-delicata> (accessed 31 August 2024).

Ineich I (1987) Description d'une nouvelle espèce du genre *Emoia* (Sauria, Scincidae) en Polynesie francaise. *Bulletin Musée National Histoire Naturelle, Paris* **9A**, 41–49.

Kelly R (2023) The mystery of the plague skinks and the small Southland they called home. <https://www.stuff.co.nz/environment/133003992/the-mystery-of-the-plague-skinks-and-the-small-southland-town-they-called-home> (accessed 18 June 2024).

Killick S, Galbraith M, Fraser D, Waipara N, Cook J, Wairepo J, Cook J, Aguliar G (2014) Detecting the spread of a known rainbow skink (*Lampropholis delicata*) population over an island. In *Proceedings of the New Zealand Biosecurity Institute (NETS) Conference*. 29 July–1 August 2014, New Plymouth, New Zealand. <https://hdl.handle.net/10652/3098> (accessed 31 August 2024).

Kolbe JJ, Glor RE, Schettino LR, Lara AC, Larson A, Losos JB (2004) Genetic variation increases during biological invasion by a Cuban lizard. *Nature* **431**(7005), 177–181. doi:10.1038/nature02807

Kraus F (2002) New records of alien reptiles in Hawai'i. *Bishop Museum Occasional Papers* 69, 48–52.

Kraus F (2003) Invasion pathways for terrestrial vertebrates. In *Invasive Species: Vectors and Management Strategies*. (Eds G Ruiz and J Carlton) pp. 68–92. Island Press, Washington, DC.

Kraus F (2009) *Alien Reptiles and Amphibians: A Scientific Compendium and Analysis*. Springer, Berlin.

Lee CE (2002) Evolutionary genetics of invasive species. *Trends in Ecology and Evolution* **17**(8), 386–391. doi:10.1016/S0169-5347(02)02554-5

Lever C (2003) *Naturalized Reptiles and Amphibians of the World*. Oxford University Press, Oxford.

Lockwood JL, Hoopes MF, Marchetti MP (2007) *Invasion Ecology*. Blackwell Publishing, Oxford.

Loveridge A (1939) A new skink (*Leiolopisma hawaiiensis*) from Honolulu. *Proceedings of the Biological Society of Washington* **52**, 1–2.

Lunney D, Ashby E, Grigg J, O'Connell M (1989) Diets of scincid lizards Lampropholis-Guichenoti (Dumeril and Bibron) and Lampropholis-Delicata (De Vis) in Mumbulla State Forest on the south coast of New South Wales. *Wildlife Research* **16**(3), 307–312. doi:10.1071/WR9890307

Mack RN, Simberloff D, Lonsdale WM, Evans H, Clout M, Bazzaz FA (2000) Biotic invasions: Cause, epidemiology, global consequences, and control. *Ecological Applications* **10**(3), 689–710. doi:10.1890/1051-0761(2000)010[0689.BICEGC]2/0.CO;2

Marbuah G, Gren I-M, McKie B (2014) Economics of harmful invasive species: a review. *Diversity* **6**(3), 500–523. doi:10.3390/d6030500

Mather PB (1986) The taxonomic status of *Lampropholis delicata* in southeast Queensland. PhD thesis. La Trobe University, Melbourne.

McElroy MT (2014) Countergradient variation in locomotor performance of two sympatric Polynesian skinks (*Emoia impar, Emoia cyanura*). *Physiological and Biochemical Zoology* **87**(2), 222–230. doi:10.1086/674920

McGeoch MA, Latombe G (2016) Characterizing common and range expanding species. *Journal of Biogeography* **43**(2), 217–228. doi:10.1111/jbi/12642

McGeoch MA, Butchart SHM, Spear D, Marais E, Kleynhans EJ, Symes A, Chanson J, Hoffmann M (2010) Global indicators of biological invasion: species numbers, biodiversity impact and policy responses. *Diversity and Distributions* **16**(1), 95–108. doi:10.1111/j.1472-4642.2009.00633.x

McKeown S (1978) *Hawaiian Reptiles and Amphibians*. Oriental Publishing Company, Honolulu.

McKeown S (1996) *A Field Guide to Reptiles and Amphibians in the Hawaiian Islands*. Diamond Head Publications, Los Osos, CA.

McKoy CJ, Busack SD (1970) The lizards *Hemidactylus frenatus* and *Leiolopisma metallica* on the island of Hawaii. *Herpetologica* **26**, 303.<https://www.jstor.org/stable/3891255>

Michaelides SN, While GM, Zajac N, Uller T (2015) Widespread primary, but geographically restricted secondary, human introductions of wall lizards, *Podarcis muralis*. *Molecular Ecology* **24**(11), 2702–2714. doi:10.1111/mec.13206

Michelangeli M, Smith CR, Wong BBM, Chapple DG (2017) Aggression mediates dispersal tendency in an invasive species. *Animal Behaviour* **133**, 29–34. doi:10.1016/j.anbehav.2017.08.027

Miller KA, Duran A, Melville J, Thompson MB, Chapple DG (2017) Sex-specific shifts in morphology and colour pattern polymorphism during range expansion of an invasive lizard. *Journal of Biogeography* **44**(12), 2778–2788. doi:10.1111/jbi.13075

Mitchell NJ, Rodriguez N, Kuchling G, Arnall G, Kearney MR (2016) Reptile embryos and climate change: modelling limits of viability to inform translocation decisions. *Biological Conservation* **204**, 134–147. doi:10.1016/j.biocon.2016.04.004

Moule H, Chaplin K, Bray RD, Miller KA, Thompson MB, Chapple DG (2015) A matter of time: temporal variation in the introduction history and population genetic structuring of an invasive lizard. *Current Zoology* **61**(3), 456–464. doi:10.1093/czoolo/61.3.456

New TR (2008) Insect conservation on islands: setting the scene and defining the needs. In *Insect Conservation and Islands*. (Ed. TR New) pp. 3–10. Springer, Dordrecht. doi:10.1007/978-1-4020-8782-0_2

New Zealand Herpetological Society (2023) Plague skink/rainbow skink. *Lampropholis delicata* (De Vis, 1888). <https://www.reptiles.org.nz/herpetofauna/non-native/lampropholis-delicata> (accessed 31 August 2024).

Nix H (1986) A biogeographic analysis of Australian elapid snakes. In *Atlas of Elapid Snakes of Australia*. Vol. 7. (Ed. R Longmore) pp. 4–15. Australian Government Printer, Canberra.

O'Dowd DJ, Green PT, Lake PS (2003) Invasional 'meltdown' on an oceanic island. *Ecology Letters* **6**(9), 812–817. doi:10.1046/j.1461-0248.2003.00512.x

OE&H (Office of Environment and Heritage) (2019a) *Lord Howe Island southern gecko – profile*. NSW Government. <https://threatenedspecies.bionet.nsw.gov.au/profile?id=10879> (accessed 1 September 2024).

OE&H (Office of Environment and Heritage) (2019b) *Lord Howe Island skink – profile*. NSW Government. <https://threatenedspecies.bionet.nsw.gov.au/profile?id=10891> (accessed 1 September 2024).

Oliver JA, Shaw CE (1953) The amphibians and reptiles of the Hawaiian Islands. *Zoologica* **38**(5), 65–95. doi:10.5962/p.203445

Peace JE (2004) Distribution, habitat use, breeding and behavioural ecology of rainbow skinks (*Lampropholis delicata*) in New Zealand. MSc thesis. University of Auckland, Auckland.

Pejchar L, Mooney HA (2009) Invasive species, ecosystem services and human well-being. *Trends in Ecology and Evolution* **24**(9), 497–504. doi:10.1016/j.tree.2009.03.016

Petford MA, Alexander GJ (2021) Potential range shifts and climatic refugia of rupicolous reptiles in a biodiversity hotspot of South Africa. *Environmental Conservation* **48**(4), 264–273. doi:10.1017/S0376892921000370

Pili AN, Tingley R, van Winkel D, Maria L, Chapple DG (2023) The escalating global problem of accidental human-mediated transport of alien species: a case study using alien herpetofauna interceptions in New Zealand. *Biological Conservation* **278**, doi:10.1016/j.biocon.2022.109860

Pyšek P, Richardson DM (2010) Invasive species, environmental change and management, and health. *Annual Review of Environmental Resources* **35**, 25–55. doi:10.1146/annurev-environ-033009-095548

Pyšek P, Hulme PE, Simberloff D, Bacher S, Blackburn TM, Carlton JT, Dawson W, Essl F, Foxcroft LC, Genovesi P, Jeschke JM, Kühn I, Liebhold AM, Mandrak NE, Meyerson LA, Pauchard A, Pergl J, Roy HE, Seebens H, van Kleunen M, Vilà M, Wingfield MJ, Richardson DM (2020) Scientists' warning on invasive alien species. *Biological Reviews* **95**(6), 1511–1534. doi:10.1111/brv.12627

Quay WB (1973) Geographic spread and habits of the metallic skink *Lygosoma metallicum*, on Kauai, Hawaiian islands. *Journal of Herpetology* **7**, 308–309. doi:10.2307/1563018

Robb J (1974) New Zealand lizards. *New Zealand's Natural Heritage* **2**, 649–656.

Robb J (1986) *New Zealand Amphibians and Reptiles in Colour*. 2nd edn. Collins, Auckland.

Schulz M (2009) Notes on the delicate skink *Lampropholis delicata* occurring on Lord Howe Island. *Herpetofauna* **39**, 48–52.

Schwaner TD, Ineich I (1998) *Emoia cyanura* and *E. impar* (Lacertilia, Scincidae are partially syntopic in American Samoa. *Copeia* **1998**(1), 247–249. https://www.jstor.org/stable/1447730

Shea GM (2010) The suburban terrestrial reptile fauna of Sydney – winners and losers. In *The Natural History of Sydney*. (Eds D Lunney, P Hutchings and D Hochuli) pp. 154–197. Royal Zoological Society of New South Wales, Sydney. doi:10.7882/FS.2010.015

Simberloff D, Schmitz DC, Brown TC (Eds) (1997) *Strangers in Paradise*. Island Press, Washington, DC.

Simbotwe MP (1985) Comparative ecology and demography of two populations of Australian reptiles (Lacertilia, Scincidae). PhD Dissertation. University of New England, Armidale, NSW.

Smith TAH (2016) Invasion ecology of the plague skink (*Lampropholis delicata*) in Hawai'i. MSc thesis. University of Hawai'i, Manoa.

Smith HM, Dickman CR, Banks PB (2017) Using effect size benchmarks to assess when alien impacts are actually alien. *Scientific Reports* **7**, 38627. doi:10.1038/srep38627

SRARNZ (2003) Rainbow skinks reach Whangarei. *Society for Research on Amphibians and Reptiles in New Zealand*, **25**, 9.

Taylor JE, Fox BJ (2001a) Disturbance effects from fire and mining produce different lizard communities in eastern Australian forests. *Austral Ecology* **26**(2), 193–204. doi:10.1046/j.1442-9993.2001.01105.x

Taylor JE, Fox BJ (2001b) Assessing the disturbance impact on vegetation and lizard communities of fluoride pollution

interacting with fire and mining in eastern Australia. *Austral Ecology* **26**(4), 321–337. doi:10.1046/j.1442-9993.2001.01117.x
Tingley R, Thompson MB, Hartley S, Chapple DG (2016) Patterns of niche filling and expansion across the invaded ranges of an Australian lizard. *Ecography* **39**(3), 270–280. doi:10.1111/ecog.01576
Towns DR (1999) Cyclodina *spp. Skink Recovery Plan 1999–2004*. Threatened Species Recovery Plan 27, Department of Conservation, New Zealand.
Viana DS, Santamaria L, Figuerola J (2016) Migratory birds as global dispersal vectors. *Trends in Ecology and Evolution* **31**(10), 763–775. doi:10.1016/j.tree.2016.07.005
Vince J (2007) Policy responses to IUU fishing in Northern Australian waters. *Ocean and Coastal Management* **50**(8), 683–698. doi:10.1016/ocecoaman.2007.05.006
Warner FJ (1995) A comparison of the dietary contents of two sympatric scincid lizards, *Lampropholis delicata* (de Vis, 1888) and *Lampropholis guichenoti* (Duméril and Bibron, 1839) (Lacertilia: Scincidae). BAppSc (Hons) thesis. University of Western Sydney, Sydney.
Webb GA (1983) Diet in a herpetofaunal community on the Hawkesbury Sandstone Formation in the Sydney area. *Herpetofauna* **14**(2), 87–91.
Wells SJ, van Winkel D, Barr BP (2023) Interference competition following a recent invasion of plague skinks (*Lampropholis delicata*) into a nationally critical native skink population. *Pacific Conservation Biology* **30**(1). doi:10.1071/PC23003
West JA (1979) The occurrence of some exotic reptiles and amphibians in New Zealand. *Herpetofauna* **10**(2), 4–9.
White AW, Burgin S (2004) Current status and future prospects of reptiles and frogs in Sydney's urban-impacted bushland reserves. In *Urban Wildlife: More Than Meets the Eye*. (Eds D Lunney and S Burgin) pp. 109–123. Royal Zoological Society, Sydney. doi:10.7882/9780958608572
White AW, Shine R (2009) The extra-limital spread of an invasive species via 'stowaway' dispersal: toad to nowhere? *Animal Conservation* **12**(1), 38–45. doi:10.1111/j.1469-1795.2008.00218.x
Whitney KD, Gabler CA (2008) Rapid evolution in introduced species, 'invasive traits' and recipient communities: challenges for predicting invasive potential. *Diversity and Distribution* **14**(4), 569–580. doi:10.1111/j.1472-4642.2008.00473.x
Wilkinson IS, Priddel D (2011) Rodent eradication on Lord Howe Island: challenges posed by people, livestock, and threatened endemics. In *Island Invasives: Eradication and Management*. (Eds CR Veitch, MN Clout and DR Towns) pp. 508–514. IUCN, Gland.
Williamson MH, Fitter A (1996) The characters of successful invasion. *Biological Conservation* **78**(1–2), 163–170. doi:10.1016/0006-3207(96)00025-0
Witmer GW, Burke PW, Pitt WC, Avery ML (2007) Management of invasive vertebrates in the United States: an overview. *Managing Vertebrate Invasive Species* **56**, 127–136. <http://digitalcommons.unl.edu/nwrcinvasive/56>
Wood R, Burney DA, Allison A, Fisher R (2013) *Emoia impar* (Squamata, Scincidae): not extinct in the Hawaiian Islands. *Oryx* **47**(3), 328. doi:10.1017/S0030605313000616
Young A, Anderson RO, Naimo A, Alton LA, Goulet CT, Chapple DG (2022) How do the physiological traits of a lizard change during its invasion of an oceanic island? *Oecologia* **198**(11), 567–578. doi:10.1007/s00442-021-05054-y
Zug GR (1991) The lizards of Fiji: natural history and systematics. *Bishop Museum Bulletin of Zoology* **2**, 1–136.
Zug GR (2013) *Reptiles and Amphibians of the Pacific Islands: A Comprehensive Guide*. University of California Press, Berkeley.

4
Behaviour

AN ETHOGRAM

Based on observation of 154 individuals of *Lampropholis guichenoti* over a total of more than 100 hours, Torr and Shine (1994) compiled an ethogram of 45 behaviours but considered the data obtained as preliminary and subject to further investigation. Evaluating observed behaviours within the context of the circumstances under which they were performed and the responses of other individuals to them, they concluded that some behaviours, such as 'head-nodding' that in various other taxa of lizards serves as warning to intruders approaching defended territories, has a different function in *Lampropholis*. *Lampropholis* do not defend territories and 'head-nodding' also takes place in non-social contexts such as focusing on potential prey; it sometimes occurs in association with 'head tilt', a behaviour with an obvious visual function. That it lacks a social function is evidenced by the fact that other lizards in clear view of the 'nodding' individual do not respond to this behaviour. Torr and Shine (1994) hypothesised that in *Lampropholis* 'head nodding' may function in visual acuity by enhancing perception of depth through estimation of parallax and that it is the evolutionary precursory stage of the derived, ritual warning expressed by other taxa of lizards.

Despite the non-social nature of the above behaviour that performs a social function in some other taxa of lizards, aggressive and submissive signals do occur in *L. guichenoti*. Examples from the ethogram of positive responses to other individuals were head-touch, allogrooming, lie on, and sexual behaviour (see section on courtship and mating below). Others were aggressive actions or responses to aggression, such as nudging, facing-off, circling, jumping toward, flinching, moving away, moving over (another lizard), chasing, fleeing, nipping, biting, and possibly gular expansion. Both the diversity and the level of such interactions were low, and even when encounters did occur, highly ritualised fixed-action patterns seldom were employed. Torr and Shine (1994) attributed the paucity of social signals to poor visibility in the cluttered structure of this species' microhabitat.

Some behaviours seemed generalised and inquisitive in nature and may have been initial responses to stimuli of an uncertain origin or context; these included lateral orientation (postural adjustment to face a stimulus) and alerting. Others were associated with maintenance activities such as feeding (e.g. foraging, stalking, food-bite, food-shake, chewing) or drinking; thermoregulation (basking; also see Chapter 6); or comfort movements (self-grooming, mouth-scraping against a hard substrate), sloughing (scraping the body against an object and removing loose skin), defecation, and perhaps eye-closing.

The ethogram of Torr and Shine (1994) brought to light two novel behaviours, perhaps related to sensory perception, that raise the possibility that pheromones may be important in the life of *Lampropholis*: licking the substrate and cloacal dragging (walking forward while dragging the cloacal region against the ground). By dragging the cloaca, lizards may leave a chemical trail that conspecific individuals could use to track them by tasting or smelling the substrate. Can *Lampropholis* track prey by licking the substrate, for example by following ant trails? Tongue-flicking in air also may be part of this syndrome. Such pheromonal (olfactory) communication could be of special importance in a habitat where the density of low vegetation inhibits visual communication. This topic is worthy of further pursuit.

Finally there were some behaviours whose function was not suggested by context or by responses of other individuals, for example strobe-motion (movement in short, rapid jerks) and gaping.

This descriptive ethogram catalogued the behaviours exhibited by one species, *L. guichenoti*. Casual observation (Burgin, pers. obs.) indicated that at least *L. delicata* exhibits similar, or perhaps identical, behaviours and Scott *et al.* (2015) found that several behaviours whose function was not evident from the ethogram of *L. guichenoti* (leg-twitching; tail-twitching; tail-undulations = tail-lashing?) were part of the early courtship of at least one population of *L. coggeri* (a population now recognised separately as *L. similis*). Another behaviour that can be added to the list for *L. delicata* is 'shakes' in which lizards shake recently captured prey in their mouths before swallowing it (Bilcke *et al.* 2006). We suggest that the function of this behaviour may be the killing or disabling of active, wriggling prey, thereby making swallowing it easier. Bilcke *et al.* (2006) also observed *L. delicata* flattening its body during basking (see Chapter 6).

Three behaviours can be added to this ethogram. Daly (1993) observed a type of behaviour in *L. delicata* that he called 'pacing'. One individual was chasing another one, and when the lizard being chased stopped and became stationary, it moved its back legs in a manner resembling walking. When the chaser moved closer to it, both lizards moved away and the encounter ceased. Elphick *et al.* (2012) noted a *L. guichenoti* that was startled by a person, first flee but after about a metre, turned toward the person and raised the distal half of its tail perpendicular to the ground and waggled the upright section. After a few seconds of 'tail waggling' the lizard escaped into nearby vegetation.

In a follow-up paper to their ethogram, Torr and Shine (1996) described an assertive display in captive *L. guichenoti* that they dubbed as 'neck-arch'. For further discussion of this behaviour see the section *Social hierarchy* below. Torr and Shine (1993) also described 'lie-on' behaviour in which one lizard would lie on top of another, either partly or completely. The possible functions of this behaviour are discussed in Chapter 6.

SOCIAL BEHAVIOUR

Chapple *et al.* (2011) explored the social tendencies of *L. delicata* and *L. guichenoti* by comparing the proportion of time these two species spent associating with other individuals when tested against different compositions of species, i.e. the same species as the one being tested (conspecific), opposite species to the one being tested (heterospecific), or mixtures of conspecifics and heterospecifics. Neither of the species showed any statistically significant preferences for basking alone or with either the same or different species (*t*-tests; *P*-values 0.05–0.93 in the six experiments of dichotomous choices).

Hamilton *et al.* (2019) modified the above experiment to test the degree of social interaction between individuals of *L. delicata* from two geographically distant populations (Sydney versus Brisbane). In three separate experiments they set up two basking sites in which the lizards could choose between a basking site with individuals from their own locality or to bask alone (treatment 1), or between a basking site with individuals from the opposite locality from their own or to bask alone (treatment 2), or between basking with individuals from their own locality or with individuals for the opposite locality. Lizards from Brisbane showed no statistically significant preferences for any of these choices and the only statistically significant choice by lizards from Sydney was for basking alone rather than with individuals from Brisbane (*t*-test, $P < 0.001$).

Are the negative results from these two studies because the lizards: (1) are seldom able to distinguish between individuals of their own species and those of the opposite species, or between conspecific individuals from their own locality from those from other localities than their own, or (2) can they recognise such differences but simply are indifferent as to which of the two species, if any, they share basking sites?

It seems unlikely that the first hypothesis is correct because if these animals cannot discriminate between members of their own species and those of the opposite one, attempts by males to copulate with females of the opposite species would be expected. Such attempts have never been reported. Although these lizards may be willing to share basking sites with individuals from other localities, perhaps they may not be willing to mate with them. Indeed, Forsman and Shine's (1995) finding that in staged mating trials males would mate with females from their own locality but not with conspecific females from a different locality suggests that *Lampropholis* has acute powers of discrimination. Perhaps sharing a basking site or merely being in the same vicinity requires only visual recognition that accepts a wider range of individuals than are acceptable for closer contact, based, say, on short-range olfactory discrimination (see Chapter 7 for a fuller discussion of this scenario).

The behaviour of individuals can be influenced by the social environment in which they find themselves. Littlewood *et al.* (2021) investigated the interactive effects of social context on individual behaviour in *L. delicata*. The behavioural phenotype of some lizards consisted of high levels of activity, boldness, and a tendency to explore (fast lizards), whereas individual characteristics of others were the opposite – relatively inactive, shy, and not prone to explore (slow lizards). They exposed both types to a series of assays measuring their characteristics in groups of one, two, four, and eight individuals and found that inactive, slow lizards increased their activity and became bolder with increasing size of the social group, whereas fast lizards did the opposite. The degree to which an individual adjusted its behaviour was contingent upon its personal phenotype, with slow lizards showing higher levels of responsiveness than did fast lizards. Nevertheless, the disparity between the two types of behavioural phenotypes persisted and the individual's personal traits remained dominant over the changes that occurred when part of a social group.

Another way that the social environment can influence individual behaviour relates to the threat of predation. Downes and Hoefer (2004) examined the effect that size of groups of conspecifics has on the antipredatory behaviour of individual *L. delicata*. They mimicked the threat of predation by applying scent of a snake known to prey on lizards (*Drysdalia coronoides*) to a tile used by the lizards for basking compared to their response when an unscented tile was substituted as a control (low risk). They found that general activity, mobility, basking, and proximity to other individuals, all were higher in tests with the unscented tile compared to those with the scented one. The frequency of all these responses increased with increases in the number of other lizards in the social group (solitary, 2, 4, 8, 12 individuals) when the perceived risk of predation was high, with that trend less pronounced in the control tests. By contrast, the number of tongue-flicks was much higher under conditions of high risk of predation than under low risk, and they decreased with progressively higher numbers of lizards in the group. These responses all are characteristic of lizards under stress and were supplemented by unquantified responses such as vibrations of the tail, fast starts, and strobe motion. Clearly the lizards were detecting the odour of the predator, were responding to it and were behaving adaptively in response to the number of individuals in their group. A predatory snake is not going to catch more than one lizard simultaneously, and if there is only one present, it will be the individual that will be targeted. If there are multiple individuals the chances that a particular lizard will be the one caught decreases with increasing number of individuals present.

Courtship and mating

There is no evidence to date of long-term pair-bonding in any *Lampropholis* species and males appear to mate with females when a chance encounter occurs while they are foraging and even such couplings (Plate 4) do not constitute prolonged contact among free-ranging individuals (Burgin, pers. obs.). When disturbed, *Lampropholis* retreat to the nearest cover (often in leaf litter; under a rock or other convenient cover). It may be assumed, therefore, that outside of the mating season or during oviposition or brumation (see below), members of this species have limited interaction except by chance.

Shine and Shine (2002) described copulation in *L. delicata* as: 'The male was slightly larger than the female, and was biting her above the right forelimb, with his body curled around above her to the left'. Scott *et al.* (2015) staged encounters between individual pairs of lizards taken from a captive population of 40 males and 40 females of *L. similis* (then called *L. coggeri*) from each of two localities and observed 39 copulations by seven pairs. Early stages of encounters involved twitching and undulation of the tail, twitching of a leg, and tongue-flicking, perhaps equivalent to the similar behaviours observed in *L. guichenoti* (see section on ethogram above); the male pursued the female. Both sexes engaged in biting during courtship but with males doing so significantly more frequently than did females. All these kinds of visual signalling were performed by both sexes with equal frequency except for undulations of the tail that were only observed in females. The most common of the visual displays was tail-twitching (57.6%), followed by leg-twitching (22.0%) and tail-undulations (20.3%). Tongue-flicks were related to sensory perception (see Chapter 7); of the 67 that occurred 64 (95.5%) were into the air and three (4.5%) were directed toward the ground; flicking the tongue against the partner was not observed. During copulation the male bit the female and held her in his mouth. Only staged encounters between individuals originating from the same locality resulted in copulation.

The courtship and mating of *L. coggeri* also has been described (Scott *et al.* 2015). Biting and visual signalling occurred in every courtship; the male bit more frequently than did the female. Visual signalling, such as tail-twitching (the most frequent visual signal) and leg-twitching were employed with equal frequency between the sexes, whereas undulation of the tail was observed only in females. In lizards, odours are detected by transferring scent from the environment via the tongue into the chemosensitive vomeronasal organ in the roof of the mouth. Thus, tongue-flicking was interpreted by the investigators as a means of identifying odour in the immediate environment. Males and females tongue-flicked with equal frequency during encounters. Most flicks of the tongue were directed into the air but some were directed at the substrate. No tongue-flicking directly against the prospective partner was observed.

Formation of 'balls' of lizards

A most unusual behaviour has been observed in *L. guichenoti*. Multiple individuals grasp another individual in their jaws and together they become entwined into a 'ball' (Fig. 4.1). Cogger (1989) described one such group as 'a tumbling, twisting, squirming "ball" of about 16 Grass Skinks' [either *L. guichenoti* or *L. delicata*, or both; Cogger, pers. comm.], each individual skink having a firm grip with its jaws on another skink. After about five minutes they scattered in all directions'. The lizards in the available photographs cannot be sexed reliably but in none of them is there any evidence of mating. Although an occasional individual in the ball may not be biting another lizard (e.g. one in seven in Fig. 4.1) most are actively doing so. It does not appear that the other six are all attempting to interact with the non-biter, but that they are biting any lizard in their vicinity. Burgin observed such aggregations in a small townhouse courtyard where *L. delicata* and *L. guichenoti* foraged year-round. Most tussles appeared to occur in late autumn and began when two individuals came in contact while foraging; their fighting attracted others to join in (Burgin, pers. obs.). The above observations led to speculation as to the function of these 'balls'. Finally, Traill (2001) discovered that it represented a group of males attempting to mate with a female. He witnessed one male in a ball achieve a partial copulation but after a union of only a few seconds, the female escaped only to be pinned down again after traveling ~70 cm. These balls thus seem to be a simultaneous attempt by multiple males to mate with a female. This is not the only context within which mating occurs as mating of one male and one female does take place, at least in some species of *Lampropholis* (Plate 4). However if other males are in the vicinity they join in and the event transforms into a 'gang rape'.

Fig. 4.1. A 'ball' of seven *Lampropholis guichenoti*. Photograph courtesy of Richard Shine. For a photograph of a 'ball' of four of *L. guichenoti* see Griffiths (1987).

Social hierarchy

The most common social structure in lizards is territoriality but dominance hierarchies can occur (Done and Heatwole 1977a, b; Heatwole and Taylor 1987). *Lampropholis guichenoti* engages in a behaviour called neck-arch (see section *An ethogram* above). It consists of an arching of the neck as an assertive display used in displacing other individuals from sites used for basking or sheltering. So far, this behaviour has not been reported from this species in the wild and in a study of 20 groups of captive individuals Torr and Shine (1996) observed it only in one. In that group of nine individuals, males tended to dominate females and larger individuals tended to dominate smaller ones. As a result, large males were more often seen sheltering with females than were smaller males and two lizards maintained top dominance over all the others. The frequency of these displays did not decrease over time and thus probably served to maintain dominance as well as to establish it.

We suggest that the dominance hierarchy observed by Torr and Shine (1996) may have been a product of the conditions of confinement. Done and Heatwole (1977a) earlier found that crowding engendered establishment of a social structure that resembled a dominance hierarchy in two other species of Australian skinks, *Eulamprus* (formerly *Sphenomorphus*) *kosciuskoi* and *Eulamprus quoyii*. In the former species the aggressive displays were accompanied by chasing and fighting, but in the latter, they were merely ritualistic without overt combat. The more dominant animals in the hierarchy fed and mated more frequently than did subordinates; also the dominant ones were better able to avail themselves of thermoregulatory sites, i.e. basking sites under cool conditions and shade under hot ones. Dominant animals obtained more food than did subordinate individuals and only the dominant male mated successfully. The animal lowest in the hierarchy died. These authors considered that the emergence of a dominance hierarchy in their experiment may have been a response to more crowded conditions than those experienced in the wild. Crowding may reduce individual's access to a finite amount of food and/or to scarce vital resources such as sites for basking or sheltering.

Done and Heatwole (1977b) found that the hormones testosterone and estradiol increased aggressiveness in male *E. kosciuskoi*, and when injected in subdominant males caused a shift in their dominance to a higher rank in the hierarchy. Adrenaline caused a temporary increase in the level of activity and aggressiveness but no change in social structure. Thyroxine did not affect social behaviour. It is likely that similar hormonal controls also operate in *Lampropholis*. Dominance hierarchies under crowded conditions are not universal among skinks, however, as similar confinement to that imposed on *Eulamprus* did not affect the behaviour of *Ctenotus robustus*.

Communal associations

Some lizards aggregate in sheltered sites, or sequentially share a common site for a particular activity. At least in some species the composition of aggregations may not be random. For example, juveniles tend to aggregate apart from adults more often than with them than expected by chance and unisexual groupings of adults are less common than both sexes aggregated together than expected at random (Kearney *et al.* 2001).

Communal nesting

Doody *et al.* (2009) reviewed communal nesting and found it to be common, especially in lizards. It has evolved multiple times in skinks; of the species whose nesting is known, 75% were recorded as nesting communally.

Loose aggregations of lizards' eggs have been observed under leaves between the buttresses of rainforest trees in Australia. The sizes of the eggs in these aggregations were the appropriate dimensions and appearance of eggs of *L. delicata* in sufficient number to be the clutches of at least two individuals. No other reptile that laid equivalent-sized/shaped eggs was present in that area (Burgin, pers. obs.).

Elsewhere, communal nesting has been documented repeatedly in *Lampropholis* (Greer 1989; Shea and Sadlier 2000) including on Lord Howe Island (Chapple *et al.* 2014), and New Zealand (Chapple *et al.* 2016). Communal nests of *L. delicata* containing up to 110 eggs contributed by as many as 36 females have been encountered at a single site (Rawlinson 1974). Mitchell (1959) reported on a communal nest of *L. guichenoti* containing 211 eggs of which 62 had just hatched and the young were still present; the remaining eggs contained fully formed embryos except for six infertile eggs. He also reported aggregation of 49, 50, 88, and 'many dozens' of eggs as well as an old nest with 29 eggshells. Other authors have found aggregations of up to 50 eggs (Clarke 1965), 132 eggs (Wells 1981), and varying between 10 and more than 100 (Burgin, pers. obs.). Cheetham *et al.* (2011) found that about two-thirds of the females of *L. delicata* nested communally. The record number of eggs in a communal nest is 497 and was recorded by Wells (1979). All these eggs had hatched except one (an *L. guichenoti*) so the identity of the others was not possible but could have been either *L. guichenoti* and/or *L. delicata* as hatchlings of both species were found within 2 metres of the nest. The large number of eggshells raises the possibility that they might represent an accumulation over multiple seasons of reproduction; however, the piece of corrugated iron under which they were located had been in place only for 18 months.

Multiple species may share communal nests. Pengilley (1972) reported a joint nest containing eggs both from *L. delicata* and *Anepischetosia maccoyi*; also mixed aggregations of eggs contributed jointly by *L. delicata* and *L. guichenoti* have been observed (Wells 1981; Burgin 1993). Doody (2019) reported a communal nest buried under 6 cm of soil in a crevice of bedrock that contained the eggs of three species of lizards. Amid a clutch of seven large eggs from *Intellagama lesueuriii* were two other clutches of smaller eggs, one of eight eggs from *Acritoscincus duperreyi* and the other of four eggs from *L. guichenoti* (species identifications verified by taking the eggs to a laboratory where they hatched). Shine (1999) reported four nests (total of 100 eggs) containing eggs of both *Acritoscincus duperreyi* and *Lampropholis guichenoti*. Shea and Sadlier (2000) found a communal nest containing 24 eggs, which they allowed to hatch. One group of 14 eggs from this nest hatched between late February to mid-March and yielded one *L. delicata* and 13 *Saproscincus mustelinus*. The remaining eggs hatched over the remaining 3 weeks; all were *L. delicata*. The extended period of hatching suggests that oviposition also occurred over a length of some weeks which in turn indicates that females were not attracted to the particular site by other ovipositing females but perhaps by the presence of the eggs themselves, or their odour.

In addition to the two widespread species, at least two of the species of *Lampropholis* with smaller ranges also nest communally. Couper and Schneider (1995) reported an aggregation of 53 eggs of *L. adonis* in a 'pocket of insect frass and decaying vegetation between sheets of corrugated iron' in a well-shaded clearing in notophyll vine forest. Several lizards were associated with the nest and it was estimated that the communal clutch was the product of at least eight females. Turner (2006) found three communal nests containing eggs both of *L. elongata* and *Saproscincus tetradactylus*, one with 16 eggs, one with 25 eggs, and another with 9 eggs. Positioning of the eggs in the nests varied from eggs being in separate clutches several centimetres apart to the eggs of *S. tetradactylus* deposited 'amongst and on top' of those of *L. coggeri*. In one case a gravid female was found on top of a cluster of eggs.

The recorded sites used for communal nests include buried 'six inches' in sawdust-like material in the centre of a grasstree stump, in loose loamy soil (Mitchell 1959), a cavity in the soil beneath a small sandstone rock (Wells 1981), 'in a soil cavity' and beneath a stone (Turner 2006), in rock walls, under logs, and in rock crevices (Burgin, pers. obs.) and under rocks and tussocks of grass on loose soil (Shea and Sadlier 2000). According to Green (1965) although *L. delicata* is 'an excellent climber in captivity, it has not been found to ascend more than a few inches above the ground in its natural habitat'. Cheetham *et al.* (2011) also found it usually to be a ground-dweller but that females in search of a nesting site in his study area resorted to climbing onto sandstone rocks that they did not visit otherwise in their daily activities. Chosen sites for nesting were narrow horizontal crevices in the rocks that were shaded from solar radiation and where humidity was higher (averaged 94%) than at random alternative sites.

Various investigators (Graves and Duvall 1995; Doody *et al.* 2009; Elphick *et al.* 2013; Domingos *et al.* 2017) have speculated as to the advantages accruing from communal nesting as compared to solitary ones, such as a saving of time and energy and ensuring the survival of mothers by copying previous females' selection of nest sites, improving individual fitness associated with thermoregulation, and protection against predators (also see Chapter 11). Cheetham *et al.* (2011), however, reported that eggs from communal nests were three times more likely to fail than those from solitary nests and suggested that the high mortality rate in communal nests likely was caused by desiccation, perhaps because eggs were displaced from their original location by subsequent gravid females seeking sites for oviposition. Such loss of eggs constitutes a clear cost of communal nesting (see Chapter 10 for greater risk of fungal infections of eggs in communal nests and of transference of ticks between communally nesting females). Radder and Shine (2007) discounted conservation of metabolic heat as an advantage of communal nesting and found that evidence for uptake of water to be equivocal but concluded that 'The sensitivity of developmental processes to relatively minor modifications of incubation conditions may help to explain an otherwise puzzling phenomenon: the widespread occurrence of communal oviposition in squamate reptiles'.

Communal brumation

Brumation in reptiles and amphibians is a form of dormancy under cold conditions that is the equivalent to hibernation in birds and mammals (Lilleywhite and Wray 2008).

Simbotwe (1985) reported aggregations of *L. delicata*, *L. guichenoti*, and *Hemiergis decresiensis* 'sheltering' communally. Powell *et al.* (1978) observed two aggregations of torpid *L. guichenoti* within crevices in two logs split open for firewood in winter (also see Chapter 6). The woodpile in which the lizards were found contained sections of logs sawn into lengths of about 0.25 m all of which had lain there for 2 years. Six sections of logs were split on 14 August but only two contained lizards. One of these logs contained nine males and nine females and the other seven males and nine females so the sexes do not seem to segregate. However, juveniles and adults did seem to occupy separate sites. In one log all lizards but one were adult (34 mm or larger snout-to-vent; juvenile to adult ratio 1:17); whereas in the other log all but three were juveniles (<34 mm snout-to-vent; j:a ratio 13:3). This could have been an artefact, however, if the size of the crevices were different in the two logs (not measured). None of the other logs from that woodpile were found to contain lizards that winter.

Hoser (2009a) found 15 *L. ronai*[1] ranging from the previous season's juveniles to adults in an aggregation under a piece of wood. All were in a single clump except for one animal 6 cm away but under the same piece of wood. The aggregation did not occur because of a lack of other choices of cover as there were other similar pieces of wood nearby, none of which sheltered any *Lampropholis*. Thus, these lizards were probably attracted by the presence of other conspecific individuals. The air temperature at the site was 1°C. Unfortunately, the temperature of the substrate from within the mass of lizards was not measured. Hoser (2009a) suggested that such aggregations might have conserved heat generated by the lizards' biological activity. However, because lizards lack any insulation such as feathers or hair, have extremely low metabolic rates that produces negligible heat at such low temperatures, and because of their small size and consequent high surface-area-to-volume ratios (see Chapter 6), small lizards such as *Lampropholis* lose heat rapidly, even if aggregated.

Shah *et al.* (2003) reviewed aggregation in large skinks and geckos that inhabit cool areas and found that huddled groups heated and cooled more slowly than did solitary individuals. They suggested that the thermal benefits arising from aggregation by large species may have been the stimulus for its evolution and that it persisted in small derivative species as a social response even after little thermal benefit accrued from its continuance. An argument against this hypothesis, however, is that even for an aggregation of large lizards the retention of heat under cool conditions would be ephemeral at best and countered by the fact that the huddled mass also warms up again more slowly than isolated individuals when favourable temperatures return.

This chapter has covered behaviour in a general descriptive way (ethogram) and dealt with social structure and communal behaviour. Chapter 6 covers behaviour in the particular context of thermoregulation.

HOME RANGE

Information on the movements of individual *Lampropholis* is scant. Simbotwe (1985) toe-clipped 86 *L. guichenoti* for individual recognition in a plot 100 by 100 m, of which only eight (9.3%) were recaptured one or more times during the annual active season. The greatest total distance known to be moved by any one individual was 70 m. The reason for the low recapture rate is unknown. In a similar study, in which marked lizards were caught more than once in a grid of pitfall traps, Milton (1980) marked individuals of *L. guichenoti* and plotted their locations of capture on a gridded map of his study site. The small size of his plot (7 × 8 m) and the fact that the number of recaptures were small meant that few animals provided sufficient data for reliable analysis. Further confounding potential analysis was the fact that four of the six species marked in the plot were found at least once either on the border of the plot or within less than half a metre from it, and two of these had all their captures at half a metre or less from the border. Nevertheless, some useful data were obtained. One animal was always found in a small area (<1 m^2) near the centre of the plot. By contrast two animals each moved 6 m between captures as estimated in reference to the grid on the map. However each of these lizards were recaptured once either on the border of the plot or within 20 cm of it and consequently their respective home ranges may have extended beyond the boundary. Accordingly the areas of home range of these two species of ~7.5 and 5.5 m^2 should be considered as minimal estimates. Of the three animals that were only captured twice, the distance travelled between captures was 1 m, 1 m, and 2.5 m, but two of these individuals were only captured very near the border. The most important finding of this study was that the two largest home ranges not only partly overlapped with each other but one almost completely enclosed the tiny home range of the third individual, and over a small area all three overlapped.

Pengilley (1972) found a slightly skewed bell-shaped curve of the distances that *L. guichenoti* moved between points of capture. Values[2] ranged from less than 1.5 m, peaked at 5 m, and extended to 14 m (with one outlier at 26 m). Unfortunately, none of the above studies were adequate to ascertain whether males and females had different sizes of home ranges or whether there are seasonal differences in the distances moved. For example, males may move greater distances during the mating season when in search for females.

The most comprehensive study of home ranges for *Lampropholis* so far was conducted by Purani (1994). He

1 A disputed name for some populations otherwise considered as *Lampropholis delicata* (see Chapter 14).
2 Pengilley (1972) reported his values in feet. We converted them to metres.

saturated a plot with pitfall traps and checked them three times per day and obtained multiple recaptures of marked individuals of both *L. delicata* and *L. guichenoti*. He mapped the type of substrate on a grid of his study area (Plate 3) and superimposed on it the locations of each of the individuals that had sufficient recaptures for assessing their home ranges. Note that some recaptures were on, or near, the edge of the grid and consequently the home ranges of those individual may have extended beyond its borders. Plate 3 shows a map of the types of substrate in the grid and one each of Purani's several maps of the home ranges of individual *L. delicata* and *L. guichenoti* connected by lines between locations in chronological order of their recaptures. One *L. delicata* was captured a total of 11 times during the 12 weeks of sampling and a *L. guichenoti* was collected nine times.

What can be gleaned collectively from the above studies is that: (1) there is great variability among individuals of *Lampropholis* in the reported sizes of their home ranges and the distances they move, and that (2) home ranges of individuals, and species, seem to broadly overlap, i.e. they constitute home ranges, not defended territories. It should also be noted that the conventional way of portraying home ranges by drawing a polygon by connecting the locations at which an individual is observed on different occasions does not necessarily outline the species' home range. Such a polygon, for example, could enclose a series of small isolated patches used alternately by an individual with other patches within the polygon not used except when the individual moved across them in transit from one favourable patch to another (see Chapter 16 for an expansion of this discussion).

ACTIVITY

Field studies

Lampropholis delicata exhibits a seasonal activity pattern with greater activity occurring during the summer months (e.g. peaks in November–December on Lord Howe Island; Chapple *et al.* 2014) and is diurnal (see Chapter 11). This daily cycle probably applies to most other species of the genus, and the seasonal one also, with the possible exception of populations in tropical areas. Greer (1997), however, found that most specimens of *L. elongata* were found 'sheltering under timber and rocks during the day' although a few were active on the surface in daytime. In Purani's (1994) study (see section immediately above) pitfall traps were cleared at 9 am, noon, and 3 pm daily and they had approximately equal numbers of *L. delicata* (30–35%, n = 168) of captures at each clearance. *Lampropholis guichenoti* had equal numbers (38% each, n = 159) at the clearances of 9 am and noon and 39% at the 3 pm clearance.

Figure 4.2 shows the activity cycle of *L. delicata* and *L. guichenoti* at three different sites near Armidale, NSW. At all sites and for both species level of activity in summer was bimodal with one peak in the morning and the second one in the afternoon, but with lesser activity occurring during the heat of late morning to midday. Presumably this pattern protects the animals from damaging heat or metabolic excess during the hottest time of day.

There was a seasonal shift in the activity cycle of these species. In the two areas in which it occurred (Pine Forest and Blue Hole), in autumn *L guichenoti* shifted to a unimodal pattern with peak activity occurring in either morning or midday, commensurate with the cooler conditions prevailing at that time of year at Puddledock. *Lampropholis delicata* exhibited a similar, but more erratic, shift.

These shifts can be interpreted in terms as adaptive; the bimodal one would protect the lizard from the heat-stress and metabolic excesses imposed by the midday heat of summer, whereas under the cooler conditions of autumn, the more moderate temperatures at midday would allow the animal to maintain body temperatures at optimal levels more easily.

There was one difference between the two species. In the one area in which they both occurred (Pine Forest), *L. delicata* maintained a bimodal peak of activity both in summer and autumn, whereas *L. guichenoti* converted from a bimodal peak in summer to a unimodal one centred between the previous morning and afternoon peaks of summer (Fig. 4.2).

Simbotwe (1985) found that in eucalypt woodland at Newholme, also near Armidale, NSW, *L. guichenoti* had a unimodal activity period peaking at about noon for adults and ~2 hours earlier for juveniles (see Fig. 11.18 in Chapter 11) but indicated in text that in summer it showed 'bimodal activity, with peaks of activity occurring early in the day when the environment is warming and again later when it is cooling down (16:00–18:00 hours)', thus confirming, for a different locality and type of vegetation, the general pattern Belmont (1977) had found for this species in Pine Forest.

Although all species of *Lampropholis* are considered to be diurnal, Schulz and Eyre (1997) reported that 'on warm nights between November and March individuals

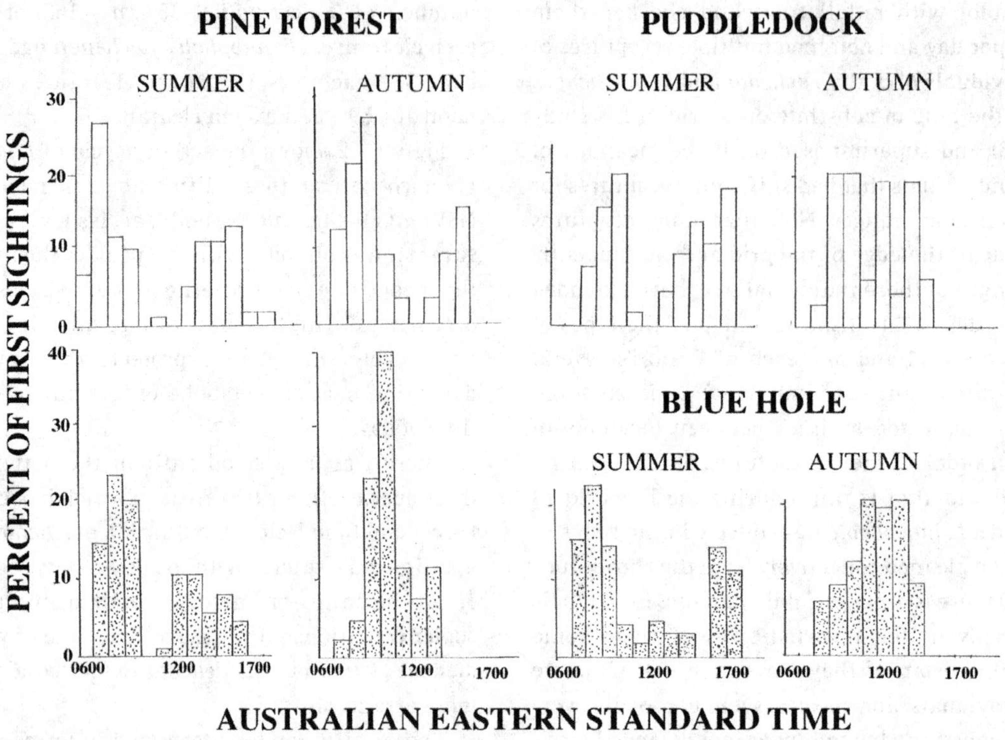

Fig. 4.2. Activity cycle of *Lampropholis delicata* (white histograms) and *Lampropholis guichenoti* (dotted histograms) in summer and autumn in three different habitats. Re-arranged and re-formatted from Belmont (1977) and published with her permission.

[of *L. delicata*] were commonly encountered active in the outbuildings and around the house to the early hours of the morning'. No mention was made as to whether lights were left on at night at those buildings; if so, that might have induced this unusual behaviour. Hoser (2009b) reported a shift in location of *L. delicata* at night after heavy rains and postulated that this unusual behaviour may have been because the rain washed them out of their habitual nocturnal retreats and their movement was to an unflooded area.

Biological rhythms

Endogenous or innate periodic behavioural cycles (circadian and circannual rhythms) may be synchronised by daily, lunar, and/or seasonal variations. These biological rhythms are ecologically important since they preadapt animals to environmental fluctuation; often photoperiod serves as an environmental cue (*Zeitgeber*) that re-sets the internal biological clock in tune with daily or seasonal environmental cycles (Golombek and Rosenstein 2010).

The pineal gland in lizards also exhibits a nocturnal-diurnal cycle in the synthesis of the hormone melatonin from serotonin peaking at night and its return to serotonin by day. The conversion of one to the other and back again is influenced by the daily cycle of light versus dark and by the fluctuation of temperature during the day–night cycle (Underwood and Calaban 1987). The pineal thus acts as a transducer that translates environmental information into an internal cue in the form of the pineal's melatonin rhythm. That rhythm, in turn, controls the phase and periodicity of 24-hour (circadian) biological rhythms, thereby ensuring that events in the lizard's daily routine occur at an appropriate time.

Lizards may alter their activity periods seasonally as environmental conditions change. Firth (1968) addressed this topic by recording the activity of individual *L. guichenoti* and *Hemiergis decresiensis* in aktographs under different conditions of light and temperature in the laboratory. Each of 10 aktographs consisted of a covered petri dish balanced on the point of an electrically connected screw. Nine slightly shorter screws were placed beneath each petri dish and evenly spaced around its periphery; every third one was connected to a copper electrode. As the lizard in the petri dish moved, the dish tilted, thereby breaking the circuit with an electrode on one side of the dish and establishing contact with a different one on the other side. These events were recorded as burnt tracings on a paper chart on the revolving drum of a 10-channel Ambrose event recorder. This apparatus was contained in a chamber in which photoperiod and temperature could be controlled and the results compared with the activity cycle under ambient conditions of light and temperature, and with the pattern of activity of *H. decresiensis*, a species sharing the same habitat with *L. guichenoti* but utilising a different microhabitat for foraging. Whereas *L. guichenoti* forages on the surface and has relatively normally sized legs. *Hemiergis decresiensis'* legs are reduced in size, a locomotory adaptation probably related to its habit of foraging in dense clumps of grass.

Firth's results confirm the diurnality of *L. guichenoti*. Under the natural cycle of light and temperature only rarely was there even a slight amount of movement in the dark and activity was concentrated in the morning at the onset of, or just after, temperature began to rise from its nocturnal low. This species maintained high diurnality (88.1–99.7% of the time) not only under natural conditions, but also under an imposed photoperiod of 12h:12h reinforcing the natural light cycle, and in complete darkness. Only under constant light (both under fluctuating and constant temperatures) did diurnality, corresponding to the natural light:dark cycle, drop to lower levels (42.8–64.8%). Thus, there was an endogenous rhythm that persisted even in constant darkness, but this rhythm was less distinct in constant light.

The activity range is defined as the temperature above which a species will not voluntarily remain active in foraging and other activities outside of their retreats and the results of Firth's (1968) study are in line with independent measures of that thermal endpoint (see Chapter 6). Animals in an aktograph in the laboratory, however, still showed some spontaneous activity at temperatures as low as 5°C, i.e. only 1–2°C above their CTMin (see Table 6.1 in Chapter 6). It is unlikely that individuals would voluntarily leave shelter and engage in activity on the surface at such low temperatures and these extremes probably merely indicate temperatures at which the animals occasionally move around within their winter refugia, rather than the lower end of the voluntary activity range for activity on the surface.

PERSONALITY

Increasingly, it has become accepted that individual animals other than humans have personal behavioural idiosyncrasies (personality) that tend to persist over time and in different contexts (Carter *et al.* 2013; Chung *et al.* 2017; Polverino *et al.* 2023) although there is some behavioural plasticity. For example, individuals may differ in their adjustment to different environments and in the consistency of their behaviour over time. Species may differ in their personalities and comparison of *L. delicata* and *L. guichenoti* revealed such differences. Whilst neither the non-directed behaviours nor the tendency of individuals to group differed between them, the former species is more exploratory than the latter and consequently is more effective in locating novel environments and finding basking sites (Chapple *et al.* 2011; also see Chapter 3 for how these differences enhance the dispersal of *L. delicata* to foreign countries).

Another aspect of the personality of a lizard is its 'approach distance', defined as 'the distance to which an animal will allow a potential predator to approach before attempting escape'. Heatwole (1968) found that lizards that were well camouflaged against the background of their habitat had shorter approach distances and remained quiescent longer upon the approach of a person than did those that were more conspicuous. For a comparison of the approach distances of *L. guichenoti* in urbanised and natural areas see Chapter 11 and Prosser *et al.* (2006).

Being temporarily caught and escaping only by autotomising its tail, either in part or in its entirely, alters the personality of the victim while the tail is regenerating (Michelangeli *et al.* 2020). The response is different depending on the extent of the loss. If only part of the trail is lost the lizard suffers little change in personality, but if the entire tail is excised the lizard becomes less active, less prone to explore, and less sociable; its capacity for differential behavioural expression is constrained. Autotomy of the complete tail has physical effects as well as those on personality. The locomotion of tailless lizards may be impaired and the animal may suffer reduced speed of sprinting and rendered less able to escape further predators, or the reverse may occur. The tail contains stores of lipids so loss of the tail deprives the animal of some of its reserves of energy. Thus, one is constrained to wonder whether it is the experience of being attacked and the trauma of autotomy that is the stimulus for the change in personality, or whether it derives from the physiological aftermath of that experience.

Personality and learning

Lizards can be divided into two different types of personalities based on their cognitive styles, i.e. whether they are: (1) bold, active, fast-learners that are neophilic (favour new and novel experiences) and that sample their environment rapidly but with less accuracy than the (2) slow learners that are shy, neophobic, inactive, and move through their environments more deliberately while paying greater attention to environmental cues (Chung *et al.* 2017). How the information is perceived and processed depends on the animals' cognitive abilities, and implementation of an appropriate response to its future fitness, such as finding prey and avoiding predators, depends on the animal's memory. Previous work on this topic examined these issues on the species level; the research of Chung *et al.* (2017) moved to the perspective of the individual. They chose *L. delicata* especially because it exhibits consistent inter-individual differences in activity, exploratory behaviour, and sociability structured into a behavioural syndrome. Their goal was: (1) to discover whether this species could learn an associative task, and (2) to assess the variation in personality among individuals. The predicted outcome was that fast individuals, having greater levels of activity, exploratory behaviour, and boldness, would exhibit greater speed in learning at the cost of accuracy and that slow individuals would take longer to become trained but by being slower and more attentive would be more accurate and flexible.

In tests for the linkage of these traits into a behavioural syndrome, they found that individuals with the highest level of activity also passed over an obstacle more quickly and spent more time basking with other lizards rather than alone. In tests of ability to learn, subjects could be trained to avoid imposed disturbance by going to safe shelter by recognising colour-cues (blue havens were safe, yellow ones not) or location (those to the right in a Y-maze were safe, those on the left were not).

Personality was found to influence the accuracy of learning but contrary to predictions, the fast lizards made fewer mistakes when learning to associate colour with a goal. Personality had little influence on learning in either the training or testing phase. During training there was no relationship between personality and the latency to reach the goal, whether the choice was correct or incorrect, or the number of trials it took to become trained. Also, in the associative learning phase, personality was not associated either with the number of incorrect turns, latency, or speed of learning. Whether a choice was correct or incorrect, however, was influenced by the combined effect of trial and personality; the fast personality type increased its accuracy at a higher rate than did the slow type. The personality of untrained individuals, trained individuals, non-learners, and learners did not vary. The conclusion of Chung *et al.* (2017) was that:

> *the expectation that fast lizards would disproportionately achieve the training criterion at a faster rate while slow lizards would achieve the learning criterion more rapidly was not observed. By contrast personality did influence learning accuracy ... but these effects were in the opposite direction. It was expected that fast personality types with their superficial sampling and inaccurate cognitive style, would not perceive the change in cue meaning during this phase and thus would make greater number of mistakes when associating the colour with the goal. However, lizards exhibiting slower personality types were less accurate.*

They anticipated that their study would throw light on the characteristics of *L. delicata* that contribute to its propensity for being transported in cargo to faraway islands, a property apparently lacking in *L. guichenoti* (see Chapter 3) and suggested that further study is needed.

Vardi *et al.* (2020) investigated the relative roles of heredity and early experience as determinants of the ability of *L. delicata* to learn by comparing the cognition of wild-caught mothers to that of their adult offspring reared in the safer, more stable environment under the care of humans. The criterion of cognition was the ability to learn to select safe conditions at one end of a Y-maze as opposed to the choice of unsafe conditions at the end of the opposite arm. They found that mothers learned the task quicker than their offspring and that the genetic impact on learning was less important than the challenges of instability during early life. By contrast, Bezzina *et al.* (2014) found that when food was the reward neither *L. delicata* nor *L. guichenoti* learned which arm of a Y-maze was the correct choice.

Kang *et al.* (2018) measured the ability of *L. delicata* to learn to choose on the basis of colour-coding between safe (lizards not disturbed) and unsafe (lizard disturbed) artificial refuges as an adaptation to urban environments. Contrary to their expectations, there was no difference in learning between populations from urban and natural environments, and it was proposed the reason for this result was that the urban and natural environments are so similar that there was no selection for enhanced cognitive function in urban lizards.

Sexual differences in personality

Males and females are subject to different selective pressures after maturation (e.g. females may be encumbered by eggs within the oviduct and consequently be awkward in locomotion). Such attributes may result in divergence between the sexes in individual adaptive behaviours and behavioural syndromes.

Lampropholis delicata

Lampropholis delicata is unusually prone to disperse overwater to distant islands (see Chapter 3) and Michelangeli *et al.* (2016a) compared five personality traits between the sexes in this species to ascertain whether they contributed to its propensity to travel overseas and invade new areas. The traits tested were activity, exploration, sociability, foraging, and boldness. They found that females were more social than males and that they also may be faster explorers. The sexes share a behavioural syndrome in which individuals of both sexes that are more active, also tend to be faster explorers and be more social than less active individuals of the same sex. Males and females did not differ in either level of activity or in foraging (voraciousness to feed).

Lampropholis similis

Goulet *et al.* (2021) tested *L. similis* for sex-specific differences in personality and behaviours (amount of activity, propensity for exploration, and sociality). They found that the sexes differed in the amount of time they spent basking with conspecifics and that females were more sociable. All other behaviours tested were similar between the two sexes and there was no evidence for a behavioural syndrome involving activity and exploration. Thus, in this regard *L. similis* differed from *L. delicata* (see above).

Personality syndromes

Combinations of personality traits can be linked into a syndrome that is repeatable across time and context (Goulet *et al.* 2021) and can influence an individual's fitness, survival, and reproductive success (Polverino *et al.* 2023). Michelangeli *et al.* (2016b) described such a syndrome in *L. delicata* in which individuals that consistently were more active also explored their environment faster and were more sociable (basked more often with other individuals).

Thus, the behaviours comprising such syndromes do not act independently but rather an adaptive change in one component of a syndrome may involve a shift in another component.

The linkage between the contributing behavioural components, however, may be decoupled under the varying selective pressures of different environments. One of these, the activity-exploratory syndrome has special application in two situations: (1) Species that are highly active and tend to explore novel situations are more likely to enter containers or crevices leading to their accidental transport in cargo to new countries and once there to expand into new habitats (see Chapter 3); (2) The second situation is adjustment to the changes occurring when their natural habitat becomes urbanised. For example, the linkage of personality traits of *L. delicata* were diverse across four Australian localities (Sydney, Coffs Harbour, Tenterfield, and Brisbane) (Michelangeli *et al.* 2019). Exploration and boldness were significantly correlated at Tenterfield and Coffs Harbour but in opposite directions (positive in Tenterfield and negative in Coffs Harbour); these traits were not significantly correlated in Sydney or Brisbane and thus lizards from these localities had somewhat different behavioural syndromes. For example, lizards from Sydney were consistently active, explorative animals compared to those from the other three localities but were less bold than those from Tenterfield and Coffs Harbour; those from Tenterfield were consistently less explorative than the lizards at any of the other localities.

Moule *et al.* (2016) addressed this issue and got somewhat different results. They found that *L. delicata*, a species that has been spectacularly successful in adjusting to the urban environment, had high activity, exploratory behaviour, and foraging behaviour linked into a behavioural syndrome. The surprising result was that there was no difference in activity or exploratory behaviour between lizards from urban populations and those from non-urban ones, although lizards from one of the urban localities were more active and more likely to forage for food than were those from three other urban regions. Taking all the above studies into consideration, it would seem that this species' activity-exploratory syndrome in its natural habitat is sufficient to serve as a pre-adaptation to an urban environment, much as it seems preadapted to overseas dispersal, but that nuances of its expression can sometimes occur locally.

Another behavioural dichotomy in *L. delicata* is the 'hot' versus 'cold' syndromes. Michelangeli *et al.* (2018) found that individuals with the hot syndrome performed optimally at higher temperatures than did those either of the cold syndrome or of an intermediate type; also that they were more active, sprinted at higher speeds, and were bolder and more explorative.

The Pace-of-Life Syndrome (POLS) is a syndrome in which behavioural traits are correlated and integrated within a fast–slow physiological continuum of which thermal physiology is a dimension (see Chapter 6 and Goulet *et al.* 2017) and to which Goulet *et al.* (2018) considered adding yet another dimension – cognition. POLS includes a fast track and a slow track. Individuals in the fast track have higher metabolic rates and are more explorative and bolder than are animals in the slow track. Goulet *et al.* (2018) predicted that the animals in the fast track would acquire a novel task faster, but with less accuracy, than those in the slow track. The animals in the slow track, however, were expected be more attentive and while responding to cues at a slower rate they were predicted to achieve greater accuracy and exhibit greater flexibility. These investigators tested their hypothesis by measuring physiological and behavioural traits in *L. delicata* and testing the lizards' ability to learn. Correlations were detected between cognition and behaviour but not with thermal physiology. Contrary to their predictions, the positioning of individuals along the axes of the fast–slow continuum were opposite to predictions. Fast lizards preferring lower body temperatures expressed higher activity, sociality, and boldness, a tendency to explore, and learned to discriminate at a slower rate and made the most errors. Also, modelling indicated that neither thermal physiology, behaviour, nor the interaction of the two, influenced cognitive performance. Because so few of their animals had the assessment of their performance completed, the authors suggested that future research be based on a larger number of animals.

GROUP-EFFECTS ON INDIVIDUALS' BEHAVIOUR

The study of animals' behaviour can be approached in two different ways: (1) from the vantage point of an individual's maintenance-activities such as foraging for food, thermoregulating, or responding to attack from predators, or (2) from the perspective of their social interactions in courtship and mating, defending territories or position in a dominance hierarchy, or caring for young. Seldom, however, has the focus been on the interface between these two emphases. Does the social context influence non-social activities of individuals? If so, do animals with different personality traits respond differently to a given social context? Brand *et al.* (2022) investigated whether boldness of *L. delicata*, as indicated by risk-taking behaviour, was different when individuals were alone compared to when they were embedded within a group of conspecifics. They measured boldness in isolation by the time it took isolated lizards to leave their single shelter to bask on a single basking-site compared to the time it took lizards in a group of 12 individuals, each in an individual shelter but with the shelters surrounding a clump of four basking sites available for use by the group.

Their results were somewhat enigmatic; they concluded that 'individual risk-taking behaviour was repeatable when measured in either social isolation or within groups both with and without food resources available' although 'lizards that were bolder during individual trials were not also bolder when in groups'. They interpreted this result to be driven by 'individual differences in social behavioural plasticity whereby individual skinks responded differently to the presence of conspecifics' and suggested that behavioural traits when measured in isolation may not reflect individual variation in behaviour when measured in more ecologically realistic social groups.

ENVIRONMENTAL DETERMINANTS OF PERSONALITY

The perpetual question attendant to any aspect of behaviour is the relevancy of environmental versus genetic forces in determining its ultimate expression, the so-called dichotomy of 'nurture versus nature'. The recent focus on the personality of *Lampropholis* and the finding that various aspects of its ecology and behaviour are affected by the temperature at which individuals were incubated in the egg (see Chapter 9, section *Effect of embryonic experience on susceptibility to predation*) has made it germane to ask whether the conditions of early development play any role in the expression of personality in the adult. de Jong *et al.* (2022) addressed this issue in two ways: (1) by comparing the activity, tendency to explore, sociability, and boldness of adult male *L. similis* that had been incubated at 23°C during their embryonic development with those that had developed during incubation at 26°C, and (2) by comparing the results of the above experiment with the equivalent behaviour of wild-caught individuals.

There were subsidiary interactive effects. Mass of the animal and time of day, for example, had a positive relationship with level of activity and there were differences between individuals subjected to the same treatment, as well as differences between their performance in response to a given treatment on different occasions. Overall, mean level of activity was not affected by the temperature at which eggs were incubated. Neither was sociability affected by the environment experienced during development and it was the only measure that was tested that was not significantly affected by the mass of the animal. For other attributes tested, however, there were significant differences between thermal treatments. Individuals experiencing higher temperatures during incubation were more explorative than either the wild-caught individuals, or those incubated at the lower temperature although the skinks incubated at the lower temperature did not differ in their propensity to explore compared to wild-caught animals.

Opposite responses to different measures of boldness occurred; there was no difference in the time that it took lizards to leave shelter to bask, but, once they emerged the time spent in basking was longer for those incubated at the higher temperature than it was for those subjected to the lower one.

The authors concluded that their results suggested that 'the potential for greater behavioural plasticity in skinks incubated at a higher temperature which may enable them to cope with environmental change, such as climate warming, in the short term. Overall we show that effects of developmental environment are complex and play a pivotal role in shaping animal personality'.

Brand *et al.* (2021) explored the activity of *L. delicata* from three populations in Hawaii within a different environmental context – the absence or presence of food (crickets). They found that individuals varied in personality; some individuals were more active than others and small skinks were more active than larger ones. There was no relationship between the level of activity in the absence of food and subsequent rates of feeding.

The repeatability of results over multiple trials did not provide consistent results among localities. At two localities there was no effect of whether food was available or not on the level of activity of the lizards; at only one location did the context of food have an effect, *viz* rate of activity was not repeatable in the absence of food but was repeatable in the presence of food. Thus, it can be concluded that the availability of food may sometimes influence the repeatability of the level of activity of *L. delicata*, at least at some localities.

REFERENCES

Belmont CM (1977) Resource partitioning and habitat selection in the ground skinks. BSc (Hons) thesis. University of New England, Armidale, NSW.

Bezzina CA, Amiel JJ, Shine R (2014) Does invasion success reflect superior cognitive ability? A case study of two congeneric lizards species (*Lampropholis*, Scincidae). *PLoS One* **9**(1), e86271. doi:10.1371/journal.pone.0086271

Bilcke J, Downes S, Büscher I (2006) Combined effect of incubation and ambient temperature on the feeding performance of a small ectotherm. *Austral Ecology* **31**(8), 937–947. doi:10.1111/j.1442-9993.2006.01663.x

Brand JA, Naimo AC, Michelangeli M, Martin JM, Sih A, Wong BBM, Chapple DG (2021) Population differences in the effect of context on personality in an invasive lizard. *Behavioral Ecology* **32**(6), 1363–1371. doi:10.1093/beheco/arab102

Brand JA, Naimo AC, Michelangeli M, Martin JM, Sih A, Wong BBM, Chapple DG (2022) Social context mediates the expression of a personality trait in a gregarious lizard. *Oecologia* **200**(3–4), 359–369. doi:10.1007/s00442-022-05269-7

Burgin S (1993) *Lampropholis*: the new 'laboratory' animals. In *Herpetology in Australia: A Diverse Discipline*. (Eds D Lunney and D Ayres) pp. 279–281. Royal Zoological Society of New South Wales, Sydney.

Carter AJ, Feeney WE, Marshall HH, Cowlishaw G, Heinsohn R (2013) Animal personality: what are behavioural ecologists measuring? *Biological Reviews* **88**(2), 465–475. doi:10.1111/brv.12007

Chapple DG, Simmonds SM, Wong BBM (2011) Know when to run, know when to hide: can behavioral differences explain the divergent invasion success of two sympatric lizards? *Ecology and Evolution* **1**(3), 278–289. doi:10.1002/ece3.22

Chapple DG, Miller KA, Chaplin K, Barnett L, Thomson MB, Bray RD (2014) Biology of the invasive delicate skink (*Lampropholis delicata*) on Lord Howe Island. *Australian Journal of Zoology* **62**(6), 498–506. doi:10.1071/ZO14098

Chapple DG, Reardon JT, Peace JE (2016) Origin, spread and biology of the invasive plague skink (*Lampropholis delicata*) in New Zealand. In *New Zealand Lizards*. (Ed. D Chapple) pp. 341–359. Springer, Cham. doi:10.1007/978-3-319-41674-8_13

Cheetham E, Doody JS, Stewart B, Harlow P (2011) Embryonic mortality as a cost of communal nesting in the delicate skink. *Journal of Zoology* **283**(4), 234–242. doi:10.1111/j.1469-7998.2010.00764.x

Chung M, Goulet CT, Michelangeli M, Melki-Wegner B, Wong BBM, Chapple DG (2017) Does personality influence learning? A case study in an invasive lizard. *Oecologia* **185**(4), 641–651. doi:10.1007/s00442-017-3975-4

Clarke CJ (1965) A comparison between some Australian five-fingered lizards of the genus *Leiolopisma* Dumeril and Bibron (Lacertilia: Scincidae). *Australia Journal of Zoology* **13**(4), 577–592. doi:10.1071/Z0965057667

Cogger HG (1989) *Australian Reptiles in Colour.* Treasure Press, Port Melbourne, Vic.

Couper PJ, Schneider CJ (1995) Communal nesting in the small skink, *Lampropholis adonis. Memoirs of the Queensland Museum* **38**, 382.

Daly G (1993) Unusual behaviour of grass skinks (*Lampropholis delicata*). *Herpetofauna* **23**(2), 41.

de Jong M, Phillips BL, Llewelyn J, Chapple DG, Wong BBM (2022) Effects of developmental environment on animal personality in a tropical skink. *Behavioral Ecology and Sociobiology* **76**(10), 137. doi:10.1007/s00265-022-03240-3

Domingos FMCB, Arantes IC, Bosue RJ, Santos MG (2017) Nesting in the lizard *Phyllopezus pollicaris* (Squamata: Phyllodactylidae) and a phylogenetic perspective on communal nesting in the family. *Phyllomedusa* **16**(2), 255–267. doi:10.11606/issn.2316-9079.v16i2p255-267

Done BS, Heatwole H (1977a) Social behavior of some Australian skinks. *Copeia* **1977**(3), 419–429. <https://www.jstor.org/stable/1443259>

Done BS, Heatwole H (1977b) Effects of hormones on the aggressive behaviour and social organization of the scincid lizard, *Sphenomorphus kosciuskoi. Zeitschrift für Tierpsychologie* **44**(1), 1–12. doi:10.1111/j.1439-0310.1977.tb00981.x

Doody JS (2019) *Intelligama lesueurii* (Australian water dragon); *Acritoscincus duperreyi* (three-lined skink); *Lampropholis guichenoti* (garden skink). Interspecific communal nesting. *Herpetological Review* **50**, 141.

Doody JS, Freedberg S, Keogh JS (2009) Communal egg-laying in reptiles and amphibians: evolutionary patterns and hypotheses. *The Quarterly Review of Biology* **84**(3), 229–252. doi:10.1086/605078

Downes S, Hoefer AM (2004) Antipredatory behaviour in lizards: interactions between group size and predation risk. *Animal Behaviour* **67**(3), 485–492. doi:10.1016/j.anbehav.2003.05.010

Elphick M, Shine R, Radder R (2012) An observation of raised tail wag behaviour in a skink. *Herpetofauna* **42**, 18–19.

Elphick MJ, Pike DA, Bezzina C, Shine R (2013) Cues for communal egg-laying in lizards (*Bassiana duperreyi*, Scincidae). *Biological Journal of the Linnean Society* **110**(4), 839–842. doi:10.1111/bij.12174

Firth BT (1968) Rhythmic activity, and the effects of photothermal acclimation and temperature on oxygen consumption in the skinks *Lampropholis guichenoti* and *Hemiergis decresiensis*. (Hons) thesis. University of New England, Armidale, NSW.

Forsman A, Shine R (1995) Parallel geographic variation in body shape and reproductive life history within the Australian scincid lizard *Lampropholis delicata. Functional Ecology* **9**(6), 818–828. doi:10.2307/2389979

Golombek DA, Rosenstein RE (2010) Physiology of circadian entrainment. *Physiological Review* **90**(3), 1063–1102. doi:10.1152/physrev.00009.2009

Goulet CT, Thompson MB, Michelangeli M, Wong BBM, Chapple DG (2017) Thermal physiology: a new dimension of the pace-of-life syndrome. *Journal of Animal Ecology* **86**(5), 1269–1280. doi:10.1111/1365-2656.12718

Goulet CT, Michelangeli M, Chung M, Riley JL, Wong BBM, Thompson MB, Chapple DG (2018) Evaluating cognition and thermal physiology as components of the pace-of-life syndrome. *Evolutionary Ecology* **32**(3), 469–488. doi:10.1007/s10682-018-9948-1

Goulet CT, Hart W, Phillips BL, Llewelyn J, Wong BBM, Chapple DG (2021) No behavioral syndromes or sex-specific personality differences in the southern rainforest sunskink (*Lampropholis similis*). *Ethology* **127**(2), 102–108. doi:10.1111/eth.13103

Graves B, Duvall D (1995) Aggregation of squamate reptiles associated with gestation, oviposition, and parturition. *Herpetological Monographs* **9**, 102–119. doi:10.2307/1466999

Green RH (1965) Two skink lizards newly recorded from Tasmania. *Records of the Queen Victoria Museum* **19**, 1–4.

Greer AE (1989) *The Biology and Evolution of Australian Lizards.* Surrey Beatty & Sons, Chipping Norton, NSW.

Greer AE (1997) A new species of *Lampropholis* (Squamata: Scincidae) with restricted, high altitude distribution in eastern Australia. *Australian Zoologist* **30**(3), 360–368. doi:107882/AZ.1997.014

Griffiths K (1987) *Reptiles of the Sydney Region.* Three Sisters Productions, Winmalee, NSW.

Hamilton K, Goulet CT, Johnstone CP, Chapple DG (2019) Does geographical isolation influence group recognition and social preference in an invasive lizard? *Journal of Zoology* **310**(3), 246–253. doi:10.1111/jzo.12738

Heatwole H (1968) Relationship of escape behavior and camouflage in anoline lizards. *Copeia* **1968**(1), 109–113. doi:10.2307/1441557

Heatwole H, Taylor J (1987) *Ecology of Reptiles.* Surrey Beatty & Sons, Chipping Norton, NSW.

Hoser R (2009a) Beating the cold: group hibernation in a species of small skink. *The Victorian Naturalist* **126**(2), 51.

Hoser R (2009b) Night movement in diurnal skinks. *The Victorian Naturalist* **126**, 27.

Kang F, Goulert CT, Chapple DG (2018) The impact of urbanization on learning ability in an invasive lizard. *Biological Journal of the Linnean Society* **123**, 55–62.

Kearney M, Shine R, Comber S, Pearson D (2001) Why do geckos group? An analysis of 'social' aggregations in two species of Australian lizards. *Herpetologica* **57**(4), 411–422. <https://www.jstor.org/stable/389048>

Lilleywhite HB, Wray KP III (2008) *Dictionary of Herpetology.* Krieger Publishing Company, Malabar, FL.

Littlewood D, Goulet CT, Chapple DG (2021) Behavioural phenotype modulates group size effects in a lizard. *Animal Behaviour* **175**(1), 181–192. doi:10.1016/j.anbehav.2021.01.022

Michelangeli M, Chapple DG, Wong BBM (2016a) Are behavioural syndromes sex specific? Personality in a widespread lizard species. *Behavioral Ecology and*

Sociobiology **70**(11), 1911–1919. doi:10.1007/s00265-016-2197-9

Michelangeli M, Wong BMB, Chapple DG (2016b) It's a trap: sampling bias due to animal personality is not always inevitable. *Behavioral Ecology* **27**(1), 62–67. doi:10.1093/beheco/arv123

Michelangeli M, Goulet CT, Kang HS, Wong BBM, Chapple DG (2018) Integrating thermal physiology within a syndrome: locomotion, personality and habitat selection in an ectotherm. *Functional Ecology*, **32**(4), 970–981. doi:10.1111/1365-2435.13034

Michelangeli M, Chapple DG, Goulet CT, Bertram MG, Wong BBM (2019) Behavioral syndromes vary among geographically distinct populations in a reptile. *Behavioral Ecology* **30**(2), 393–401. doi:10.1093/beheco/ary178

Michelangeli M, Melki-Wegner B, Laskowski K, Wong BBM, Chapple DG (2020) Impacts of caudal autotomy on personality. *Animal Behaviour* **162**, 67–78. doi:10.1016/j.anbehav.2020.02.001

Milton D (1980) Some aspects of the population dynamics of *Lampropholis guchenoti* in Toohey Forest near Brisbane. *Herpetofauna* **8**, 19–23.

Mitchell FJ (1959) Communal egg-laying in the lizard *Leiolopisma guichenoti* (Dumeril and Bibron). *Transactions of the South Australian Museum* **82**, 121–122.

Moule H, Michelangeli M, Thompson MB, Chapple DG (2016) The influence of urbanization on the behaviour of an Australian lizard and the presence of an activity-exploratory behavioural syndrome. *Journal of Zoology* **298**(2), 103–111. doi:10.1111/jzo.12288

Pengilley R (1972) Systematic relationships and ecology of some lygosomine lizards from southeastern Australia (2 Vols). PhD thesis. Australian National University, Canberra.

Polverino G, Buchholz KM, Goulet CT, Michelangeli M, Chapple DG (2023) Temporal repeatability of behaviour in a lizard: implications for behavioural syndrome studies. *Evolutionary Biology* **37**(3), 401–418. doi:10.1007/s10682-023-10232-w

Powell HP, Heatwole H, Heatwole M (1978) Winter aggregation of *Leiolopisma guichenoti*. *British Journal of Herpetology* **5**, 789–791.

Prosser C, Hudson S, Thonpson MB (2006) Effects of urbanization on behavior, performance, and morphology of the garden skink, *Lampropholis guichenoti*. *Journal of Herpetology* **40**(2), 151–159. doi:10.1670/38-05A.1

Purani K (1994) Population dynamics of *Lampropholis guichenoti* (Dumeril and Bibron) and *L. delicata* (De Vis) in an open sclerophyll forest. MSc thesis. Environmental Science, University of Western Sydney – Hawkesbury, Richmond.

Radder RS, Shine R (2007) Why do female lizards lay their eggs in communal nests? *Journal of Animal Ecology* **76**(5), 881–887. doi:10.1111/j.1365-2656.2007.01279.x

Rawlinson PA (1974) Biogeography and ecology of the reptiles of Tasmania and the Bass Strait area. In *Biogeography and Ecology of Tasmania*. (Ed. WD Williams) pp. 291–339. Dr. W. Junk, The Hague.

Schulz M, Eyre T (1997) Observations on some reptiles from Minnie Water, north-eastern New South Wales. *Herpetofauna* **27**(1), 41–42.

Scott ML, Llewelyn J, Higgie M, Hoskin CJ, Pike K, Phillips BL (2015) Chemoreception and mating behaviour of a tropical Australia skink. *Acta Ethologica* **18**(3), 283–293. doi:10.1007/s10211-015-0213-0

Shah B, Shine R, Hudson S, Kearney M (2003) Sociality in lizards: why do thick-tailed geckos (*Nephrurus milii*) aggregate? *Behaviour* **140**(8–9), 1039–1052. doi:10.1163/156853903322589632

Shea GM, Sadlier RA (2000) A mixed communal nest of skink eggs. *Herpetofauna* **30**(2), 46–47.

Shine R (1999) Egg-laying reptiles in cold climates: determinants and consequences of nest temperatures in montane lizards. *Journal of Evolutionary Biology* **12**(5), 918–926. doi:10.1046/j.1420-9101.1999.00093.x

Shine R, Shine T (2002) An observation of copulation by garden skinks, *Lampropholis delicata*. *Herpetofauna* **32**(1), 55.

Simbotwe MP (1985) Comparative ecology and demography of two populations of Australian reptiles (Lacertilia, Scincidae). PhD Dissertation. University of New England, Armidale, NSW.

Torr GA, Shine R (1993) Experimental analysis of thermally dependent behavior patterns in the scincid lizard *Lampropholis guichenoti*. *Copeia* **1993**, 850–854. doi:10.2307/1447250

Torr GA, Shine R (1994) An ethogram for the small scincid lizard *Lampropholis guichenoti*. *Amphibia-Reptilia* **15**(1), 21–34. doi:10.1163/156853894X00524

Torr GA, Shine R (1996) Patterns of dominance in the small scincid lizard, *Lampropholis guichenoti*. *Journal of Herpetology* **30**(2), 230–237. doi:10.2307/1565514

Traill A (2001) Observations of skink mating behaviour. *The Victorian Naturalist* **118**(2), 58.

Turner G (2006) Communal egg-laying in *Lampropholis coggeri* and *Saproscincus tetradactyla* (Scincidae). *Herpetofauna* **36**(2), 88–92.

Underwood H, Calaban M (1987) Pineal melatonin rhythms in the lizard *Anolis carolinensis*: I. Response to light and temperature cycles. *Journal of Biological Rhythms* **2**, 179–193.

Vardi R, Goulet CT, Matthews G, Berger-Tal O, Wong BBM, Chapple DG (2020) Spatial learning in captive and wild-born lizards: heritability and environmental effects. *Behavioral Ecology and Sociobiology* **74**(2), 23. doi:10.1007/s00265-020-2805-6

Wells R (1979) A large aggregation of skink eggs. *Herpetofauna* **11**(1), 19–20.

Wells R (1981) Utilisation of the same site for communal egg-laying by *Lampropholis delicata* and *L. guichenoti*. *Australian Journal of Herpetology* **1**(1), 35–36.

5 Natural history and life cycles

INTRODUCTION

The life histories of animals have evolved under the agency of natural selection in ways that adapt the vulnerable young to appear at optimal seasons of food supply and favourable levels of environmental conditions. The timing and occurrence of such essentials vary geographically and for the two widespread species of *Lampropholis* one might expect regional adjustments in adaptive life-history traits. This chapter reviews what is known about the life histories of *Lampropholis* in hope that it will stimulate research leading to a fuller understanding of the relationships of these animals with their environment.

GONADAL CYCLES

Females

Although publishing their results in the same year, Joss and Minard (1985) and Simbotwe (1985a, b) unknowingly conducted similar investigations into the life cycle of *Lampropholis*, the former during 1975–1980 in the Sydney area on both *L. delicata* and *L. guichenoti* and the latter during 1981–1984 in New England, NSW, only on *L. guichenoti*. In the Sydney area during normal moist conditions (1975–1976 and 1976–1977) mean monthly rainfall was 102 mm and 106 mm respectively. *Lampropholis guichenoti* and *L. delicata* had different cycles of ovarian weight than in the drier reproductive season of 1979–1980 (mean rainfall only 52.3 mm) (Joss and Minard 1985).

During the wetter season *L. guichenoti* had a bimodal cycle in which follicular growth was rapid, with the first set of vitellogenic follicles being ovulated in mid-spring (September–October) and the second one in summer (December–January) (Fig. 5.1a). The peak of *L. delicata* was juxtaposed between the two peaks (late spring to early summer) (Fig. 5.1b). From August to October some follicles began to enlarge and become vitellogenic and a few were ovulated by October–November, followed by formation of *corpora lutea* that persisted until oviposition occurred about a month later. Follicles that became vitellogenic later did not ovulate but merely regressed.

In 1975–1976, a moist year in Sydney, the ovarian cycle *of L. guichenoti* was bimodal (Fig. 5.1a). The mean weight of the ovaries increased from a low in winter (July) to a peak in spring (September–October); then, upon the completion of ovulation, it declined precipitously in October to rise again to a second peak in summer (December–January) followed by a second abrupt decline to a very low level in January that was sustained until the end of observations in June. In that same year and locality, *L. delicata* exhibited a very different cycle. It was unimodal with the single peak spaced between the twin peaks of *L. guichenoti* (mid- to late spring; October–November) (Fig. 5.1a).

In the drier year of 1979–1980 at Sydney the cycle of ovarian weight of *L. guichenoti* was unimodal (Fig. 5.1b); thus, its own cycle differed between wet and dry years at the same locality, with its wet-year cycle resembling that of *L. delicata* during a dry-year cycle at the same place. The unimodal peak of *L. delicata* was slightly earlier in the season (October–November) than the one for *L. guichenoti* (November–December) and declined somewhat more slowly, thereby reaching the low level of ovarian weight only in late summer and maintaining it until observations ceased in the following June when the study ended. Unfortunately, data were not obtained for

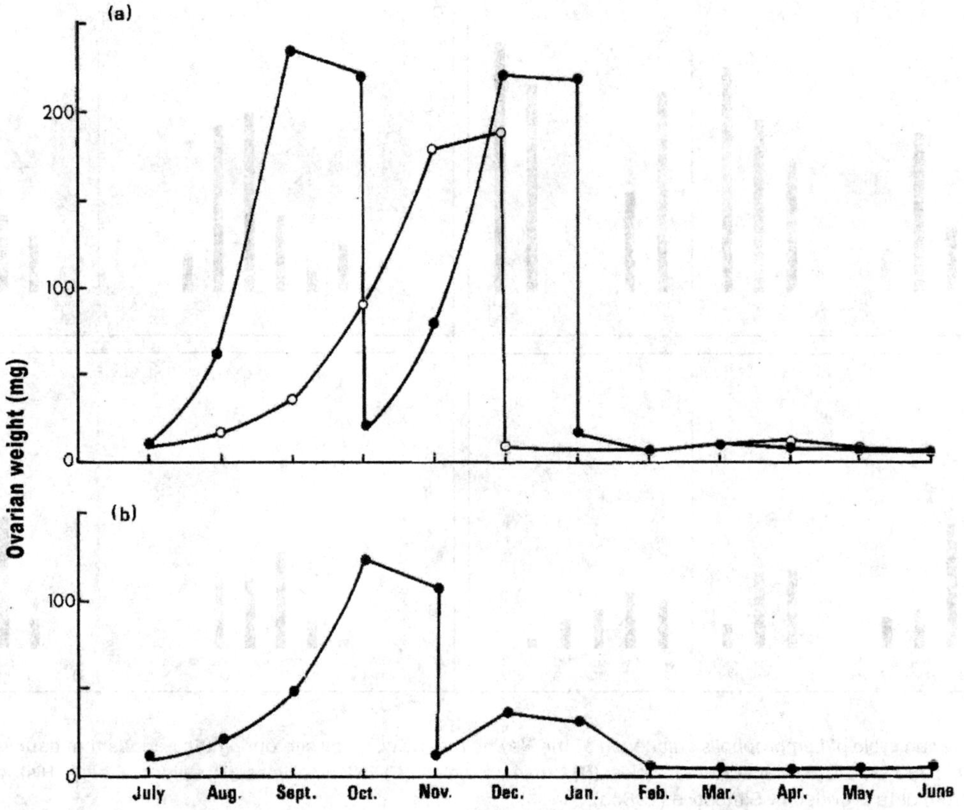

Fig. 5.1. The mean ovarian weights of *Lampropholis guichenoti* (**a**) during 1975–1976 (dots) and 1979–1980 (circles) and of *Lampropholis delicata* (**b**) during 1975–1976. From Joss and Minard (1985). Printed with permission.

L. delicata during the wet year to ascertain whether its cycle would change in relation to hydric conditions so, unfortunately, at present it is not possible to tell whether it has a flexible ovarian cycle like *L. guichenoti* or not.

Pengilley (1972) also demonstrated a unimodal ovarian cycle in *L. guichenoti*: at Coree Flats in the Brindabella Range, NSW, vitellogenesis began in spring with ova rapidly increasing from a mean diameter of 1.5 mm in October to 4.5 mm in early November when ovulation occurred. Finally, Simbotwe (1985a) found a unimodal ovarian cycle in *L. guichenoti* in the Newholme area of New England, NSW, during the reproductive seasons of 1981–1982 and 1983–1984. He classified the progressive stages of development of ovarian follicles according to colour and size: translucent (<1.9 mm in diameter); creamy (2.0–3.0 mm); yellowing (4.0–5.8 mm); and finally oviductal eggs (5.9–10.2 mm). In the 1981–1982 season the early translucent stage predominated in early to mid-spring (September–October; Fig. 5.2A) although some had already begun to develop to the second stage (creamy) (fig. 5.2b). The third stage (yellowing) first appeared in October, peaked in late spring to early summer, declined to low levels by mid-summer, and disappeared by summer's end (Fig. 5.2C). Oviductal eggs were first observed in mid-spring (October) and peaked in early summer (December), with few remaining by mid-summer because most females had oviposited by then (Fig. 5.2D).

In general, oviposition began at the warmest and wettest time of year as predicted by the long-term average weather conditions. Thus, the peak in the laying of eggs occurred at the time when environmental temperatures would be at their highest yearly average and hence the newly laid eggs developed under optimal thermal conditions and hatched when food was maximally available.

Three studies, each at a different location, agree that *L. guichenoti*, like *L. delicata* at the single place it was studied, has a unimodal ovarian cycle but at least at one site modified its cycle to a bimodal one. Joss and Minard (1985) noted an annual bimodal reproductive cycle in female *L. delicata* but, at least in 1 year, *L. guichenoti* laid eggs in mid-November to early December that hatched

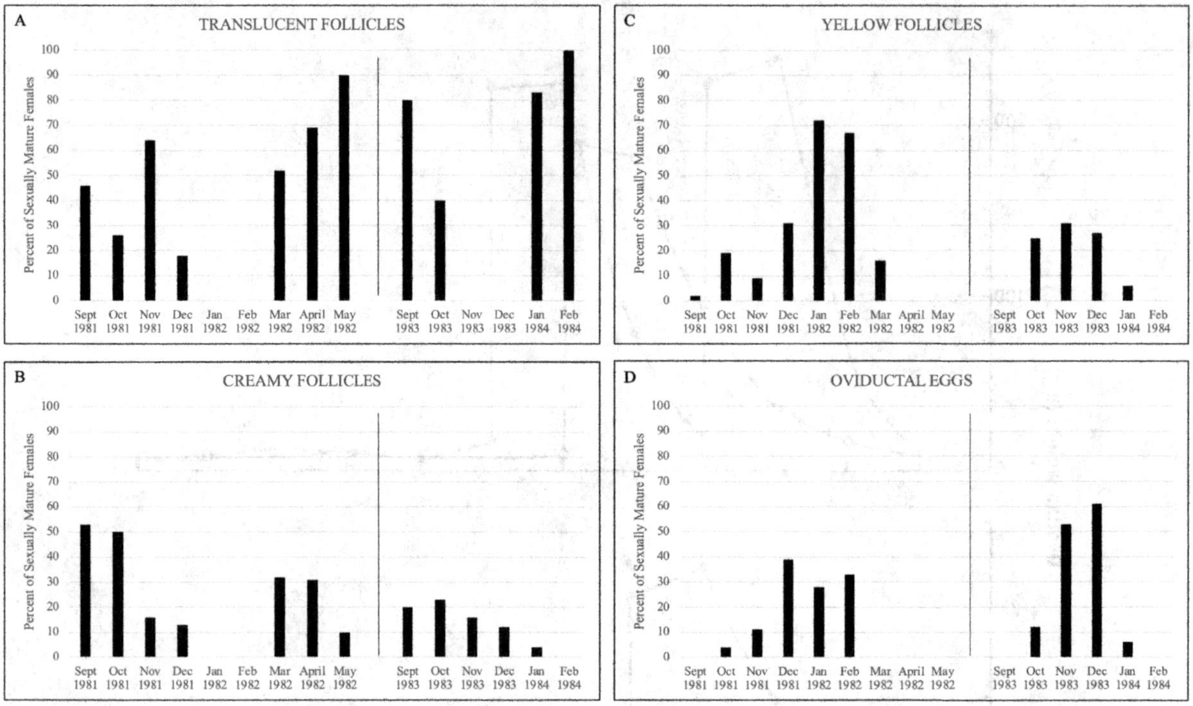

Fig. 5.2. The ovarian cycle of *Lampropholis guichenoti* at the Newholme, NSW, study site during the periods that data were available for 1981–1982 and 1983–1984. **(A)** Translucent follicles, **(B)** Creamy follicles, **(C)** Yellow follicles, **(D)** Oviductal eggs. Histograms prepared by Susan King from data provided by Simbotwe (1985a).

in late January or early February. The second clutch was laid as the first juveniles were hatching (i.e. late January to early February); the eggs of the second clutch hatch in late March to early April. Thus, this species can vary between unimodal and bimodal patterns in the production of eggs but responds somewhat differently to the environmental regimen because at the same locality in the same year *L. guichenoti* had a bimodal cycle and *L. delicata* a unimodal one (Fig. 5.1a). Although Wells (2002) commented that sometimes two clutches may be produced in a single season for *L. delicata*, Shea (2010), based on 30 years of data, found no evidence supporting more than one clutch per season.

No data on rainfall were available for Pengilley's study area during the time of his research but a weather station at a similar elevation 2 kilometres away indicated precipitation to be 'fairly uniform throughout the year, though summer droughts are not infrequent'. Thus, it is possible that the observed unimodal ovarian cycle could have occurred because of insufficient rain to trigger a second cycle, but a different cause altogether cannot be ignored. Elevation is 1,040 m at the Brindabella Range at Coree Flats (Pengilley 1972) and 900–1,300 m on the New England Plateau where Newholme is located (Debus *et al.* 2006) compared to only 80–180 m at Sydney. Thus, even if unimodality of the ovarian cycle was in response to drier conditions when they occurred in the Sydney area, it is not requisite that the same limitation applied to the other two sites at higher elevations. Perhaps there the unimodal ovarian cycle was in response to a shorter duration of the breeding season at higher elevations, not to dryness. Indeed, at Newholme, the reproductive season, when Simbotwe (1985a) studied the ovarian cycle (1983–1984), was considerably wetter than the 30-year average for that time of year (Fig. 5.3) and wetter than that for the preceding year (1981–1982; Fig. 5.4). Finally, it may be that using a bimodal ovarian cycle as the default condition and the reduction to a unimodal one as an adaptation to dryness is the reverse of reality. Perhaps the unimodal cycle is the default one, and the bimodal one an adaptation that increases the number of viable offspring under exceptionally permissive conditions.

In Queensland enlarged follicles first appeared in September and 40% of females became gravid then; from

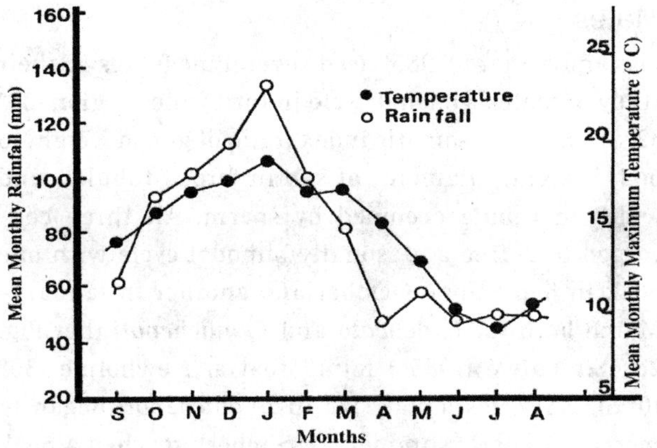

Fig. 5.3. Mean monthly rainfall (mm) and mean monthly maximum temperatures (°C) over 30 years at the Newholme, NSW, study site. From Simbotwe (1985a).

October to February all mature females that were collected were gravid (Clerke and Alford 1993).

The ability of a species to adjust the number of times that eggs are produced per year in response to hydric conditions, as seen in *L. guichenoti* would seem to be adaptive. In dry years when food is scarce, having only one ovarian cycle not only would improve the mothers' condition, but would save wastage of energy and materials occasioned by producing a second group of offspring that would have low chances of survival. By contrast in wet years when food is more abundant, two clutches per year would be less demanding on the mother than in dry years and a greater number of offspring, with better chances of survival, may result.

Another important consideration is whether the degree of flexibility within a species varies geographically. For example, high flexibility would seem to have greater value in drier habitats than in wetter ones because it would allow maximising the number of viable offspring produced in the few favourable wet years that occur and reduce wastage of energy and materials in the more frequent dry ones. Important initiatives in research would seem to be to: (1) compare populations of the same species from different parts of their geographic range and ascertain whether the degree of flexibility varies in relation to the variability of rainfall at different localities, and (2) compare the degree of flexibility between *L. guichenoti* and *L. delicata* at a variety of localities where they are syntopic. As similar as these two species are, ecologically, it is already known that at some localities, one is far more abundant than the other (Pengilley 1972). There must be some reason why.

Follicular development and ovulation are controlled by steroids secreted by the ovaries (see section *Endocrinology* in Chapter 7). Perhaps the seasonality of

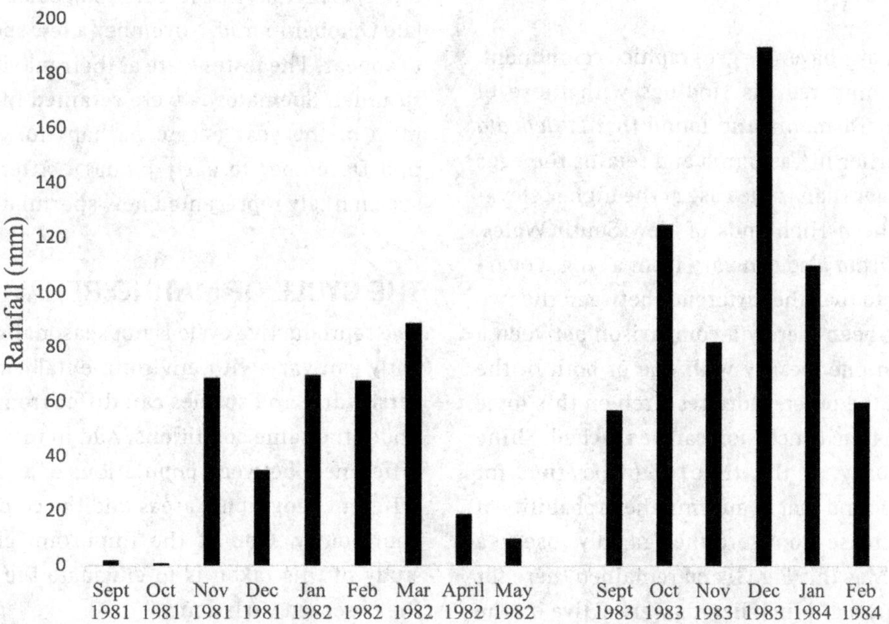

Fig. 5.4. Monthly rainfall (mm) at the Newholme, NSW, study site for the years 1981 through 1984 during the times that data were available for the ovarian cycle of *Lampropholis guichenoti*. Figure prepared by Susan King from data from results provided by Simbotwe (1985a).

oogenesis in these species is mediated by the operation of various environmental influences upon the endocrine system.

Females of *L. elongata* were found with oviducal eggs in mid- to late spring (24 October to 7 December); the embryos are in developmental stage 33 at the time of oviposition, i.e. slightly later in the year than either *L. delicata* or *L. guichenoti* (Greer 1997).

There may be a loss of ova between ovulation and the egg reaching the oviduct. The *corpora lutea* essentially represent scars remaining on the ovary after an egg has been ovulated. Pengilley (1972) compared the number of *corpora lutea* in gravid females of *L. guichenoti* with the number of eggs in their oviducts over three consecutive activity seasons. His estimate of mean loss of eggs for these years was reported slightly discrepantly for the same dataset for the three active seasons (his Tables 48 and 50 and page 178); re-calculation from the raw data indicated that the number of ovulated eggs lost before forming a clutch in the oviduct was 3.5% for 1968–1969, and zero for both 1967–1968 and 1969–1970.

Oviposition

Lampropholis deposit their eggs in various kinds of sites including under logs or slabs of bark (Rawlinson 1974), in shallow depressions in soil under rock or stone (Bilcke *et al.* 2006) and even in burrows of spiders (Humphreys 1976).

Oviposition may have a geographic component. Pengilley (1972) compared his findings with those of Green (1965) from Tasmania and found that *L. delicata* ovulates much earlier in Tasmania and retains the eggs in the oviduct longer than is the case at the higher elevations of the Southern Highlands of New South Wales. However, if *L. delicata* also can vary from a single ovarian cycle per year to two, the difference between the two regions may have been merely a comparison between a unimodal cycle in one locality with one or both of the bimodal peaks in the other. More research on this topic is needed before a firm conclusion can be reached. Shine (1985) listed January as the time of oviposition for *L. guichenoti* and found that in autumn the probability of frost began to increase from zero then rapidly rose to a certainty by mid-May (his Fig. 3) and remained there for the rest of the season at which lizards were active on the surface of the ground. See section *Oviparity versus viviparity* below for further discussion about the timing of oviposition at different elevations

Males

Joss and Minard (1985) used several measures in their study of males' gonadal cycle in the Sydney region (33° 52′ S): the goniosomatic index (ratio of gonad weight to body weight), diameter of seminiferous tubules, and width of tubule occupied by sperm. All three converged to define a seasonally bimodal cycle with one peak in September–October and another in January–March both for *L. delicata* and *L. guichenoti* (her Fig. 2c). Simbotwe (1985a) found that at Newholme (30° 10′ S), NSW, testicular size in *L. guichenoti* began to increase in early summer (December), reached a peak in January, and then gradually declined until late autumn (May) when it reached the small size maintained throughout the winter. There seems to be a latitudinal difference in the testicular cycle as Clerke and Alford (1993) on the basis of the ratio of testis volume to SVL, reported that farther north in the seasonal tropics in the vicinity of Townsville (19° 15′ S), Queensland, males of *L. delicata* were in reproductive condition in all months.

Pengilley (1972) used the size of the testes and the cytology of the seminiferous tubule as his criteria for staging males' reproductive cycle in *L. guichenoti*. He found that the testes were at their minimum size in autumn; at that time the seminiferous tubules had few spermatogonia, many secondary spermatocytes, and a few spermatids in the early stages of development. By late October to mid-November a few spermatozoa began to appear. The testes were at their maximum size during summer. Spematozoa were retained in the vas deferens most of the year except perhaps for a brief period in mid-December to early January. After that period any sperm likely represented new spermiation.

THE CYCLE OF MATING BEHAVIOUR

The reproductive cycle is not seasonally fixed but apparently can vary with environmental conditions (e.g. wet versus dry) and species can differ from each other even under the same conditions. Add to that the possibility of differences between populations of a given species from different geographic areas and the complexity increases enormously. One of the important challenges in the study of this taxon is to elucidate the mechanisms for the observed variation.

It is quite clear that, at least at some latitudes, the ovarian and testicular cycles are not in chronological synchrony for either *L. delicata* or *L. guichenoti*. How, under such circumstances, can reproduction be successful for either of these species? The answer lies in the storage of sperm in the spermatheca of the female at the time of mating, with fertilisation taking place later, following ovulation.

Burgin (1993) reported that both species have similar seasonal patterns of mating in the Sydney region, although, as noted above, in seasons of high rainfall, *L. guichenoti* may produce a second clutch of eggs. Both *L. delicata* and *L. guichenoti* mate in late summer. The following spring, the females lay their eggs. The gonadal cycle of males shows two peaks of spermatogeneis, one occurring in spring and the other in summer. The follicles show increasing maturation with some becoming vitellogenic between August and October. Oviposition takes place in late summer, after which copulation occurs. At the time of mating, females lack mature ova receptive to fertilisation. Instead, they store sperm with fertilisation only occurring during the spring of the following year. Perhaps the storage of sperm is an adaptation that obviates the necessity of coordinating the reproductive cycles of males and females. It does not really matter whether the complexities of the physiological (e.g. hormonal) and ecological cycles of both sexes are synchronous provided that the sperm can be stored viably from whenever the males' sperm matures until the females have receptive ova. Only the behavioural cycle of mating needs to be in synchrony. It is immaterial when females produce eggs, or whether they do so in one or two peaks during the year (see section on gonadal cycles of females above) because stored sperm ensures fertilisation regardless of the seasonality of females' production of eggs (Couper and Schneider 1995).

There is limited information about the other species of *Lampropholis*; however, *L. elongata* have been reported to lay 2–6 eggs, with well-developed embryos occurring by November and December. Incubation is shorter than for other species of *Lampropholis* and Greer (1997) suggested that this may be an adaptation to the cool conditions to which the eggs are exposed during incubation at high elevations.

DEVELOPMENT

Eggs

At oviposition, the eggs are at similar stages of development for both widespread species: *L. delicata* stages 29–32 (Shine 1983a), 29–30 (Thompson *et al.* 2001); *L. guichenoti* stages 25–31 (Shine 1983a), 29–31 (Thompson *et al.* 2001), and 30–32 (Pengilley 1972). The dimensions of the oviductal eggs of *L. guichenoti* have been variously reported as 7 × 4 mm (Mitchell 1959) and 7.3 ± 2.0 mm long and 4.4 ± 0.7 mm wide (Simbotwe 1985a). Recently laid eggs of *L. guichenoti* on average were 9.4 × 5.4 mm and those of *L. delicata* from the same locality measured 8.6 × 6.4 mm (Pengilley 1972); oviductal eggs of *L. adonis* were 9.5–10.8 mm long by 7.8–7.9 mm wide (mean 9.9 × 7.6 mm and weighed 0.30–0.38 g (mean 0.33 g). The dimensions of eggs of *L. guichenoti* in nests were 9 × 6 mm (8 × 5 mm for six infertile ones) (Mitchell 1959). Green (1965) reported oviductal eggs in *L. delicata* to vary from a diameter of 1.7 to 4.5 mm wide × 5.5 mm long and that just after oviposition they had a dimension averaging 5 × 7.5 mm that at the point of hatching were 6 × 11 mm. Wapstra and Wapstra (1986) found clusters of eggs of *L. delicata* measuring 9–13 mm in length. Thompson *et al.* (2002) reported the dry mass of eggs of *L. delicata* and *L. guichenoti* as 27.1 and 34.9 mg respectively.

The greater size of eggs already laid compared to oviducal ones probably is attributable to the absorption of water from the external environment after deposition. Consequently, eggs that have already been laid are variable in weight due to the amount of water they have taken up; in some cases the volume of absorbed fluid increases the volume as much as four-fold, with the eggs measuring 13 × 8 mm just before hatching (Mitchell 1959). Pengilley (1972) reported that the proportional weights of clutches were 22–32% of the mothers' gross body weight; note, however, that these values are over-estimates as the weight of the oviduct was included in the weight of the clutch.

Eggs of *L. adonis* measured 9.5–10.8 mm long by 7.8–7.9 mm wide (mean 9.9 × 7.6 mm) and weighed 0.30–0.38 g (mean 0.33 g) and those in nests were 9.54–10.7 mm long by 7.07–7.94 wide (Schneider and Couper 1995). Data extrapolated from a graph by Uller *et al.* (2009) give the mean weight of eggs of *L. mirabilis* as ~0.13–0.18 g depending on the number of eggs in the clutches.

Eggshell

The eggshell is an important structure that protects the developing embryo and mediates gaseous exchange with the external environment. It also provides a source of calcium for the developing embryo (Hallmann and Griebeler

2015). Thompson *et al.* (2001) reported that eggs of *L. guichenoti* and *L. delicata* take up water and some ions from the environment during incubation and that hatchlings contain almost twice the amount of calcium as occurs in freshly deposited eggs, part of which is due to resorption of calcium from the eggshell. By contrast, there is no significant difference between eggs and neonates of either potassium or magnesium (Osborne and Thompson 2005). The levels of calcium and potassium also are significantly higher both in the eggs and hatchlings of *L. guichenoti* than they are for *L. delicata*. For example, the content of calcium in *L. guichenoti*'s eggs was 0.303 ± 0.051 mg compared to 0.641 ± 0.047 mg in neonates, whereas those of *L. delicata* contained 0.187 ± 0.013 mg in eggs (neonates 0.435 ± 0.033 mg) (Osborne and Thompson 2005). Sodium was present in significantly higher levels in the hatchlings of *L. guichenoti* than in those of *L. delicata* but there was no significant difference between the levels of these ions in hatchlings.

The amount of protein in the form of nitrogen is only slightly lower in eggs than in neonates; however, there is a significant decline in total lipids due to its use as a protein source for metabolic energy during embryogenesis; this source of energy is not as critical as lipids (Osborne and Thompson 2005).

There are three kinds of eggshell in squamates: parchment (flexible), calcified (rigid), and reduced or missing (shell-less) (Hallmann and Griebeler 2015). The eggs of *Lampropholis* are of the flexible, parchment type and are composed of three layers: an inner boundary layer, a middle fibrous membrane, and an outer calcareous layer. As with all such shells, the outer layer is the thickest of the three (Osborne and Thompson 2005).

Yolk

Yolk is composed primarily of lipids and proteins, which serve as a source of energy and constitute the essential components to build embryonic tissues (Thompson and Speake 2002). In addition, embryos contain many other nutrients, including inorganic ions, vitamins, carotenoids, water, and hormones. Apart from water and oxygen, which are taken up by eggs, and some inorganic ions that can come from the eggshell or even from outside the egg, in oviparous species everything required by the embryo must be in the egg when it is laid. Species with complex placentae (see section *Oviparity versus viviparity* below) have reduced volumes of yolk, with the mother augmenting embryonic nutrition by provision across the placenta (Stewart and Thompson 2000; Thompson and Speake 2002; Thompson *et al.* 2002). This avenue is unavailable to strictly oviparous forms such as *Lampropholis*.

Yolk not only supplies nutrients and energy required during development within the egg but residual amounts stored in the gut may contribute significantly to the sustenance of young lizards for a while after hatching. Mitchell (1959) reported that *L. guichenoti* measuring 14 mm in total length at hatching grew to 47–48 mm without external food.

Roles of the liver and fat-bodies

The vitellogenin used in the deposition of yolk is produced by the liver and hence the condition of the liver might be reflected in the reproductive cycle of females. Indeed, Joss and Minard (1985) showed that seasonal changes in the weights of livers of *L. guichenoti* decreased in winter and early spring (September) but then increased markedly during the rest of spring and much of the summer (October to December) when vitellogenin was being produced. After December when vitellogenesis and ovulation had been completed there was a dramatic drop in the weights of livers.

The fat-body stores energy as fat that can be mobilised in times of need. This stored energy is used in two ways: (1) to sustain life during brumation over winter, and (2) to produce eggs. Pengilley (1972) found that in females in April, prior to brumation, the weights of fat-bodies were significantly higher than in females in October following quiescence. There also was a decrease in the fat-bodies of males during brumation. Males and females, however, differed in respect to the role of the fat-body in reproduction. In contrast to females' seasonal changes in fat-bodies during vitellogenesis, in male *L. guichenoti* there was no decrease in size of the fat-body concomitant with the period of testicular growth. Thus, it appears that in this species both sexes rely on their fat-bodies for energy to sustain them overwinter, but only the females draw upon that source for the demands of reproduction.

Clutch size

Skinks generally have smaller clutch/litter sizes than do other families of lizards (Chapple *et al.* 2023). Sometimes, because of communal nesting (see Chapter 4), it is not always possible to identify the separate contributions of individual females to the combined clutches of eggs. However, in the laboratory, Bilcke *et al.* (2006)

found that individual clutches of 52 female *L. delicata* averaged 3.2 ± 0.12 SE eggs. Similarly, Rawlinson (1974) reported an average clutch size of three (range = 1–4). Wells (2002) reported clutch sizes for *L. delicata* as 3–4 but up to 8 and Du and Shine (2008) gave a value of 2–8; the upper limit of these values was higher than reported elsewhere and one wonders whether it represents combined clutches from more than one female. For *L. delicata* in Tasmania, Green (1965) reported clutch sizes of 2–4 and 2–5 and Wapstra and Wapstra (1986) reported 3–5. Simbotwe (1985a, b) reported mean clutch sizes for *L. guichenoti* of 2.26 ± 0.55 (1981–1982) and 2.61 ± 0.55 (1983–1984), with the difference between these two periods being significant ($t = 2.58$; $P < 0.02$). The prospective clutch sizes, as indicated by the follicles developing for the following season, were 2.41 ± 0.81 (for 1982–1983) and 2.50 ± 0.58 (for 1984–1985) and were not significantly different ($P > 0.05$).

At Coree Flats, Pengilley (1972) found mean clutch sizes for *L. guichenoti* to be 2.5 ± 2.1, 3.2 ± 2.7, and 3.0 ± 1.4 respectively in the active seasons of 1967–1968, 1968–1969, and 1969–1970. Clutch size, however, as estimated both by the number of embryos produced and by the number of *corpora lutea*, increases with increasing size of the female. Wells (2002) gave the clutch size for *L. delicata* as usually 3–4, but up to 6. Variation in clutch size may be influenced by the size-structure of the population at a particular time as Forsman and Shine (1995a) and Chapple *et al.* (2014) found that clutch size increased in proportion to increasing size of the females' bodies. In Townsville, Qld., clutch size of *L. delicata* was 1–4 (mean 2.67) (Clerke and Alford 1993).

Joss and Minard (1985) estimated that in times of good rainfall the size of clutches produced by females of *L. delicata* and *L. guichenoti* are identical (2–4 eggs) with *L. guichenoti* ovipositing twice per year, but *L. delicata* only once. Simbotwe (1985a) compared the sizes of clutches of *L. guichenoti* in two different years of reproductive activity. The number of oviductal eggs was significantly higher ($t = 2.58$; $P < 0.02$) in 1983–1984 (2.61 ± 0.55) than it was in 1981–1982 (2.26 ± 0.55). By contrast, Thompson *et al.* (2001) found clutch size to be larger in *L. delicata* (3.6 ± 1.0; range = 2–5) than in *L. guichenoti* (1.8 ± 0.2; range = 1–3); females of both species had the same body size and there was no significant relationship between clutch size and body size within the range of the sizes of females in their sample of either species. However, Pengilley (1972) did find that the number of the *corpora lutea* ($y = -4.913 + 0.180X$; $r = 0.62$) and the number of eggs per clutch ($y = -8.56 \pm 0.252X$; $r = 0.68$) both increased significantly with increase in the body size of the mother. Forty-five per cent of the gravid females had equal numbers of *corpora lutea* in both ovaries, 26% had a greater number in the right ovary than in the left one and 29% had more in the left one than in the right. Loss of eggs during ovulation was only 1.7%. Forsman and Shine (1995b) also found that clutch size increased with the size of the female.

If Joss and Minard's (1985) estimate of the number of eggs per clutch being the same in both species is consistent in time and place, and if the number of eggs is used as the criterion for reproductive output, then *L. guichenoti* would seem to have double the reproductive output of *L. delicata* when the former oviposits twice per year instead of just once. If, however, the estimate by Thompson *et al.* (2001) that *L. guichenoti* lays smaller clutches (see above) but of larger eggs (mean dry weight 41.6 ± 1.2 mg) than in the clutches of *L. delicata* (26.6 ± 2.8 mg), and if the *weight of eggs* produced is used to express reproductive output, then the annual value for *L. guichenoti* becomes 150 mg dry weight (mean dry weight per egg 41.6 × mean clutch size 1.8 × 2 clutches per year) compared to 96 mg (26.6 × 3.6 × 1) for *L. delicata*, thereby making the ratio between the two species 1.6:1 for *L. guichenoti* relative to 2:1 for *L. delicata* if the size of the clutch were the sole criterion for annual reproductive output and the difference between the two species was dependent solely on the number of clutches produced per year.

Baker (1979) reported on the clutch sizes of *L. delicata* introduced into Hawaii. They were 3.3 ± 1.8 for the island of Oahu, 4.1 ± 1.0 for Kauai, and 4.7 ± 1.3 for Hawaii. The sizes of clutches were significantly different between Hawaii and Oahu ($P < 0.01$) but not between Oahu and Kauai. The values from Kauai and Hawaii were higher than those recorded from populations of this species within Australia, their native land.

Clutch sizes reported for the species with smaller geographic ranges are: *L. adonis* up to 3 eggs; *L. amicula* 2–3 eggs; *L. caligula* 2 eggs; *L. coggeri* up to 3 eggs; *L. colossus* up to 5 eggs; *L. mirabilis* 2–4 eggs (Uller *et al.* 2009); *L. elongata* 4–6 eggs, mean 3.7 (Greer 1997); *L. robertsi* up to 3 eggs (Wells 2002). Turner (2006) found 2–3 eggs in gravid females of *L. coggeri*; again, it is possible that some of the higher values represent contribution to the clutch by multiple females.

Larger females have more room inside for eggs than do smaller individuals and Baker (1976) reported that clutch sizes increased significantly with increase in the snout-to-vent length of female *L. delicata*. By contrast Greer (1997) found no correlation between size of female *L. elongata* and the size of their clutches. The effect of body size on the size of clutches may stem from a complicated array of unexpected sources. For example, Downes (2001) found that females of *L. guichenoti* subjected to the odour of one of its predatory snakes spent less time foraging outside of refugia and as a result obtained less food and grew more slowly, which resulted in reaching sexual maturity at a smaller size; when mature these smaller females produced clutches of eggs that were lighter in weight than those produced by larger females; hatchlings from those clutches also were smaller than those hatching from eggs produced by larger females. Thus, indirectly the smell of a predator was responsible for smaller offspring in a lizard, not a chain of events that is intuitively obvious.

REPRODUCTIVE OUTPUT

The energy stored in eggs can be used as a measure of reproductive output. Thompson *et al.* (2002) addressed this issue and discovered that the eggs of *L. guichenoti* contained significantly more energy (939.0 ± 74.6 joules) than did those of *L. delicata* (523.7 ± 63.5 joules), a ratio of 1.8:1. Surprisingly, unlike the eggs, the hatchlings of the two species did not differ significantly in their content of energy. Deeming (1989) noted that residues of semi-gelatinous fluid were left behind in the eggshells of lizards after hatching and that the amount of this material seldom was measured. Perhaps the discrepancy in differences between eggs and hatchlings may have resulted from differences in the relative amounts of residue left behind by these two species.

In retrospect, and in looking toward the future, allocation of energy and/or materials to produce either a few large eggs or a large number of smaller eggs may require equivalent physiological effort but the ecological impact may differ greatly and represent different evolutionary strategies. Thus far, the status of such allocation is not clear for the genus *Lampropholis* and the data are somewhat contradictory. It well may be that the responses may differ not only between species, but among populations within a species inhabiting different geographical locations, different habitats, or embedded within the varying matrices of interacting species in their assemblages and biotic communities. To fully understand the dynamics of reproductive output it will be necessary to examine its different manifestations over a range of localities, habitats, and biotic communities. Environmental conditions are undergoing flux under the spectre of worsening climatic instability and it is likely that a temporal element will be imposed that will change the suitability of present strategies and influence the future reproductive ecology of this genus.

Qualls and Shine (1997) examined the costs of reproduction in two neighbouring populations of *L. guichenoti* that were similar in the sizes and shapes of their bodies, sizes of eggs, numbers of eggs in their clutches, and relative clutch-mass. Despite all these similarities, there still was a cost to the reproductive output that differed between these two populations. When not gravid, females from these two populations ran at equal speeds while being chased along a racetrack. When, gravid, however, the locomotory performance of females from these two populations differed greatly. At one location, gravid females speed on the racetrack was faster than unmated ones, whereas at another the benefit was reversed and gravid females ran more slowly. This study emphasises the complexity of the topic and illustrates the need to avoid accepting blanket conclusions to cover all contingencies.

The above study was followed up by an experimental investigation in which female *L. guichenoti* were injected with sterile Reptile Ringer's solution equivalent at doses up to 30% of their body weight to mimic the physical burden that gravidity imposes (Shine 2003a). Animals burdened in this way ran significantly slower than non-burdened controls with the reduction in speed increasing in proportion to the weight of the load. Injected males did not differ in reduction of speed compared to females so any difference between the sexes in relative shape of the body was discounted as influencing speed. There was a significant effect, however, between individual lizards. Some were faster than others in their non-burdened state and their reduction of speed when burdened was greater than it was for 'slow' lizards. Thus, in terms of running speed, females that are intrinsically swift runners would benefit more from carrying fewer eggs (smaller clutch size) than would slower females. This would seem to be a counter effect to the advantage otherwise of greater reproductive output. The relative advantage of these two opposite influences would probably be tempered by the abundance and efficiency of would-be predators.

In a further study, Shine (2003b) found that factors such as a moderate decrease in the temperature of the body, a large meal, or autotomy of the tail reduced the speed of locomotion even more that did gravidity. Thus, the reproductive state of the female causes only a minor proportion of the total variation in her locomotory ability and may not impose as significant a pressure on her reproductive investment a previously thought.

OVIPARITY VERSUS VIVIPARITY

Oviparity (shelled eggs laid externally) is the primitive mode of reproduction in skinks but viviparity (embryos developing within the mother; eggshells much reduced or lost) has arisen more than 100 times independently within lizards and snakes; skinks have taken this evolutionary route more often than any other family (Murray and Thompson 2011; Chapple *et al.* 2023). It has arisen at least 19 times in Lygosominae (Blackburn 1982), the subfamily of skinks to which *Lampropholis* belongs. Complex placentae have rarely evolved in squamates but all five lineages in which this has occurred are skinks (see Thompson *et al.* 2002).

The species of *Lampropholis* lay eggs and have served in various studies as the oviparous model against which viviparous species are compared as a means of understanding the fundamental steps required when viviparity, and ultimately complex placentae, evolved from oviparous ancestors.

Shine (1985) emphasised that viviparity and oviparity each have costs and benefits. Most attention has centred around the benefit of viviparity in cold climates where behavioural thermoregulation by the mother can maintain the embryos more consistently at higher favourable temperatures, thereby speeding up development, accelerating hatching, enhancing successful hatching, and increasing locomotory speeds of hatchlings. The temperature at which skinks develop may have important effects on a variety of adaptive attributes of offspring. For example, Shine (1992; and references reviewed therein) compared the attributes of hatchlings from eggs incubated at temperatures simulating those of nests in the field with those incubated at higher temperatures simulating the body temperatures of females in the field in two species of skinks (*Acritoscincus duperreyi* and *Anepischetosia maccoyi*). He found that the phenotypes of their hatchlings were affected by the temperatures at which they were incubated, including their body size, relative length of the tail, running speed in a laboratory raceway, frequency of basking, and antipredator tactics; the running speeds of *A. duperreyi* were also altered by brief retention at simulated body temperatures of females and by body temperatures of females prior to oviposition. Thus, such thermally induced plasticity alone plays a role in adaptive modification of life-history phenomena, including viviparity. Genetic factors and environmentally induced plasticity are probably complex and their interaction in the evolution of viviparity requires further study.

Other advantages of viviparity are avoidance of predators by the mother and prevention of internal embryos from contact with infectious fungi or with desiccating air, earlier hatching because of development at higher temperature and consequently being able to emerge earlier and seek shelter prior to lethally low temperatures in autumn (see review by Shine 1992). These are all benefits to the embryos or hatchlings but the cost to the mother is that she is more cumbersome when gravid and therefore probably more vulnerable to predators and less efficient in capturing prey. Thus, oviparity may improve her chances of survival to breed another year and increase her long-term production of progeny. Neither reproductive mode is necessarily 'better' under all circumstances, but rather their relative advantages and disadvantages will depend on other attributes of the species and on environmental conditions. Viviparity would seem to be most advantageous in cool environments.

In oviparous lizards the oviducts function in fertilisation, storage of sperm, transport of eggs, deposition of shell, maintenance of the early embryo, and expulsion of the egg; in viviparous forms they also provide placentae responsible for gaseous exchange and provision of nutrients for the developing embryo (Blackburn 1998), but have lost, or diminished, the deposition of shell. Thus, in viviparous lizards the oviduct develops a uterine function.

During the formation of the placenta in mammals dome-like protrusions on the cells of the uterine epithelium, called uterodomes, develop (see Thompson *et al.* 2005) and these also occur in some viviparous skinks. Adams *et al.* (2004) used electron microscopy to describe the changes in the oviductal epithelium of *L. guichenoti* during the production and deposition of eggs. They identified three morphologically distinct regions of the oviduct: an anterior 'uterine tube', a central 'uterus', and a terminal posterior 'vagina'. Each of these developed a greater morphological complexity in gravid females than in those undergoing vitellogenesis. The uterine tube undergoes a transition from a lush ciliation during vitellogenesis to an epithelium composed of well-defined ciliated cells, microvillous cells, and rosettes composed of central gland cells surrounded by cells densely covered by microvilli. After the eggs are laid an apical secretory layer masks the surface of the epithelium, thereby disguising cellular differences among the three sections.

Although changes in the morphology of the oviductal epithelium take place in *L. guichenoti* during its reproductive cycle, no uterodomes are formed in any section of the tract at any stage of the cycle. Hence, this oviparous species does not undergo the typical transformation of the plasma membrane occurring in mammals and viviparous skinks (Adams *et al.* 2004).

There are intermediates between strict oviparity and strict viviparity (sometimes referred to a 'ovoviviparity') in terms of the retention of the eggs in the oviduct; species differ in the stage of development reached before oviposition take place. Andrews and Mathies (2000) posited that the length of time, and hence stage of development to which development can proceed in eggs retained in the oviduct, is limited by hypoxia in the absence of placental supply of oxygen by the mother. The lack of uterodomes in *L. guichenoti*, such as those seen in some truly viviparous skinks (e.g. Hosie *et al.* 2003) suggested that retention of eggs in the oviduct in *Lampropholis* is unlikely and that this genus probably lies toward strict oviparity on the continuum in mode of reproduction (but see section *Eggs* above). The scincid genus *Eulamprus* illustrates an intermediate condition. All are viviparous and possess a placenta but their extraembryonic membranes share features both with other viviparous skinks and with oviparous ones like *Lampropholis* (Murphy *et al.* 2012). Since viviparity seems to arise easily in skinks, sometimes even occurring in some populations of an otherwise oviparous species (Thompson *et al.* 2002), the question arises as to why some genera inhabiting cool regions, such as *Lampropholis*, remained oviparous. The path to viviparity involves the progressive shortening of the period of incubation under cool external conditions by increasing the duration of retention within a behaviourally thermoregulating mother. Shine *et al.* (2018) obtained data suggesting that an oviparous European skink in a cold environment achieved reduction of the duration of exposure of eggs to an inclement external environment by producing smaller eggs that developed at an accelerated rate, thereby allowing the embryos to develop externally within the brief seasonal window of sufficiently warm conditions in the soil. On the basis of their study they proposed the following hypotheses: (1) These same shifts in size of eggs (and in shortened external incubation periods and earlier stage of development at hatching) likely occurred in numerous lineages of oviparous squamates as they extended their geographic ranges into cooler regions, or as climate cooled in their existing one. Note that *L. delicata* and *L. guichenoti* may be examples of species undergoing this process. (2) Prolonged retention of the eggs within the mother may have conferred a selective advantage by extending the seasonal duration of ovulation that could result in viable offspring. (3) The transition toward viviparity may confer the additional advantage of allowing reproducing females to escape the need to reduce the size of offspring in late-season clutches.

Shine and Thompson (2006) and Shine *et al.* (2018) recognised, however, that the transition from oviparity to viviparity may not be restricted to a single trajectory and in the latter paper summarised their views as:

> *Our data suggest that as an oviparous squamate species extends its range into cooler climates, it experiences a suite of selective forces. Maternal fitness can be enhanced not only by judicious selection of warm nest-sites, but also by prolonged uterine retention of developing eggs, by bringing forward the seasonal timing of ovulation, by reducing egg size, by accelerating developmental rate for any given size of egg, and by stage of development at hatching. Shifts in these traits can enable hatching to occur sooner after oviposition than would otherwise be possible. Hence, selection on that duration likely favoured concurrent shifts in all those traits. From this perspective, the oviparity-viviparity continuum in reproductive modes ... may be multidimensional and season-specific, rather than a simple linear transition based on incremental increases in uterine retention of eggs.*

How far along the trajectory from their perceived strict oviparity toward viviparity the various species of *Lampropholis* have progressed has not been widely accessible. Although some relevant data are available, such as sizes of eggs, stage of development at oviposition, seasonality of the various stages of gonadal cycles and reproductive activities, and size of hatchlings and the effect of environmental variables upon their phenotypes, much of it resided in unpublished parts of two PhD theses until presented above (this chapter). One of these

(Pengilley 1972) contains a comprehensive treatment of the ecology of an entire assemblage of lizards in the Brindabella Range, NSW, mainly at Coree Flats (elevation 1,040 m). The other (Simbotwe 1985a) was an intensive comparison of two syntopic species of skinks, *L. guichenoti* and *Hemiergis decresiensis* at Newholme, NSW (elevation 144 m). The collective data from all sources are geographically spotty and alone are of limited value in assessing the evolution of viviparity. However, they serve as models of the kinds of detailed comparative studies needed for the species of *Lampropholis* at the centre and at the geographic and elevational limits of their distributions in order to assess the variability within the taxon and especially within the two widely distributed species and the extent to which they variously have adapted to the constrictions of oviparity under different climatic regimens. Blackburn (2006) warned that viviparity in mammals and reptiles may not have evolved in the same way and that the value of the squamate model may reside in the insights it provides into physiological problems rather than in the universality of specific mechanisms that have evolved in meeting those problems.

Shine (2002) addressed the theory that viviparity developed as a result of selection for retention of fertilised eggs in the oviducts until seasonal conditions became better indicators of the suitability of potential nest sites for the incubation of eggs (the predictability hypothesis). For 7 years he monitored temperatures at natural nest sites at three different elevations. He found that early prediction of subsequent temperatures at nest sites was high at all three elevations and that longer delay decreased, rather than increased, the predictability of future temperatures, thereby nullifying the predictability hypothesis.

DEVELOPMENT OF THE EMBRYO

There is no comprehensive treatise on the details of the overall embryology of any species of *Lampropholis*. There are, however, descriptions of the development of particular structures and these are treated immediately below. For a treatment of the effect of different temperatures during incubation on attributes of the individuals during their life post-hatching see section *Pre-hatching environment* below.

Development of scales

Alibardi and Thompson (1999) described the development of the scales covering the body of *L. guichenoti*. The outer of the embryonic scale (peridermis) is partially lost during development and a second layer develops beneath it. Two or three layers of this type develop in the outer surface near the hinges of the scales. These layers form cells that secrete a lipid-like material into the spaces between the scales and eventually the whole epidermis becomes coated with it. This material may reduce friction and reduce accumulation of dirt between adjacent, overlapping scales. At the end of their differentiation, the modified inner peridermis turns into extremely thin cornified cells. The layer beneath it accumulates granules and forms a complex with the overlying layer. Typically, aggregations of small filaments form into characteristically pointed cytoplasmic outgrowths. During differentiation, the outer layer merges with keratin produced underneath, so that a typical layer of keratin eventually is formed before hatching. Near the tip of the scale, the dermis under the epidermal cells is scarce or absent so that the mature tip of the scale consists of a solid rod of keratinised cells. At the time of hatching, differentiation of a middle layer is well advanced, and the epidermal histology of scales corresponds to Stage 5 of an adult's shedding cycle.

Development of claws

The development of claws of *L. guichenoti* was investigated by Alibardi (2008a, b, 2009) who concluded that the claws evolutionarily originated from scales on the tips of the digits (fingers and toes). He found that during development claws formed by a thickening of the mesenchyme under the epidermis covering the terminal joint of the digits. There, beneath the alpha-keratin cells of the scale, beta-keratin cells differentiate and spread over the underside of the digits to form a corneous layer. These corneocytes form a row of cells parallel to the surface of the unguis (newly forming claw).

The developing claw has an internal region with 20 amino acids in the core extending into a beta-keratin sheath with 32 amino acids. The keratins are rich in glycine and stabilise the beta-keratin polymer (Alibardi and Toni 2009). The N-regions and C-regions contain cysteine that form disulphide bonds. The claws contain higher amounts of cysteine and glycine than do proteins in the scales and are suited to produce a very hard corneous material.

The upperside of the unguis becomes a hard, horny layer thickened by keratin but its underside (subunguis) remains softer.

HATCHING

Wells (1981) collected an aggregation of clutches from a communal nest on 25 February; of the original 132 eggs, 44 (33%) had already hatched and there were some hatchlings in the leaf litter near the nest site. He collected the remaining 85 eggs and took them into captivity (47 hatched to *L. delicata,* 23 to *L. guichenoti,* and 15 died). Thus, the time of hatching in both species of *Lampropholis* ranges from the latter part of February to early March in Central Coastal NSW. Couper and Schneider (1995) found eggs of *L. adonis* on 12 November, some of which were transported to the laboratory where they hatched on 25–26 January. Recall that in sections *Introduction* and *Gonadal cycles* above it was noted that hatching occurs at an optimal season. A clutch of eggs of *L. elongata* hatched at 33–34 days after oviposition (Greer 1997). Greer speculated that the advanced stage of embryonic development and more rapid rate of development of this species might be an adaptation to the conditions at the high elevations it inhabits.

Doody (2011) reviewed environmentally cued hatching (ECH) in reptiles and noted that there were three types: (1) early hatching, (2) delayed hatching, and (3) synchronous hatching, and indicated that *L. delicata* and *L. guichenoti* exhibited the last of these types.

Disturbance of the nest by vibration, such as may be caused by a predator, or perhaps by a human transporting eggs to the laboratory causes early hatching (see Chapter 9). Thus, the above dates of hatching may be slightly earlier than those of undisturbed eggs.

Temperature affects the duration of incubation. Eggs of *L. delicata* and *L. guichenoti* incubated at different, controlled temperatures in the laboratory took significantly longer at lower temperatures to reach hatching than they did at higher ones (Table 5.1). Incubation took slightly longer for *L. guichenoti* than for *L. delicata* at the moderate temperatures of 20°C and 26°C but the two species differed greatly in their sensitivity to temperatures at more extreme levels. *Lampropholis delicata* had a much narrower range of thermal tolerance than did *L. guichenoti*. The survival of the eggs of *L. delicata* to hatching was high in the mid-range of temperatures (91% at 20°C and 95% at 27°C; Downes and Shine 1999); none of the eggs survived to hatching at either the lowest temperature tested or the highest one tested (Table 5.1), whereas there was no mortality of the eggs of *L. guichenoti* at those extremes. In one respect, this result is the reverse of expectations as *L. delicata* has a much wider latitudinal range, both northward and southward, than does *L. guichenoti*. The former species extends into warm tropical Queensland, well beyond the northern limit of *L. guichenoti* and to cool Tasmania, beyond *L. guichenoti*'s southern limit in the south (see Chapter 2). However, the observed difference in thermal sensitivity makes more sense if viewed from the perspective of the western limits of the ranges of these two species rather than their latitudinal distribution. The range of *L. guichenoti* extends farther west into the hotter interior of New South Wales than does that of *L. delicata*. It would be instructive to compare the thermal sensitivity of these two species throughout their entire geographic ranges.

HATCHLINGS

Size of hatchlings

The lengths of the young at the time of hatching are similar for the five species of *Lampropholis* that have been measured (Table 5.2). There was great overlap between species and almost as much variation within a species at different times, different localities, and different temperatures of incubation as there was between species. Downes and Shine (1999) extended their comparisons to skinks in other genera (*Saproscincus mustelina*; previously included in the genus *Lampropholis*, and *Anepischetosia* [formerly *Nannoscincus*] *maccoyi*) and found that 'cold-incubated' individuals of all species investigated were heavier at hatching than were their 'hot-incubated' siblings. The thermal regime experienced during incubation also influenced hatchlings' SVL and

Table 5.1. Effect of temperature on the duration of incubation in *Lampropholis delicata* and *Lampropholis guichenoti*.

Species	No. of clutches	No. of eggs	Mean days from oviposition to hatching (range) at: 15°C	20°C	26°C	27°C	30°C
Lampropholis delicata[1]	5	10	All died	65.3 (64–66)	36.8 (35–42)	-	All died
Lampropholis delicata[2]	-	-	-	54.9 ± 0.67	-	30.5 ± 0.28	-
Lampropholis guichenoti[1]	42	83	150 (all 150)	72.0 (66–73)	38.6 (35–45)	-	27.8 (26–29)

[1]Data from Shine (1983b). [2]Data from Downes and Shine (1999).

Table 5.2. Mass and length of the body and relative tail lengths of the species of *Lampropholis* at the time of hatching.

N = Number of animals measured; SVL = length from the tip of the snout to the vent. Dashes indicate no data available.

		Mean (range)			
Species of *Lampropholis*	**N**	**Mass of body in g**	**SVL in mm**	**Length of tail in % of SVL**	**Authority**
L. guichenoti	–	–	22.4	~150*	Duméril and Bibron (1839)
L. guichenoti	23	–	18.2 ± 0.5 (17–19)	133.3 (121.1–155.6)	Wells (1981)
L. guichenoti in 1968	24	0.18 ± 0.06 (0.12–0.29)	20.8 ± 2.22 (19–24)	–	Pengilley (1972)
L. guichenoti in 1970	26	–	20.8 ± 1.06 (19–21)	–	Pengilley (1972)
L. guichenoti (1 week old)	10	–	19.24 (17–20)	–	Clarke (1965)
L. guichenoti (incubated at 25°C)	7	0.16	19.1	–	Booth *et al.* (2000)
L. guichenoti (incubated at 30°C)	6	0.16	19.0	–	Booth *et al.* (2000)
Overall	**96**	**0.18 ± 0.06 (0.12–0.29)**	**18.2–22.4 (17–24)**	**133–150* (121.1–155.6)**	
L. delicata	–	–	38.0	123.7	De Vis (1888)
L. delicata in 1968	26	0.13 ± 0.046 (0.09–0.22)	18.8 ± 1.56 (17–20)	– –	Pengilley (1972)
L. delicata in 1970	8	–	20.0 ± 1.06 (19–21)	– –	Pengilley (1972)
L. delicata in 1976	45	–	16.1 ± 0.6 (14–17)	133.6 (117.6–152.9)	Wells (1981)
L. delicata (incubated at 20 ± 4°C)	41	0.15 ± 0.01	21.0 ± 0.10	121	Downes and Shine (1999)
L. delicata (incubated at 27 ± 4°C)	41	0.14 ± 0.00	20.3 ± 0.10	130	Downes and Shine (1999)
Overall	**161**	**0.13–0.15 (0.09–0.22)**	**16.1–38.0 (14–38)**	**121–130 (117.7–152.9)**	
L. adonis	7	0.18 (0.17–0.20) Preserved weight	18.86 (17.12–20.36)	228.9 –	Couper and Schneider (1995)
L. elongata	3	–	20.0 19.5–20.5	109 (105–113)	Greer (1997)
L. mirabilis	34	0.16 ± 0.01	–	–	Uller *et al.* (2009)

*Based on the statement by Duméril and Bibron (1839) that its tail is half times longer than the rest of the body [translated from the French: '*sa queue est une demi-foi plus longue que le reste du corps*']. The length they provided for the tail otherwise would make it 326% of the SVL, clearly an error. They provide several other dimensions; Neck = 7 mm; head = 9 mm; trunk = 26 mm.

the length of their tails, but the direction of change varied among the three genera.

The hatchlings of *L. guichenoti* are heavier and longer than those of *L. delicata* (Table 5.2). Thompson *et al.* (2002) reported the dry mass of hatchlings of *L. guichenoti* as 34.9 mg and that of hatchlings of *L. delicata* as 27.1 mg, and weighing 75% and 89% of the dry mass of freshly ovulated eggs. The mean SVL of the hatchlings of *L. guichenoti* were 44% of that of adult females and their masses were 15% of the females' masses. Green (1965, 1981) reported total length (SVL plus tail) of hatchlings to be 37–43 mm on one occasion and 32 mm on another. See Table 5.3 for an account of the influence of the temperature at which incubation occurs on the subsequent mass of the hatchlings of *L. delicata*.

Composition of hatchlings compared to that of eggs

The amount of nitrogen (protein) in hatchlings is only slightly lower than in the eggs of *L. delicata* and *L. guichenoti*, whereas lipids show a marked decline between these two stages. Thus, lipids are a much more

Table 5.3. Summary of the effects of various temperatures during incubation (22°C cool, 26°C mild, 30°C hot) on the phenotypes of hatchlings of *Lampropholis delicata*.

For multivariate analyses, mass of hatchlings was selected as the covariate. Data from de Jong *et al.* (2022).

Characteristic affected	Effect of temperature of incubation	Effect of other variables (tested independently)
Univariate analyses		
Hatching success	No effect.	–
Sex ratio of hatchlings	No effect.	–
Mass of hatchlings	Hot-incubated significantly lighter in mass than those incubated under mild or cool conditions.	There was a positive correlation between the mass of eggs and the mass of their resulting hatchlings.
Growth (also see Fig. 5.5)	Growth rate of mild-incubated and hot-incubated lizards was slower than those of cool-incubated ones. Lizards that were cool-incubated or mild-incubated grew faster than those that were hot-incubated.	Growth rate of hatchlings from cool treatment not affected by hatchlings' SVL. Growth rates of mild-incubation and hot-incubation treatments were higher in lizards with small SVLs.
Preferred temperature	Lizards from the hot-incubation treatment selected a significantly narrower range of temperatures than did those from the mild-incubation treatment but not from those incubated under cool conditions; there was no significant differences between cool and mild treatments.	There was a negative relationship between the mass of lizards and their average selected temperatures.
Selection and use of habitat	Neither the frequency of movement from one habitat to another, nor the times spent in particular microhabitats was affected by the temperature of incubation. Hatchlings from all three treatments spent more time in vegetated areas, followed by rocky ones, and least on logs.	There was no significant relationship between the mass of hatchlings and the number of times they moved between different microhabitats, either for males or females.
Dispersal	Cool-incubated and hot-incubated males were less likely to disperse, compared to mild-incubated ones.	–
Metabolic rate	Resting metabolic rate measured at a given temperature was progressively higher in offspring hatching from cool-incubated to mild-incubated to hot-incubated eggs.	Resting metabolic rate increased with increasing mass of the eggs and with temperature of the hatchlings.
Multivariate analyses		
Mass of hatchlings	Had an effect.	
Principal components 1	Separated cool-components groups from the mild and hot treatments. Individuals separated along PC1 differed in developmental, performance, and physiological traits, as well as in one behavioural trait – dispersal.	
Principal components 2	Separated hot components from the mild and cool treatments. Incubation at the cool temperature resulted in individuals characterised by higher resting metabolic rate, mass-specific growth rates, size-specific growth rates, and maximum speed of sprints, but a narrower breadth of thermal performance, and a reduced propensity for dispersal.	

important source of metabolic energy during embryogenesis than is protein (Thompson *et al.* 2001). The loss of lipids during embryonic development is due mainly to the disproportionate use of triglycerol (80%). Phospholipids accounted for 10% of the lipids and the total level did not differ between eggs and hatchlings, although the relative contributions of the component constituents (arachidonic acid, eicosapentaenoic acid, docosahexaenoic acid) did vary.

Levels of free cholesterol are low (0.7% for *L. guichenoti*; 1.3% for *L. delicata*) in the eggs but were significantly higher in hatchlings, indicating that additional cholesterol was synthesised during embryonic development (Thompson *et al.* 2001). For levels of ions in hatchlings see section *Eggshell* above.

Factors affecting the phenotypes and performance of hatchlings

It has long been known that the sex of reptiles is determined by the temperature they experience during development and that, opposite to the effect in turtles, embryonic lizards usually differentiate as males at higher temperatures and as females at lower ones (Packard and Packard 1988). The type of sex-differentiation, however, has not been verified for *Lampropholis*.

Pre-hatching environment

Incubating eggs at different temperatures had no effect on the mass or length of the resulting hatchlings, or at best only by a trivial amount, in either *L. delicata* or *L. guichenoti* (Table 5.3). Similarly, under laboratory conditions, Du and Shine (2008) found that the per cent of eggs that successfully produced live hatchlings exceeded 88% independently of the level of substrate moisture ranging from zero to −500 kPA ($P > 0.05$). Similarly, the length of time from oviposition to hatching (duration of incubation) was not affected by hydric conditions ($P > 0.05$) or the initial mass of the eggs ($P > 0.9$); the sex ratio of the hatchlings remained at the expected 1:1 under all these conditions ($P > 0.05$).

Hatching success and phenotypes (e.g. size and locomotory performance) of some lizards are known to be affected by even short-term episodes of elevated temperatures during incubation (Shine and Elphick 2001). Du and Shine (2008) found that both temperature and moisture influenced hatchlings' phenotypes. When eggs of *L. guichenoti* were incubated under four different moisture contents between −500 and zero kPA, those incubated on drier substates had a lower uptake of water and the hatchlings were smaller than those exposed to moister substrates. However, there was no difference among hatchlings from eggs incubated under different hydric conditions in terms of other phenotypic traits such as duration of intraoval development, sex of hatchlings, bodily proportions, running speed, and growth rate after hatching. Under the range of moisture contents investigated, *L. guichenoti* were less sensitive to hydric conditions than to temperature.

Bilcke *et al.* (2006) incubated the eggs of *L. delicata* at two different temperatures and compared various aspects of their ecology and performance. One group was 'hot-incubated' at 28°C, thereby mimicking the natural warm conditions in one of its microhabitats (open grassland) and the other 'cold-incubated' at 20°C, mimicking conditions in another of its microhabitats (an area invaded by *Vinca major*, blue periwinkle). After hatching, cold-incubated lizards were more mobile (crossed gridlines more often) than were hot-incubated ones when tested at 20°C, but when tested at 28°C the hot-incubated lizards exhibited the higher mobility. Thus, the activity of the hatchlings had adjusted to the temperature at which they had been incubated. When placed in a thermal gradient equipped with a basking perch the cold-incubated lizards maintained their body temperature at a slightly lower level than did the hot-incubated ones.

The question arises as to whether the conditions under which eggs are incubated in the laboratory mimic those in the field sufficiently to provide valid conclusions. This issue has not been addressed for any species of *Lampropholis* but Shine *et al.* (1997) have done so for a different species of small scincid lizard, *Bassiana* (now *Acritoscincus*) *duperreyi*. They divided clutches into two parts and left half of the eggs from each clutch at their original site but reciprocally exchanged the other half with similarly divided clutches from other locations. They concluded that hatchlings' phenotypes were influenced both by the conditions of incubation in the field as well as by the location of their nest-of-origin, that male and female hatchlings respond differently to thermal conditions during incubation, but that there was no evidence that females select nest sites that match the reaction norms displayed by their offspring. The extent to which these attributes apply to *Lampropholis* remains to be seen but it seems likely that they would be similar.

Qualls and Shine (1998) examined the role of the environment of incubating embryos in affecting the subsequent hatchlings' phenotypes and assessed the relative influence of proximate (environmental) factors from the underlying genetic ones. In the laboratory they incubated eggs of *L. guichenoti* from two different geographic localities (warmer, lower elevation at Sydney and cooler conditions at the higher elevation of the Blue Mountains, 120 km apart) each reciprocally at the thermal regime characteristic of their home area as well as the one at the non-resident locality. They found that when both were incubated under the same thermal regimen, those from Blue Mountain (cooler area) hatched sooner than those from Sydney. Statistical analysis indicated that this result was not attributable either to differences in the stage of development at which the eggs were laid or to their mass.

There was great variation in the way the hatchlings' morphology and performance were affected by genetic and environmental factors. For example, neither SVL nor maximum speed over short distances differed between individuals from the two different populations or between the temperatures of incubation. Mass and stockiness (ratio of mass to SVL) differed between populations but were not influenced by the temperature of incubation. Finally, features such as period of incubation, length of the tail, shape of the body, inter-leg distance, and various ratios of different dimensions of body parts, as well as running speeds under two different conditions, were affected both by geographic location of origin and temperature during incubation. These various relationships provide an indication of the complexity of the number of interacting factors that affect the phenotypes of developing embryos up to the time of hatching. The authors concluded that differences in temperatures of the nest were sufficient alone to induce significant geographic variation in hatchlings' phenotypes. As complex as these relationships are, the described relationships are an oversimplification. As the authors indicated, other possible influences such as the hygric conditions of nests, were not evaluated.

de Jong *et al.* (2022) assessed the importance of additional factors during incubation in influencing the subsequent hatchlings' phenotypes. They incubated eggs of *L. delicata* at three different levels of temperature: cool (22°C), mild (26°C), and hot (30°C), and performed two types of experiments. One set involved univariate analysis (testing the individual effect of one variable alone, without exploring its interactive effect with other variables) and the other set involving multivariate analysis (testing the relationship of two or more independent variables in their simultaneous combined effect). Table 5.3 summarises their findings. In general, the suites of traits affected by incubation at different temperatures differed depending on whether temperature during incubation was cool or hot relative to the mild treatment, with impacts of the cool treatment primarily involving development, physiology, and performance (sprinting), whereas incubation under hotter conditions had greater effects on behaviour as well as upon performance.

de Jong *et al.* (2023) extended their previous analyses of the effects of the prehatching thermal environment upon *L. delicata* hatchlings to include study of lasting effects persisting into adulthood. They used the same three test temperatures as before and assessed the results on juveniles 28–42 days old, subadults ~200 days old, and adults ~2 years old. Their results are summarised in Table 5.4.

The long-lasting and multifarious effects that the thermal conditions under which *Lampropholis*' eggs are incubated have on so many facets of the hatchlings' lives emphasise the importance of the type of environments females select as sites for nesting and lends relevance to the observation that female *L. delicata* may range afield and oviposit in thermally favorable structures not available in their traditional home range (Cheetham *et al.* 2011).

Post-hatching environment

To assess the effect of post-hatching environments upon lizards' phenotypes Qualls and Shine (2000) extended their previous study in Sydney and the Blue Mountains to include effects of the environment lizards experience after they hatch. The thermal regimes of the two areas were closely simulated in the laboratory as a 'warm' regime mimicking conditions in Sydney and a 'cool' one mimicking those in the Blue Mountains. When eggs from both localities were incubated under the same conditions, those from the Blue Mountains hatched sooner than did those from Sydney. Mortality of eggs did not differ between localities or thermal regimes. The prehatching effects were similar to those obtained under their previous (1998) study although the attributes tested were not all identical.

Hatchlings were released into outdoor enclosures at both Sydney and the Blue Mountains and recaptured at intervals. Because incubation regime influenced the dates of hatching, warm-incubated juveniles were older than cool-incubated ones but these variations did not result in different rates of mortality; recaptured warm-incubated juveniles were longer and heavier than cool-incubated ones, but this difference disappeared when the two groups were age-corrected. The only difference in locomotory performance between warm-incubated and cool-incubated recaptured juveniles was that the former ran faster than the latter at distances greater than 1 metre.

The juveniles recaptured in the enclosures at the two sites did not suffer different rates of mortality but those recaptured at the site in Sydney were longer and heavier than those recaptured in the enclosures in the Blue Mountains, despite the former being younger than the latter. Despite their younger age, however, the recaptured lizards reared in the enclosures in Sydney ran faster than their counterparts who were reared in the Blue Mountains. This difference was not related to their differences in body size.

In terms of the magnitudes of the effects that took place, geographic origin was the least important and analyses showed that the phenotypes of young skinks are influenced by regimes they experienced during embryogenesis as well as by their geographic origin and the subsequent environment in which they are reared. Qualls and Shine (2000) suggested two alternative hypotheses to

Table 5.4. Summary of the effects of various temperatures during incubation (22°C cool, 26°C mild, 30°C hot) on the attributes of hatchlings of *Lampropholis delicata*.

Data from de Jong *et al.* (2023).

Characteristic affected	Effect of temperature of incubation	Interactive effects
Maximum size of hatchlings	Lizards incubated under cool and hot conditions grew to larger maximum size than those incubated under mild conditions.	–
Survival of hatchlings to maturity	Cool-incubated = 67.7% survival; mild-incubated = 84.0% survival; and hot-incubated = 70.4% survival. These differences were not statistically significant.	–
Speed of sprinting	Hot-incubated skinks were slower sprinters than mild-incubated or cool-incubated ones.	Effect of temperature of incubation on speed of sprinting changed with age. The increase in speed of sprinting with increase in temperature was smaller in juveniles than in subadults or adults. Both under cool and mild conditions there was an effect of mass and of temperature of incubation on predicted maximum speed of sprinting; neither measure was significantly affected by temperature of incubation.
Locomotory performance	Incubation under hot conditions reduced locomotory performance compared to that of animals subjected to either mild or cold treatments during incubation.	No effect of the interaction between temperature of incubation and age on the shape of the performance curve. There was a significant effect of mass and the main effect of the temperature of incubation. Neither the temperature of optimal performance nor breadth of performance was significantly affected by mass.
Resting metabolic rate	Subadults had a higher resting metabolic rate than did other categories of age.	Under cool conditions, animals both of average and low mass had a higher resting metabolic rate than those of greater mass; the latter had a lower resting metabolic rate both from hot and mild temperatures of incubation.
Growth rate (also see Fig. 5.5)	Individuals incubated at mild temperatures grew slower than those that were either cool-incubated or hot-incubated.	–
Mass of testes	Temperature during incubation had no effect on subsequent size of testes in adults.	Mass of the testes had a positive effect on the sperm count.
Sperm count	Temperature during incubation had no effect on subsequent sperm count in adult males.	
Morphology of sperm	Cool-incubated and hot-incubated adult males had sperm with shorter mid-pieces and heads compared to that from adult males incubated under mild conditions, although there is no significant differences on total length.	–

account for these results: either (1) there is a divergence in developmental pathways induced during embryogenesis, or (2) these phenomena are a by-product of differences in the date of hatching and the concomitant environmental conditions they experience post-hatching. To decide between these alternative hypotheses, they conducted seasonally adjusted, age-corrected statistical analyses and found that the environments the young experienced after hatching deteriorated as the season progressed. Thus, lizards that hatched later started their

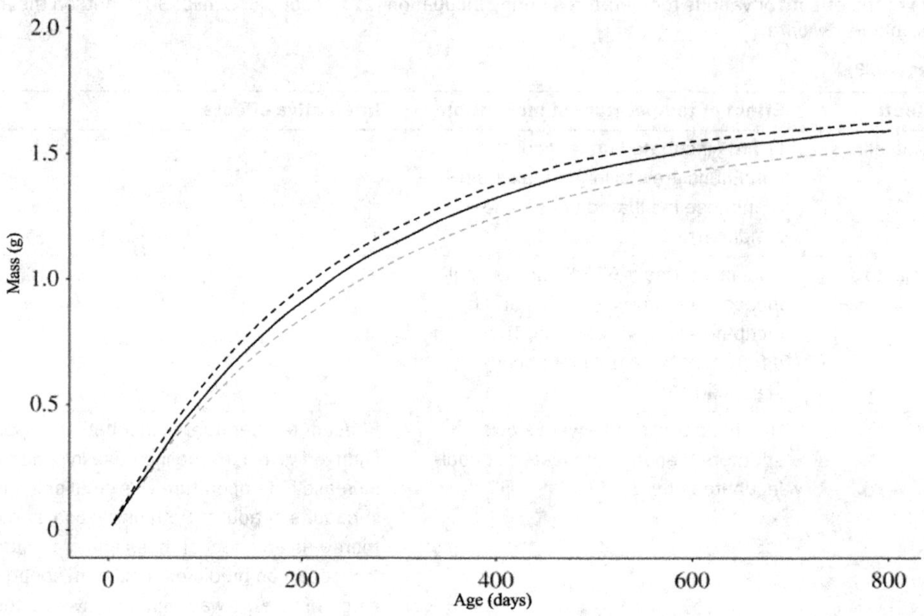

Fig. 5.5. Growth of *Lampropholis delicata*. The three lines portray the growth curves of animals hatching from eggs subjected to different temperatures during incubation: upper line (incubated at 22°C), central line (incubated at 30°C), and lower line (incubated at 26°C). Simplified from de Jong *et al.* (2023) by Susan King.

post-hatching life under less favourable conditions than did those that hatched earlier. When this was taken into account statistically, the effect upon the phenotypes of the locality of origin or of the conditions experienced during incubation disappeared. Instead, it was the conditions that juveniles experienced after hatching that was important and that depended directly upon the time at which hatching occurred. In summary, the determining factor influencing the differences in phenotypes of these lizards was neither genetic nor the conditions experienced during incubation, but rather the quality of the environment experienced after hatching, and that was dependent directly on the time that hatching occurred, which thus was the critical influence.

GROWTH

Immediately after hatching, the hatchlings initially depend partly or entirely on residual yolk in the gut rather than completely on prey that they capture (see section *Yolk* above). Mitchell (1959) found that some *L. guichenoti* measuring 41 mm in total length (body plus tail) at hatching and grew to 47–48 mm without external food.

Clarke (1965) found that 10 *L. guichenoti* about a week old on 16 March had a mean SVL of 19.24 mm (range 17–20 mm). Periodic samples taken over the following year indicated growth to sizes of 27–35 mm SVL, an increase of 10–15 mm in SVL during their first year of life. Simbotwe (1985a) marked individual *L. guichenoti* by coded toe-clipping and measuring SVL upon recapture. Growth rates were low and, as expected, the rate of growth (slope of the lines) decreased as the animals grew. Clarke (1965) calculated the coefficients of growth between various body parts relative to the SVL and to the length of the head. These coefficients differed between the sexes, for some parts higher for males, for others higher for females, and therefore the general shape of the body varied slightly between males and females as growth proceeded (Table 5.5). See Tables 5.3 and 5.4 for a summary of the effect that temperature during incubation had on the subsequent growth of the hatchlings of *L. delicata*.

Some effects on growth can emanate from unexpected sources. For example, Downes (2001) found that during their first 6 months of life, the young of *L. guichenoti* subjected to the odour of one of their natural predators (the yellow-faced whip snake, *Demansia psammophis*) became active later in the day, were less mobile, and selected safer (more secluded) microhabitats. These responses resulted in less time spent foraging for surface-active prey (crickets) and consequently in less food eaten, and lower growth rates.

Table 5.5 The coefficients of growth of various body parts of *Lampropholis guichenoti* and *Lampropholis delicata* to the SVL and the length of the head.

Boldface indicates values higher than the equivalent for the opposite sex. Data from Clarke (1965).

	Lampropholis guichenoti		*Lampropholis delicata*	
Coefficient of growth	**Males**	**Females**	**Males**	**Females**
Relative to SVL				
Length of trunk (axilla to groin)	1.0450	**1.5403**	1.0640	**1.2200**
Length from tip of snout to forelimb	0.6281	**0.7450**	**0.7764**	0.6523
Length of head	**0.8061**	0.6813	**0.7224**	0.5695
Width of head	**0.7997**	0.5203	**0.6386**	0.3829
Length of forelimb	0.7972	**0.7796**	**0.8198**	0.4806
Length of hind limb	**0.8400**	0.7013	**0.8934**	0.5546
Relative to length of head				
Length of trunk (axilla to groin)	1.0091	**1.8574**	1.3723	**1.6253**
Length from tip of snout to forelimb	0.6336	**0.8981**	**1.0583**	1.0534
Length of forelimb	0.5686	**0.9969**	**1.1648**	0.9107
Length of hind limb	**0.9419**	0.9112	**1.2220**	1.0284

SEXUAL MATURITY

The smallest male *L. guichenoti* (*N* = 103) with enlarged testes and convoluted semeniferous tubules, and thus capable of producing sperm, had a SVL of 27.2 mm, whereas the largest one still lacking that capability measured 32.8 mm. Thus, most males probably reach sexual maturity somewhere within this range (Simbotwe 1985b). Pengilley's (1972) estimate was higher (34–45 mm; age 1 year) and weight of testes 0.6–1.6 g. The mean dimension of mature testes was 4.27 ± 1.60 mm long and 2.7 ± 0.94 wide (Simbotwe 1985b[1]). Clarke (1965) reported that in *L. guichenoti* sexual maturity was reached at a mean SVL length of 35 mm in males and 37 mm in females and based on her estimates of growth rates (see section *Growth* above). She concluded that it would probably take this species more than 1 year to grow to sexual maturity.

Simbotwe (1985b) used the creamy-colour of follicles and the smallest female with oviductal eggs to determine that sexual maturity of females was reached between mean SVL of 26.5 and 34.0 mm (range 27–43). The age at first reproduction was 8.9 months (Simbotwe (1985a). Pengilley's (1972) estimate was higher: sexual maturity being reached at 35–49 mm SVL in 1 year.

Immature *L. delicata* ranged from 18 mm to 29 mm SVL (mean = 21.2 mm) and reached sexual maturity the following summer at 35 mm SVL for males and 36 mm SVL for females (Clarke 1965). Bell *et al.* (2010) reported a minimum SVL of 31.5 mm for male *L. coggeri* and one of 39 mm for male *L. robertsi*. Greer (1997) found that all females of *L. elongata* were immature until they were at least 28 mm SVL and that all were mature by the time they had grown to 41 mm. These observations indicate that this species probably reaches maturity at ~2 years of age.

It may well be that the time it takes to reach sexual maturity may differ among species or vary geographically depending on seasonality and local weather conditions, but summarising the above studies, it appears that it may be reached within 1 year, and certainly by 2 years of age.

Size at sexual maturity may not be controlled solely by genetic determination. Downes (2001) found that animals that had lower intake of food grew more slowly and reached sexual maturity at a smaller body size.

SIZE OF THE BODY AND PROPORTION OF BODILY REGIONS

The two eurytopic species

Various investigators have compared the dimensions of adult *L. guichenoti* with somewhat different results (Tables 5.6 and 5.7). Simbotwe (1985a, b) found females of *L. guichenoti* to have a longer snout-to-vent length

1 These values differ between the summary in his Table 46 and his summary on page 175 and it appears that the value for the sexes may have been reversed in one of these sources. However. the difference between the sexes was slight, being identical for the year 1968–1969 and very similar in 1967–1968 (1.1 and 1.2 g).

Table 5.6. Comparison of body sizes of adult males and females of *Lampropholis guichenoti* at Newholme, NSW, during the active seasons of 1981–1982 and 1983–1984.

Dashes indicate no data available. Values of SVL are means ± standard deviation. S = statistically significant. NS = not statistically significant. Modified from Simbotwe (1985a). **Boldface** indicates a higher value than for the opposite sex during the equivalent year.

	S-V length (mm)		Tail length (mm)		Mass of the body (g)	
Year	**Males**	**Females**	**Males**	**Females**	**Males**	**Females**
1981–1982	35.93 ± 3.58	**36.86 ± 4.24**	**48.27 ± 7.62**	47.13 ± 10.90	0.97 ± 0.32	**1.06 ± 0.36**
1983–1984	37.27 ± 3.21	**38.72 ± 3.60**	**52.96 ± 5.28**	50.72 ± 7.90	0.97 ± 0.26	**1.11 ± 0.32**
t	2.69	3.68	3.28	1.84	Identical	1.15
P	<0.01 S	<0.001 S	<0.01 S	>0.10 NS	NS	>0.4 NS

Table 5.7. Snout-to-vent length of mature adult individuals of *Lampropholis*.

N = number of observations. SE = standard error of the mean. SD = standard deviation of the mean. **Boldface** indicates that the measurement is higher than the one for the opposite sex. All distances and lengths are in millimetres. – indicates that no data were given. * = regenerated tail.

	SVL (mm)							
Authority	***Lampropholis guichenoti***				***Lampropholis delicata***			
Years								
Sex	**N**	**mean**	**range**	**SE**	**N**	**mean**	**range**	**SE**
Clarke (1965)								
Males	28	39.5	35–46	0.48	14	39.7	35–47	0.78
Females	31	**42.1**	37-51	0.71	10	**42.0**	36–48	1.37
Green (1965)								
Males	–	–	–	–	5	38	32-41	–
Females	–	–	–	–	4	37	29–40	–
Pengilley (1972)								
1967–1968								
Males	34	**42.6**	35.0–49.0	3.14	–	–	–	–
Females	73	40.7	34.0–45.0	3.06	–	–	–	–
1968–1969								
Males	43	**44.2**	37.0–51.0	3.67	–	–	–	–
Females	45	42.6	35.0–45.6	2.82	–	–	–	–
Simbotwe (1985a)				**SD**				
1981–1982								
Males	103	**39.95**	27.0–43.1	3.58	–	–	–	–
Females	103	36.86	26.0–45.0	4.24	–	–	–	–
1983–1984								
Males	89	37.27	25.2–42.6	3.21	–	–	–	–
Females	136	**38.72**	26.8–46.0	3.60	–	–	–	–
Shine and Greer (1991)				**SE**				
Females	–	43.9	38–49	2.5	–	37.6	34–42	?
Thompson *et al.* (2001)								
Females	13	39.5	35–42	2.5	9	40.0	36–43	2.1
Matthews *et al.* (2023)								
Males (Plain)	–	–	–	–	371	37.7	30.2–44.3	–
Males (Striped)	–	–	–	–	81	38.5	30.3–43.9	–
Females (Plain)	–	–	–	–	25	37.3	30–45	–
Females (Striped)	–	–	–	–	333	37.7	30.3–42.7	–
Qualls and Shine (1998)								
Sexes combined	–	–	up to 51	–	–	–	–	–
Cogger (2018)								
Sexes combined	–	40	–	–	–	45	–	–

Authority Years Sex	SVL (mm)							
	Lampropholis guichenoti				*Lampropholis delicata*			
	N	mean	range	SE	N	mean	range	SE
Baker (1979)[1]								
Females								
Oahu	–	–	–	–	45	38.6	–	1.7
Kaui	–	–	–	–	9	41.8	–	3.5
Hawaii	–	–	–	–	10	41.2	–	2.8

[1]Introduced into Hawaii.

than males, but males had longer tails; these differences were small, only by 1–2 mm. Females were heavier than males in both years, probably because their weights would have included the weight of oviductal eggs they were carrying. Clarke (1965) also found adult female *L. guichenoti* to have a greater SVL than males (Table 5.7), but Pengilley (1972) found just the opposite in two separate seasons.

Du and Shine (2008) found no significant differences between the sexes of hatchlings of *L. guichenoti* in SVL (P = 0.91), weight (P = 0.81), length of head (P = 0.66), relative length of the head to that of the body (P = 0.49), tail length (P = 0.24), or ratio of the length of the tail to that of the body (P = 0.26).

There are fewer data for *L. delicata* than for *L. guichenoti*, but the SVL measured longer in males than in females in each of two seasons in one study and the opposite for a single season in another study (Table 5.7).

Differences between the two eurytopic species (Table 5.7) also gave contrary results with Shine and Greer (1991) showing *L. guichenoti* to have greater SVL than *L. delicata* but Clarke (1965) and Thompson *et al.* (2001) indicating trivial and probably statistically insignificant differences between the two species (0.5 mm or less); Cogger's (2018) summary indicated that *L. delicata* were larger (43 mm SVL) than *L. guichenoti* (40 mm SVL).

Some of these perceived differences may have arisen perhaps because some studies included some subadults in their samples. A further confounding factor is that Simbotwe (1985a) found the 1-year-old females were significantly larger than males ($P < 0.001$), whereas in a different year the reverse occurred ($P < 0.01$) (see Table 5.7).

Of the *L. delicata* introduced into Hawaii, the invasive population on Oahu were larger than those on the other two islands for which data were available (Table 5.7).

Table 5.8 compares the sizes of different parts of the body of *L. guichenoti* and *L delicata*. Female *L. guichenoti* were the larger in all characteristics of the body and limbs but smaller in those of the head, indicating that the sexes differ somewhat in shape, with the females having relatively smaller heads relative to the body than do males. The results were somewhat different in *L. delicata*. In this species the females exceeded the males in all measures of the head and trunk of the body, but had smaller limbs.

The species with small distributions

Bell *et al.* (2010) presented box diagrams indicating the intraspecific differences in the size of the body in genetic lineages both of *L. coggeri* and *L. robertsi*. They found that in *L. coggeri* there were statistically significant, regional differences in the sizes of the bodies of males. Those from the southern lineage were larger than the ones from the northern lineage. Females did not differ in size between these two lineages.

Lampropholis robertsi had a different pattern. Males from a lineage from Atherton–Bartle Frere were significantly smaller than their counterparts in lineages from Bellenden Ker, Carbine Uplands, and Thornton Uplands but did not differ from each other. Females of this species were significantly larger in the lineage on the Carbine Uplands than in other places.

Epilogue

The fact that species of *Lampropholis* in general are smaller than most skinks, including those she considered the taxon's nearest relatives, led Clarke (1965) to hypothesise that *Lampropholis*' miniature status may have evolved via neoteny, with the adults ceasing to grow earlier than previously reported and adults accordingly never exceeding their previous juvenile size. However, she also recognised that the opposite could have occurred with *Lampropholis* representing the ancestral condition and its nearest relatives evolving extended growth beyond the previous maximum.

The varying conclusions reached by different authors about the sizes of the body in the two eurytopic species

Table 5.8. Sizes of various parts of the bodies of adult *Lampropholis guichenoti* and *Lampropholis delicata*.

N = number of observations. SE = standard error of the mean. Ratio = the ratio of the measurement to the snout-to-vent length. **Boldface** indicates that the measurement is higher than the one for the opposite sex. All distances and lengths are in millimetres. - indicates that no data were provided. Data from Clarke (1965).

Lampropholis guichenoti						***Lampropholis delicata***				
Characteristc	**N**	**Mean**	**Range**	**SE**	**Ratio**	**N**	**Mean**	**Range**	**SE**	**Ratio**
Length of trunk (axilla to groin)										
Males	28	21.51	18.0–27.0	0.35	0.550	14	21.96	18.3–28.0	0.90	0.55
Females	31	**24.11**	20.2–31.0	0.59	**0.570**	10	**24.67**	21.0–30.9	0.99	**0.59**
Distance from tip of snout to forelimb										
Males	28	14.14	12.8–16.0	0.16	**0.360**	14	14.20	12.2–15.4	0.26	**0.36**
Females	31	**14.38**	12.6–16.0	0.26	0.340	10	**14.57**	13.0–16.4	0.37	0.35
Distance from tip of snout to ear										
Males	28	**7.91**	7.2–8.7	0.08	**0.200**	14	7.54	5.8–8.2	0.16	**0.190**
Females	27	7.80	6.7–9.3	0.12	0.185	10	**7.57**	6.9–8.2	0.18	0.180
Width of head, between the eyes										
Males	28	**4.55**	3.9–5.2	0.06	**0.580**	13	4.17	3.4–4.9	0.10	**0.55**
Females	26	4.45	3.8–5.0	0.06	0.570	10	**3.90**	3.4–4.5	0.14	0.52
Length of forelimb										
Males	28	10.83	9.2–12.7	0.15	**0.272**	14	**10.32**	8.3–11.5	0.24	0.26
Females	26	**10.63**	9.3–12.5	0.16	0.253	9	9.94	8.1–10.8	0.27	**0.24**
Length of hind limb										
Males	28	15.25	13.6–16.8	0.19	**0.390**	14	**14.64**	13.4–16.4	0.23	**0.37**
Females	27	**14.76**	12.7–17.9	0.24	0.350	9	13.93	12.8–15.1	0.27	0.33

probably are attributable to operation of a number of uncontrolled factors that could confound the results, such as nutritional condition varying from one season and/or locality to another, and regional or elevational differences in the environmentally and/or genetically determined sizes of the species. All these influences need to be assessed before a full understanding to the observed differences is attained. The research by Bell *et al.* (2010) represents an important contribution in that direction.

SEXUAL DIMORPHISM

Sexual dimorphism in size and shape of the body has been reported to occur in *L. delicata,* not only at several locations in its native geographic range in Australia, but also in some locations on the Pacific islands to which it had been introduced (Miller *et al.* 2017; see Chapter 3). For example Chapple *et al.* (2014) found that in introduced populations of *L. delicata* on Lord Howe Island males had smaller bodies than females (except in the northern parts of the island), shorter abdomens, and longer and broader heads than females. By contrast, Simbotwe (1985a, b) found no statistically significant difference in SVL between the sexes of *L. guichenoti* ($t = 1.66$; $P > 0.10$), nor was the difference in length of the tail significant (mean for males = 48.27 ± 7.62 mm; mean for females 47.13 ± 10.94 mm) ($t = 0.61$; $P > 0.90$) or of the head (mean for males = 8.60 ± 0.06, mean for females 8.55 ± 0.07 mm) ($t = 0.57$; $P > 0.57$). Du and Shine (2008) found that hatchlings of *L. guichenoti* did not show significant differences either in SVL ($P = 0.91$), body weight ($P = 0.08$), length of the head ($P = 0.66$), length of the head relative to that of the body ($P = 0.49$), or length of the tail relative to that of the head ($P = 0.08$). However, Simbotwe (1985a, b) found that the heads of males of this species were significantly narrower (mean 3.74 ± 0.05 mm) than those of females (mean 3.60 ± 0.05 mm; $t = 2.12$; $P < 0.036$) and significantly thicker (4.90 ± 0.04; $t = 2.37$; $P < 0.019$) than those of females (mean 4.72 ± 0.04 mm). There also was a difference in that the lengths of the body in both sexes and in the tails of males differed between years at a given site and thus the extent, or even the existence, of dimorphism may not be consistent over time. Surprisingly, even though bodily dimensions may show significant year-to-year differences, weight of the body did not.

The significantly wider gape (as indicated by width of the head) in adult females may allow them to eat prey larger than the upper size manipulatable by adult males of equivalent length, thereby partially reducing overlap between the sexes in utilisation of resources. Simbotwe (1985a) found that the mean body length of prey consumed by *L. guichenoti* increased significantly with increasing size of the lizard's head ($r = 0.83$; $P < 0.01$; $N = 76$). Also as hatchlings grow into adults the size of their heads increases (Simbotwe 1985a) and hence so does their gape, thereby allowing consumption of progressively larger prey.

The significance of the narrower, but thicker heads of males is not clear. Combat does occur in *Lampropholis* (see Chapter 4) and one would expect a thicker head to be indicative of greater size of the jaw muscles and hence of a stronger bite in subduing prey or opponents but its narrower head would suggest the opposite.

Cooper and Vitt (1987) have shown that in the skink, *Eumeces laticeps,* larger males have an advantage over smaller ones in fights and accordingly young males tend to avoid aggressive contests with older males and, in their presence, to defer attempting to mate with females. Under such circumstances one would expect there to be selection for larger size in males than in females, as has been noted for snakes that have male-to-male combat (Shine 1978).

What is the role of dimorphism in colour pattern?

There is genetically coded dimorphism in striping in both sexes of *L. delicata* (see Chapter 14). *Lampropholis delicata* is dimorphic in relation to the presence (striped morph) or absence (plain morph) of a lateral white stripe on the lower side of the body (see Chapter 14) and the colour pattern of the body is believed to play a role in camouflaging lizards from potential predators, influence the behavioural responses of other conspecific individuals, and affect thermoregulation.

Forsman and Shine (1995a) found that in *L. delicata* choice of mates showed no relationship to colour pattern so dimorphism seemingly lacks a social function. They (1995b) extended their investigation on the adaptative value of dimorphism in colour in this species and demonstrated a north-to-south decline in the incidence of the white stripe in both sexes with its disappearance altogether somewhere between Sydney and Melbourne. At all localities, the incidence of the striped morph was much lower in males than in females. For example, north of Townsville the incidence in females was 84%, whereas that in males was only 35%; this relative discrepancy became even more pronounced at distances farther south. Across populations the relative frequency of striped individuals decreased commensurately with increasing proportion of regenerating tails; however, there was no significant difference between striped and non-striped individuals in this regard, nor between the sexes or individuals of different sizes. They divided a large number of specimens from one locality into two groups collected at intervals many years apart and there was no significant change in the incidence of striping over time. They also painted white stripes on non-striped individuals and found no change in their probability of being recaptured by humans at a later time.

These authors favoured the hypothesis that the relationship between incidence of colour pattern between the sexes was due to differences in activity patterns between males and females, with the lateral stripe increasing predation on males but not on females. Females were less active than males on cold days, suggesting that males were active at lower body temperatures, on average, than were females. The authors posited that relative risk of predation of striped and non-striped individuals depends on sprint speed and anti-predator behaviour (and hence on body temperature). Accordingly, the effect of the stripe on survival may have differed between sexes simply because males and females were active at different body temperatures. The fact that male and female *L. delicata* selected different microhabitats, and that females moved twice as far between captures than did males may also have contributed to the different response by the two sexes. They considered the disadvantage of the white stripe in males to be consistent with their observations that the striped morph was less common in males than in females in all polymorphic populations, and that the stripe was less distinct in males than in females. A maladaptive stripe in males could be retained as a less distinct form simply because of its associated advantage to females.

Matthews *et al.* (2023) tested heating and cooling rate, sprint speed, thermal preference, body temperatures in the field, and metabolic rate in both morphs and sexes to see if there was any link to dimorphism in colour pattern. Plain individuals heated more quickly than striped ones but other thermal traits showed little variation. Matthews and Chapple (2023) also tested the conspicuousness of the pattern of *L. delicata* and found that plain skinks were more conspicuous than striped ones both to predatory birds and predatory lizards. The *Lampropholis* did not select microhabitats on the basis of matching colour so it is likely that the advantage of the stripe is that it disrupts the outline of the lizard against the structure of various variegated backgrounds whether of litter or of vegetation.

Females reflected more light from their dorsum than did males and plain individuals reflected more light dorsally than did striped ones. Reflection laterally from the body differed between left and right sides; females reflected more light from the left side than did males but on the right side the sexes reflected equally. In general the stripe is paler in males than it is in females. Lateral reflectance was greater in striped animals than in plain ones. These reflectivities were related to models of the visual systems of three groups of potential predators: (1) birds whose colour-detecting cone cells in the eyes were sensitive to ultraviolet light, (2) birds whose cone cells were less sensitive to ultraviolet light, and (3) lizards. For left-angle contrast, plain females were more conspicuous than either striped females or striped males; plain males were more conspicuous than striped females, and striped males were more conspicuous than striped females.

Eventually, elimination of the alternate alleles can lead to extirpation of specific disadvantageous colour morphs. There are, however, exceptions such as heterozygote advantage, frequency-dependent selection, and differential selection between the sexes, that support the maintenance of dimorphisms in many populations.

For a discussion of the relationship between colour pattern and thermal biology see Chapter 6.

SEX RATIO

Pengilley (1972) determined the sex ratios of *L. guichenoti* of different age-groups (juveniles, subadults, and adults) captured by two methods (hand-collecting and pitfall traps) at different months in 1967 and 1968. Samples for some months were too small for testing but for all of those that were the sex ratio did not differ significantly from a ratio of 1:1 between the sexes. Similarly Chapple *et al.* (2014) found an equal sex ratio in *L. delicata* at all regions of Lord Howe Island where this species was introduced.

REDUCTION OF LIMBS

Skinks vary in the size of their limbs in relation to the size of their torso from fully limbed (well-developed legs similar to those of most lizards), to proportionately tiny limbs for the size of their body, to completely limbless (snake-like) (Chapple *et al.* 2023). Although Camaiti *et al.* (2022) in their review of limblessness in lizards, portrayed *Lampropholis* as 'fully limbed' (as opposed to 'limb reduced' or 'limbless'), they included only four of the species of this genus in their assessment. In fact, various species of *Lampropholis* have some degree of reduction in the size of the limbs in comparison with *L. guichenoti* and *L. delicata* (see Table 14.4 in Chapter 14 and photographs on the cover and title pages of this book). Of these, *L. elongata* is the species in the genus with the greatest reduction in limbs; it has lost entire phalanges, specifically from the interior of digit 4, both from the hands and feet, thereby giving it the phalangeal formula of 2.3.4.**4**.3/2.3.4.**4**.3 instead of 2.3.4.5.3/2.3.4.5.3 (Greer 1997); bold highlight indicates the digit from which the phalanx was lost.

LONGEVITY

Maximum longevity in skinks is similar to that in other families of lizards (Chapple *et al.* 2023). Two estimates of the longevity of *L. guichenoti* are 'at least 2 yr, and probably more' (Clarke 1965) and greater than 4 years (Pengilley 1972).

REFERENCES

Adams SM, Hosie MJ, Murphy CR, Thompson MB (2004) Changes in oviductal morphology of the skink, *Lampropholis guichenoti*, associated with egg production. *Journal of Morphology* **262**(1), 536–544. doi:10.1002/jmor.10262

Alibardi L (2008a) Cornification of developing claws of the common Australian skink (*Lampropholis guichenoti*) (Squamata, Lacertidae). *Italian Journal of Zoology* **75**(4), 327–336. doi:10.1080/11250000801973334

Alibardi L (2008b) Microscopic analysis of lizard claw morphogenesis and hypothesis on its evolution. *Acta Zoologica* **89**(2), 169–178. doi:10.1111/j.1463-6395.2007.00312.x

Alibardi L (2009) Development, comparative morphology and cornification of reptilian claws in relation to claws evolution in tetrapods. *Contributions to Zoology* **78**(1), 25–42. doi:10.1163/18759866-07801003

Alibardi L, Thompson MB (1999) Epidermal differentiation in the developing scales of embryos of the Australian scincid lizard *Lampropholis guichenoti*. *Journal of Morphology* **241**(2), 139–152. doi:10.1002/SICI1097-4687(199908)-241:2>3.0.C0;2;2-H

Alibardi L, Toni M (2009) Immunocytochemistry and protein analysis suggest that reptilian claws contain small high cysteine-glycine proteins. *Tissue and Cell* **41**(3), 180–192. doi:10.1016/j.tice.2008.10.002

Andrews RM, Mathies M (2000) Natural history of reptilian development: constraints on the evolution of viviparity. *Bioscience* **50**(3), 227–238. doi:10.1641/0006-3568(2000)050[0227:NHORDC]2.3.CO;2

Baker K (1976) The occurrence and ecological significance of metallic skinks on the islands of Hawai'I and Kaua'i. In *Proceedings of the First Conference in Natural Sciences, Hawaii Volcanoes National Park, August 19-20, 1976*. (Ed. CW Smith) pp. 11–17. Cooperative National Park Studies Unit, Department of Botany, University of Hawaii, Manoa.

Baker KJ (1979) The rainbow skink, *Lampropholis delicata*, in Hawaii. *Pacific Science* **33**(2), 207–212. <http://hdl.handle.net/10125/1470>

Bell RC, Parra JL, Tonione M, Hoskin CJ, Mackenzie JB, Williams SE, Moritz C (2010) Patterns of persistence and isolation indicate resilience to climate change in montane rainforest lizards. *Molecular Ecology* **19**(12), 2531–2544. doi:10.1111/j.1365-294X.2010.04676.x

Bilcke J, Downes S, Büscher I (2006) Combined effect of incubation and ambient temperature on the feeding performance of a small ectotherm. *Austral Ecology* **31**(8), 937–947. doi:10.1111/j.1442-9993.2006.01663.x

Blackburn DG (1982) Evolutionary origins of viviparity in the Reptilia. I. Sauria. *Amphibia-Reptilia* **3**, 185–291. doi:10.1163/156853882X00419

Blackburn DG (1998) Structure, function, and evolution of the oviducts of squamate reptiles with special reference to viviparity and placentation. *Journal of Experimental Zoology* **282**(4–5), 560–617. doi:10.1002/(SICI)1097-010X(199811/12)282:4/5<560::JEZ10>3.0.CO;2-J

Blackburn DG (2006) Squamate reptiles as model organisms for the evolution of viviparity. *Herpetological Monographs* **20**(1), 131–146. doi:10.1655/0733-1347(2007)20[131:SRAMOF]2.0.CO;2

Booth DT, Thompson MB, Herring S (2000) How incubation temperature influences the physiology and growth of embryonic lizards. *Journal of Comparative Physiology B* **170**(4), 269–276. doi:10.1007/s003600000097

Burgin S (1993) *Lampropholis*: the new 'laboratory' animals for teaching. In *Herpetology in Australia: A Diverse Discipline*. (Eds D Lunney, D Ayers) pp. 279–282. Surrey Beatty & Sons, Chipping Norton, NSW.

Camaiti M, Evans AR, Hipsley CA, Hutchinson MN, Meiri S, Anderson RO, Slavenko A, Chapple DG (2022) A database of the morphology, ecology and literature of the world's limb-reduced skinks. *Journal of Biogeography* **49**(7), 1397–1406. doi:10.1111/jbi.14392

Chapple DG, Miller KA, Chaplin K, Barnett L, Thompson MB, Bray RD (2014) Biology of the invasive delicate skink (*Lampropholis delicata*) on Lord Howe Island. *Australian Journal of Zoology* **62**(6), 498–506. doi:10.1071/ZO14098

Chapple DG, Slavenko A, Tingley R, Farquhar JE, Camaiti M, Roll U, Meiri S (2023) Built for success: distribution, morphology, ecology and life history of the world's skinks. *Ecology and Evolution* **13**(12). doi:10.1002/ece3.10791

Cheetham E, Doody JS, Stewart B, Harlow P (2011) Embryonic mortality as a cost of communal nesting in the delicate skink. *Journal of Zoology* **283**(4), 234–242. doi:10.1111/j.1469-7998.2010.00764.x

Clarke CJ (1965) A comparison between some Australian five-fingered lizards of the genus *Leiolopisma* Duméril & Bibron (Lacertilia: Scincidae. *Australian Journal of Zoology* **13**(4), 577–592. doi:10.1071/ZO9650577

Clerke RB, Alford RA (1993) Reproductive biology of four species of tropical Australian lizards and comments on the factors regulating lizard reproductive cycles. *Journal of Herpetology* **27**(4), 400–406. doi:10.2307/1564826

Cogger HG (2018) *Reptiles and Amphibians of Australia*. Updated 7th edn. CSIRO Publishing, Clayton South, Vic.

Cooper WE, Vitt LJ (1987) Deferred agonistic behavior in a long-lived scincid lizard *Eumeces laticeps*: field and laboratory data on the roles of body size and residence in agonistic strategy. *Oecologia* **72**(3), 321–326. doi:10.1007/BF00377558

Couper PJ, Schneider CJ (1995) Communal nesting in the small skink, *Lampropholis adonis*. *Memoirs of the Queensland Museum* **38**, 382.

Debus SJS, Ford HA, Page D (2006) Bird communities in remnant woodland on the New England Tablelands, New South Wales. *Pacific Conservation Biology* **12**(1), 50–63. doi:10.1071/PC060050

Deeming DC (1989) The residues in the eggs of squamate reptiles at hatching. *Herpetological Journal* **1**, 381–385.

de Jong M, White CR, Wong BBM, Chapple DG (2022) Univariate and multivariate plasticity in response to incubation temperature in an Australian lizard. *Journal of Experimental Biology* **225**(22), jeb244352. doi:10.1242/jeb.244352

de Jong MJ, Alton, LA, White CR, O'Bryan MK, Chapple DG, Wong BBM (2023) Long-term effects of incubation temperature on growth and thermal physiology in a small ectotherm. *Philosophical Transactions of the Royal Society B: Biological Sciences* **378**(1884), 20220137. doi:10.1098/rstb.2022.0137

De Vis (1988) A contribution to the herpetology of Queensland. *Proceedings of the Linnean Society of New South Wales* **2**, 811–826.

Doody JS (2011) Environmentally cued hatching in reptiles. *Integrative and Comparative Biology* **51**(1), 49–61. doi:10.1093/icb/icr043

Downes S (2001) Trading heat and food for safety: costs of predator avoidance in a lizard. *Ecology* **82**(10), 2870–2881. doi:10.1890/0012-9658(2001)082[2870:THAFFS]2.0.CO;2

Downes SJ, Shine R (1999) Do incubation-induced change sin a lizards' phenotype influence its vulnerability to predators? *Oecologia* **120**(1), 9–18. doi:10.1007/s004420050827

Du W-G, Shine R (2008) The influence of hydric environments during egg incubation on embryonic heart rates and offspring phenotypes in a scincid lizard (*Lampropholis guichenoti*). *Comparative Biochemistry and Physiology Part

A: Molecular and Integrative Physiology **151**(1), 102–107. doi:10.1016/j.cbpa.2008.06.005

Duméril AMC, Bibron G (1839) *Erpétologie Général ou Histoire Naturelle Complète des Reptiles*. Roret, Paris.

Forsman A, Shine R (1995a) Parallel geographic variation in body shape and reproductive life history within the Australian scincid lizard *Lampropholis delicata*. *Functional Ecology* **9**(6), 818–828. doi:10.2307/2389979

Forsman A, Shine R (1995b) The adaptive significance of colour pattern polymorphism in the Australian scincid lizard *Lampropholis delicata*. *Biological Journal of the Linnean Society* **55**(4), 273–291. doi:10.1016/S0024-4066(05)80001-X

Green RH (1965) Two skink lizards newly recorded from Tasmania. *Records of the Queen Victoria Museum* **19**, 1–4.

Green RH (1981) Distribution of the delicate skink. *Tasmanian Naturalist* **68**, 8.

Greer AE (1997) A new species of *Lampropholis* (Squamata: Scincidae) with a restricted, high altitude distribution in eastern Australia. *Australian Zoologist* **30**(3), 360–368. doi:10.7882/AZ.1997.014

Hallmann K, Griebeler EM (2015) Eggshell types and their evolutionary correlation with life-history strategies in squamates. *PloS One* **10**(9), e0138785. doi:10.1371/journal.pone.0138785

Hosie MJ, Adams SM, Thompson MB, Murphy CR (2003) Viviparous lizard, *Eulamprus tympanum*, shows changes in the uterine surface epithelium during early pregnancy that are similar to the plasma membrane transformation of mammals. *Journal of Morphology* **258**(3), 346–357. doi:10.1002/jmor.10163

Humphreys WF (1976) Spider induced egg mortality in a skink population. *Copeia* **1976**, 404. doi:10.2307/1443980

Joss JMP, Minard JA (1985) On the reproductive cycles of *Lampropholis guichenoti* and *L. delicata* (Squamata: Scincidae) in the Sydney region. *Australian Journal of Zoology* **33**(5), 699–704. doi:10.1071/ZO9850699

Matthews G, Chapple DG (2023) Testing the function of a colour pattern polymorphism in a lizard. *Biological Journal of the Linnean Society* **142**(2), 197–207. doi:10.1093/biolinnean/blad124

Matthews G, Farquhar JE, White CR, Chapple DG (2023) Does thermal biology differ between two colour pattern morphs of a widespread Australian lizard? *Journal of Thermal Biology* **114**, 103579. doi:10.1016/j.jtherbio.2023.103579

Miller KA, Duran A, Melville J, Thompson MB, Chapple DG (2017) Sex-specific shifts in morphology and colour pattern polymorphism during range expansion of an invasive lizard. *Journal of Biogeography* **44**(12), 2778–2788. doi:10.1111/jbi.13075

Mitchell FJ (1959) Communal egg-laying in the lizard *Leiolopisma guichenoti* (Dumeril and Bibron). *Transactions of the Royal Society of South Australia* **82**, 121–122.

Murphy BF, Brandley MC, Murphy CR, Thompson MB (2012) Morphology and development of the placentae in *Eulamprus quoyii* group skinks (Squamata: Scincidae). *Journal of Anatomy* **220**(5), 454–471. doi:10.1111/j.1469-7580.2012.01492.x

Murray BF, Thompson MB (2011) A review of the evolution of viviparity in squamate reptiles: the past, present and future role of molecular biology and genomics. *Journal of Comparative Physiology B* **181**(5), 575–594. doi:10.1007/s00360-011-0584-0

Osborne L, Thompson MB (2005) Chemical composition and structure of the eggshell of three oviparous lizards. *Ichthyology and Herpetology* **2005**(3), 683–692. doi:10.1643/CH-04-280R1

Packard GC, Packard MJ (1988) The physiological ecology of reptilian eggs and embryos. Chapter 8. In *Biology of the Reptilia*. (Eds C Gans and RB Huey) Vol. 16, Ecology B. pp. 523–605. Alan R Liss, New York.

Pengilley R (1972) Systematic relationships and ecology of some lygosomine lizards from southeastern Australia (2 Vols). PhD thesis. Australian National University, Canberra.

Qualls FJ, Shine R (1997) Geographic variation in 'costs of reproduction' in the scincid lizard *Lampropholis guichenoti*. *Functional Ecology* **11**(6), 757–763. doi:10.1046/j.1365-2435.1997.00150.x

Qualls FJ, Shine R (1998) Geographic variation in lizard phenotypes: importance of the incubation environment. *Biological Journal of the Linnean Society* **64**(4), 477–491. doi:10.1111/j.1095-8312.1998.tb00345.x

Qualls FJ, Shine R (2000) Post-hatching environment contributes greatly to phenotypic variation between two populations of the Australian garden skink, *Lampropholis guichenoti*. *Biological Journal of the Linnean Society* **71**(2), 315–341. doi:10.1111/j.1095-8312.2000.tb01260.x

Rawlinson PA (1974) Biogeography and ecology of the reptiles of Tasmania and the Bass Strait area. In *Biogeography and Ecology of Tasmania*. (Ed. WD Williams) pp. 291–339. Dr. W. Junk, The Hague.

Schneider CJ, Couper PJ (1995) Communal nesting in the small skink, *Lampropholis adonis*. *Memoirs of the Queensland Museum* **38**(2), 382.

Shea GM (2010) The suburban terrestrial reptile fauna of Sydney – winners and losers. In *The Natural History of Sydney*. (Eds D Lunney, P Hutchings, D Hochuli) pp. 154–197. Royal Zoological Society of NSW, Sydney. doi:10.7882/FS.2010.015

Shine R (1978) Sexual size dimorphism and male combat in snakes. *Oecologia* **33**(3), 269–277. doi:10.1007/BF00348113

Shine R (1983a) Reptilian reproductive modes: the oviparity-viviparity continuum. *Herpetologica* **39**(1), 1–8. <https://www.jstor.org/stable/3892470>

Shine R (1983b) Reptilian viviparity in cold climates: testing the assumptions of an evolutionary hypothesis. *Oecologia* **57**(3), 397–405. doi:10.1007/BF00377186

Shine R (1985) The evolution of viviparity in reptiles: an ecological analysis. Chapter 8. In *Biology of the Reptilia*. (Eds C Gans and F Billett) Vol. 15, Development B. pp. 605–694. John Wiley & Sons, New York.

Shine R (1992) A new hypothesis for the evolution of viviparity in reptiles. *The American Naturalist* **145**(5), 809–823. doi:10.1086/285769

Shine R (2002) An empirical test of the 'predictability' hypothesis for the evolution of viviparity in reptiles. *Journal of Evolutionary Biology* **15**(4), 553–560. doi:10.1046/j.1420-9101.2002.00420.x

Shine R (2003a) Effects of pregnancy on locomotor performance: an experimental study on lizards. *Oecologia* **136**(3), 450–456. doi:10.1007/s00442-003-1281-9

Shine R (2003b) Locomotor speeds of gravid lizards: placing 'costs of reproduction' within an ecological context. *Functional Ecology* **17**(4), 526–533. doi:10.1046/j.1365-2435.2003.00756.x

Shine R, Elphick MJ (2001) The effect of short-term weather fluctuations on temperatures inside lizard nests, and on the phenotypic traits of hatchling lizards. *Biological Journal of the Linnean Society* **72**(4), 555–565. doi:10.1111/j.1095-8312.2001.tb01338.x

Shine R, Greer AE (1991) Why are clutch sizes more variable in some species than in others? *Evolution* **45**(7), 1696–1706. doi:10.2307/2409790

Shine R, Thompson MB (2006) Did embryonic responses to incubation conditions drive the evolution of reproductive modes in squamate reptiles? *Ecological Monographs* **20**(1), 159–171. doi:10.1655/0733-1347(2007)20[159:DERTIC]2.0.CO;2

Shine R, Elphick MJ, Harlow PS (1997) The influence of natural incubation environments on the phenotypic traits of hatchling lizards. *Ecology* **78**(8), 2559–2568. doi:10.2307/2265914

Shine R, Wapstra E, Olsson MM (2018) Seasonal shifts along the viviparity-viviparity continuum in a cold-climate lizard population. University of Wollongong. Faculty of Science, Medicine and Health - Papers: Part B. <https://ro.uow.edu.au/smhpapers1/984>

Simbotwe MP (1985a) Comparative ecology and demography of two populations of Australian reptiles (Lacertilia, Scincidae). PhD Dissertation. University of New England, Armidale, NSW.

Simbotwe MP (1985b) Sexual dimorphism and reproduction of *Lampropholis guichenoti* (Lacertilia: Scincidae). In *The Biology of Australasian Frogs and Reptiles*. (Eds G Grigg, R Shine and H Ehmann) pp. 11–16. Surrey Betty & Sons and Royal Zoological Society of New South Wales, Chipping Norton, NSW.

Stewart JR, Thompson MB (2000) Evolution of placentation among squamate reptiles: recent research and future directions. *Comparative Biochemistry and Physiology Part A: Molecular & Integrative Physiology* **127**(4), 411–431. doi:10.1016/S1095-6433(00)00273-7

Thompson M, Herbert J, Murphy CR, Thompson MB (2005) HoxA10-like proteins in the reproductive tract of the viviparous lizard *Eulamprus tympanum* and the oviparous lizard *Lampropholis guichenoti*. *Comparative Biochemistry and Physiology Part B: Biochemistry and Molecular Biology* **142**(1), 123–127. doi:10.1016/j.cbpc.2005.07.001

Thompson MB, Speake BK (2002) Energy and nutrient utilisation by embryonic reptiles. *Comparative Biochemistry and Physiology Part A: Molecular & Integrative Physiology* **133**(3), 529–538. doi:10.1016/S1095-6433(02)00188-5

Thompson MB, Speake BK, Russell KJ, McCartney RJ (2001) Utilisation of lipids, protein, ions and energy during embryonic development of Australian oviparous skinks in the genus *Lampropholis*. *Comparative Biochemistry and Physiology Part A: Molecular & Integrative Physiology* **129**(2–3), 313–326. doi:10.1016/S1095-6433(00)00349-4

Thompson MB, Stewart JR, Speake BK, Hosie MJ, Murphy CR (2002) Evolution of viviparity: what can Australian lizards tell us? *Comparative Biochemistry and Physiology Part B: Biochemistry and Molecular Biology* **131**(4), 631–643. doi:10.1016/S1096-4959(02)00013-1

Turner G (2006) Communal egg-laying in *Lampropholis coggeri* and *Saproscincus tetradactyla* (Scincidae). *Herpetofauna* **36**(2), 88–92.

Uller T, While GM, Wapstra E, Warner DA, Goodman BA, Schwarzkopf L Langkilde T, Doughty P, Radder RS, Rohr DH, Bull CM, Shine R, Olsson M (2009) Evaluation of offspring size-number in 12 species of lizard. *Journal of Evolutionary Biology* **22**(1), 143–151. doi:10.1111/j.1420-9101.2008.01629.x

Wapstra M, Wapstra E (1986) Breeding record of the delicate skink *Leiolopisma delicata* in southern Tasmania. *Tasmanian Naturalist* **87**, 5–6.

Wells R (1981) Utilisation of the same site for communal egg-laying by *Lampropholis delicata* and *L. guichenoti*. *Australian Journal of Herpetology* **1**(1), 35–36.

Wells RW (2002) Some taxonomic changes to the genus *Lampropholis* (Reptilia: Scincidae) from Australia. *Australian Biodiversity Record* **8**, 1–24.

6
Thermal biology

INTRODUCTION

There are three major categories of animals in terms of their thermal biology. *Poikilotherms* are animals that passively accept and track the temperature of their environment and consequently have body temperatures that do not depart significantly from environmental ones. In stark contrast are *endotherms,* like birds and mammals, whose body heat is produced and regulated internally by metabolic means. Accordingly, they can maintain optimal, nearly constant, levels of body temperature even when environmental ones fluctuate markedly. Until the seminal work of Cowles and Bogert (1944) reptiles were considered to be poikilotherms. Few, however, conform to that designation and many reptiles exert a level of control of their body temperature between that of poikilotherms and that of endotherms. They achieve this primarily by behavioural means that alter their exchange of heat with their external environment, and hence are classified as *ectotherms* (meaning external heat). Ectotherms, in turn, are classified into several subcategories, depending on the medium that serves as the vehicle of their primary exchanges of heat. Those that depend mainly upon solar radiation are called *heliotherms,* whereas those that depend mainly on exchange of heat with the substrate with which they are in contact are classified as *thigmotherms.* Even for ectotherms, regulation of body temperature incurs a variety of costs for its execution as well as benefits received once attained (Meek 1995) and is an exceedingly complex phenomenon (see Box 6.1).

Effect of size and shape of the body on ectothermic exchange of heat

Within lizards, there is an enormous range in size, from up to 1.54 m snout-to-vent and 81.5 kg for an adult Komodo Dragon, the world's largest living lizard (Wood 1982) to the smallest one, a chameleon, *Brookesia nana* (Glaw *et al.* 2021), measuring <1.4 cm snout-to-vent (adult male) and 1.9 cm (adult female). *Lampropholis* are closer to the lower end of the range in body size known for lizards (maximum reported SVL 5.1 cm for adults of both sexes of *L. guichenoti* and 4.7 and 4.8 cm respectively for males and females of *L. delicata*; see Chapter 5, section *Size of the body and proportion of bodily regions*).

Body size has a profound effect on the thermal biology of lizards because of the relationship of differences in the ratios of surface area to volume between large and small entities. This relationship is illustrated by different-sized cubes in Fig. 6.1. The amount of heat contained in the body of an ectotherm, such as a lizard, depends on the animal's volume. For example, at a given temperature it may hold a certain number of calories per gram, or per cubic centimetre, of tissue. By contrast, the rate of loss or gain of heat to or from the environment depends on the surface area (number of square centimetres of skin exposed to the air) through which heat can pass. The larger the reptile, the more stable is its body temperature because for the great amount of heat stored in its body, whereas there are relatively few square centimetres of skin through which heat can be either gained or lost (Heatwole and Taylor 1987). Thus, small lizards, because of their high surface-area-to-volume ratio, rapidly absorb heat, and even a small gain contributes a large amount of heat in proportion of the total amount of heat already in the tiny body. The heat being gained through a given square centimetre of skin only has to warm a small amount of tissue, whereas in a very large animal, the heat gained through a unit area of skin has an immense volume of tissue to heat, and consequently body temperature rises only slightly. The same applies to the loss of heat; a small ectotherm has only a few cubic centimetres of tissue with its contained heat, but a relatively large surface area through which that heat can pass to a cooler environment. Because of their great surface area to volume ratio, very small ectotherms rapidly lose heat to, and gain heat from, their environment as they perform their daily activities and move through patches of sunlight and shade or pass between cooler and warmer substrates.

Shape of the body is also important for thermal biology. A long, slender object, such as a lizard or snake, has a much greater SA:V than does a squat, rotund object of the same volume (Fig. 6.1). Consequently, a long slender lizard, such as *Lampropholis*, gains or loses heat to the environment more rapidly than does a more squat, rotund animal (such as a frog) of equal weight or volume.

Behavioural thermoregulation

Lizards can alter their absorption and loss of heat by alternately shuttling between sunlit and shaded areas, resting on warmer or cooler substrates, basking in sunlight (sometimes accompanied by changing their angle of orientation to the sun's rays by flattening their bodies to expose its greatest surface area and tilting their backs toward the sun versus facing the sun and exposing the least possible area), and perching at different heights above the heated ground (Heatwole 1970; Spellerberg 1972b; Heatwole and Taylor 1987).

Different objects, by virtue of their chemical composition, colour, and reflective and absorptive properties may heat at different rates under the same burden of solar radiation and thus a lizard might have a different body temperature than its surrounding substrate. Complete congruence in such properties could mean that similarity in body and substrate temperatures merely denote

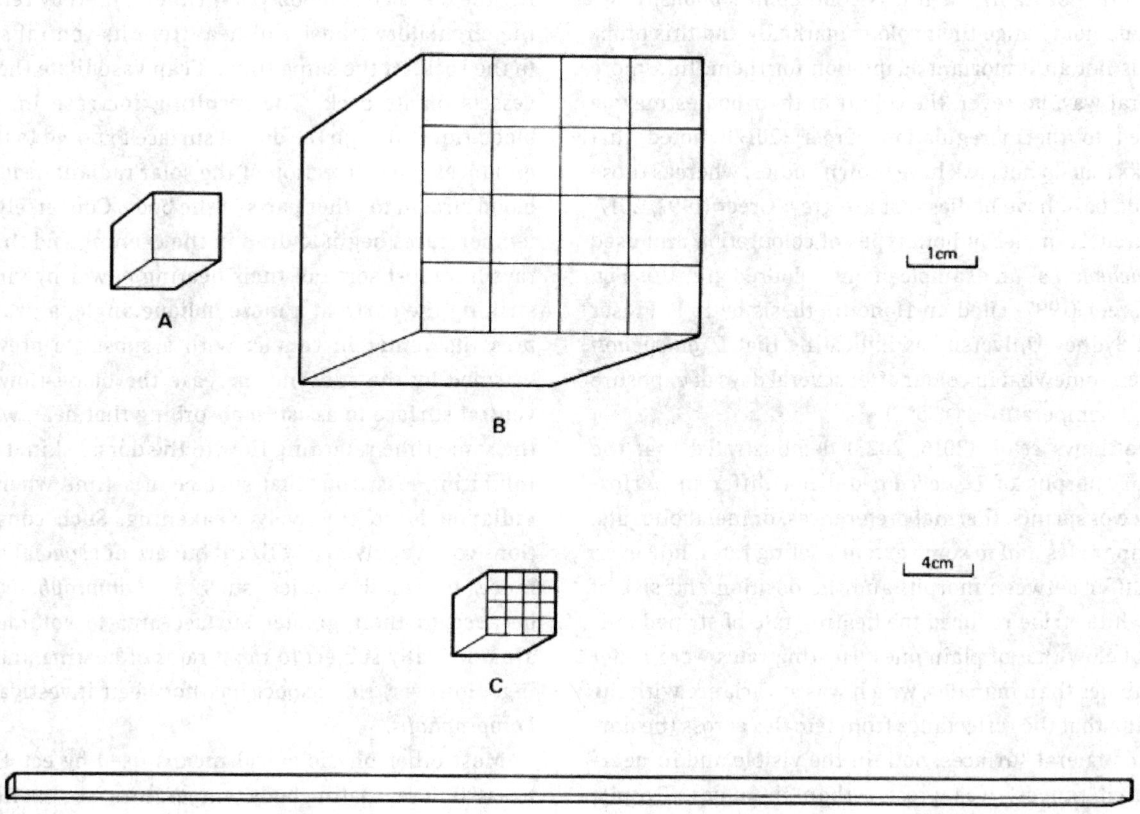

Fig. 6.1. The relationship of size and shape to surface-area to volume ratio (SA:V). Objects A and B are the same shape (cubes) but of different sizes. The SA:V ratio of object A is 6 cm^2 (six sides, each of 1 cm^2) of surface area to 1 cm^3 of volume (SA:V = 6:1). Object B has a surface area of 16 cm^2 on each of six sides for a total of 96 cm^2 and a volume of 64 cm^3 for a SA:V ratio on only 1.5:1. Thus, object A has four times as much surface area for each cubic centimetre of its volume as does object B. Compare objects C and D (note that they are drawn at a different scale than A and B). C and D are of the same volume (64 cm^3) but of a very different shape. Object C has a SA:V ratio of 1.5:1 and D one of more than 4:1. From Heatwole and Taylor (1987).

similar responses to solar radiation, rather than that body temperature was directly dependent upon conduction of heat from the substrate. If a cold lizard, just emerging in the morning and basking on a cold rock, heats faster than the rock, it could actually conduct heat to the rock; conversely, if the rock heats faster it would transfer heat to the lizard. Bartlett and Gates (1967) compared the reflective properties of lizards and their substrates separately by different wavelengths of light. They found that within the visible range of light the two objects tended to have similar reflectivities and be similarly coloured, thereby being camouflaged from potential predators. Beyond the range of wavelengths of light visible to predators, however, the reflectance curves diverged in ways that enhanced the basking lizards' ability to absorb solar energy and heat faster. In addition, some lizards can change colour as temperature changes, becoming lighter (more reflective) under hot conditions, and darker (more absorptive of light) when it is cool. *Lampropholis*, however, do not change their colour markedly and this probably is not an important adaptation for them. In a more general way, however, the colour of their bodies may be related to thermoregulation. Greer (2017)[1] noted that skinks that do not bask have brown bodies, whereas those that do bask have bodies that are grey; Greer (1997, 2017) pictured examples of both types of colouration and used *L. guichenoti* as an example of one coloured grey (his Fig. 19). Greer (1997) cited an Honours thesis by S. P. Fraser from Sydney University as indicating that *L. guichenoti* lighten somewhat in colour after several days of exposure to high temperatures (~35°C).

Matthews *et al.* (2016, 2023) demonstrated that the colour morphs of *L. delicata* did not differ in performance of sprints, thermal preferences, or metabolic rate. Heating rates and to some extent cooling rates, however, did differ between morphs and the position and size of the white stripe reduced the heating rate of striped animals below that of plain ones. Heating rates were faster in females than in males, which was at variance with the finding that the reflectance from females across the dorsal and lateral surfaces, both in the visible and in near-infrared ranges, was greater than in males. Despite these differences males and females maintained similar body temperatures and the authors concluded that behavioural thermoregulation compensated for any effect of colour pattern.

For a detailed flow-chart of the dynamics of the exchange of energy between a lizard and its environment see Fei *et al.* (2012) and for a discussion of other possible functions of the colour pattern see Chapter 5.

Physiological thermoregulation

In addition to the behavioural means of ameliorating temperature, there are a few physiological ways that ectothermic lizards can use to alter their body temperatures in the field and keep them in a favourable range when environmental temperature fluctuates beyond optimally favourable limits. These include metabolic and cardiac adjustments (Heatwole and Taylor 1987).

One of these is a local response of peripheral blood vessels to dilate under warm conditions and constrict under cool ones. Thus, a lizard basking on a cold rock can restrict flow of blood to the venter by vasoconstricting the cutaneous blood vessels there, thereby retarding the circulatory transfer of heat from its ventral surface to the rock. At the same time, it can vasodilate the blood vessels on its back. The resulting increase in rate of blood flow through the dorsal surface exposed to the sun enhances the convection of the solar radiant heat by the blood stream to other parts of the body. Conversely as air temperatures begin to drop in the evening and the sun's rays have lost some of their heating power by virtue of striking the earth at a more oblique angle, a lizard can press its venter in contact with a substrate previously warmed by the sun and increase the blood flow to its ventral surface to assist in absorbing that heat, while at the same time retarding flow to the dorsal skin, thereby inhibiting loss from that surface at a time when solar radiation is progressively weakening. Such considerations could apply to any lizard but are of especial importance to small species such as *Lampropholis* who, because to their greater surface-area-to-volume ratio are unusually subject to rapid rates of heating and cooling. However, this aspect has not been investigated in *Lampropholis*.

Most other physiological means used by ectotherms to assist in regulating body temperature, are not likely to be available to *Lampropholis* because of their small body size and their large surface area to volume ratios. For assessment of the role of metabolism on thermoregulation of ectotherms see Chapter 7.

1 Published only online in 2017 although written in 2007.

Ecologically important temperatures

There are various critical temperatures that mark the limits of behavioural and physiological activities of lizards. One of the most important ones is the preferred temperature, or better designated as a preferred range of temperatures, above or below which the animal attempts to bring its body temperature back within the preferred range.

Ectothermic lizards, during their periods of daily activity, tend to maintain their body temperature at or near a preferred level. The preferred level depends on the type of activity in which the lizard is engaged; higher temperatures are maintained while digesting than they are during foraging, and when lizards are suffering infection, they seek higher temperatures than when they are well ('behavioural fever') (see review by Heatwole and Taylor 1987). The behavioural means of regulating body temperature are basking in sunlight, posturing in ways that either enhance or inhibit absorption of solar energy, perching at different heights above a heated substrate, flattening against a warm substrate, and shuttling between sun and shade. Although if conditions are such that they are unable to regulate body temperature at the preferred level for the activity in which they are engaged, they will accept lower or higher temperatures over a certain range (the activity range) above or below which they cease overt activity and seek shelter. The preferred temperature is much closer to the upper boundary of the activity range, than it is to the lower limit; indeed, it seems dangerously close to the upper limit (Fig. 6.2). For heliothermic animals, the activity range can be subdivided into two sections: (1) the part at the lower end is the basking range over which the animal raises its temperature by basking in sunlight and (2) the activity range proper at the upper part over which the animal engages in maintenance activities, such as foraging for food, or engaging in social activities (Fig. 6.2). Finally, most taxa of lizards begin to pant at a temperature at the upper end of the activity range – the panting threshold. Panting selectively cools the head (including the brain, the part of the body most sensitive to damage by heat) and retards further rate of heating generally. Panting is a very old, primitive trait in the squamates and goes back at least to the tuatara (Heatwole 1982) but subsequently was lost by snakes and most skinks. Heatwole *et al.* (1973) reviewed the literature on the heating to high temperatures of 74 species of lizards in 41 genera and nine families, of which the number of species of skinks was 20 in 11 genera; *L. guichenoti* was the only species of *Lampropholis* tested. All of the non-skinks tested panted except two of three varanid individuals and the sole individual of pygopodids that was tested, whereas none of the skinks panted except for *Tiliqua rugosa* and *T. scincoides*. The *Tiliqua* panted discontinuously and opened the mouth only during exhalation, whereas all other taxa that panted maintained the mouth open continuously. Greer (1980) extended the above study by testing 44 species of lizards in 16 genera and added a third species of *Tiliqua* (*T. branchialis*) that panted in the same manner as described by Heatwole *et al.* (1973) for the other species in that genus. In addition, Greer found a third skink that panted (*Menetia greyi*)

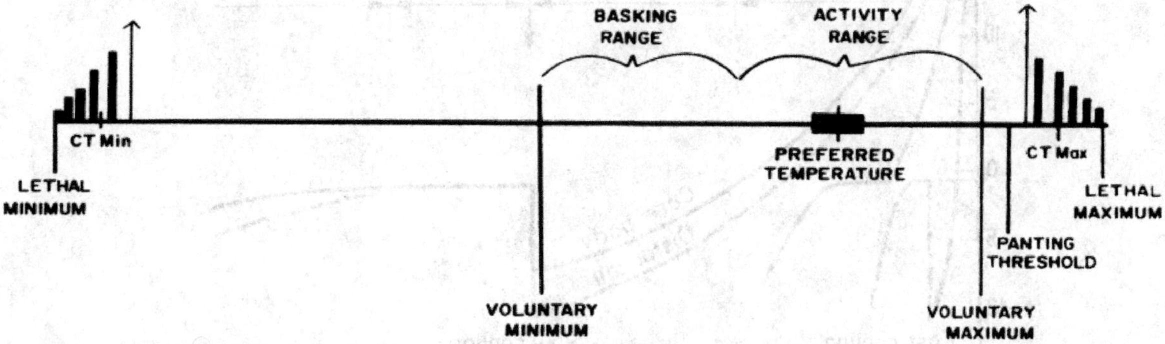

Fig. 6.2. The ecologically important body temperatures of lizards. The black vertical bars indicate decreasing survival times as temperatures approach the nearly instantaneously lethal maximum temperature (Lethal Maximum) and nearly instantaneously lethal minimum temperature (Temperature of Spontaneous Freezing; Tsf) after having passed through a period of supercooling (Fig. 6.3). The vertically pointing arrows indicate indefinite survival. The range in preferred temperature is indicated by a horizontal black bar within the voluntary range and enclosing a short vertical line (the mean value). This is an oversimplification in that there should be a series of preferred temperatures, one for each of different activities. Note also that many of these critical values are affected by acclimation (see below). From Heatwole and Taylor (1987). Printed with permission.

and it panted in a different way from non-scincid lizards or the species of *Tiliqua*; it opened the mouth only slightly but rapidly flicked the tongue in and out of the mouth.

The presence of panting in a few skinks apparently is a re-evolved characteristic independently achieved twice, once by *Tiliqua* and once by *Menetia* (Heatwole *et al.* 1973; Greer 1980). Some other skinks that have been tested, however, do open their mouths at high temperatures but not until after both their heart rate and breathing rate have dropped precipitously, thereby no longer being effective in depressing further rise in temperature. This response is merely gaping, not panting (Heatwole *et al.* 1973), and sometimes is accompanied by spasms.

Either the heating or the cooling of skinks at some point beyond the limits of their activity range causes them to lose coordination and the ability to right themselves when turned on their backs, thus rendering them unable to escape dangerous thermal levels. At the upper end of the scale this point is called the Critical Thermal Maximum (CTMax) and that at the lower end is the Critical Thermal Minimum (CTMin). The animal usually recovers unharmed if immediately returned to favourable temperatures unless there are accompanying spasms. These two endpoints are not symmetrically spaced around the preferred temperature; rather the CTMax is only a short distance above the Voluntary Maximum (VMax), compared to the much greater span between the Voluntary Minimum (VMin) and the CTMin (Fig. 6.2).

Finally, above the Voluntary Maximum at the upper end of the scale and below the Voluntary Minimum at the lower end are the temperatures at which the lizard's survival time is progressively reduced until the lethal minimum (also called Temperature of Spontaneous Freezing or Tsf), the point at which death is nearly immediate, is reached.

Under some conditions, some lizards can supercool. Supercooling is the cooling of body-fluids below their normal physical freezing point without formation of ice crystals (Lillywhite 2008). If supercooled animals (or supercooled solutions of salt), are moved or shaken, they spontaneously freeze and their temperature immediately rises to the Tsf (Fig. 6.3).

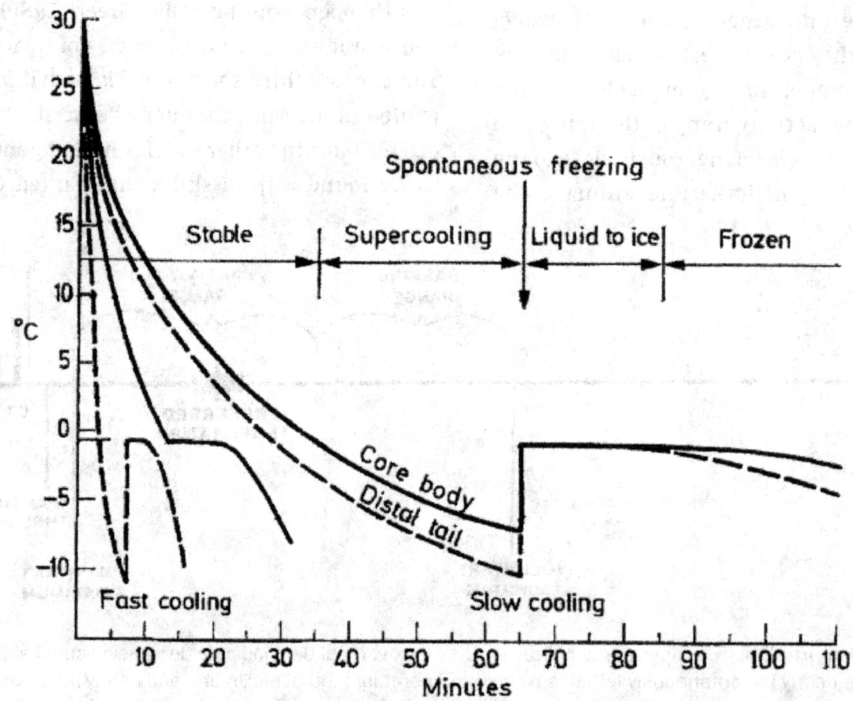

Fig. 6.3. Diagrammatic representation of rapid and slow progressive cooling of the core and tail of a lizard from favourable temperatures through supercooling to death by freezing. Although freezing probably represents the instantaneous death point for most ectotherms, there are some that can survive freezing of their tissues and body fluids. From Spellerberg (1972a). Printed with permission.

The Temperature of Spontaneous Freezing (Tsf) is near the instantaneous death point (lethal minimum) for many species although some ectotherms normally survive freezing in winter (Spellerberg 1972a).

At any point before the animal reaches its lethal minimum, returning it to warm conditions revives it, even though the animal may have survived lower body temperatures during supercooling (Fig. 6.3).

Many, or perhaps all, of the endpoints discussed above are not as precise as they may initially appear. They are subject to alteration by various environmental influences (Spellerberg 1973); also see section immediately below).

ADAPTATION TO CHANGING THERMAL CONDITIONS

Acclimatisation, acclimation, and heat-hardening are temporal adjustments to environmental changes on three different time-scales (Phillips *et al.* 2016). Whereas acclimatisation is an evolutionary change in morphology, behaviour, or physiology in response to natural selection, acclimation is a shorter-term response to temporary but somewhat prolonged (e.g. seasonal) environmental changes (Lillywhite 2008); heat-hardening, sometimes called 'heat shock' is a response to more immediate, unusual, environmental emergencies (Phillips *et al.* 2016).

Various researchers have assessed the relative importance of genetic ancestry (phylogeny) of lizards versus adaptation to local physical environments or interaction with species sharing their niche (i.e. which is the most important, phylogeny or geography and ecology?). Not all facets of adaptation respond in the same way, nor do all taxa of lizards follow the same patterns. For example, Grigg and Buckley (2013) found that phylogeny accounts for most variation in body temperatures, probably because behavioural thermoregulation allows lizards to maintain preferred levels over a range of environmental conditions. By contrast, local conditions have a greater impact on the variation in the CTMax and neither space nor genetic ancestry has much effect on the CTMin. However, Spellerberg and Hoffman (1972) found that the CTMin in a European lizard showed significant circadian variation under the natural rhythm of light and dark and that this rhythm was governed not only by an endogenous cycle, but that it also was influenced by short-term acclimation during the cooler conditions at night.

BOX 6.1 SUPPLEMENTARY LITERATURE ON THERMAL BIOLOGY

The brief review of the thermal biology of reptiles above highlights the background necessary for an understanding of the following discussion. The literature in this field, however, is burgeoning and cannot be reviewed in detail herein. In addition to the papers discussed in the present chapter the following references are a small representative sample of a wider range of literature on this topic, and perusal of them, and the papers they cite, can provide a more detailed and complete coverage of the topic: Huey and Slatkin (1976), Garland *et al.* (1991), Huey and Kingsolver (1993), Tosini and Avery (1993), Meek (1995), Belliure *et al.* (1996), Angilletta *et al.* (2002), Díaz and Cabezas-Díaz (2004), Herczeg *et al.* (2006), Labra *et al.* (2009), and Kong *et al.* (2022).

THE THERMAL BIOLOGY OF *LAMPROPHOLIS*

Heliothermy: Basking

Lampropholis position themselves in sunlight and raise their body temperature through absorption of radiant energy. Basking was observed at three different sites near Armidale in New South Wales, a Pine Forest and eucalypt woodlands at Puddledock and Blue Hole. In summer, basking of both species was observed mainly in the early morning. A few individuals of *L. delicata*, also were seen basking in the mid-afternoon, but this was not observed for *L. guichenoti*. In autumn, basking had a bimodal pattern that was sustained by both species for much longer during the day, with only a brief interlude at midday. In neither species did basking extend into the evening period of cooling and the latest occurrence of basking at any season was the hour between 15:00 and 16:00. Thereafter, a thigmothermic response to warm substrate, perhaps accompanied by some absorption of solar energy, superseded ordinary basking (see section *Thigmothermy* below).

Basking at Puddledock (*L. delicata*) and Blue Hole (*L. guichenoti*) showed a somewhat different pattern; in both cases basking was observed less often in summer than in the Pine Forest. In addition, at Blue Hole one individual of *L. guichenoti* was seen to bask in the afternoon; that was not observed for *L. delicata* at Puddledock. In autumn, basking was unimodal, first observed at 07:00 for *L. delicata* and at 09:00 hours for *L. guichenoti* and ending for both species during the hour between 14:00 and 15:00. Bilcke *et al.* (2006) observed that sometimes a basking *L. delicata* spread its ribs laterally, thereby flattening its body and increasing the surface area exposed to the sun.

There are seasonal and latitudinal differences in the opportunity for basking. Clearly, less time would need to be spent in summer when conditions generally are hotter than in either spring or autumn. Also, the generally hotter conditions at lower latitudes would reduce the time spent in basking there and in the extreme presumably could obviate it altogether. Basking probably exposes the animal to a greater risk of predation and expends time that otherwise could be used in foraging for food or engaging in other vital activities. Environmental changes, such as global warming, likely will alter the basking behaviour of many ectotherms and affect adjustment of physiological functions that are thermally dependent.

Anderson *et al.* (2023b) investigated the effect of basking regime on various physiological traits in *L. delicata*. They first ascertained the thermal preferences and speed of sprinting in animals brought into the laboratory and then assigned them randomly into two groups based on the thermal conditions at the latitudinal extremes of this widely distributed species: one group assigned to a regime of 'high basking' mimicking the greater daily and annual opportunities for basking occurring at the lower latitudes, and the other to a regime of 'low basking' mimicking the briefer occurrence of such opportunities experienced at the higher latitudes. The high-basking regime provided opportunity to bask for 14 hours per day, whereas the low-basking regime provided only 6 hours per day. After acclimation at these regimes for 6 weeks the animals' thermal preference and locomotory performance were assessed, followed by re-assessment after 12 weeks.

These acclimations did not invoke any change in the animals' preferred temperatures but did produce changes in their locomotory performance. Acclimation elevated the maximum speed of sprints in both treatments, with the greatest change occurring between the 6-week and 12-week intervals. The extent of acclimation, however, did not differ between the two regimes.

Downes (2001) tested the relative importance of thermoregulation and safety from predators to *L. guichenoti* by comparing the behaviour of lizards in enclosures covered with the scent of one of their natural predators, the snake *Demansia psammophis* to lizards in unscented (control) enclosures. Those in scented environments had fewer opportunities to bask because they emerged from shelter later in the day than were those from control enclosures. They thus sacrificed effective thermoregulation for a gain in perceived safety (see Chapter 8 for a fuller account of these experiments).

Thigmothermy

Belmont (1977) noted a peculiar stance utilised by *L. delicata*, in which the animal folds its forelimbs, and sometimes the hind limbs as well, backwards and up over the back (sometimes with the palms uppermost), thereby lowering the entire venter against the substrate (Fig. 6.4), a response subsequently termed 'adpress behaviour' by Torr and Shine (1993). In the field, this posturing occurs on autumn afternoons when the substrate is still warm from heating by the sun during the day. Accordingly, it is considered to be a thigmothermic response. It maximises contact of the body with the warmest substrate available as the sun is declining. This behaviour differs from the typical basking posture in the morning when the contact of the venter with the still-cold substrate is minimised; it may also confer some of the advantages of basking as the lizards position themselves on the western side of rocks where at that time of day the sun's rays strike the tilted substrate (and lizards' backs) more nearly at right angles. Although this response occurs in autumn in the field, it can be induced at other times in the laboratory, as described by Belmont (1977, p. 81):

> *The behaviour was first noticed in some animals in the laboratory provided with a 60-watt basking light with a metal shade, placed just above the substrate of sticks and small branches .... A few lizards crawled up onto the metal shade, flattened their bodies and lowered their heads against it and at the same time lifted their front limbs, or both front and hind limbs off the substrate and backwards along their bodies. Sometimes the back legs would be held up over the dorsal surface of the animal so that they almost met. Black pieces of paper or foil, first heated under a lamp, elicited the same response. It was noticed that the behaviour only occurred in animals recently brought into the laboratory during autumn and winter but a few minutes in a refrigerator seemed to lower the threshold of response for acclimatised animals.*

There were variations in the details of such posturing. For example, since it usually occurred on a surface tilted toward the source of light, maintaining position could be a problem. Belmont noticed that sometimes the front and back leg on the uphill side would be used to grasp the substrate and maintain position, whereas the legs on the downhill side would be folded against the body in the thigmothermic position.

Adpression maximises ventral contact of the undersurface of the body with the substrate, but wouldn't merely splaying the legs against the substrate have the same effect, or even increase the total surface employed in thigmothermy?

Lampropholis guichenoti also exhibits adpression when basking and Torr and Shine (1993) conducted laboratory experiments in which its basking behaviour at different temperatures of the basking site was controlled by the wattage of the light bulb serving as the source of heat (25, 60, or 100 watts). The experiments were designed to distinguish among alternative hypotheses previously advanced to explain the following functions of adpression: (1) Adpression facilitates the uptake of heat, either by maximising the surface area of the feet exposed to the source, by increasing the surface area of the animals' venters in contact with the warm substrate, and/or by increasing the flow of blood through the appendages (Cogger 1974; Fraser 1980). (2) Adpression is an avoidance reaction to uncomfortably high temperatures impinging on the feet (Bradshaw and Main 1968; Mitchell 1973; Cogger 1974). Torr and Shine (1993) found that the results favoured the heat-avoidance hypothesis as this response occurred earlier under the 100-watt heat source than it did under lower wattages. This response is surprising as one would expect that heat uncomfortable for the feet would be even more uncomfortable for the venter of the body. However, the conclusion was also supported by the fact that only the feet that were directly illuminated were lifted. The answer to this conundrum may be that heat conduction laterally to substrate beneath the body would take longer than the almost immediate delivery beneath the smaller feet.

Torr and Shine (1993) also studied another response in basking *L. guichenoti* that they termed 'lie-on behaviour'. It consisted of one lizard lying on top of another one with the extent of coverage ranging from only the head of one lizard overlapping the other lizard (usually the middle but sometimes only over the tail or head) to nearly complete coverage of one by the other. Again, there were several hypotheses as to the function of this behaviour: (1) members of this species often aggregate during basking, so perhaps lie-on results from insufficient space to accommodate all lizards seeking a place to bask; (2) it may have a social function with territorial or sexual connotations; or (3) it may function in thermoregulation

Fig. 6.4. Photograph (A) and sketch (B) of a *Lampropholis delicata* in the adpressing thigmothermic position. Drawing and photograph courtesy of Caroline Belmont.

if another lizard, say, either has a higher thermal conductance, or is occupying an especially effective basking site. The observation that lie-on occurs even when unoccupied space is readily available at basking sites falsifies the first of these hypotheses. There was a sexual bias in the frequency of cross-gender examples of lie-on, i.e. males tended more often than random to lie on females, and females tended more often to lie on males, so social factors may be involved. Thermoregulation also seemed to be important, however, as lie-on occurred significantly more often at the lower wattage of the heat-producing lamps, suggesting either that lie-on may enhance heating rate, or that at the higher wattages, animals may reach high levels of body temperature quickly and then dispense with lie-on behaviour.

Torr and Shine (1993) concluded that 'adpress behavior ... is probably due to avoidance of hot substrates. Lie-on behavior ... has thermoregulatory significance also, but frequency is modified by social factors'.

In summary, *Lampropholis* employ exposure to sunlight (heliothermism) in the morning) but in the afternoon as the day cools, they switch to maximising contact with warm substrates previously heated by the sun (thigmothermism). They thus are neither strict heliotherms nor strict thigmotherms but rather employ both approaches sequentially, and even simultaneously, during the day. When their body temperatures are well within the activity range, they do not overtly thermoregulate in this way, but rather forage for prey.

Substrate temperatures in the field

Belmont (1977) recorded the temperature of the substrates where *L. delicata* and *L. guichenoti* were first sighted as an indication of the thermal environment that they experience during their activity periods and the extremes of environmental temperatures they voluntarily tolerate. The lower limit of substrate temperatures at which these lizards were active was 18°C for both species in all habitats from which temperatures were taken (Figs 6.5 and 6.6), which is also the lowest body temperature of active *L. guichenoti* that Simbotwe (1985) recorded. At Pine Forest both species of *Lampropholis* had very similar profiles of substrate temperature. The unimodal distribution of substrate temperatures for active individuals in Pine Forest began at 18°C, peaked at 25°C for both species, and then trailed off to a maximum temperature of 34°C for *L. delicata* and 35°C for *L. guichenoti*. Thus, in the same habitat these species were remarkably similar with respect to the substrate temperatures they experienced.

Lampropholis delicata had a practically identical profile in Puddledock as it did in Pine Forest. The lowest substrate temperature at the site of a lizard was 18°C and the peak records were at 25°C in both places. The upper limit, however, was marginally different (34°C in Pine Forest and 36°C at Puddledock). The results for *L. guichenoti* showed a greater discrepancy among areas as temperatures of substrates it occupied extended from 35°C in Pine Forest (Fig. 6.5) to 42°C at Blue Hole (Fig. 6.6).

Allusion has already been made to the difficulty of accurately measuring the body temperatures of ectotherms as small as *Lampropholis* (see section *Physiological thermoregulation* above). Simbotwe (1985), in a study of *L. guichenoti*, was aware of these limitations and attempted to minimise the problem by not taking a reading if a lizard attempted to flee before capture (and thus moving into a different thermal environment), and

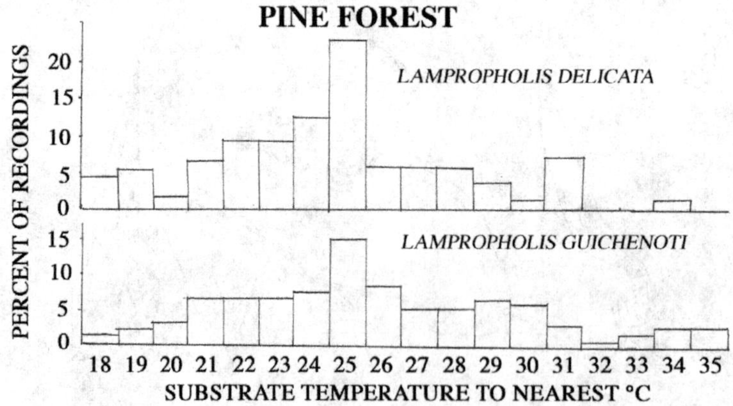

Fig. 6.5. Substrate temperatures associated with active *Lampropholis delicata* and *Lampropholis guichenoti* in Pine Forest during summer and autumn. Re-formatted from Belmont (1977) and published with her permission.

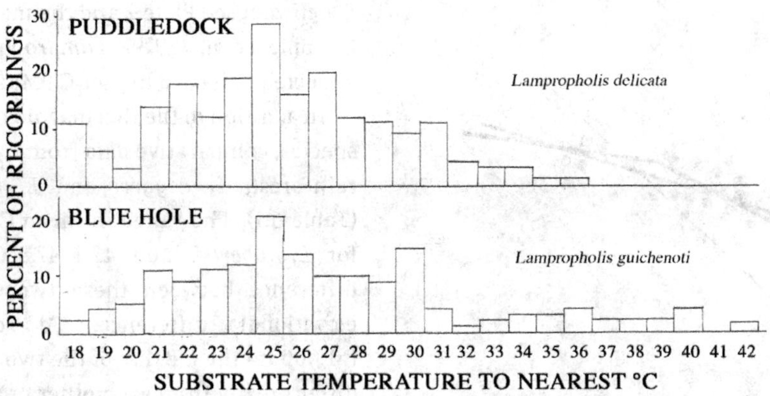

Fig. 6.6. Substrate temperatures associated with active *Lampropholis delicata* at Puddledock and *Lampropholis guichenoti* at Blue Hole. Re-formatted from Belmont (1977) and published with her permission.

by working as quickly as possible. His measurements were taken by a quick-recording copper-constantan thermocouple probe immediately inserted 6 mm into the cloaca, technology that was not available to Belmont in the study described above. Also, he recorded body temperatures in summer when the thermal gradient between the investigator's hand and the lizard was not as great as would be true under colder conditions. Accordingly, the body temperatures he recorded were probably as accurate as possible with the technology available at that time. After measurement of the lizard's body temperature, temperatures of the air and substrate were taken at the site of the lizard's capture.

Simbotwe (1985) found that body temperatures of active *L. guichenoti* in the field did not show a significant correlation with air temperature. By contrast, a positive relationship of body temperature with substrate temperature was significant. In other words, substrate temperature had a greater influence on lizards' temperatures overall than did air temperature. This result is understandable because air is a poor conductor of heat, whereas many solid substrates are good ones.

Figure 6.7 shows that, in contrast to air temperatures, those of the substrate have a marked relationship to substrate temperatures, at least over part of the thermal range. The bold straight line is the best straight line that fits all the data relating substrate and cloacal temperatures. However, a straight line is not the best form of line to give the best possible fit. Rather it is the bold bent line which more accurately describes the relationship. At the lower range of temperatures, the cloacal temperatures track those of the substrate. At the higher end of the thermal range, however, cloacal temperatures become less dependent upon those of the substrate and level off, often being lower than those of the substrate.

The preferred body temperature may be subject to acclimation, even of embryos developing within the egg, as Bilcke *et al.* (2006) found that hatchlings of *L. delicata* from eggs incubated at 20°C had slightly lower preferred temperatures than those from eggs incubated at 28°C (Table 6.1; also see Chapter 5).

Critical temperatures

Range for voluntary activity

The range in body temperatures that *L. delicata* and *L. guichenoti* voluntarily accept during their activity period were similar, a span of 15.0 degrees Celsius (mean VMin 18.6°C–mean VMax 33.6°C) for *L. delicata* and 13 degrees Celsius (mean VMin ~18°C–VMax ~31°C) for *L. guichenoti* (see Table 6.1); the slightly greater span for *L. delicata*, if statistically valid and if constant throughout these species' geographic range, would probably allow it to have a longer period of daily activity period than *L. guichenoti*.

See Chapter 5, Table 5.3 for information on the effect that temperature at which eggs are incubated has on the preferred temperature of the resulting hatchlings.

CTMin and CTMax

The CTMin and CTMax are vitally important values as they bracket the range of temperatures over which a species can escape inclement conditions. The greater the span between these two points the greater is the versatility of the animal and the more expansive are the latitudinal and elevational ranges it could occupy.

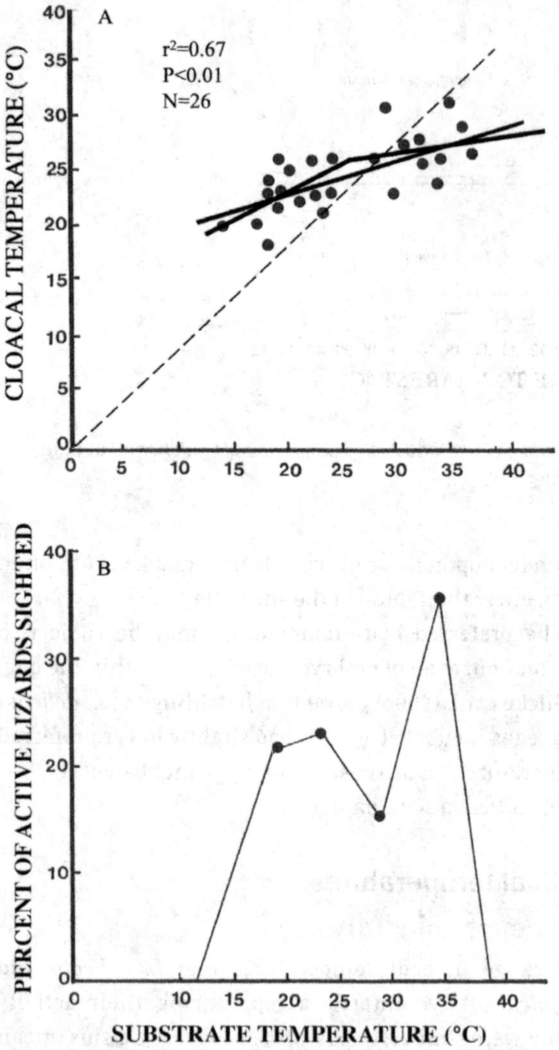

Fig. 6.7. The relation of cloacal temperature of *Lampropholis guichenoti* to substrate temperature (A) and the change in the percentage of individuals on the surface that were 'active' as substrate temperature changed during the day (B) at Newholme, near Armidale, NSW. Note that, opposed to its use elsewhere, in this figure the author used the word 'active' to refer to the total percentage of lizards abroad on the surface of the ground that were involved in such functions as maintenance, foraging, fighting, or mating, as opposed to quiescently basking. Modified from Simbotwe (1985).

The mean CTMin of *L. delicata* and *L. guichenoti* are slightly different, which means that the thermal ranges over which they remain ambulatory also differ. *Lampropholis guichenoti* has both a lower CTMin and a higher CT Max than does *L. delicata* (Tables 6.2 and 6.3). Its higher CTMax is in disaccord with the fact that its geographic range does not extend nearly as far north into the warmer tropical latitudes as does that of *L. delicata*; see Plate 2 and the maps in Cogger (2018) and Chapple *et al.* (2019). *Lampropholis delicata* would be expected to have a higher CTMax than *L. guichenoti*.

In addition to the thermal biology of the two eurytopic species, comparative data from two species from tropical rainforest, *L. coggeri* and *L. robertsi*, are available (Table 6.1). The range in mean CTMax was 43.1–43.4°C for *L. robertsi*, and 42.1–47.8°C for *L. coggeri*. The difference between these two may partly reflect an elevational difference. At equivalent elevations (~1,000 m) the CTMax of the two tropical species did not differ greatly from each other or from the two eurytopic species. *Lampropholis coggeri*, however, was tested over a wide range of elevations and it had a higher CTMax in the warm tropical lowlands (100 m) than it did at a higher elevation (1,000 m); *Lampropholis robertsi* showed no elevational effect between the two high elevations where it was tested (Table 6.1).

The important feature of this comparison was the extreme difference in the CTMin of the tropical forms versus the widespread ones. The CTMin both of *L. coggeri* and *L. robertsi* were much higher than those of *L. delicata* and *L. guichenoti*, which means that the tropical forms are poorly adapted for life in cold areas relative to the widespread species because the former lose coordinated movement well above the temperatures that the eurytopic species do.

In summary: (1) *L. delicata* has a slightly greater range of temperatures at which it is voluntarily active on the surface of the ground than does *L. guichenoti* but contrariwise has a shorter span of temperatures between those that are incapacitatingly high and those that are incapacitatingly low); (2) the most striking result is the extremely high CTMin of the tropical species and their resulting inability to escape even moderately cold conditions, compared to the two eurytopic species; (3) there may be an elevational adjustment of some of these endpoints in some species; and finally (4) some of these comparisons are based on small numbers of observations from non-equivalent latitudinal or elevational conditions and may not be statistically significant. Therefore, these summaries should be viewed as inadequately tested hypotheses, rather than as firm conclusions.

Either latitudinal, elevational, or clinal variation in any of these endpoints could confound the above patterns. Hence, until the intraspecific patterns over a wide range of populations at different latitudes and elevations are compared, especially at the edges of species'

Table 6.1. Important body temperatures (°C) of species of *Lampropholis*.

T_B = temperature of the body; T_S = temperature of the substrate.
Data from Hastie (1966), Pengilley (1972), Spellerberg (1972a), Rawlinson 1974), Powell *et al.* (1978), Greer (1980), Simbotwe (1985a), Bilcke *et al.* (2006), Downes and Hoefer (2007), Muñoz *et al.* (2016), Goulet *et al.* (2017a), Llewelyn *et al.* (2017), Farquhar *et al.* (2023), Mitchell (unpubl. data), Webb (unpubl. data).

				Preferred temperature			
Season Species Elevation (m)	**Mean T_B of torpid animals in the field (range)**	**Mean CTMin (range)**	**Mean Voluntary Minimum**	**Mean T_B in the field (range)**	**Mean T_S selected in laboratory gradient**	**Mean CTMax (range)**	**Mean Voluntary Maximum**
IN SUMMER							
L. delicata	–	4.7	18.6	26.1	25.8	40.8	33.6
–	–	(4.6–5.1)	–	–	(24.9–26.9)	(40.6–41.3)	–
–	–	–	–	–	26.4	–	–
–	–	–	–	–	(25.4–27.6)	–	–
–	–	–	–	–	30.5 ± 0.1	–	–
L. delicata	–	–	–	20.7	–	–	–
Inactive under cover	–	–	–	(19.5–23.5)	–	–	–
L. guichenoti	–	3.3	~18	–	–	42.0	–
–	–	(3.0–3.6)	–	(24.0–27.0)	–	(41.3–42.1)	~31
–	–	–	–	–	–	42.6 ± 0.4	–
–	–	–	–	–	–	42.1 ± 41.7	–
L. guichenoti	–	–	–	17.4	–	–	–
Inactive under cover	–	–	–	(14.5–20.2)	–	–	–
L. coggeri							
100 m	–	14.4 ± 0.3[1,2]	–	–	24.9[2,3]	–	–
100 m	–	14.6 ± 0.9[1,2]	–	–	29.3 ± 1.7[2,3]	47.8 ± 0.4[2]	–
1,000 m	–	13.5 ± 0.4[1,2]	–	–	28.8 ± 2.1[2,3]	42.6 ± 0.4[2]	–
1,000 m	–	12.6 ± 0.7[1,2]	–	–	28.2 ± 1.9[2,3]	42.1 ± 0.6[2]	–
L. robertsi							
1,000 m	–	13.1 ± 0.8[1,2]	–	–	32.0[2,3]	43.1 ± 0.3[2]	–
1,000 m	–	10.1[1,2]	–	–	36.1[2,3]	–	–
1,600 m	–	10.0 ± 0.6[1,2]	–	–	31.4[2,3]	43.4 ± 0.4[2]	–
IN WINTER							
L. guichenoti [2]	1.9	1.4–3.6	–	–	–	40.7	–
–	(1.4–2.1)	–	–	–	–	(40.6–41.3)	–
–	3.6	–	–	–	–	–	–
–	(1.9–4.7)	–	–	–	–	–	–
–	–	–	–	–	–	(36.7–38.8)	–
–	–	–	–	–	–	(40.6–42.1)	–
–	–	–	–	–	–	(35.9 ± 1.0)	–
L. guichenoti	**Temperatures of crevices**						
–	3.5; 5.0; 5.8; 7.6	–	–	–	–	–	–

[1]These values are exceptionally high. However, they are the values that were reported and high values were obtained for all seven species in this study (all were from rainforest).
[2]Values are from different localities.
[3]These are the mean values of the optimal temperatures for sprinting as an approximation of the mean temperature preferred in a laboratory gradient.

geographic limits, unequivocal interpretations are not possible. Table 6.3 provides a crude summary of the values of CTMin and CTMax.

For an account of the effect of temperature during incubation upon the phenotype of hatchlings see Chapter 5.

Temporal and spatial scope

The temporal scope of adaptation to changes in temperature has two components. One is acclimation to repetitive seasonal changes whereby animals adjust their critical endpoints upward or downward in conformity with the annual progression of summer to winter and back to summer again. The stimuli for acclimation can be such phenomena as repeated changes in the annual cycle of daylength. The other component is not cyclic, but rather progressive longer-term, unidirectional alteration in thermal conditions such as the global anthropogenic changes in temperature we currently are experiencing. Successful adaptation to global warming may require expanding the range of temperatures achievable by acclimation at higher than previous limits – metaphorically to move the goalposts farther apart.

The spatial scope encompasses geographical differences in where those goalposts are in geographically separate populations of the same species. Progressive, longer-term changes. such as global warming, may not return to the starting point during the lifetime of an individual as the goalposts keep on receding. The capacity of a species to survive changes in its thermal environment depends in part on whether it can adapt by altering such critical thermal limits as the CTMin and CTMax to track the extent of climatic change and the rate at which it occurs.

Having noted that *L. guichenoti* had a lower CTMin in winter than in summer Spellerberg (1972a) investigated this species' rate of acclimation. He subjected animals to a temperature 1°C below the mean CTMin in summer. At first lowering of the CTMin was rapid after which the rate of change declined. The summer level of ~3°C was ~0.2°C within ~48 hours, then acclimated more slowly for the next 48 hours to about −0.2°C, and thereafter nearly levelling off; it was still only at about −0.4 degrees at the termination of the experiment 240 hours later.

Llewelyn *et al.* (2016) held temporal scope constant for *L. coggeri* and studied the spatial one separately. The acclimation of four different traits, CTMin, CTMax, performance (thermal optimum for sprinting), and desiccation (rate of evaporative loss of water), were compared at the same time among populations from different localities. All four of these traits showed marked differences among localities. The CTMin varied from 8.2 ± 0.3°C to 16.3 ± 0.3°C and the CTMax from 36.6 ± 0.2°C to 41.3 ± 0.2°C in different populations measured at the same time; optimal temperature for sprinting ranged from 27.1 ± 0.8°C to 37.1 ± 1.1°C, and the desiccation rate from 1.8 ± 0.2 mg/hr to 6.9 ± 0.6 mg/hr. In addition to the spatial scope there was a temporal one with values varying over time within the same locality and, finally, there also were other differences; for example, large skinks tested at the higher temperatures lost more weight than did smaller ones. In summary, there were local adjustments to environmental conditions independent of seasonal ones.

The most surprising finding was counter-intuitive: the relationship between CTMax and temperature of the environment was negative, that is, the CTMax was higher in areas with lower prevailing temperatures. In a later paper (Llewelyn *et al.* 2017), however, the reverse was found. The paradox was resolved by analysing the lizards' thermoregulatory behaviour: animals from hot environments chose cooler microenvironments than did those from habitats that were cool and the physiology of the animals from hot environments acclimated to the chosen temperature, rather than to the prevailing one of the habitat in general. Such nuances need to be assessed in predicting the response of lizards to impending hotter environments.

Finally, in a triad of papers, Llewelyn *et al.* (2016, 2017, 2018) compared the traits of populations of *L. coggeri* (known to differ physiologically) from two areas, Paluma (cooler, wetter site) and Hervey Range (hotter, drier site). In addition to the traits used in previous studies, maximum speed of sprinting, breadth of the range of temperatures permitting sprinting, and heat shock (see section *Adaptation to changing thermal conditions* above) were added. Results were compared between animals recently after capture, again after acclimation, and finally a genetic component was added by studying the subjects' offspring.

Initially, there were small, but statistically significant, differences between the two sites in most of the traits. In comparison to Hervey Range, lizards from Paluma had lower CTMin, higher performances in several measures of the speed of sprinting, and higher desiccation rates. Acclimation also affected most of the traits, bringing them closer to laboratory conditions; an exception was the response to heat shock; it was significantly higher for animals from Paluma than for those from Hervey Range.

Eggs from the subjects of the experiments described above were incubated at two different temperatures and the progeny tested for some of the same attributes as those to which their parents had been subjected. CTMin did not differ significantly between locations when the second-generation individuals were tested at the age of 1 month (Paluma 13.12°C, Hervey Range 13.12°C) or as adults. CTMax also did not differ significantly between populations at 1 month of age (Paluma 38.73°C, Hervey Range 38.51°C). However, animals had higher CTMax at Paluma than at Hervey Range when retested as adults.

This array of physiological, behavioural, and genetic responses by so many traits from different localities within a small area attests to a surprising versatility of this species. Even if one combination of traits fails to adapt to changing conditions, others may succeed – to use a familiar notation in a different context, the situation denotes 'the survival of the fittest'. The authors suggest that because of this versatility the species with small geographic ranges may not be as vulnerable to climatic change as originally thought. A critical question remains, however, as to whether the breadth of any of these combinations of responses are sufficient to cope with the extent and rapidity of the impending march of climatic changes.

Phillips *et al.* (2014) contrasted two hypotheses on the evolutionary constraints on an organism's thermal sensitivity: (1) the 'hotter-is-better expectation' (organisms adapted to higher temperatures should be able to attain higher maximum performance than organisms adapted to cooler temperatures), and (2) the 'specialist-generalist trade off' (organisms have a limited allocation of energy, and therefore, gains anywhere entail losses somewhere else) and proposed that these constraints should operate at different levels of organisation (species > population > individual). They measured the thermal sensitivity of sprinting in 440 individuals of *L. coggeri* from 10 different populations and found that the hotter-is-better expectation was evident only between populations, whereas the specialist-generalist trade-off was evident within, as well as between, populations.

Macdonald *et al.* (2018) examined the relative effect of two avenues of adaptation to environmental conditions: (1) genetic adaptation to specific environments via natu ral selection ('local adaptation'), and (2) within the context of a given genetic constitution, adaptation to a wide range of different environments ('plasticity'). They found that various physiological characteristics of *L. coggeri*, but especially its CTMin and CTMax, exhibited stronger local adaptation than did morphological characteristics.

Clinal differences

Thermal conditions change both latitudinally and elevationally and for widespread eurytopic species spatial adjustments in thermoregulation along such gradients are expected. Anderson *et al.* (2022) analysed elevational changes in preferred temperature, CTMin, and CTMax for *L. guichenoti* at four elevations along a cline from lowlands in Victoria (11 m asl) through two intermediate sites (elevation 228 and 772 m asl) to the Alpine National Park (1,546 m asl). They found that all three of these thermal attributes varied significantly with elevation. Skinks from the high elevations had higher thermal preferences and lower CTMin than at lower elevations, thereby behaviourally matching body temperature with optimum temperatures for performance and maximising the time they could remain operative at the lower temperatures prevailing in the mountains. In contrast, lowland lizards had narrower breadth of thermal performance and higher optimum temperatures that maximised their performance at the warmer conditions in the lowland. CTMax did not vary significantly across elevations.

They also examined the effect of temperature on locomotory ability and found that the speed of sprinting was affected by temperature and was different at different elevations; the sensitivity of the speed of sprints varied among elevations. Animals at higher elevations could support activity over a broader range of temperatures than could animals from the lowlands.

Anderson *et al.* (2022) calculated the number of hours per day that the lizards' temperatures would be above their CTMin with and without the intervention of their behavioural responses. At all elevational levels that number was higher with behavioural regulation. They also calculated the number of hours of daily activity that would accrue with and without those behaviours. In that case the number of hours was lower without the behavioural effect only at the lowest elevation tested (11 m asl); at all higher elevations the length of the daily activity period was increased by behaviour thermoregulation.

Thermal safety margins

Heatwole and Taylor (1987) defined several comparisons of the spans between various thermal endpoints that they termed Thermal Safety Margins that, like the span between the CTMax and CTMin, are expressions of the breadth of the thermal ranges to which a species is adapted. They fall into three categories: (1) *Physiological Thermal Safety Margins* that indicate the number of degrees Celsius between the mean temperature of an active animal's gradient and the CTMin (loss of coordination) or the lethal temperature, (2) *Ecological Thermal Safety Margins*, the span between the mean temperature in a laboratory gradient and either the CTMin or lethal temperature, and (3) *Minimum Thermal Safety Margin*, the temperature between maximum T_B in the field and the upper tolerances, either CTMax or lethal temperature. The greater the span of these safety margins, the more capable is a species of using longer periods of the day or the season for activity.

Table 6.2. The spans between the critical endpoints of CTMin and CTMax to each other and to the preferred temperatures of active animals (as indicated either by direct measurements in the field or in laboratory gradients) where there are available data from all three categories.

For evaluation of the trauma to ectothermic animals during assessments of various endpoints, see Box 6.2.
Calculated from data in Table 6.1.
The values in **boldface** indicate the span between the two values flanking them.

	Degrees Celsius					
Species **Elevation**	**CTMin**		**Preferred temperature**		**CTMax**	**Span between CTMax and CTMin**
Lampropholis guichenoti	3.3		24.0		42.0	
1,000 m		**20.7**		**18.0**		**38.7**
Lampropholis delicata	4.7		26.1		40.8	
–		**21.4**		**14.7**		**36.1**
Lampropholis coggeri	14.6		29.3		47.8	
100 m		**14.7**		**18.5**		**33.2**
1,000 m	13.5		28.8		42.6	
		15.3		**13.8**		**39.5**
1,000 m	12.6		28.2		42.1	
		15.6		**13.9**		**29.5**
Lampropholis robertsi	13.1		32.0		43.1	
1,000 m		**18.9**		**11.1**		**30.0**
1,600 m	10.0		31.4		43.4	
		21.4		**12.0**		**33.4**

Table 6.3. Effect of cold on hatching success of eggs of *Lampropholis guichenoti* nearly at full term near Penrith, NSW.

Data from Shine (1983a). $N = 20$ for each treatment.

Treatment	**Duration of cold (hours)**	**Hatching success (%)**
Controls	0	100
-0.6°C	5	95
-1.0°C	5	100
-1.8°C	5	35
-0.6°C	15	50
-1.8°C	15	0

Several such spans can be calculated approximately from some of the data in Table 6.1; these are shown in Table 6.2. A caveat applies to the interpretation of these safety margins. Research by Llewelyn *et al.* (2016) has shown remarkable spatial and temporal variation in the various endpoints discussed above. This means that both endpoints used in calculating the spans may be subject to considerable variation and accordingly those previously reported are mere approximations and not as definitive as originally conceived by Heatwole and Taylor (1987) (see section *Temporal and spatial scope* above). For this reason, the data from Tables 6.2 and 6.3 should not be considered precise designations of spans, but rather as approximations of the order of magnitude of the ranges between different endpoints. Allusion has already been made to the difficulty of accurately measuring the body temperatures of such small ectotherms (see section *Physiological thermoregulation* above).

With this brief background, we can now consider the actual thermoregulation of *Lampropholis* in the field.

The daily course of body temperatures of *L. guichenoti* soon after morning emergence and during early basking were between 17°C and 19°C when substrate temperatures in the area were between 10°C and 15°C (Fig. 6.7) (Simbotwe 1985) but rose quickly during a period devoted primarily to basking to between 21°C and 31°C (Mean = 24°C ± 2.77 SD) during the period of activity. Over this range fewer than 25% of the lizards on the surface of the ground were actively engaged in maintenance or social activities, as opposed to basking, and the mean substrate temperature was slightly higher than the line of equality (compare the left-hand part of the bold bent line with the dashed line of equality between substrate and body temperature in Fig. 6.7). Once the substrate's temperatures reach ~24°C, the body temperature levels off at that mean level (compare the right-hand part of the bold bent line with the dashed line of equality) but with fluctuation around the mean greater than at lower temperatures, and with T_B lower than T_S. The upper level of body temperature of animals on the surface of the ground was ~31°C. Interpretation of these results allow an estimate of some of the temperatures of ecological importance indicated in Fig. 6.2. The mean Voluntary Minimum thus appears to be ~18°C and the mean Voluntary Maximum just above 25°C, although individual values were much higher.

Belmont (1977) used a somewhat different approach. She recorded the temperature of the substrates where lizards were first sighted as an indication of the thermal environment that *L. delicata* and *L. guichenoti* experience during their activity periods and the extremes of environmental temperatures they voluntarily tolerate. The lower limit of substrate temperatures at which lizards were active was 18°C for both species in all habitats from which temperatures were taken (Figs 6.5 and 6.6), which is also the lowest body temperature of active *L. guichenoti* that Simbotwe (1985) recorded (see paragraph above). At Pine Forest both species of *Lampropholis* had very similar profiles of substrate temperature. The unimodal distribution of substrate temperatures for active individuals in Pine Forest, began at 18°C and peaked at 25°C for both species, and then trailed off to a maximum temperature of 34°C for *L. delicata* and 35°C for *L. guichenoti*. In the same habitat, these species were thus remarkably similar with respect to the substrate temperatures they experienced.

Lampropholis delicata had a practically identical profile in Puddledock as it did in Pine Forest. The lowest substrate temperature at the site of a lizard was 18°C and the peak was 25°C in both places. The upper limit was marginally different, however (34°C in Pine Forest, 36°C at Puddledock).

The results for *L. guichenoti* showed a greater discrepancy between areas than did *L. delicata*. Its relation to substrate temperatures was studied at three localities near Armidale, NSW: at Pine Forest and Blue Hole by Belmont (1977) and at Newholme by Simbotwe (1985). At Pine Forest no animals were found active on the surface when substrate temperatures exceeded 35°C (Fig. 6.5) but at Blue Hole, where *L. delicata* did not occur, there were a few *L. guichenoti* still on the surface when substrate temperatures there reached as high as 45°C (Fig. 6.6).

Based on the CTMax of other species of lizards (see Heatwole and Taylor 1987), the highest substrate temperatures experienced by *L. guichenoti* at Blue Hole, (see Fig. 6.7), must be near the upper limit of its activity range and probably approaches its CTMax. This finding also suggests that *L. guichenoti* may sometimes operate at higher temperatures than does *L. delicata*, and that the absence of the latter from Blue Hole may be because high substrate temperatures there exceed its tolerance limits. Thus, these results pose an enigma for the pattern of distribution. If *L. delicata* cannot support the higher substrate temperatures at the Blue Hole site in temperate NSW, but *L. guichenoti* can, how does one explain the extension of the geographic range of *L. delicata* into the subtropics and tropics, beyond the northern limits of the distribution of *L. guichenoti*? To resolve this paradox, an important approach would be to compare the thermal biology of *L. delicata* among populations located in different parts of its geographic range to ascertain whether there is a latitudinal cline in thermal adaptation and relate the results to similar studies of *L. guichenoti*.

In summary, both widespread species initiate activity outside of their nocturnal shelters only when they can attain temperatures of 18°C or higher, i.e. the lower limit of their activity ranges. Both species have identical preferred temperatures for daily activity of ~25°C, and the upper thermal limit of the activity range of *L. delicata* is ~36°C but seems to be higher (to 42°C) for *L. guichenoti*.

Returning now to Fig. 6.7A, we can begin to relate the findings of Belmont (1977) and Simbotwe (1985) to the mode of thermoregulation of *L. delicata* and *L. guichenoti*. Note that although there was a significant positive relationship between the temperature of the substrate and the body temperatures of *L. guichenoti*, that the two seldom were equal. In other words, although body temperatures increased with increasing substrate temperatures in an orderly way, the two temperatures were not the same, and thus one cannot use substrate temperature as a direct surrogate for body temperature, as would be the case for a true poikilotherm (except for lizards brumating under cover and hence in equilibrium with substrate temperatures). Seldom were the temperatures of the substrate and the body the same as one would expect if body temperatures were determined solely by direct conduction from the substrate. Remember that because of surface-area-to-volume ratios in such small lizards, equilibration would be expected to be very rapid. It is clear from Fig. 6.7A that at the lower environmental temperatures, lizards' temperatures were higher than either the temperature of the substrate or that of the air, but at the higher ones, body temperatures were lower than environmental ones. Clearly, the lizards are doing something that enhances their gain in heat at low temperatures but retards it at higher ones.

There are alternative methods for measuring this attribute, such as recording the body temperatures of lizards subjected to a thermal gradient in the laboratory (see Veron and Heatwole 1970), but when field data are available construction of graphs such as those in Fig. 6.7 are an easy way of determining preferred temperatures. Recall the caveat, however, that there are different preferred temperatures for different activities. For example, lizards' preferred temperatures are somewhat higher during digestion, then while foraging for food. The intersection method probably gives preferred temperatures somewhere between these values, as during the early part of the day the lizards are mainly foraging for prey but later in the day when they have full stomachs, they are digesting it.

The convergence between the body temperature of *L. guichenoti* at the intersection of the bold bent line and line of equality (Fig. 6.7A), the mean body temperature of active individuals of this species in the field recorded by Simbotwe (1985) (24°C), and the substrate temperature most often selected by both species of *Lampropholis* recorded by Belmont (1977) (25°C) lends confidence to this method as an estimator of their preferred temperature. Now that it is clear that *Lampropholis* are not poikilotherms, i.e. do not passively accept and track whatever temperatures impinge upon them, it is germane to consider their mechanism of thermoregulation.

The scale of critical temperatures for *Lampropholis*

In the field *L. coggeri* and *L. robertsi* both maintain their body temperature closer to the higher end of the range of temperatures at which they accept than to the lower end of that range. The span between the temperature at which *L. coggeri* is maximally operative for activity (sprinting) is about equally distant from its CTMax and its CTMin (~13 degrees Celsius; calculated from Muñoz *et al.* 2016); the comparable respective value for *L. robertsi* is ~21–23 degrees Celsius). This span is unusually small for both species because of the high level of their high CTMin. In the other species of *Lampropholis* investigated, the CTMin were much lower.

As indicated above (section *Ecologically important temperatures*) there are various factors that can modify the preferred temperature of lizards (one being that there may be an elevation in their preferred temperature when they are ill – 'behavioural fever'). It is not only important, however, to know which environmental parameters affect behavioural or physiological attributes, but also to understand which ones do *not* do so, and whether this varies among different taxa of lizards. For example, Lee *et al.* (2023) found that neither infestation by endoparasitic nematodes nor their ectoparasitic mites, nor the combined action of both, had any effect on the preferred body temperature of *L. guichenoti*.

For additional discussion of critical temperatures, see section *Temporal and spatial scope* above.

Body temperatures in winter

Ascertaining the values of the critical ecological temperatures in the species of *Lampropholis* is difficult because of the small size of these species, the rapidity with which their body temperature changes to approximate momentary environmental temperatures, and the technological difficulties in measuring body temperatures in such small lizards. Using the temperatures of the substrates occupied by torpid lizards at the time of first sighting may, nevertheless, serve as a surrogate for their body temperatures.

Few data are available for the body temperatures of lizards in the field while they are brumating. Cloacal temperatures of brumating *L. delicata* under logs and rocks during winter near Armidale, NSW, registered 4.3 degrees Celsius above substrate temperatures, and 7.1°C above air temperature at the site as well as temperatures of the air and substrate at the lizards' locations (Table 6.2).

Equivalent values for adult *L. guichenoti* were 3.9 and 2.7 degrees respectively. Ectotherms as small as *Lampropholis* would not be able to sustain such elevations above body temperature by metabolic means even if highly active, and certainly not in the state of torpor in which they were found. The explanation probably lies in the conduction of heat from the investigator's hand to the lizard being measured. Recall that the surface-area-to-volume ratio of such a small lizard is that they exchange heat with their environment extremely rapidly (see section *Effect of size and shape of the body on ectothermic exchange of heat*) and traditional methods of measuring body temperature applicable to larger lizards are difficult to apply accurately even in summer when the gradient of temperatures between the hand of the endothermic investigator and the ectothermic subject is not so great. In winter when the temperature of the investigator's hand is much higher than the body temperature of the lizard, it is exceedingly difficult to obtain accurate data.

Despite every precaution being taken to minimise conduction of heat to the lizard while measuring its temperature, for example, holding it only by one hind limb, conduction of heat to the lizard probably accounts for the perceived difference in temperature between the lizard and its environment. Nevertheless, these data are extremely valuable as not only do they illustrate the extreme difficulty in recording accurate body temperatures for cold lizards of this size, but they do indicate, indirectly, the body temperatures of these lizards in the field in winter. Small lizards, such as *L. delicata* and *L. guichenoti*, equilibrate rapidly to the temperature of solid substrates with which they were in intimate contact. Hence, we believe the temperature of the substrate with which they were in full-body contact during brumation (the soil on which they lay, and the rock or log covering them), provides an accurate surrogate for their body temperatures. Thus, using substrate temperature as such a surrogate, we know that brumating *Lampropholis* (including juveniles who are too small to be subjected to cloacal probes) survive low temperatures within the range of 13.5–22.0°C (mean 17.5°C) for *L. delicata*, and 10.0–16.9°C (mean 13.6°C) for *L. guichenoti* because when all these animals were taken to the laboratory and warmed; all recovered full activity. The actual range of survivable temperatures may, of course, extend even lower.

The air temperatures in winter also make an important point. They were consistently lower than those of the substrate by as much as 0.2–6.3 degrees (mean 3.0 degrees) for *L. delicata*. Thus, by brumating in a secluded place surrounded by solid substrate, these animals were protected from the thermal extremes of more exposed locations. This again illustrates the importance of logs and rocks in the ecology of *Lampropholis* (see Chapter 11). Recall that air temperature is not a good surrogate for the body temperature of lizards.

The results were different for *L. guichenoti* although the data are too few to draw any definite conclusions. An important research initiative would be to install automatic temperature-recording devices at brumation sites and monitor substrate and air temperatures throughout an entire winter. With improved miniaturisation, perhaps even body temperatures could be monitored directly.

BOX 6.2 HUMANE TREATMENT OF ANIMALS

Most of the endpoints discussed in this chapter are experienced in the life of lizards on a regular basis and do not cause trauma to the animals. The practice of chilling reptiles as a means of anaesthetising them for experimental or surgical purposes, or for euthanising them humanely has been criticised, mostly by persons interpreting the relevant data from the perspective of homeotherms rather than a true understanding of the physiology of ectotherms. However, a number of veterinarians and biologists reviewed the relevant physiological data and concluded that small ectothermic vertebrates do not experience pain attributable to low temperature (and even to the freezing of their bodily fluids or tissues in some species that normally survive freezing during winter) and that whole-body cooling, followed by freezing, is a humane form of euthanasia and that cooling offers a humane and useful means of anaesthetising reptiles for surgery (Lillywhite *et al.* 2017). Studies that involve comparisons with the lower endpoints can thus be conducted without distress to the animals. The same is not true, however, of the endpoints above the CTMax; those cause obvious pain and distress and should be avoided in the absence of a compelling need for such information.

Tolerance of eggs to cold

Shine (1983b) subjected eggs of *L. guichenoti* that were nearly ready to hatch to a series of subzero temperatures and compared their hatching success to controls that were not cooled. There was a significant decrease in the survival of the embryos both with decreasing temperature ($F = 21.9$; $P < 0.05$) and increasing duration of exposure ($F = 12.5$; two-factor ANOVA) (Table 6.3). The temperatures of nests measured in the field did not reach subzero values in Pengilley's (1972) study at Coree Flats and ranged from ~3°C to 27°C during the period from October through March (his Fig. 2). Nests both of *Anepischetosia* (formerly *Anotis*) *maccoyi* and *Lampropholis* were both included in that figure but were not separately designated nor was the species of *Lampropholis* identified.

The temperatures in those nests did not, however, reach temperatures either low enough or high enough to reduce the viability of the eggs of either species of *Lampropholis* during that period (Table 6.3). Shine (1985) wanted to see how well the air temperature estimated temperatures in the nest and found that the two were significantly related ($N = 27$; $r = 0.97$; $P < 0.001$; his Fig. 2). The correlation was not, however, a straight-line one; at temperatures of nests below ~15°C eight of the 10 points were above the line of equality 80%, whereas above that temperature, 16 of the 17 points were below that line; thus the temperatures of nests are underestimated by air temperatures at low values and overestimated at high ones.

Shine (1999) found that although viviparity allowed females to elevate the temperature of their developing embryos by basking, *Nannoscincus* (now *Anepischetosia*) *maccoyi* from high-elevation areas of south-eastern Australia follows the alternative strategy and stated that '***Bassiana* eggs tolerate higher temperatures than do *Nannoscincus* eggs, but do not** develop as rapidly at low temperatures. **Incubation at lower temperatures produces a larger, faster hatchling** in *Nannoscincus*, whereas the reverse is true in *Bassiana*' [emphasis added]. Although detailed data were not presented, Shine (1999) cited a personal communication as '*Lampropholis delicata* from the Brindabella Range are similar to those of *Bassiana* (*e.g.* **higher-temperature incubation yields hatchlings that are larger and faster** than are those from cool incubation' [emphasis added]. One of these conflicting statements must be an inadvertent misprint so the status of *L. delicata* remains unresolved.

EFFECT OF TEMPERATURE ON PERFORMANCE AND ON TEMPERAMENT (PERSONALITY)

Performance

Muñoz *et al.* (2016) examined assemblages collectively containing seven species of lizards from Australian rainforests and compared their performances as measured by their: (1) capability of sprinting at different temperatures, and (2) the range of temperatures at which they were capable of righting themselves if turned over and thus with enough celerity to be able to escape threatening thermal extremes. *Lampropholis coggeri* and *L. robertsi* were the two closely related species involved in their study. Like the other species from rainforests, they reached their CTMin when temperatures had dropped only to 10–15°C, whereas *L. delicata* and *L. guichenoti* could chill to 5–10°C before reaching their CTMin (Tables 6.1 and 6.2). By contrast their CTMax were on a par with those of *L. delicata* and *L. guichenoti*.

For further discussion of performance see Chapters 4, 5, and 7.

Temperament (personality)

Michelangeli *et al.* (2018) investigated the role of temperature in 'animal personality' (*i.e.* consistent individual behavioural differences, also referred to as temperament, coping style, or behavioural syndrome) in *L. delicata*. They confirmed that traits associated with thermal physiology corresponded with the individual's placement along a cold–hot continuum. 'Hot' thermal phenotypes were more 'proactive'. They exhibited superior performance at higher temperatures and were more active, bold, explorative, and had higher sprint speed than did 'cold' phenotypes. 'Hot' individuals also had superior performance within a narrower temperature range, at higher optimum body temperatures. Despite the different thermal types spending most time in the vegetated microhabitat of the constructed laboratory-based landscape, 'hot' thermal-types spent less time in the cool microhabitats than did 'cold' thermal-types.

A cold ectotherm's ability to conduct its daily activities increases as its temperature rises to an optimal peak and then wanes with further increments of temperature. Anderson *et al.* (2023a) measured eight species of *Lampropholis'* maximum speed of sprinting along a race-track at temperatures ranging from 15°C to 35°C as an indication of the thermally driven change in their capacity for performance. The resulting 'thermal performance' curves indicated that of the species tested, the optimal temperature for peak performance was significantly lower for the widespread species (*L. delicata* and *L. guichenoti*) than for those with more restricted geographic ranges (*L. mirabilis*, *L. amicula*, *L. couperi*, *L. adonis*, and *L. caligula*). The range of temperatures over which sprints occurred was broader for the widespread species. However, this shortening of the span of temperatures at which physical effort is possible is not as disadvantageous as it would seem; the span of temperatures over which they were capable of active, coordinated movement is shortened by truncation on the lower end of the scale (Muñoz *et al.* 2016; their Table 1). It would seem therefore, that they would not suffer from global warming to a greater extent than the widespread species as they are adapted to heat (share their CTMax) but would not suffer from rising temperatures so much because they would not need the protection against cold conditions to which *L. delicata* and *L. guichenoti* are adapted.

The Pace-of-Life Syndrome (POLS) is a syndrome in which behavioural traits are correlated and integrated within a fast–slow physiological continuum. Goulet *et al.* (2017a, b) added thermal physiology as a new dimension to the cold–hot axis and their research confirmed that thermal physiology and behaviour covaried along the thermal continuum. They measured thermal traits of *L. delicata* and used principal components analysis and structural equation modelling to test whether behavioural traits were structured into the syndrome and found that behavioural traits either constrained or promoted activity, exploration, boldness, and social behaviour. They found that in the syndrome there were 'hot' lizards that sprinted at higher speeds, were more active and exploratory and more social than were 'cold' individuals. A physiologically driven model was strongly supported and there were various pathways that integrated these variables but in general their results supported the theory that high thermal traits promoted greater capacity for performance as well as greater boldness, greater tendencies to explore, and higher levels of activity. Counter to their predictions, however, sociability also exhibited a positive relationship with thermal physiology.

REFERENCES

Anderson RO, Alton LA, White CR, Chapple DG (2022) Ecophysiology of a small ectotherm tracks environmental variation along an elevational cline. *Journal of Biogeography* **49**(2), 405–415. doi:10.1111/jbi.14311

Anderson RO, Tingley R, Hoskin CJ, White CR, Chapple DG (2023a) Linking physiology and climate to infer distributions in Australian skinks. *Journal of Animal Ecology* **92**(10), 2094–2108. doi:10.1111/1365-2656.14000

Anderson RO, Goulet CT, Chapple DG (2023b) Acclimation of thermal physiology to new basking regimes in a widespread Australian skink. *Journal of Thermal Biology* **113**. doi:10.1016/j.jtherbio.2023.103530

Angilletta MJ Jr, Niewiarowski PH, Navas CA (2002) The evolution of thermal physiology in ectotherms. *Journal of Thermal Biology* **27**(4), 249–268. doi:10.1016/S0306-4565(01)00094-8

Bartlett PN, Gates DM (1967) The energy budget of a lizard on a tree trunk. *Ecology* **48**(2), 315–322. doi:10.2307/1933120

Belliure J, Carrascal LM, Díaz JAS (1996) Covariation of thermal biology and foraging mode in two Mediterranean lacertid lizards. *Ecology* **77**(4), 1163–1173. doi:10.2307/2265585

Belmont C (1977) Resource partitioning and habitat selection in ground skinks. BSc (Hons) thesis. University of New England, Armidale, NSW.

Bilcke J, Downes S, Büscher I (2006) Combined effect of incubation and ambient temperature on the feeding performance of a small ectotherm. *Austral Ecology* **31**, 937–947. doi:10.1111/j.1442-9993.2006.01663.x

Bradshaw SD, Main AR (1968) Behavioural attitudes and regulation of temperature in *Amphibolurus* lizards. *Journal of Zoology (London)* **154**(2), 193–221. doi:10.1111/j.1469-7998.1968.tb01659.x

Chapple DG, Tingley R, Mitchell NJ, McDonald SL, Keogh JS, Shea GM, Bowles P, Cox NA, Woinarski JCZ (2019) *The Action Plan for Australian Lizards and Snakes 2017.* CSIRO Publishing, Clayton South, Vic.

Cogger HG (1974) Thermal relations of the Mallee dragon, *Amphibolurus fordi* (Lacertilia: Agamidae). *Australian Journal of Zoology* **22**(3), 319–339. doi:10.1071/ZO9740319

Cogger HG (2018) *Reptiles and Amphibians of Australia.* Updated 7th edn. CSIRO Publishing, Clayton South, Vic.

Cowles RB, Bogert CM (1944) A preliminary study of the thermal requirements of desert reptiles. *Bulletin of the American Museum of Natural History* **83**, 261–296.

Díaz JA, Cabezas-Díaz S (2004) Seasonal variation in the contribution of different behavioural mechanisms to lizard thermoregulation. *Functional Ecology* **18**(6), 867–875. doi:10.1111/j.0269-8463.2004.00916.x

Downes S (2001) Trading heat and food for safety: costs of predator avoidance in a lizard. *Ecology* **82**(10), 2870–2881. doi:10.1890/0012-9658(2001)082[2870:THAFFS]2.0.CO;2

Downes S, Hoefer A-M (2007) An experimental study of the effects of weed invasion on lizard phenotypes. *Oecologia* **153**(3), 775–785. doi:10.1007/s00442-007-0775-2

Farquhar JE, Russell W, Chapple DG (2023) Identifying the abiotic factors that determine the inland range limits of a mesic-adapted lizard species. *Integrative and Comparative Biology* **64**(1), 55–66. doi:10.1093/icb/icad124

Fei T, Skidmore AK, Venus V, Wang T, Schlerf M, Toxopeus B, van Overjijk S, Bian M, Liu Y (2012) A body temperature model for lizards as estimated from the thermal environment. *Journal of Thermal Biology* **37**(1), 56–64. doi:10.1016/j.therbio.211.10.013

Fraser S (1980) Insights into the physiological and behavioural thermoregulatory abilities of small lizards. (Hons) thesis. University of Sydney, Sydney.

Garland T Jr, Huey RB, Bennett AF (1991) Phylogeny and co-adaptation of thermal physiology in lizards: a reanalysis. *Evolution* **45**(8), 1969–1975. doi:10.1111/j.1558-5646.1991.tb02703.x

Glaw FG, Köhler J, Hawlitschek O, Ratsoavina FM, Rakotoarison A, Scherz MD, Vences M (2021) Extreme miniaturization of a new amniote vertebrate and insights into the evolution of genital size in chameleons. *Nature. Scientific Reports* **11**(1), 2522. doi:10.1038/s41598-020-80955-1

Goulet CT, Thompson MB, Chapple DG (2017a) Repeatability and correlation of physiological traits: do ectotherms have a 'thermal type'? *Ecology and Evolution* **7**(2), 710–719. doi:10.1002/ece3.2632

Goulet CT, Thompson MB, Michelangeli M, Wong BBM, Chapple DG (2017b) Thermal physiology: a new dimension of the pace-of-life syndrome. *Journal of Animal Ecology* **86**(5), 1269–1280. doi:10.1111/1365-2656.12718

Greer AE (1980) Critical thermal maximum temperatures in Australian scincid lizards: their ecological and evolutionary significance. *Australian Journal of Zoology* **29**, 91–102.

Greer AE (1997) *The Biology and Evolution of Australian Snakes*. Surrey Beatty & Sons, Chipping Norton, NSW.

Greer A (2017) *The Biology and Evolution of Scincid Lizards*. Published online only. <www.academia.edu/35305801/The_Biology_and_Evolution_of_Scincid_Lizards_doc>

Grigg JW, Buckley LB (2013) Conservatism of lizard thermal tolerances and body temperatures across evolutionary history and geography. *Biology Letters* **9**(2), 20121056. doi:10.1098/rsbl.2012.1056

Hastie A (1966) Studies of critical thermal maxima in some southeastern Australian 'skinks'. Unpublished Zool. III project. University of Melbourne.

Heatwole H (1970) Thermal ecology of the desert dragon, *Amphibolurus inermis*. *Ecological Monographs* **40**(4), 425–457. doi:10.2307/1942339

Heatwole H (1982) Panting and other responses to high temperature in the tuatara, *Sphenodon punctatus*. *New Zealand Wildlife Service Occasional Publications* No. **2**: 251–269.

Heatwole H, Taylor J (1987) *Ecology of Reptiles*. Surrey Beatty & Sons, Chipping Norton, NSW.

Heatwole H, Firth BT, Webb GJW (1973) Panting thresholds of lizards. I. Some methodological and internal influences on the panting threshold of an agamid, *Amphibolurus muricatus*. *Comparative Biochemistry and Physiology Part A: Physiology* **46**(4), 799–826. doi:10.1016/0300-9629(73)90130-8

Herczeg G, Gonda A, Saarikivi J, Merilä J (2006) Experimental support for the cost-benefit model of lizard thermoregulation. *Behavioral Ecology and Sociobiology* **60**(3), 405–414. doi:10.1007/s00265-006-0180-6

Huey RB, Kingsolver JG (1993) Evolution of resistance to high temperature in ectotherms. *The American Naturalist* **142**, S21–S46. doi:10.1086/285521

Huey RB, Slatkin M (1976) Cost and benefits of lizard thermoregulation. *The Quarterly Review of Biology* **51**(3), 363–384. doi:10.1086/409470

Kong JD, Arnoldi J-F, Jackson AL, Bates AE, Morley SA, Smith JA, Payne NL (2022) Heating tolerance of ectotherms is explained by temperature's non-linear influence on biological rates. *bioRxiv*. doi:10.1101/2022.12.06.519315

Labra A, Pienaar J, Hansen TF (2009) Evolution of thermal physiology in *Liolaemus* lizards: adaptation, phylogenetic inertia, and niche tracking. *American Naturalist* **174**(2), 204–220. doi:10.1086/600088

Lee K-H, Whiting MJ, Leu ST (2023) Are whole-organism performance and thermal preference linked to endo- and ectoparasites in a short-lived lizard? *Ethology* **129**(1), 432–443. doi:10.1111/eth.13379

Lillywhite HB (2008) *Dictionary of Herpetology*. Krieger Publishing Company, Malabar, FL.

Lillywhite H, Shine R, Jacobson E, Denardo D, Gordon M, Navas C, Wang T, Seymour R, Storey KB, Heatwole H, Heard D, Brattstrom B, Burghardt GM (2017) Anesthesia and euthanasia of amphibians and reptiles used in scientific research. Should hypothermia and freezing be prohibited? *Bioscience* **67**(1), 53–61. doi:10.1093/biosc/biw143

Llewelyn J, Macdonald S, Hatcher A, Moritz C, Phillips BL (2016) Intraspecific variation in climate-relevant traits in a tropical rainforest lizard. *Diversity and Distributions* **22**(10), 1000–1012. doi:10.1111/ddi.12466

Llewelyn J, Macdonald S, Hatcher A, Moritz C, Phillips BL (2017) Thermoregulatory behaviour explains countercurrent variation in the upper thermal limit of a rainforest skink. *Oikos* **126**(5), 748–757. doi:10.1111/oik.03933

Llewelyn J, Macdonald SL, Moritz C, Martins F, Hatcher A, Phillips BL (2018) Adjusting to climate: acclimation, adaptation and developmental plasticity in physiological traits of a tropical rainforest lizard. *Integrative Zoology* **13**, 411–427. doi:10.1111/1749-4877.12309

Macdonald SL, Llewelyn J, Phillips BL (2018) Using connectivity to identify climatic drivers of local adaptation. *Ecology Letters* **21**(2), 207–216. doi:10.1111/ele.12883

Matthews G, Goulet CT, Delhey K, Chapple DG (2016) The effect of skin reflectance on thermal traits in a small heliothermic ectotherm. *Journal of Thermal Biology* **60**, 109–124. doi:10.1016/j.jtherbio.2016.06.013

Matthews G, Farquhar JE, White CR, Chapple DG (2023) Does thermal biology differ between two colour pattern morphs of a widespread Australian lizard? *Journal of Thermal Biology* **114**, 103579. doi:10.1016/j.jtherbio.2023.103579

Meek R (1995) Reptiles, thermoregulation and the environment. *Testudo* **4**, 56–78.

Michelangeli M, Goulet CT, Kang HS, Wong BM, Chapple DG (2018) Integrating thermal physiology within a syndrome: locomotion, personality and habitat selection in an ectotherm. *Functional Ecology* **32**(4), 970–981. doi:10.1111/1365-2435.13034

Mitchell FJ (1973) Studies on the ecology of the agamid lizard, *Amphibolurus maculosus* (Mitchell). *Transactions of the Royal Society of South Australia* **97**, 47–76.

Muñoz MM, Langham GM, Brandley MC, Rosauer DF, Williams SE, Moritz C (2016) Basking behavior predicts the evolution of heat tolerance in Australian rainforest lizards. *Evolution* **70**(11), 2537–2549. doi:10.111/evo.13064

Pengilley R (1972) Systematic relationships and ecology of some lygosomine lizards from southeastern Australia (2 Vols). PhD thesis. Australian National University, Canberra.

Phillips BL, Llewelyn J, Hatcher A, Macdonald S, Moritz C (2014) Do evolutionary constraints on thermal performances manifest at different organizational scales? *Journal of Evolutionary Biology* **27**(12), 2687–2694. doi:10.1111/jeb.12526

Phillips BL, Muñoz MM, Hatcher A, Macdonald SL, Llewelyn J, Lucy V, Moritz C (2016) Heat hardening in a tropical lizard: geographic variation explained by the predictability and variance in environmental temperatures. *Functional Ecology* **30**(7), 1161–1168. doi:10.1111/1365-2435.12609

Powell H, Heatwole H, Heatwole M (1978) Winter aggregation of *Leiolopisma guichenoti*. *British Journal of Herpetology* **5**, 789–791.

Rawlinson PA (1974) Biogeography and ecology of the reptiles of Tasmania and the Bass Strait area. In *Biogeography and Ecology of Tasmania*. (Ed. WD Williams) pp. 291–339. Dr. W. Junk, The Hague.

Shine R (1983a) Reptilian viviparity in cold climates: testing the assumptions of an evolutionary hypothesis. *Oecologia* **57**(3), 397–405. doi:10.1007/BF00377186

Shine R (1983b) Reptilian reproductive modes: the oviparity-viviparity continuum. *Herpetologica* **39**(1), 1–8. <https://www.jstor.org/stable/3892470>

Shine R (1985) The evolution of viviparity in reptiles: an ecological analysis. In *Biology of the Reptilia*. Vol. 15. (Eds C Gans, F Billett) pp. 805–894. John Wiley & Sons, New York.

Shine R (1999) Egg-laying reptiles in cold climates: determinants and consequences of nest temperatures in montane lizards. *Journal of Evolutionary Biology* **12**, 918–926. doi:10.1046/j.1420-9101.1999.00093.x

Simbotwe MP (1985) Comparative ecology and demography of two populations of Australian reptiles (Lacertilia, Scincidae). PhD Dissertation. University of New England, Armidale, NSW.

Spellerberg IF (1972a) Temperature tolerances of southeast Australian reptiles examined in relation to reptile thermoregulatory behaviour and distribution. *Oecologia* **9**(1), 23–46. doi:10.1007/BF00345241

Spellerberg IF (1972b) Thermal ecology of allopatric lizards (*Sphenomorphus*) in southeast Australia. II: Physiological aspects of thermoregulation. *Oecologia* **9**(4), 385–398. doi:10.1007/BF00345341

Spellerberg IF (1973) Critical minimum temperatures of reptiles. In *Effects of Temperature on Ectothermic Organisms, Ecological Implications and Mechanisms of Compensation*. (Ed. W Wagner) pp. 239–247. Springer-Verlag, Heidelberg.

Spellerberg IF, Hoffmann K (1972) Circadian rhythm in lizard critical minimum temperature. *Naturwissenschaften* **59**(11), 517–518. doi:10.1007/Bf00609830

Torr GA, Shine R (1993) Experimental analysis of thermally dependent behavior patterns in the scincid lizard *Lampropholis guichenoti*. *Copeia* **1993**(3), 850–854. doi:10.2307/1447250

Tosini G, Avery R (1993) Intraspecific variation in lizard thermoregulatory: a thermographic study in *Podarcis muralis*. *Journal of Thermal Biology* **18**(1), 19–23. doi:10.1016/0306-4565(93)90037-T

Veron J, Heatwole H (1970) Temperature relations of the water skink, *Sphenomorphus quoyii*. *Journal of Herpetology* **4**(3/4), 141–153. doi:10.2307/1562889

Wood G (1982) *The Guinness Book of Animals Facts and Feats*. Guinness Superlatives, Enfield, Middlesex.

7 Physiology

INTRODUCTION

Physiology is the branch of biology that deals with the physical and chemical functioning of living organisms and their organs, tissues, and cells, and as such it abuts the field of ecology, which is the study of the relationship of organisms with their external environment. Physiology thus underpins all animal behaviour; however, vertebrate ectotherms (including *Lampropholis*) are particularly influenced by environmental effects on their physiology since they rely on the external environment to maintain a balanced energy budget (Lillywhite 2016).

The present chapter lies at the nexus of the disciplines of physiology and ecology in that it treats how the external environment influences physiological processes in *Lampropholis*, and how they diverge from, or conform to, patterns seen in other lizards.

METABOLISM

Metabolism encompasses the totality of the body's chemical processes and thus deals with the assimilation of energy and its role in the rates and kinds of transformations of materials within an individual. It is the basis of sustaining living organisms and is indicative of the energy expended in vital activities. Metabolism is of two basic types: (1) anaerobic, which utilises chemical pathways that do not involve oxygen, and (2) aerobic, which is based on oxidative processes and hence its intensity can be measured by the rate at which oxygen is consumed. Aerobic organisms, such as vertebrates, can undergo anaerobic metabolism temporarily, but in the long term, need to oxidise its by-products (e.g. lactic acid) and hence pay back the 'oxygen debt' incurred during anaerobiosis. Endotherms, such as mammals and birds, exert a greater physiological control over metabolic rates than do ectotherms such as *Lampropholis*.

Endotherms physiologically control body temperature within rather narrow limits and use metabolic means to elevate body temperature (T_B) above ambient temperatures under extremely cold conditions; in very hot environments metabolic rate is also elevated as a consequent of expending energy for cooling (e.g. sweating and panting). By contrast, in ectotherms, such as *Lampropholis*, metabolic rate is directly dependent upon the temperature of the animal's body, the higher the T_B the higher is the metabolic rate. Thus, most ectotherms can only indirectly control metabolic rate by behaviourally keeping the T_B within certain limits. A somewhat oversimplified[1] generalisation of the differences between these two groups of animals is that ectotherms use behaviour to regulate body temperatures during their activity period (see Chapter 6), with metabolic rate tracking the thermal changes that occur over the diel cycle, whereas endotherms primarily use metabolic means to maintain more precise body temperatures under varying environmental conditions throughout the entire day.

1 This summary ignores mammalian and avian *heterotherms*, who alternate between the endothermic and ectothermic relationship between metabolism and temperature, such as the daily torpor of some small mammals and birds (Heldmaier *et al.* 2004; Lyman 2013; Ruf and Geiser 2015) and the seasonal torpor of hibernators (Brown 1991; Murphy *et al.* 2006). Also, some larger reptiles have more metabolic control over body temperature than do small species such as *Lampropholis* (Spotila et al. 1973).

Small ectotherms, such as *Lampropholis*, have such a high ratio of surface area to volume that they lose or gain heat from the environment very rapidly and consequently, their T_B varies even from minute to minute as they move from sun to shade in a dappled environment such as a forest floor (see Chapter 6). It is clear, therefore, that their metabolic rate will depend greatly on the thermal conditions under which they are active, and the environmental factors that influence their body temperatures and activity cycle. An individual lizard that is active during the day when temperatures are higher than at night, *ceteris paribus*, would be expected to have a higher metabolic rate than one active nocturnally. In turn, then, factors affecting the timing of activity, such as light or an internal activity rhythm might be expected to affect the daily regimen of the animal's metabolism.

Not only does the immediate thermal environment affect the metabolic rate of most ectotherms, but so do antecedent ones. Lizards, including *Lampropholis*, kept at a given temperature for an extended time, then moved to another one, adjust their metabolic rate, so that when it is measured at any given test temperature the results from the two sets of tests usually optimise the metabolic rate relative to the conditioning environment. This process is called acclimation.

In addition, there are various other environmental influences that affect metabolic rate. Some, such as exercise, have an immediate stimulatory effect, whilst others, such as photoperiod, take longer to exert their influence. The present chapter addresses how various factors interact in influencing the metabolic rates of *Lampropholis*.

Effect of temperature

The metabolic rate of ectotherms increases with progressive elevation of temperature and *L. guichenoti* (Firth 1968; Anderson *et al.* 2022) and *L. delicata* (Merritt *et al.* 2013; Kar *et al.* 2021) are no exception to this generality. Kar *et al.* (2021) also demonstrated how metabolic rate changes with temperature (a reaction norm they termed 'metabolic thermal plasticity') was 'repeatable to a certain extent' and that individual differences in this attribute were 'somewhat consistent over time'. After taking account of various sources of variation, however, they found that scaling for mass did not change with temperature.

In addition to the expected increase in metabolic rate with rising temperature Firth (1968) demonstrated that when *L. guichenoti* was acclimated at 32°C for 2 weeks prior to measuring metabolic rates over a range of temperatures, the rate at the higher end of the range was lower than was the case for the same animals acclimated at 3°C (Plate 5). Animals experiencing several weeks of only winter-like conditions in nature would be unlikely to be subjected suddenly to exceptionally high temperatures and accordingly their acclimation would not need to prepare them for such an extreme, unlikely event. Acclimation at the higher temperature depressed the metabolic rate under hot conditions, thereby keeping metabolic rate closer to optimal levels, even at high temperatures, and protecting the animals from episodic, thermally induced metabolic excesses and wastage of energy in the upper range of temperature at which they were active.

The full import of Firth's (1968) findings on *L. guichenoti* can be appreciated by comparing his results with his identical treatment of *Hemiergis decresiensis*, another species of skink syntopic with it. *Lampropholis guichenoti* and *H. decresiensis* are active at the same time of day and even known to shelter together under the same object when inactive (see Chapter 4). However, the metabolism differed between these species in two ways: (1) the metabolic rate at all temperatures was much lower in *H. decresiensis* than in *L. guichenoti*, and (2) acclimation had a greater effect on oxygen consumption at low temperatures than it did at high temperatures in *H. decresiensis* whereas the reverse was true for *L. guichenoti* (Plate 5). This difference may reflect the differences in foraging behaviour of these two species, *L. guichenoti* being a large-legged forager on the surface in direct sunlight where it actively pursues prey, whereas *H. decresiensis* has minute legs in comparison to its body size and forages by slithering with serpentine undulations through the crevices in leaf litter and grass where conditions are shaded and likely cooler. Thus, *H. decresiensis* is less likely to reach high body temperatures during foraging and therefore has little need of acclimating to high temperatures but in order to maintain celerity at the lower end of its active temperatures would benefit from raising its oxygen consumption via acclimation during periods of low temperatures. Consequently, the difference in metabolism between these two species may relate to the discrepancy of their lifestyles. They may be within centimetres of each other but under different thermal regimens, an example of ecological segregation on a microscale. Such nuances of ecological and physiological differences between species in the same assemblage illustrates the complexity of the web of interactions between syntopic species and constitutes an example of point-diversity.

Small species, like these two, because of their higher surface area to volume ratios, have much less ability to physiologically influence body temperature than do larger species (see Chapter 6). Anderson *et al.* (2022) found that in *L. guichenoti* the metabolic rate varied depending on body weight and that in an elevational cline the animals from the higher, colder sites had higher metabolic rates than did those from lower elevations tested at the same temperature. Therefore, it seems that the animals from higher elevations adjust their metabolic rate in ways that compensate for the colder conditions in their montane habitat.

Merritt *et al.* (2013) took the study of metabolism and its relationship to performance a step further by studying the inter-individual differences in the aerobic resting metabolic rates (metabolic rates of fasting animals at rest) in *L. delicata* and how that affected the animals' other attributes. They verified first of all that even animals of similar size consistently differed significantly in their resting metabolic rates and in the maximum rate of aerobic metabolism of which they were capable (aerobic capacity), the former by as much as seven-fold; both of these attributes increased with increase the mass of the animals tested. Aerobic scope (aerobic capacity divided by resting metabolic rate) was 4.6, a value lower than for most animals, but was attributable to the small size of *Lampropholis* because smaller animals generally tend to have lower scopes than do larger ones. In summary, this research demonstrated that: (1) individuals of equivalent size consistently differed in their resting metabolic rate, but that (2) the only variables tested that were significantly correlated with each other were resting metabolic rate and aerobic capacity, but even that correlation was not a direct one; rather it merely reflected the fact that both were similarly dependent upon the mass of the body. Thus body mass and individual idiosyncrasy seem to be the only determinants of metabolic differences identified thus far.

Fraser and Grigg (1984) using step-function changes in temperature, found that in five small (<20 g) species of Australian skinks, including *L. delicata* and *L. guichenoti*, individuals cooled more slowly than they heated, surprisingly suggesting some physiological influence that either retarded cooling and/or enhanced heating at least within the activity range of temperatures. This aspects needs confirmation and further study.

For information about the effect that the temperature at which they were incubated has on *L. delicata* hatchlings' metabolic rate, see Chapter 5.

Comparison of widespread and range-restricted species of *Lampropholis*

Anderson *et al.* (2023) compared the sensitivity of metabolic rates to temperature of widespread and range-restricted species of *Lampropholis* and found no significant differences between any of two widespread species (*L. delicata* and *L. guichenoti*) and six species with restricted ranges (*L. caligula, L. amicula, L. couperi, L. adonis, L. mirabilis, L. similis*) that they tested. However, they did find a latitudinal effect in that populations from southern Queensland had significantly lower sensitivity of metabolic rate to temperature than those located elsewhere.

Q_{10}

An important physiological measure is the relative effect of a given change in temperature, often expressed as Q_{10}, the rate of change of a physiological or chemical system as a consequence of changing the temperature by 10°C. It indicates the progression of efficiency between the increments in temperatures toward the physiological optimal temperature for a given process. In the case of metabolism, then, it refers to the change in metabolic rate relative to a 10-degree change in temperature which varies over different ranges of the thermal scale. The classic view of ectotherms' metabolism did not recognise such differences but rather that it was best represented by a scatter of points through which an upward-trending straight line relating oxygen consumption to temperature was calculated, and whose slope indicated the increase in metabolic rate with increasing temperature. The scatter of points around such lines was interpreted as statistical 'noise'. However, more recent analysis revealed that the points around this line sometimes did not merely represent random scatter, but rather those at the lower end of the scale more often tended to be above the straight line rather than below, with the opposite trend exhibited at the upper end of the curve. Thus, at least for some lizards, the best fit seems not to be a straight line, but rather one with a lightly more sigmoid shape (Rojas *et al.* 2012). At the low end of the thermal scale Q_{10} may be higher than an overall straight line would suggest but at the upper end of the scale may be lower. The metabolic implications are that at low temperatures lizards maintain a higher metabolic rate than expected and hence physiological activity is closer to optimum than would otherwise be the case, whereas at high temperatures a depressed Q_{10} dampens metabolic excesses and wastage of energy. Thus, this variable effect of temperature on metabolism can have the effect of prolonging the diel period of effective metabolic activity. In Firth's (1968) comparison the values of Q_{10} for the curves overall were quite different for the two species.

Metabolism of developing embryos

As the eggs of *L. guichenoti* develop and the embryos become larger and more complex, their initially low metabolic rate rises during incubation to a peak several days before hatching, after which it then drops somewhat (Booth *et al.* 2000) (Fig. 7.1). Eggs acclimated at 30°C consumed significantly more oxygen than did those incubated at 25°C. The group incubated at 25°C consistently had a much lower metabolic rate and required about 12 more days of incubation before hatching (Fig. 7.1) than did those acclimated at 30°C. However, the group incubated at 25°C incurred a lower metabolic cost of production (3.47 kJ × g^{-1}) than did the group incubated at 30°C (4.00 kJ × g^{-1}). Accordingly, one would expect that the hatchlings from eggs incubated at 25°C would have a greater reserve of yolk remaining in the gut to sustain them during early life until they were successful in capturing prey than would be the case for those acclimated at a higher temperature.

Kar *et al.* (2022) incubated eggs *of L. delicata* under two regimes of fluctuating temperatures: 'cold' (23°C ± 3°C) and 'hot' (29°C ± 3°C) and found that variation among individuals' metabolic rates changed consistently with short-term changes in temperature, although credible intervals were quite broad. The repeatability of the reaction did not depend on the temperatures experienced during early development. These authors concluded that the implications are that 'thermal metabolic plasticity' is robust to early developmental temperatures and has the capacity to evolve, despite there being less consistent variation in metabolic rate under hot environments.

OSMOREGULATION, IONS, WATER BALANCE, AND EXCRETION

These aspects of the biology of *Lampropholis* scarcely have been investigated and then often only in an indirect way, ancillary to other topics. It can be assumed, however, that *Lampropholis*, like other terrestrial reptiles, excrete their nitrogenous wastes as uric acid and since neither nasal salt glands nor other extra-renal salt glands have been reported from any other skinks, osmoregulation and water balance in this genus probably is mediated by the kidneys and by selection of habitats

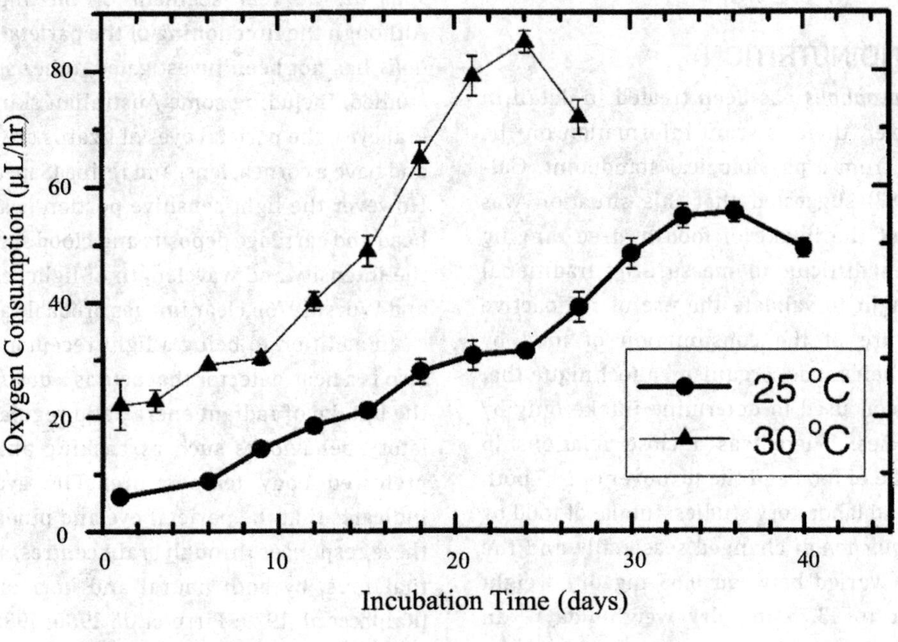

Fig. 7.1. Effect of temperature on the oxygen consumption of developing embryos of *L. guichenoti*. Modified from Booth *et al.* (2000) with permission.

and microhabitats with suitable hydric conditions (Lourdais *et al.* 2017). Booth *et al.* (2000) noted that the eggs of *L. guichenoti* incubated at different temperatures (25°C and 30°C) absorbed equivalent amounts of water and the embryos maintained the same percentage of water content at these two temperatures.

The study of Anderson *et al.* (2023) on eight species of *Lampropholis*, including both widespread species as well as six with restricted ranges, revealed no significant interspecific differences in the thermal sensitivity of metabolic rate to evaporative loss of water, nor were there any significant latitudinal differences among species.

In a study of the water balance of *L. guichenoti* along an elevational gradient from lowland Victoria to an elevation of 1,546 m above sea level Anderson *et al.* (2022) found that evaporative loss of water was greater from lizards from the higher elevations than it was from those from lower elevations tested under equivalent conditions. This probably is validly interpreted as an adaptation on the part of lowland populations in response to the drier conditions there compared to the wetter montane habitats.

Skinks lay parchment-shelled eggs and Thompson *et al.* (2001) reported that those of *Lampropholis'* take up water and ions during incubation and that this may result in hatchlings containing almost twice as much sodium at hatching as is present in freshly deposited eggs. The shell also provides a source of calcium for the developing embryo (Hallmann and Griebeler 2015).

DIGESTION AND NUTRITION

The diet of *Lampropholis* has been treated in detail in Chapter 8; however, there is scant information on the digestive system from a physiological standpoint. Gallagher *et al.* (1983) suggested that this situation was because studies of the intake of food by free-ranging animals has been difficult to measure by traditional means. They sought to validate the use of radioactive ^{22}Na as a measure of the consumption of food by *L. guichenoti* as their model organism, a technique that had previously been used to determine intake only by mammalian species. There was a close relationship between the intake of food and the turnover of Na^+ both in field surveys and laboratory studies. Intake of food by free-ranging *L. guichenoti* changed seasonally and the turnover of ^{22}Na varied between 0.65 mg dry weight (day)$^{-1}$ in winter to 13.75 mg dry weight (day)$^{-1}$ in summer. Estimated assimilated and expended energy agreed closely with previous research.

Bilcke *et al.* (2006) found that, as expected, digestion by *L. delicata* was faster at 28°C than at 20°C; however, unlike the capture and retention of prey it was not influenced by the temperature to which the lizard had been subjected during incubation.

SENSORY PERCEPTION

The eyes and the pineal complex

Most lizards have well developed sense organs and nervous systems (Huntingford and Turner 1987) and most Australian skinks have large, fully developed eyes and likely have colour vision (Hutchinson 1993). They certainly can distinguish blue from yellow as Chung *et al.* (2017) was able to train *L. delicata* to retreat to havens marked blue rather than unsafe ones marked yellow. The pupil is typically round with a distinct iris, moveable lower eyelid with an associated nictitating membrane. The eyes are lubricated by glands and the eyeball is surrounded by bony plates which provide protection. Ingram (1990) used his observations that *Lampropholis* are diurnal and employ visual cues in their behaviour and breeding displays to justify using colour pattern as a basis for describing new species in the genus when there were few other distinctive differences.

In addition to the ordinary paired eyes, all Australian skinks have a single parietal eye positioned in the parietal region of the head. It is externally visible as a circular spot in the rear segment of the interparietal scale. Although the functioning of the parietal eye in *Lampropholis* has not been investigated, other species have been studied, including some Australian skinks. Like the lateral eyes, the parietal eyes of lizards are sensitive to light and have a cornea, lens, and retina (Sapède and Cau 2013). However the light-sensitive portion is subsurface in the head and cartilage deposits and blood sinuses may modify the intensity and wavelength of light reaching it (Gundy and Wurst 1976); clear images probably are not formed.

In addition to being a light receptor, the parietal eye also is a heat-detector that acts as a dosimeter measuring the uptake of radiant energy and regulates thermoregulatory behaviours such as basking and influences the preferred body temperature. The available evidence indicates that the parietal eye and pineal organ mediate these responses through brain centres, such as the hypothalamus, by both neural and hormonal mechanisms (Ralph *et al.* 1979; Firth *et al.* 1980, 1988). Shielding the parietal eye can lead to lizards becoming more assertive (Phillips and Howes 1988).

Olfaction

Lizards use their tongues to transfer scent from their environment to the chemosensitive vomeronasal organ in the roof of their mouth and flicking of their tongues is an indication they are actively engaged in detecting odours. There is indirect evidence that *Lampropholis* possess sensitive olfactory apparatus as they respond to the odour of snakes that prey upon them (see Chapter 9). Scott *et al.* (2015) conducted experiments on the ability of *L. similis* (then called *L. coggeri*) to respond to the odours of conspecific individuals by counting the number of flicks of the tongue made by subjects to cotton swabs impregnated with the odour of other lizards in comparison to their response to control swabs soaked in distilled water. They found that although both sexes could distinguish between the scent of conspecific lizards (of either sex) and the controls, males responded to female scent more strongly than to that of another male, whereas females did not respond differently to the scents of males versus those of females. Further testing to see whether this response might be confounded by whether or not the test-female had mated previously was undertaken to determine the response of mated versus non-mated females. Although mated females showed stronger responses than did non-mated ones overall, they still did not distinguish between scents of males and females. Similarly, males responded equally to the scent of mated versus unmated females.

The subjects came from two separate localities, Paluma and Hervey ranges. Both sexes from Hervey Range in general gave somewhat stronger responses to odours than did those from Paluma Range. When tested to see whether lizards could differentiate between the odours of animals from these different localities, it was found that both sexes tongue-flicked in response to swabs from lizards from their own locality more often than they did to those from the unfamiliar one, but the response was sufficiently weak (probability of not being due to chance only 0.057). However, no copulation took place between animals from these different localities (see Chapter 4).

There are two tests that were not carried out that would be of interest. Do these lizards distinguish between their own personal scent as opposed to that of any other individual? The functional significance of this is that lack of such discrimination would make it difficult for a lizard to use olfaction to track another lizard in the search for a mate. The second test is to check whether *Lampropholis* responds to odours of prospective prey. They do respond visually to prey, but in their habitat close to the ground in the midst of leaf litter, low plants, logs, and other barriers to long-distance vision, olfactory tracking would seem to be of value. The fact that most tongue-flicking was in air suggests that these lizards were using volatile airborne chemicals to assess a potential mate; however, some tongue-flicks also were in contact with the ground which raises the possibility that, like many other lizards (Scott *et al.* 2015), *Lampropholis* may also track via odour trails left on the ground. See Chapter 4 for a discussion of the role of odour in the social behaviour of *Lampropholis* and Chapter 9 for its role in detecting predators.

Hearing

Wever (1978) suggested that some Australian skinks have a better quality of hearing than most other lizards, and Hutchinson (1993) reported that the ears of skinks are sensitive even to faint sounds although discrimination among frequencies is poor.

CIRCULATION

Larger species of lizards have various circulatory adaptations that aid in thermoregulation (Heatwole and Taylor 1987) but because of the small size of *Lampropholis*, and the consequent overwhelming immediate effect of their large surface area to volume ratio on heat exchange with their environment, circulatory adjustments would not be expected to exert very great control of body temperature in this genus (see Chapter 6). Similarly, Bartholomew (1982) reported that physiological modification of rates of heat exchange was of negligible importance in the temperature regulation of small reptiles, compared with the effect of behavioural responses. However, Fraser and Grigg (1984) found that *L. delicata* and *L. guichenoti* cooled more slowly than they heated under equivalent gradients and thus tended to maintain more equable favourable temperatures. This finding suggests a possible circulatory mechanism.

Du and Shine (2008) tested the hypothesis that more viscous blood arising from lower availability of water during incubation would alter the heartbeat of developing embryos of *L. guichenoti*. They found that lower uptake of water did not affect embryos' heart rates and rejected the hypothesis. Also, heart rates were similar on days 10 and 20 of incubation. Du *et al.*

(2010) found that embryos of a different species of lizard that occupied deep nests (*Sceloporus undulatus*) exhibited thermal acclimation of heart rate but that one that occupied shallow nests, and hence subject to more variable temperatures in nature (*Bassiana duperreyi*) did not. *Lampropholis* remains to be tested in this regard.

SEROLOGY

Baverstock and Donnellan (1990) used disc electrophoresis to purify the plasma of various taxa of lizards and injected it into rabbits. They reported the cross-reactions as Albumin Immunologic Distances (AIDs), equivalent approximately to the number of substitutions of amino acids. They found the genus *Lampropholis* (as understood at that time) to be highly diverse with AIDs among species as high as 29, i.e. as diverse as for the entire amphibolurid radiation. However, this high diversity was partly an artifact as later, some of the species then classified as *Lampropholis* were later assessed as belonging to other genera.

ENDOCRINOLOGY

Hormones influence almost all physiological, reproductive, and behavioural processes in vertebrate animals, yet very little research on the endocrinology of *Lampropholis* has been performed. Joss (1977, 1985), however, compared the production of ovarian steroids in four scincid species including *L. delicata* and *L. guichenoti*. She used radioimmunoassay to ascertain the levels of progesterone and 17β-estradiol in the plasma coupled with ovarian histochemical localisation of the enzymes 3β-steroid dehydrogenase and 17β-steroid dehydrogenase in the ovaries. Ultrastructural examination of the ovaries demonstrated that in these species there are two distinct types of thecal cells and one type of luteal cell responsible for the production of ovarian steroids correlated with follicular development and with ovulation. A related, but non-congeneric skink, *Saproscincus challengeri*, lacked one of the types of thecal cells found in the two species of *Lampropholis*.

Crino *et al.* (2024) tested how corticosterone (often known as a 'stress' hormone) and high temperatures during embryonic development of *L. delicata* affect growth, physiology, and survival. They exposed developing eggs of *L. delicata* to different doses of corticosterone (untreated controls, low level of hormone, high level of hormone) at several different temperatures and found that temperature during incubation and exposure to corticosterone had some sustained effects, as well as some short-term ones. They found that topical application of corticosterone to the eggs increased corticosterone levels in the yolks within a biologically relevant range, suggesting that topical application of hormones is an effective method to manipulate embryonic exposure to corticosterone.

They found that both corticosterone and temperature had independent effects on growth and physiology. Treatment with corticosterone during development did not affect the duration of incubation. There were no significant differences in mass between the controls and those treated with low doses of corticosterone, but those treated at the high dosage weighed less than the controls but did not differ in this respect from the ones treated with a low dose. There were no effects of corticosterone on the SVL or body condition of juveniles. However, adults that received high doses of the hormone as embryos had shorter SVLs than did controls or those that received low doses of treatment. Corticosterone administered to embryos had no significant effect on adults' subsequent mass or body condition.

Individuals developing under warmer conditions had shorter duration of incubation and hatched at a smaller body size compared to those incubated under cooler conditions. Temperatures during incubation had sustained but not permanent effects on the mass of the body and SVL such that juveniles incubated at the higher temperature were smaller than those incubated at lower ones, but these differences did not extend into adulthood.

Treatment with corticosterone during development affected adults' baseline level of corticosterone such that embryos subjected to low doses of corticosterone had higher levels of baseline corticosterone once they became adults compared to control animals. There was no difference in baseline corticosterone between males and females. There was, however, a positive association between adults' baseline levels of corticosterone and growth, suggesting that conditions during development can have sustained effects on growth through changes in endocrine function. It was also observed that treatment with corticosterone during development did not affect mitochondrial bioenergetics in the liver. Males had more efficient mitochondrial respiration than did females.

REGENERATION

Many species of lizards are capable of autotomy, after which the dismembered tail then wriggles conspicuously and attracts the attention of the predator who attacks it and may consume it (see Chapter 9). Whilst the predator is thus distracted the mutilated lizard often can escape and survive. In time, tails that have been lost in this way can regenerate and eventually regain their original size. The mechanism of this extraordinary extent of healing has received attention and the genus *Lampropholis* has been one of the taxa investigated.

The spinal cord

Regenerated sections of tails are different in appearance from original parts and Alibardi and Meyer-Rochow (1988) captured *L. delicata* from the field in New Zealand and compared regenerated tails of a length indicating at least a year of regrowth with material from intervals of varying lengths post-autotomy in the laboratory. The regenerated spinal cords had a single layer of cells interspersed with axons and cellular processes lining the central canal filled with cerebrospinal fluid. Special neurons called 'cerebrospinal fluid contacting neurons' (CSF-cNs) were distinguishable from other types of cells and there were vesicles rupturing on their luminal face. This configuration is similar to that seen in sensory organs such as the inner ear of vertebrates and the lateral line of fish. The authors postulated three possible functions for these vesicles: (1) a secretory one, (2) a sensory one that monitors the flow of cerebrospinal fluid in the central canal, and (3) a mechanical one of causing the cerebrospinal fluid to flow. Whatever the specific function, these authors established that the CSF-cNs contribute to the neuropil in regenerating the spinal cord. Particular chemicals located in the CSF-cNs of the older regenerate suggest that the organisation of the cells in the regenerating spinal cord varies with the stage of regeneration. There is much still to be learned before the mechanism of the regeneration of the spinal cord in the tails of lizards is understood.

The muscles

At the site where the break occurred that led to shedding of the tail during autotomy, a mass of undifferentiated cells, called a blastema, forms. The blastema then undergoes changes that eventually lead to the growth of a new tail. The progression of regeneration of the tail in *L. delicata* passes through a series of stages (Alibardi 1995) (Fig. 7.2): (1) the undifferentiated cells of the blastema form myoblasts, the precursors of muscle cells and fibroblasts, the precursors of the connective tissue; (2) the myoblasts and firbroblasts aggregate; then (3) myoblasts form bundles in the centre with the fibroblasts at either end; (4) the myoblasts fuse to form long cells (called myotubes) with multiple nuclei; and finally (5) the myotubes aggregate into bundles called myotomes and the fibroblasts fuse to form the myosepta (connective tissue) that partitions the myotomes into segments.

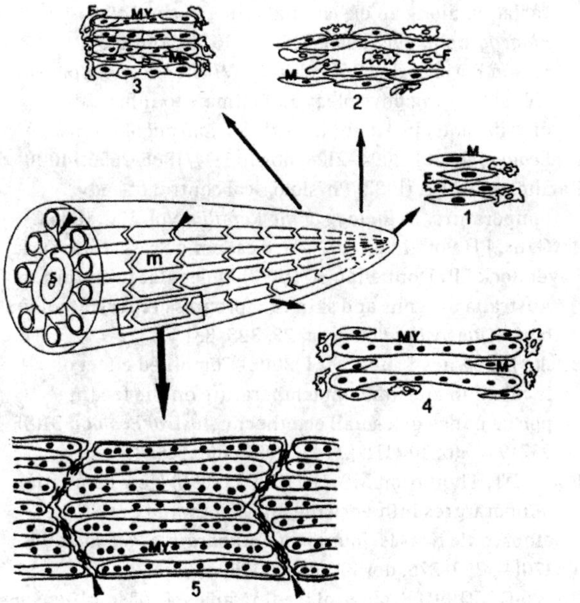

Fig. 7.2. Stages in the regeneration of the tail of *Lampropholis delicata* following autotomy. 1. Formation of myoblasts and fibroblasts. 2. Aggregation of myoblasts and fibroblasts. 3. Orientation of myoblasts in bundles in the centre and fibroblasts at the end. 4. Formation of myotubes. 5. Regenerated myotome. F = fibroblasts; M = myoblast; MY = myotube; m = myotome; Arrows indicate the progressive locations of the various stages along the length of the regenerating tail from the tip of the new tail in the earlier stages to the fully regenerated parts. Arrowheads indicate the myosepta. From Alibardi (1995), with permission.

REFERENCES

Alibardi L (1995) Muscle differentiation and morphogenesis in the regenerating tail of lizards. *Journal of Anatomy* **186**(Pt 1), 143–151.

Alibardi L, Meyer-Rochow B (1988) Ultrastructure of the neural component of the regenerating spinal cord in the tails of three species of New Zealand lizards (*Leiolopisma nigriplantare maccanni, Lampropholis delicata*, and *Hoplodactylus maculatum*). *New Zealand Journal of Zoology* **15**(4), 535–550. doi:10.1080/03014223.1988.10422633

Anderson RO, Alton LA, White CR, Chapple DG (2022) Ecophysiology of a small ectotherm tracks environmental variation along an elevational cline. *Journal of Biogeography* **49**(2), 405–415. doi:10.1111/jbi.14311

Anderson RO, Tingley R, Hoskin CJ, White CR, Chapple DG (2023) Linking physiology and climate to infer distributions in Australian skinks. *Journal of Animal Ecology* **92**(10), 2094–2108. doi:10.1111/1365-2656.14000

Bartholomew GA (1982) Physiological control of body temperature. In *Biology of the Reptilia*. Vol. 12C. (Eds C Gans, FH Pough) pp. 167–211. Academic Press, New York.

Baverstock PR, Donnellan SC (1990) Molecular evolution in Australia dragons and skinks: a progress report. *Memoirs of the Queensland Museum* **29**, 323–331.

Bilcke J, Downes S, Büscher I (2006) Combined effect of incubation and ambient temperature on the feeding performance of a small ectotherm. *Austral Ecology* **31**(8), 937–947. doi:10.1111/j.1442-9993.2006.01663.x

Booth DT, Thompson MB, Herring S (2000) How incubation temperatures influence the physiology and growth of embryonic lizards. *Journal of Comparative Physiology B* **170**(4), 269–276. doi:10.1007/s003600000097

Brown GW (1991) Ecological feeding analysis of south-eastern Australian scincids (Reptilia, Lacertilia). *Australian Journal of Zoology* **39**(1), 9–29. doi:10.1071/ZO9910009

Chung M, Goulet CT, Michelangeli M, Melki-Wegner B, Wong BBM, Chapple DG (2017) Does personality influence learning? A case study in an invasive lizard. *Oecologia* **185**(4), 641–651. doi:10.1007/s00442-017-3975-4

Crino OL, Wild KH, Friesen CR, Leibold D, Laven N, Peardon AY, Recio P, Salin K, Noble DWA (2024) From eggs to adulthood: sustained effects of early developmental temperature and corticosterone exposure on physiology and body size in an Australian lizard. *Journal of Experimental Biology* **227**, jeb249234.

Du W-G, Shine R (2008) The influence of hydric environments during egg incubation on embryonic heart rates and offspring phenotypes in a scincid lizard (*Lampropholis guichenoti*). *Comparative Biochemistry and Physiology Part A: Molecular and Integrative Physiology* **151**(1), 102–107. doi:10.1016/cbpa.2008.06.005

Du W-G, Ye H, Zhao B, Warner SA, Shine R (2010) Thermal acclimation of heart rates in reptilian embryos. *PLoS One* **5**(12), e15308. doi:10.1371/journal.pone.0015308

Firth BT (1968) Rhythmic activity, and the effects of photothermal acclimation and temperature on oxygen consumption in skinks *Lampropholis guichenoti* and *Hemiergis decresiensis*. (Hons) thesis. University of New England, Armidale, NSW.

Firth BT, Mauldin RE, Ralph CL (1988) The role of the pineal complex in behavioral thermoregulation in the collared lizard *Crotaphytus collaris* under seminatural conditions. *Ecological and Evolutionary Physiology* **61**(2). doi:10.1086/physzool.61.2.30156149

Firth BT, Ralph CL, Boardman TJ (1980) Independent effects of the pineal and a bacterial pyrogen in behavioural thermoregulation in lizards. *Nature* **285**(5764), 399–400. doi:10.1038/285399a0

Fraser S, Grigg G (1984) Control of thermal conductance is insignificant to thermoregulation in small reptiles. *Physiological Zoology* **57**(4), 392–400. doi:10.1086/physzool.57.4.30163341

Gallagher KJ, Morrison DA, Shine R, Grigg GC (1983) Validation and use of ^{22}Na turnover to measure food intake in free-ranging lizards. *Oecologia* **60**(1), 76–82. doi:10.1007/BF00379323

Gundy GC, Wurst GZ (1976) Parietal eye-pineal morphology in lizards and its physiological implications. *Anatomical Record* **185**(4), 419–431. doi:10.1002/ar.1091850404

Hallmann K, Griebeler EM (2015) Eggshell types and their evolutionary correlation with life-history strategies in squamates. *PLoS One* **10**(9), e0138785. doi:10.1371/journal.pone.0138785

Heatwole H, Taylor JA (1987) *Ecology of Reptiles*. Surrey Beatty & Sons, Chipping Norton, NSW.

Heldmaier G, Ortmann S, Elvert R (2004) Natural hypometabolism during hibernation and daily torpor in mammals. *Respiratory Physiology and Neurobiology* **141**(3), 317–329. doi:10.1016/j.resp.2004.03.0154

Huntingford FA, Turner AK (1987) *Animal Conflict*. Chapman & Hall, London.

Hutchinson MN (1993) Chapter 31. Family Scincidae. In *Fauna of Australia*. (Eds CG Glasby, GJB Ross and PL Beesley) pp. 261–279. Australian Government Publishing Service, Canberra.

Ingram GJ (1990) Five new skinks from Queensland rainforest. *Queensland Museum* **30**(3), 443–453.

Joss JMP (1977) Pituitary ovarian relationships during the breeding cycle of the Australian scincid lizard *Lampropholis* (*Leiolopisma*). In *Reproduction and Evolution*. (Eds JH Calaby and CH Tyndale-Biscoe) pp. 211–212. Australian Academy of Science, Canberra.

Joss JMP (1985) Ovarian steroid production in oviparous lizards of the genus *Lampropholis* (Scincidae). In *The Biology of Australasian Frogs and Reptiles*. (Eds G Grigg, R Shine and H Ehmann) pp. 319–326. Surrey Beatty & Sons, Chipping Norton, NSW.

Kar F, Nakagawa S, Friesen CR, Noble DWA (2021) Individual variation in thermal plasticity and its impact on mass-scaling. *Oikos* **130**(7), 1131–1142. doi:10.1111/oik.08122

Kar F, Nakagawa S, Noble DWA (2022) Impact of developmental temperatures on thermal plasticity and repeatability of metabolic rate. *Evolutionary Ecology* **36**, 199–216. doi:10.1007/s10682-022-10160-1

Lillywhite HB (2016) Behavior and physiology: an ecological and evolutionary viewpoint on the energy and water relations of ectothermic amphibians and reptiles. In *Amphibians and Reptile Adaptations to the Environment: Interplay Between Physiology and Behavior*. (Eds SV Andrade, CR de Bevier and JE de Carvalho) pp. 1–40. CRC Press, Boca Raton.

Lourdais O, Dupoué D, Guillon M, Guiller G, Michaud B, DeNardo DF (2017) Hydric 'costs' of reproduction: pregnancy increases evaporative water loss in the snake *Vipera aspis*. *Physiological and Biochemical Zoology* **90**(6), 663–672. doi:10.1086/694848

Lyman CP (2013) *Hibernation and Torpor in Mammals and Birds*. Elsevier Science, New York.

Merritt L, Matthews PGD, White CR (2013) Performance correlates of resting metabolic rate in garden skinks *Lampropholis delicata*. *Journal of Comparative Physiology B* **183**(5), 663–673. doi:10.1007/s00360-012-0736-x

Murphy K, Husaon S, Shea G (2006) Reproductive seasonality of three cold-temperate viviparous skinks from southeastern Australia. *Journal of Herpetology* **40**(4), 454–464. doi:10.1670/0022-1511(2006)40[454-RSOTCV]2.0.CO:2

Phillips JA, Howes KA (1988) The pineal complex, aggressive behavior and thermoregulation in curly-tailed lizards, *Leiocephalus carinatus*. *Physiology & Behavior* **42**(1), 103–108. doi:10.1016/0031-9384(88)90268-5

Ralph CL, Firth BT, Turner JS (1979) The role of the pineal body in ectotherm thermoregulation. *American Zoologist* **19**, 273–293.

Rojas AD, Körtner G, Geiser F (2012) Cool running: locomotor performance at low body temperature in mammals. *Biological Letters* **8**(5), 868–870. doi:10.1098/rsbl.2012.0269

Ruf T, Geiser F (2015) Daily torpor and hibernation in birds and mammals. *Biological Reviews* **90**(3), 891–926. doi:10.1111/brv.12137

Sapède D, Cau E (2013) Chapter Five – The pineal gland from development to function. *Current Topics in Developmental Biology* **106**, 172–215. doi:10.1016/B978-0-12-416021-7.00005-5

Scott ML, Llewelyn J, Higgie M, Hoskin CJ, Pike K, Phillips BL (2015) Chemoreception and mating behaviour of a tropical Australian skink. *Acta Ethologica* **18**(3), 283–293. doi:10.1007/s10211-015-0213-0

Spotila JR, Lommen PW, Bakken GS, Gates DM (1973) A mathematical model for body temperatures of large reptiles: implications for dinosaur ecology. *The American Naturalist* **107**(955), 391–404. doi:10.1086/282842

Thompson MN, Speake BK, Russell KJ, McCartney RJ (2001) Utilisation of lipids, proteins, ions and energy during embryonic development of Australian oviparous skinks in the genus *Lampropholis*. *Comparative Biochemistry and Physiology A* **129**(2–3), 313–326. doi:10.1016/S1095-6433(00)00349-4

Wever EF (1978) *The Reptile Ear*. Princeton University Press, Princeton.

8
What's on the menu? The diet of *Lampropholis*

INTRODUCTION

There have been more studies of the diet of *Lampropholis* perhaps than of any other aspect of this taxon's biology. Some papers merely list in a qualitative way what was eaten. Most papers, however, quantified the intake of prey, albeit in different ways by different authors, either as the percentages of the stomachs (or of the entire guts) in which the various taxa of prey occurred, or as the proportional contribution of each component taxon of prey to the total number of items and/or total volume of food ingested. In some cases, several habitats were lumped together instead of treated separately, and in others the sizes of available samples were too small for confident comparisons to be made.

The present chapter provides data on the diets of *Lampropholis*, as expressed in these various ways. Treatment of this information as metadata can improve confidence in the collective results and aid in interpreting trends and in constructing testable hypotheses, even when some of the individual reports are not adequate for drawing firm conclusions. This body of literature is reviewed, conclusions drawn when warranted, and hypotheses formulated.

FORAGING

Lampropholis delicata and *L. guichenoti* have very broad diets and seem to eat anything of the appropriate size that they encounter. This does not mean, however, that merely because they will eat almost anything that they lack preferences if faced with a choice. Another variable is that they may only eat prey when they are specifically foraging for food. For example, do they feed on insects that they encounter while sequestered under cover. It is conceivable that they could share cover with a semi-fossorial species that seldom or never leaves cover but preys only on insects that seek shelter under the same objects, whereas *Lampropholis* might obtain most of their food while actively pursuing a different kind of prey on the surface. Martin (2015) observed a group of *L. guichenoti* feeding on a swarm of ants (*Iridomyrmex rufoniger* group). The small workers greatly outnumbered the alates who individually were almost twice as large. Despite that, the lizards attacked far more alates than workers and on some occasions a worker was avoided; at other times several lizards would pursue the same alate or try to take one away from another lizard.

Iridomyrmex workers emit a strong odour of formic acid so the lizards' responses to them may be a combination of an avoidance of a noxious stimulus from the workers coupled with a positive response to the larger, softer alates, especially the females packed with nutritious eggs. Whether selectivity also applies to other taxa of prey remains to be tested.

Webb (1995) also noted ants as a prominent item in the diet of *L. guichenoti* but didn't indicate whether they were workers, soldier, or alates.

TAXONOMIC BREADTH OF *LAMPROPHOLIS*' DIETS

Most skinks are predators. Of the species whose diet is known, 90.1% are carnivorous, 8.8% are omnivorous, and 1.1% are herbivorous; the smaller ones, tend more toward carnivory than the larger ones (Chapple *et al.* 2023), and *Lampropholis*, being small, conforms to that assessment. In an analysis of 936 skinks from 15 species of skinks from south-eastern Australia, Brown (1991) found that the above generalisation held true not only for body size but for taxonomic affiliation. The smaller skinks such as those in the genus *Lampropholis,* mostly belong in the Eugongyline group or the *Sphenomorphus* group and are primarily insectivorous, whereas the larger, more herbivorous species are members of the *Egernia* group.

The most striking feature arising from multiple studies is the wide range of invertebrates eaten by *L. delicata* and *L. guichenoti*. Except for the more-detailed studies by Belmont (1977), Brown (1983) and Warner (1995), which include some identifications to the familial level, most of the literature on *Lampropholis* shown in Plate 10 were identified only to Order or higher, which means that the breadth of trophic niches would be underestimated and the overlap among taxa would be overestimated (Greene and Jaksić 1983). Note also that not all taxa are necessarily available, or eaten, in all habitats. The salient message of Plate 10 is the extreme flexibility of the diet of these lizards and the remarkable range of taxa consumed. Especially notable is (the rare) inclusion in the diet of even some freshwater items, such as amphipods (Rose 1974), 'freshwater Crustacean' (Crome 1981), and the nymphs of dragonflies (Simbotwe 1985). Green (1965) reported that although the diet of most *L. delicata* consisted mainly of hymenopterans and dipterans, one individual collected near the highwater mark of a beach had its alimentary canal distended with the remains of small crustaceans. Laboratory experimentation leading to discovery of how such items are detected and processed where the habitat borders on bodies of water, might expand our perception of the feeding responses of these remarkable lizards. Even such relatively immobile prey as the pupae of beetles are eaten, which suggests that the lizards may use chemical recognition, in addition to the perception of motion, to detect prey.

The collective list of prey for *L. delicata* and *L. guichenoti* contains invertebrates in two phyla (Mollusca and Arthropoda) and six classes (Appendix 4); 26 Orders of invertebrates have been identified. Eight families of spiders and two of hexapods are known from the diet of *Lampropholis*. The main bulk of the diet, however, consists of the Class Insecta, which is represented in stomachs by 16 Orders and 59 families. This is, of course, a minimal list. The lower the taxonomic scale, the less complete it is, because each entry not identified beyond the level of Order, suborder, or superfamily may contain various families, genera, and species not otherwise included in this list. The number of species involved is incalculable, and given the diversity of forms, shapes, colours, modes of locomotion, and microhabitat of the known prey, these results suggest, but do not prove conclusively, that *Lampropholis* will eat almost any invertebrate of a manageable size they encounter (but note a few exceptions below and in Chapter 11).

There is one example of an atypical item in the diet of *L guichenoti*. Metcalfe (2013) observed an individual with a dried scat of a tree frog (*Litoria dentata*) in its jaws. Skeletal elements of arthropods could be seen in the scat. When disturbed the lizard fled without relinquishing its meal.

Appendices 4 and 5 and Fig. 8.1 suggest that *Lampropholis* (except for the exceptions noted above) may not feed selectively but rather mainly eat whatever they chance upon during foraging. If that hypothesis is true, then it may be more important to examine the lizards' foraging behaviour and the abundances and attributes of potential prey that might affect the frequency of encounters between these two groups, rather than to provide an exhaustive catalogue of what is eaten. These topics are

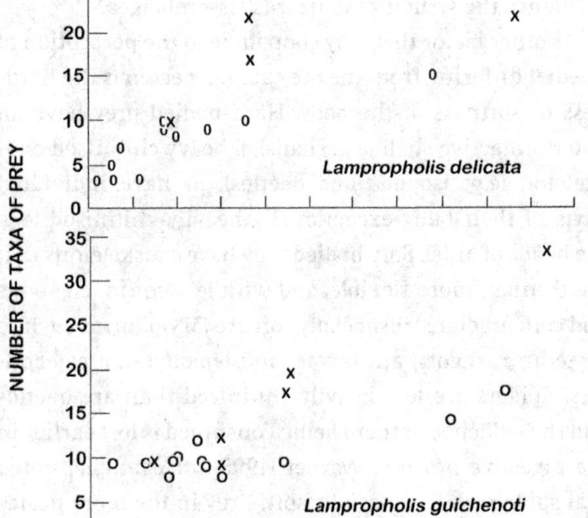

Fig. 8.1. The number of taxa identified only to ordinal, or higher, taxonomic level (circles) and the number of families (Xs) represented in the gut contents of *Lampropholis delicata* and *Lampropholis guichenoti* from different studies. Based on data from Rose (1974), Belmont (1977), Crome (1981), Webb (1983), Simbotwe (1985), Lunney *et al.* (1989), and Warner (1995, including separate data for males and females).

Table 8.1. Relative proportions of hard-bodied and soft-bodied arthropods in the stomachs of *Lampropholis delicata* and *Lampropholis guichenoti*.

N = Number of lizards. Data from Warner (1995).

	Per cent of prey items		Per cent of volume of prey	
Species	**Hard-bodied**	**Soft-bodied**	**Hard-bodied**	**Soft-bodied**
Lampropholis delicata				
Males: *N* = 13	15.0	85.0	15.7	84.3
Females: *N* = 36	11.1	88.9	43.4	56.6
Lampropholis guichenoti				
Males: *N* = 30	14.6	85.4	10.9	89.1
Females: *N* = 43	9.7	90.3	6.8	93.2

addressed in the following sections. Specifically, we review the relative abundances of prey in the stomachs to ascertain which are the common prey and which are the rarities, and whether they occur there in the same relative proportions as in the habitat at large. This approach provides a means of detecting the staple dietary items that are of greater importance in the lizards' ecologies, and which merit more-detailed study. Then we enquire into the attributes of prey that may influence their frequency of encounters with the lizards and consequently their appearance as major dietary constituents. Finally, we examine whether there are sexual or interspecific differences in diet and later (Chapter 11), how these may influence the structure of lizards' assemblages.

Another factor that may contribute to the perception of the diet differing from the prey actually eaten is the hardness or softness of the body. Hard-bodied prey have an outer protective shell (e.g. snails), a heavy chitinised exoskeleton (e.g. isopods and beetles), or have individual parts of their body exceptionally heavily chitinised (e.g. the heads of ants). Soft-bodied prey have exoskeletons that are thinner, more flexible, and with less chitin (e.g. bugs and collembolans). Especially soft are larval forms such as dipteran maggots, ant larvae, and lepidopteran caterpillars. Spiders are less heavily chitinised than are beetles and the evidence of them being consumed is lost earlier in the digestive process. Warner (1995) for example, noted that spiders were especially soft. Prey in the more posterior region of the gut will be further along in the digestive process and therefore subject to greater error. If one compares, for example, collembolans (soft-bodied) and equally small ants (hard-bodied), the former may be fully digested by the time the bolus reaches the lower intestine, but the heavily chitinised heads of the ants may be fully intact. Consequently, examining the entire gut, rather than just the stomach alone, increases the distortion of the perceived diet, even though it augments the number of individuals in the sample. One way of correcting for this discrepancy is to compare the taxonomic composition of the stomach contents with that of the intestine as a measure of the disappearance of certain types of prey during passage through the gut. Another way is to feed captive lizards a known quantity of both types of prey and compare the persistence of each type after measured intervals of time in the digestive tract.

Because of the more rapid disintegration of soft-bodied prey during digestion, their contribution to lizards' diets probably is underestimated. Even though spiders are among the most abundant items in diets of *Lampropholis delicata* and *Lampropholis guichenoti* they probably are more important, proportionally, than the data suggest. Similarly, the incidences of soft-bodied larvae probably are underestimated in comparison to adults of the same taxa. By contrast, the proportions of hard-bodied taxa such as beetles and ants may be overestimated.

Warner (1995) compared the relative proportions of soft-bodied and hard-bodied prey in the stomachs of both common species of *Lampropholis* (Table 8.1). For both sexes in both species of lizards, soft-bodied prey contributed a much higher percentage of the total numbers of prey, as well as of the total volume of prey consumed. The raw data are no longer available for testing, but it seems unlikely that the slight differences observed would approach significance.

COMMONNESS AND RARITY OF DIETARY ITEMS

There are three ways of estimating the importance of various prey species in lizards' diets: (1) the number of studies in which they are recorded as dietary items, (2) the numbers of individuals eaten, and (3) their proportional volumetric contribution to stomach contents. These are addressed in turn.

Table 8.2. Consistency of occurrence of the prey of *Lampropholis* in different dietary studies.

There were nine studies in total (Rose 1974; Belmont 1977; Crome 1981; Brown 1983; Webb 1983, 1995; Simbotwe 1985; Lunney *et al.* 1989; Warner 1995), with six involving *Lampropholis delicata* and seven involving *Lampropholis guichenoti.* Both species of *Lampropholis* co-occurred in some of the studies, but in others, only one or the other of the two occurred singly.

	Studies containing the taxon			
	Number	Per cent	Number	Per cent
Taxon of prey	*Lampropholis delicata*		*Lampropholis guichenoti*	
Snails	3	50	5	71
Amphipods	1	17	0	0
Isopods	3	50	4	57
Acari (mites)	1	17	3	48
Araneae (spiders)	5	83	6	86
Chelonethida	0	0	1	14
Opiliones	0	0	1	14
Scorpionida	0	0	1	14
Centipedes	2	33	3	43
Collembola	3	50	2	29
Odonata	0	0	1	14
Orthoptera	2	33	6	86
Isoptera	1	17	5	71
Blattodea	1	17	3	48
Mantodea	0	0	2	29
Dermaptera	2	33	1	14
Psocoptera	1	17	1	14
Thysanoptera	1	17	0	0
Hemiptera	4	67	5	71
Neuroptera	1	17	2	29
Diptera	3	50	6	86
Coleoptera	4	67	6	86
Mecoptera	0	0	1	14
Trichoptera	1	17	2	29
Lepidoptera	3	50	6	86
Hymenoptera				
Ants	4	67	7	100
Non-ants	3	50	3	43

Various authors have indicated either qualitatively or quantitatively that certain taxa are relatively abundant in the stomachs of both common species of *Lampropholis.* In multiple studies at different sites, certain taxa of prey occur more often than at others. By the criterion of consistency of occurrence in different studies, spiders, beetles, ants, and Hemiptera ranked highest and were the only taxa of invertebrates to occur in more than half of the studies for both species (Table 8.2; Fig. 8.2). It is also clear that a greater number of taxa of prey consistently occur among different studies of *L. guichenoti* (20 cases) than they do among those of *L. delicata* (6 cases).

Large prey may be more difficult to subdue and consume than an equal volume of smaller, but more numerous, prey, but each individual large prey item, *ceteris paribus*, yields a greater energetic and nutritional return. Numerical and volumetric analyses thus provide different perspectives on the feeding process and on the relative contributions of individual species of prey. Several authors

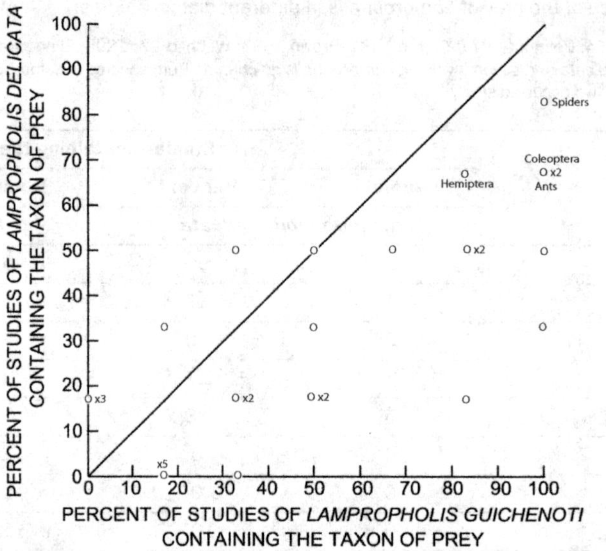

Fig. 8.2. Consistency of occurrence of higher Orders of taxa among studies of the diets of *Lampropholis delicata* and *Lampropholis guichenoti*. There were eight studies in total (Rose 1974, Belmont 1977, Crome 1981, Brown 1983, Webb 1983, Simbotwe 1985, Lunney *et al*. 1989, and Warner 1995) with six involving *Lampropholis delicata* and six involving *Lampropholis guichenoti*. In some of the studies, both species of *Lampropholis* co-occurred, but in other studies only one, or the other, species occurred singly. Numbers accompanying circles indicate the number of identical values represented by that circle, e.g. x2 indicates two identical values at the same point. The diagonal line represents equality in consistency of occurrence of prey taxa in *Lampropholis delicata* and *Lampropholis guichenoti*. For other details, see the text and Table 8.1.

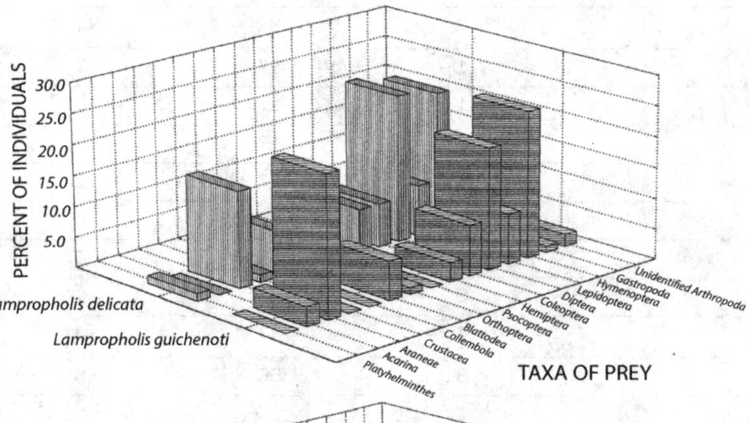

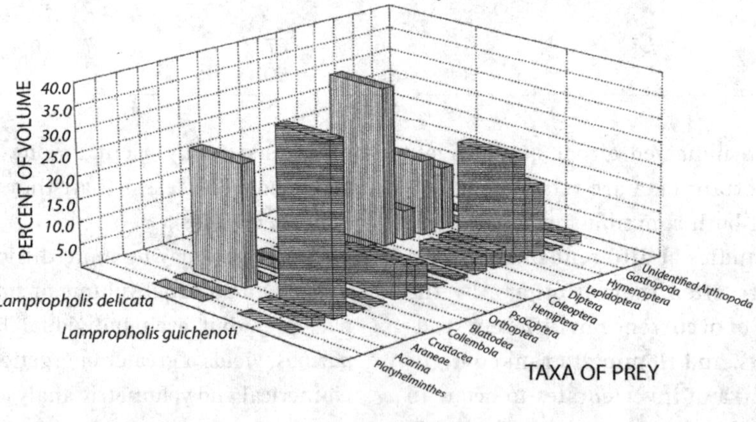

Fig. 8.3. Proportional representation of different taxa of prey in the stomachs of *Lampropholis delicata* and *Lampropholis guichenoti* from open woodland, based on number (upper) and volume (lower) of prey ingested. Modified from Warner (1995).

have expressed their results in both ways. Figure 8.3 compares numerical and volumetric expressions of the stomach contents of samples of *L. delicata* and *L. guichenoti* from an open woodland near Hawkesbury NSW (Warner 1995). Although spiders, beetles, flies, and hymenopterans were the prominent items in the diet by both means of expression, their relative prominence differed. For *L. delicata*, flies constituted the most numerous items in stomachs, followed closely by hymenopterans and then spiders. Volumetrically, the order changed to beetles, followed by spiders and then Lepidoptera. The most notable differences were in the Hymenoptera and Diptera, which were numerically abundant but, by including small-bodied taxa, e.g. ants and gnats (Appendix 4), they collectively accounted for a relatively smaller volume, and were superseded volumetrically by larger-bodied Lepidoptera and an increase in representation by spiders, some of which are large-bodied, at least as adults. In general, Simbotwe (1985) found that the size of prey eaten increased with increase in the body size of *L. guichenoti*.

For *L. guichenoti*, the three numerically dominant taxa in the stomachs were spiders and Hymenoptera subequally, followed by flies, with all other taxa providing much lower contributions. Volumetrically, there was a marked shift, resembling that occurring in *L. delicata*, and reflecting a hymenopterous diet of more numerous, but smaller, items and a lepidopterous one consisting of fewer, but larger, individuals.

DOES THE TAXONOMIC COMPOSITION OF THE DIET REFLECT THE RELATIVE ABUNDANCES OF PREY IN THE HABITAT?

Unless the relative abundances of potential prey in the habitat are known, it is difficult merely from dietary analysis, to assess whether lizards target certain prey, or alternatively, selectively exclude some items from their diet. Simbotwe (1985) addressed this problem by relating the stomach contents of *L. guichenoti* to the components of the fauna in its habitat. He supplemented general collecting by using hand-held nets and 10 pitfall traps in various microhabitats occupied by *L guichenoti*, including fallen bark and timber, granite outcrops, and litter under shrubs and trees. The ranks of species in descending order of abundance, both in stomachs and in samples from the environment, were Coleoptera first, ants second, and spiders equal third with termites. Among the less common taxa, for which presence or absence of one or a few individuals could unduly affect rankings, the two kinds of sampling did not always correspond.

Warner (1995) also related the types of prey consumed to the array of prey available in the habitat. She used Tullgren funnels to extract prey from the litter, sweep-netting to capture volant forms, and pitfall traps to sample surface-active species. From this, she assembled a list of potential prey species and compared it to the actual gut contents of the animals in six different microhabitats in her study area (Table 8.3). The habitats

Table 8.3. Prey available to *Lampropholis delicata* and *Lampropholis guichenoti* in six different microhabitats compared to the prey in the lizards' stomachs.

Data from Warner (1995).

			Prey consumed by *Lampropholis delicata*			Prey consumed by *Lampropholis guichenoti*		
Microhabitat	Area (m^2)	Total known taxa of potential prey	Number of taxa	Per cent of potential taxa	Taxa not present in sample of potential food	Number of taxa	Per cent of potential taxa	Taxa not present in sample of potential food
1	387	41	7	17.1	1	0	17.1	1
2	532	40	5	12.5	0	1	17.5	1
3	1,371	50	12	24.0	6	6	20.0	10
4	371	31	2	6.5	2	1	12.9	3
5	121	35	3	8.6	2	0	8.6	2
6	484	32	4	12.5	1	1	6.3	2

1. Shrubland: *Melaleuca* and *Leptospermum*; leaf litter cover 5–35%, 0.5 cm deep; moss >10%; bare ground >30%.
2. Woodland: *Eucalyptus, Angophora*; leaf litter cover 15–60%, 0.7–1.0 cm deep; grass >20%; bare ground <10%.
3. Woodland: *Angophora, Eucalyptus*; leaf litter cover >60%, 1–2 cm deep; herb cover; <40% grass.
4. Woodland: *Eucalyptus, Angophora*; *Acacia* shrubs; leaf litter cover >60%, 0.20–0.60 cm deep; 10–20% grass.
5. Woodland: *Angophora*; leaf litter cover, 2 cm deep; >70% grass.
6. Woodland: *Angophora*; low shrubs; litter cover 30–40%; 0.5 cm deep; >20% grass, >25% moss.

differed in the proportion of the potential prey species consumed by the lizards. These microhabitats differed in the extent and depth of leaf litter, the coverage of grass and mosses, and the floristic composition of the canopy and ground layer, any one of which potentially could affect the proportion of the total list of potential prey that might be available in that specific microhabitat. A finer resolution by laboriously sampling the potential prey present separately in each microhabitat would doubtless restrict the list for each individual habitat and raise the percentage of that list eaten by the lizards. Nevertheless, it is impressive that these species consume up to almost a quarter of the known taxa of invertebrate fauna present in the area at large. The proportion consumed varied among the microhabitats from 6.3–20% of the potential prey species for *L. guichenoti* to 6.5–24% for *L. delicata* (Table 8.3). Part of this effect may be attributable to the size of the microhabitats. The largest areal extent of the microhabitats sampled was for microhabitat number 3; it was 1,371 m^2 and yielded the highest percentage consumption of species on the list of potential prey for both species, The correlation of the number of taxa of potential prey consumed per areal extent of the microhabitat was significant (Spearman Rank Correlation test, one-tailed: $r_s = 0.771$, $P < 0.05$, identical for both species). From these data, it is difficult to separate the effect of area from other attributes of the habitat (but see the section *Spatial niche* in Chapter 11).

A reasonable hypothesis is that species capable of flight might escape predation by *Lampropholis* more easily than non-volant species by flying out of reach, and thus volant species should be underrepresented relative to non-volant ones relative to their respective abundances in the habitat. The relevant abundances of populations of volant versus non-volent species is not available yet for testing this hypothesis.

INTRASPECIFIC DIFFERENCES IN DIET

Warner (1995) collected 122 *Lampropholis* over 4 months (May to August 1995) – 22.0% of *L. delicata* and 21.9% of *L. guichenoti* had empty stomachs. There was not a significant difference (by volume) consumed by the other 95 individuals ($X^2 = 0.013$; $P > 0.90$). *Lampropholis guichenoti* consumed a significantly greater volume of prey than did *L. delicata*. There were both intraspecific and interspecific differences in the diet between males and females and between adults and juveniles of *L. guichenoti*. There also were significant differences between the sexes in the ranking of the relative abundance of different taxa of prey in the stomachs of *L. delicata* ($r_s = 0.66$; $0.05 > P > 0.02$), although not in their volume. The top five contributors in this ranking for males were: ants > spiders > sciarid fungus-gnats > beetle larvae > all others. For females, the corresponding rankings were: spiders > sciarid fungus-gnats > ants > roaches > mites = dascilloid beetles = geometrid moths = caterpillars.

When rankings were carried out on a volumetric basis (Table 8.4), rather than on the number of individuals eaten, a somewhat different picture emerged. As with the abundance of prey, the volumetric contribution of the taxa of prey differed significantly between male and female *L. delicata* ($r_s = 0.62$; $0.05 > P > 0.02$). The top five values of the volumetric rankings were: spiders > unidentified caterpillars > roaches > ants > pseudococcid bugs for males, and sciarid fungus-gnats > ants > spiders > hepialid moths > collembolans for females. Similarly, *L guichenoti* did not exhibit a significant difference between the sexes in the rankings of the volumes of prey ($r_s = 0.14$; $P > 0.50$). The top five of these ranks for males were: crickets > ants > spiders > beetle larvae > geometrid moths.

Several things emerge from this analysis that could not be obtained from a similar exercise conducted solely on prey identified only to the ordinal level. A classification based on Orders alone would not have recognised the predominance of Diptera as prey for these lizards. At least at this locality Diptera may have been represented by a single family (Sciaridae) or perhaps even one species, rather than upon a wide variety of taxa of flies. In fact, at this site, only four families of dipterans were recorded from the diet of the lizards, but except for one instance when mosquitoes were eaten in relatively large numbers, only the Sciaridae reached high ranking. Was this because: (1) sciarids are smaller than most other dipterans and hence lie within a more suitable range in size of prey for the lizards, (2) because they occupy the litter (where they feed on fungi) and thus are more accessible to the lizards, (3) because they are extremely abundant in comparison to other dipterans in the habitat (4) because the lizards have a behavioural predilection for this species, or (5) some combination of these factors? Brown (1983) listed a minimum of 18 families of dipterans from the diet of *L. guichenoti*, collectively from five different habitats covering all the major zoological subregions in the state of Victoria. These families were rare (1–4 individuals each) in the diets of this species except

Table 8.4. Volume (% of total volume of contents) of different taxa of animals in the digestive tracts of males and females of *Lampropholis delicata* and *Lampropholis guichenoti*.

'Unidentified' signifies that the prey were not identified in any category below that indicated. For example, 'ARTHROPODA (Unidentified)' means that the data in that row included only animals that were not identified to any taxonomic category lower than the Phylum Arthropoda, i.e. that line does not include any of the animals in the arthropod Classes Crustacea, Arachnida or Insecta that are listed in the lines below. Data from Warner (1995).

	Lampropholis delicata		***Lampropholis guichenoti***	
Taxa of known prey	**Males**	**Females**	**Males**	**Females**
PHYLUM PLATYHELMINTHES*	0	0.04 (0.01)	0	0
PHYLUM MOLLUSCA				
Family Mesogastropoda	0	0.2 (0.02)	1.4 (0.2)	0
PHYLUM ARTHROPODA (Unidentified)	0	0	0	84.7 (2.2)
CLASS CRUSTACEA	1.1 (0.5)	0	0	0
CLASS ARACHNIDA				
Order Acarina	0	0	0	74.9 (2.0)
Order Araneae	119.7 (55.1)	105.2 (16.0)	157.1 (19.4)	1,568.1 (40.8)
CLASS ENTOGNATHA				
Order Collembola**				
Family Entomobryidae	3.4 (1.6)	3.9 (0.6)	0	0
CLASS INSECTA				
Order Blattodea				
Family Blattidae	31.2 (14.4)	1.9 (0.3)	0	313.8 (8.2)
Family Blaberidae	0	0	0	48.0 (1.3)
Order Orthoptera				
Family Gryllidae	0	0	274.6 (33.9)	0
Order Psocoptera				
Suborder Psocomorpha	0	0.4 (0.1)	0	0
Order Hemiptera				
Suborder Sternorhybcha				
Family Pseudococcidae	5.0 (2.3)	0	0	0
Suborder Auchenorrhyncha	0.3 (0.2)	0	0	0
Suborder Heteroptera	0	2.7 (0.4)	0	1.4 (0.04)
Family (Pyrrhocoridae)	0	0	15.5 (1.9)	0
Order Coleoptera (Unidentified)	0	0	7.4 (0.9)	0
Unidentified beetle larvae	0	128.5 (19.5)	76.0 (9.4)	0
Suborder Polyphaga	1.8 (0.8)	1.5 (0.2)	0	0
Family Dascilloidae	0	160.0 (24.3)	0	73.0 (1.9)
Family Curculionidae	0	0	0	4.3 (0.1)
Order Diptera				
Suborder Nematocera				
Family Culicidae	0	0	0	90.9 (2.4)
Family Sciaridae	2.3 (1.1)	4.3 (0.7)	5.4 (0.7)	43.5 (1.1)
Suborder Brachycera				
Family Xenasteiidae	0	0	0.3 (0.03)	0
Family Phoridae	0	0.3 (0.04)	0	0
Family Calliphoridae	0	48.6 (7.4)	0	34.7 (0.9)
Order Lepidoptera				
Family Hepialidae	0	85.9 (13.1)	13.0 (1.6)	0
Family Geometridae	0	8.0 (1.2)	15.7 (1.9)	951.2 (24.8)
Caterpillars (Unidentified)	43.9 (20.2)***	0	0	80.6 (2.1)
Order Hymenoptera				
Unidentified adults	0	3.7 (0.6)	0	7.8 (0.2)
Unidentified larvae	0	0	0	11.5 (0.3)
Suborder Symphyta				
Family Tenthredinidae	0	0	0	60.0 (1.6)
Suborder Apocrita (Unidentified)	0	5.0 (0.8)	0	6.4 (0.2)

continued

Table 8.4. continued

	Lampropholis delicata		*Lampropholis guichenoti*	
Taxa of known prey	**Males**	**Females**	**Males**	**Females**
Family Diapriiade	0	1.6 (0.3)	0	0
Family Platygastridae	0	0.1 (0.01)	0	0
Family Eurytomidae	0	0.2 (0.03)	0	0
Family Cynipidae	0	0	0	0.7 (0.02)
Family Formicidae	8.4 (3.9)	95.7 (14.6)	244.4 (30.2)	384.8 (10.0)
Number of stomachs	**13**	**36**	**30**	**43**
Sexes combined	**49**		**73**	
Total volume (mm^3)	**217.1**	**657.7**	**810.6**	**3,840.5**
Sexes combined	**874.8**		**4,651.1**	
Mean volume/Stomach (mm^3)	**16.7**	**18.3**	**27.1**	**89.3**
Sexes combined	**17.9**		**63.7**	
Range in volume (mm^3)	**0.3–119.7**	**0.1–160.0**	**0.3–274.6**	**0.7–1,568.1**
Sexes combined	**0.1–160.0**		**0.3–1,568.1**	

*Excluding parasitic taxa.
**No longer considered as insects but rather as a subphylum under the class Hexapoda.
***Sex indeterminate.

for the Sciaridae (60 individuals). The prevalence of this family in two different states suggests that perhaps it is a staple dietary item over wide areas and that it alone may account for a great share of the prevalence and consistency of the Diptera in the diet of *Lampropholis*. It also provides a strong link between fungi and *Lampropholis* in the food web of woodland communities (litter → fungi → sciarid fungus-gnats → lizards).

Perhaps higher taxa prevalent in the diet of *Lampropholis*, other than Diptera, also are disproportionately represented, with a few families, genera, or species, constituting the bulk of the diet, with the remainder of the vast list of taxa made up primarily of rare species that collectively play a minor role in the trophic ecology of the lizards. The extent to which this phenomenon extends to non-dipteran prey is not known but might be a fruitful line of research.

Assemblages and communities can be analysed by the equitable or inequitable distribution of individuals within species. Many communities are highly inequitable with a relatively few common species accounting for large proportions of the individuals in the assemblage and a much greater number of rare species that play a minor role in community dynamics. *Lampropholis* consume a highly diverse body of prey, and even if they are non-selective and eat nearly anything they encounter, their diets could still be highly inequitable if it merely reflected the relative abundances of the spectrum of taxa in the community in general. It would be instructive if an analysis of the equitability of prey in the diets of *Lampropholis* were carried out at the species level of prey. This not only would allow identification of the species of prey most important to the lizards, but also would enable one to concentrate on such species and how they, and the lizards, are structured into the community. In this context, a comparison of the equitability of prey in lizards' stomach contents in comparison with that of the populations from which the prey were derived, would be highly instructive.

The study by Warner (1995) examined sexual differences in diet at the familial level of prey (Tables 8.4 and 8.5). For males of *L. delicata*, the number of stomachs was low (only 11) and consequently so was the number of taxa of prey. A Spearman Rank Correlation test indicated that the numbers of individuals in different taxa of prey (abundance) in the stomachs was significantly different between the sexes for *L. delicata* ($r_s = 0.64$; $0.05 > P > 0.02$). The ranks for males, in descending order, were: spiders > collembolans = pseudococcids (mealybugs) = sciarid fungus-gnats > ants > low numbers of all others (Table 8.5). The order for the five highest ranks for female *L. delicata* was: sciarid fungus-gnats > ants > spiders > hepialid moths (ghost moths) > collembolans.

In Warner's (1995) study, the three most abundant taxa of prey present in the stomachs of *L. guichenoti* and *L. delicata* were the same. The ranking, however, altered slightly, with the proportion of Hymenoptera nearly the same in both, but in *L. delicata* the proportion of flies

Table 8.5. Number and per cent (in parentheses) of different taxa of prey in the stomachs of males and females of *Lampropholis delicata* and *Lampropholis guichenoti*.

Empty stomachs were excluded from the analysis, as were parasites (they were not prey). Data from Warner (1995).

	Lampropholis delicata		***Lampropholis guichenoti***	
Taxa of known prey	**Males**	**Females**	**Males**	**Females**
Arthropoda (Unidentified)	0	0	0	2 (3.1)
Crustacea (Unidentified)	1 (5.0)	0	0	0
Family Mesogastropoda	0	1 (1.6)	1 (3.3)	0
Arachnida				
Order Acarina (Unidentified)	0	0	0	3 (4.6)
Order Araneae (Unidentified)	4 (20.0)	9 (14.3)	8 (25.8)	14 (21.5)
Insecta				
Order Collembola*				
Family Entomobryidae	3 (15.0)	3 (4.8)	0	0
Order Blattodea (Roaches)				
Family Blattidae	1 (5.0)	1 (1.6)	0	5 (7.7)
Family Blaberidae	0	0	0	1 (1.5)
Order Orthoptera				
Family Gryllidae	0	0	1 (3.3)	0
Order Psocoptera				
Suborder Psocomorpha	0	1 (1.6)	0	0
Order Hemiptera (Unidentified)	0	0	0	1 (1.5)
Suborder Sternorhycha	3 (15.0)	0	0	0
Pseudococcidae	1 (5.0)	0	0	0
Suborder (Unidentified)	0	2 (3.2)	0	1 (1.5)
Suborder Heteroptera	0	0	1 (3.3)	0
Family (Pyrrhocoridae)	0	0	1 (3.3)	0
Order Coleoptera (Unidentified)	0	2 (3.2)	3 (9.7)	0
Unidentified beetle larvae	1 (5.0)	1 (1.6)	0	0
Suborder Polyphaga (Unidentified)	0	2 (3.2)	0	3 (4.6)
Family Dascilloidae	0	0	0	1 (1.5)
Family Curculionidae	0	2 (3.2)	0	0
Order Diptera				
Suborder Nematocera				
Family Culicidae	0	0	0	1 (1.5)
Family Sciaridae	3 (15.0)	13 (20.6)	4 (12.9)	11 (16.9)
Suborder Brachycera				
Family Xenasteiidae	0	0	1 (3.3)	0
Family Phoridae	0	1 (1.6)	0	0
Family Calliphoridae	0	1 (1.6)	0	2 (3.1)
Order Lepidoptera				
Family Hepialidae	0	5 (7.9)	1 (3.3)	0
Family Geometridae	0	1 (1.6)	1 (3.3)	3 (4.6)
Unidentified caterpillars	1 (5.0)	0	0	3 (4.6)
Order Hymenoptera (Unidentified)	0	2 (3.2)	0	1 (1.5)
Unidentified larvae	0	0	0	1 (1.5)
Suborder Symphyta				
Family Tenthredinidae	0	0	0	2 (3.1)
Suborder Apocrita (Unidentified)	0	1 (1.6)	0	1 (1.5)
Family Diapriidae	0	2 (3.2)	0	0
Family Platygastridae	0	1 (1.6)	0	0
Family Eurytomidae	0	1 (1.6)	0	0
Family Cynipidae	0	0	0	1 (1.5)
Family Formicidae**	2 (10.0)	10 (15.9)	9 (29.0)	8 (12.3)

continued

Table 8.5. continued

Taxa of known prey	*Lampropholis delicata* Males	*Lampropholis delicata* Females	*Lampropholis guichenoti* Males	*Lampropholis guichenoti* Females
Number of stomachs	13	36	30	43
Total items in stomachs	20	63	31	65
Mean items per stomach	1.54	1.75	1.03	1.51
Number of higher taxa	10	22	11	20
Total combined taxa of both species	26		27	
Number (% of combined taxa) shared	6 (23)		4 (15)	
Number (% of combined taxa) not shared	20 (77)		23 (85)	
Number (% of own prey taxa) shared with opposite sex	6 (60)	6 (27)	4 (36)	4 (20)
Number (% of own prey taxa), that are exclusive	4 (40)	16 (73)	7 (64)	16 (80)

*No longer considered as insects but rather as a subphylum under the class Hexapoda.
**The category 'Females' includes workers (genetically females) as well as reproductive females.

was slightly higher, and that of spiders lower, than for *L. guichenoti*, thereby altering the ranking from beetles > Hymenoptera > spiders in *L. delicata* to spiders > Hymenoptera > flies in *L. guichenoti* (Fig. 8.2).

It was noted in the section *Commonness and rarity of dietary items* above that the taxonomic composition of prey in the stomach contents of *L. delicata* and *L. guichenoti* were similar on a numerical basis but that because of differences in body size among the various taxa, on a volumetric basis there were reductions in proportional representation of Diptera and Hymenoptera and a corresponding increase in that of Lepidoptera, and to a certain extent also of spiders. Although both species had high volumetric representation by spiders and moderately high ones for Lepidoptera and Hymenoptera (Fig. 8.3), the volumetric representation by Coleoptera differed markedly between the two species. In *L. delicata*, beetles accounted for the greatest volume of prey of any taxon (about 35%), whereas in *L. guichenoti* they accounted for less than 5%.

Even if both species eat any manageable prey they encounter, it does not mean that their diets will be identical throughout their ranges. For example, distributional data (see Chapter 2) indicates that *L. delicata* does not occur in some of the drier inland areas occupied by *L. guichenoti*, and even where the two species are geographically sympatric, *L. delicata* is absent from some of the more open sites inhabited by *L. guichenoti*, e.g. Blue Hole (Belmont 1977). *Lampropholis* of both common species regularly eat crickets in various wooded habitats, but grasshoppers have been recorded only in the diet of *L. guichenoti*. This again reflects the difference in type of habitat occupied by these two species of *Lampropholis*. Both species of lizards occur in wooded habitats where crickets occur, but only *L. guichenoti* occurs in the more open grassy ones inhabited by grasshoppers. Thus, this interspecific difference in the diet of the lizards may not indicate a difference in choice of prey but merely differences in the type of orthopteran prey available in different habitats. A simple food-choice experiment testing whether *L. delicata* would eat suitably sized grasshoppers would be a critical test of this hypothesis.

In a study of lizards' diets, including that of *L. guichenoti*, Brown (1991) listed five taxa not included in Appendix 4: vertebrates (one individual of unspecified species), millipedes, and three taxa of arachnids (pseudoscorpions, harvestmen, and scorpions). Unfortunately, he did not list dietary items by individual species of lizards separately. Rather, he lumped the data by Greer's (1979) phylogenetic lineages. This meant that the detailed data from *L. guichenoti* were combined with those of six other species from several vegetation types in five different localities. Consequently, it is not known whether those five taxa were eaten by *L. guichenoti* or only by one or more of the other species. Conversely, three Orders of insects, Mantodea, Mecoptera, and Trichoptera, all in very small numbers, occur in Appendix 4 but were not included in Brown's list. All three were present in very small numbers, however, and it is possible that their absence from diets of *Lampropholis* in other studies was merely stochastic. Some of these taxa are rather large and appear unlikely dietary items for such small animals as *Lampropholis*. The occurrence of scorpions in the list, however, does indicate that at least some species of small skinks are either immune to, or can counteract, the noxious qualities of some prey. Also, the consumption of harvestmen suggests that at least some species in this group of lizards must have ways of manipulating prey with unusually long legs relative to the size of the lizard.

Because only a few studies identified prey below the ordinal level, it is at that level that comparisons among different sites and species need to be (Plate 10: colour-coded for easy comparison among species). Despite the diversity of habitats and geographic regions involved, not only is the diet of each species of *Lampropholis* relatively constant at the ordinal level of prey, but the two species also are remarkably similar. Some differences in the prey consumed that do exist are among the rarer species whose absences could be attributable to sampling error, especially in the studies involving low numbers of stomachs.

The data in Appendix 4 were used to tally the times that lizards had similar prey (at the ordinal level or above) in their stomachs/guts, as opposed to dissimilar contents (taxon absent in one but present in the other). Table 8.6 presents comparisons of the similarity in all available pairs of studies, separately for *L. delicata* and *L. guichenoti*, in which the studies were carried out by different investigators and at different locations and times. The gut contents of both species were always more similar than dissimilar between pairs of studies (Table 8.6), except for one study in which only 12 (46%) of the taxa of prey were similar between one pair of studies and 14 (54%) were dissimilar (*L. guichenoti*: data from Simbotwe [1985] versus those from Brown [1983]). Indeed, the mean Similarity for all the paired comparisons were practically identical for *L. delicata* and *L. guichenoti*: 69% and 67% respectively. Thus, studies carried out in different locations at different times and by different authors nearly always show greater similarity than dissimilarity in the taxonomic composition of the gut contents, both within each of the two species individually and between them; on average about two-thirds of the taxa at ordinal level are shared (Miller *et al.* 2017).

To interpret this finding, several caveats and one source of error must be taken into consideration. The latter is that the results are dependent on sample size (number of stomachs examined). Figure 8.2 shows the sample sizes of members of a paired study with the values for *L. guichenoti* plotted against those of *L. delicata* and each point depicting the value of the percentage similarity of that comparison. Note that three points fall above the line of commonality and 20 below, with one occurring directly on the line. This means that more species from the combined list of prey were found in the stomachs of *L. guichenoti* more often than they were in in the stomachs of *L. delicata*. Only in one study were the same number of prey species from the combined list equally represented in the stomach of the two species of *Lampropholis*. Before accepting the hypothesis that this indicates that *L. guichenoti* consumes a broader spectrum of prey than does *L. delicata*, one should consider the caveat that the number of stomachs examined is important in analysing breadth of diet. It is known that where these two species are syntopic they are not always equally abundant (see Chapter 2).

Also, the number of variables make precise interpretation difficult. For example: (1) the fact that different locations and times are involved, (2) the taxonomic composition of the gut contents may reflect either temporal or spatial differences in the relative abundance of prey taxa, and that (3) rare taxa would be less likely to be encountered at a particular time or place, whether or not the animal would prey upon them if the opportunity arose. Alternately some populations may not prey upon temporally or spatially rare species merely because their search image at that time or place has been conditioned by more abundant species. Because of differences in activity period, camouflage, or microhabitat of the prey, most of the studies reviewed here did not attempt to ascertain the relative abundance of prey taxa in nature, or to ascertain whether those present varied in their availability to the lizards. More-detailed, experimental approaches are required to separate the individual effects of these variables.

When the analysis was focused on comparisons between *L. delicata* and *L. guichenoti* studied at the same time and place, the spatial and temporal variables were minimised, and accordingly, the mean correspondence between the two species at the same site, 85% were higher than the individual means of 67–69% (Tables 8.6 and 8.7) for these two species at different sites and times. Again, the departure from random expectation was significant for all comparisons (Table 8.7). The problem with low numbers of stomachs examined was reduced in this analysis as one of the two studies with low sample size did not include the two widespread species (Rose 1974).

Table 8.6. Comparison, separately by species, of the similarity and dissimilarity of the prey in the stomach/guts of the two common species of *Lampropholis* (*L. delicata* and *L. guichenoti*), studied by different authors at different locations and times.

Numbers in **boldface** indicate summaries in order to distinguish them from individual values.
Data from Rose (1974), Belmont (1977), Crome (1981), Brown (1983, 1991), Lunney *et al.* (1989), and Warner (1995), for prey at the ordinal level or above.

	Stomachs/guts	
Authors	**Number**	**Per cent**
Lampropholis delicata		
Rose vs Belmont		
Similar prey	16	62
Dissimilar prey	10	39
Rose vs Crome		
Similar prey	19	73
Dissimilar prey	7	27
Rose vs Lunney *et al.*		
Similar prey	16	62
Dissimilar prey	10	39
Rose vs Warner		
Similar prey	17	65
Dissimilar prey	9	35
Belmont vs Crome		
Similar prey	15	58
Dissimilar prey	11	42
Belmont vs Lunney *et al.*		
Similar prey	18	69
Dissimilar prey	8	31
Belmont vs Warner		
Similar prey	21	81
Dissimilar prey	5	19
Crome vs Lunney *et al.*		
Similar prey	19	73
Dissimilar prey	7	27
Crome vs Warner		
Similar prey	18	69
Dissimilar prey	8	31
Lunney *et al.* vs Warner		
Similar prey	20	77
Dissimilar prey	6	23
Per cent similarity		
Mean	**68.9**	
Range	**58–81**	
Per cent dissimilarity		
Mean	**31.3**	
Range	**19–42**	
Lampropholis guichenoti		
Belmont vs Crome		
Similar prey	18	69
Dissimilar prey	8	31
Belmont vs Simbotwe		
Similar prey	19	73
Dissimilar prey	7	27
Belmont vs Lunney *et al.*		
Similar prey	16	62
Dissimilar prey	10	39
Belmont vs Warner		
Similar prey	19	73
Dissimilar prey	7	27
Belmont vs Brown		
Similar prey	18	69
Dissimilar prey	8	31
Crome vs Simbotwe		
Similar prey	17	65
Dissimilar prey	9	35
Crome vs Lunney *et al.*		
Similar prey	16	62
Dissimilar prey	10	39
Crome vs Warner		
Similar prey	21	81
Dissimilar prey	5	19
Crome vs Brown		
Similar prey	16	62
Dissimilar prey	10	39
Simbotwe vs Lunney *et al.*		
Similar prey	15	58
Dissimilar prey	11	42
Simbotwe vs Warner		
Similar prey	20	77
Dissimilar prey	6	23
Simbotwe vs Brown		
Similar prey	12	46
Dissimilar prey	14	54
Lunney *et al.* vs Warner		
Similar prey	19	73
Dissimilar prey	7	27
Lunney *et al.* vs Brown		
Similar prey	18	69
Dissimilar prey	8	31
Warner vs Brown		
Similar prey	17	65
Dissimilar prey	9	35
Per cent similarity		
Mean	**66.9**	
Range	**46–81**	
Per cent dissimilarity		
Mean	**33.3**	
Range	**19–54**	

Table 8.7. Percentage of stomachs with food and the number of prey per stomach in *Lampropholis delicata*, *Lampropholis guichenoti*, and some other small Australian ground-skinks from various habitats.

N = number of stomachs examined. Data from Belmont (1977), Crome (1981), Lunney *et al.* (1989), Simbotwe (1985a), Warner (1995), and Webb (1983).

Habitat	N	Per cent of stomachs		Mean prey per stomach with food	Authority
		Empty	With food		
Lampropholis delicata					
Hawkesbury Sandstone, Sydney	8	25	75	1.2	Webb (1983)
Three habitats combined	99	21	79	2.8	Belmont (1977)
Angophora-Eucalyptus woodland	49	22	78	2.2	Warner (1995)
Eucalypt woodland, Eastwood State Forest; healthy, abundant leaf litter	5	0	100	2.8	Crome (1981)[1]
Eucalypt woodland, Newholme Site 1; moderate dieback with slight epicormic growth	–	–	–	–	Crome (1981)[1]
Mumbulla State Forest; healthy forest, 1980, pre-fire, pre-drought	20	–	–	3.1	Lunney *et al.* (1989)[2]
Mumbulla State Forest; 1980, post-fire, pre-drought	11	–	–	0.5	Lunney *et al.* (1989)[2]
Mumbulla State Forest	16	–	–	4.2	Lunney *et al.* (1989)[2]
Mumbulla State Forest	17	–	–	5.3	Lunney *et al.* (1989)[2]
Mumbulla State Forest	7	–	–	3.7	Lunney *et al.* (1989)[2]
Eucalypt woodland, Newholme Site N-3; mostly dead trees open, grassy, sparse litter	2	0	100	2.0	Crome (1981)[1]
Lampropholis guichenoti					
Three habitats combined	94	9.0	90	3.4	Belmont (1977)
Angophora-Eucalyptus woodland	73	22	78	1.7	Warner (1995)
Eucalypt woodland, Eastwood State Forest; healthy	12	35	65	2.3	Crome (1981)[1]
Eucalypt woodland, Newholme Site 1; moderate dieback with slight epicormic growth	20	30	70	1.5	Crome (1981)[1]
Eucalypt woodland, Newholme Site 2; severe dieback; mostly dead trees; open, grassy, sparse litter	24	8	92	3.4	Crome (1981)[1]
Eucalypt woodland, Newholme Site 3, Armidale; grazed by cattle and sheep	82	0	100	1.2	Simbotwe (1985)
Mumbulla State Forest; healthy forest, 1980, pre-fire, pre-drought	17	–	–	4.5	Lunney *et al.* (1989)[2]
Mumbulla State Forest; 1980, post-fire, pre-drought	5	–	–	0.4	Lunney *et al.* (1989)[2]
Mumbulla State Forest	25	–	–	4.3	Lunney *et al.* (1989)[2]
Mumbulla State Forest	26	–	–	6.4	Lunney *et al.* (1989)[2]
Mumbulla State Forest	17	–	–	–	Lunney *et al.* (1989)[2]
Hemiergis decresiensis					
Eucalypt woodland, Eastwood State Forest, Site E-1; healthy with abundant leaf litter	2	0	100	9.5	Crome (1981)[1]
Eucalypt woodland, Newholme Site 1; moderate dieback	10	0	100	7.6	Crome (1981)[1]

continued

Table 8.7. continued

Habitat	N	Per cent of stomachs		Mean prey per stomach with food	Authority
		Empty	With food		
Eucalypt woodland, Newholme Site N-2; severe dieback; mostly dead trees; open, grassy, sparse litter	20	5	95	4.6	Crome (1981)[1]
Eucalypt woodland, Newholme Site N-3; grazed by cattle and sheep	34	66	34	2.1	Simbotwe (1985)
Morethia boulengeri					
Eucalypt woodland, Newholme Site N-2; mostly dead trees	3	33	67	1.5	Crome (1981)[1]

[1]Values of percentages for each taxonomic category were obtained by projecting the histograms of Figures 1 and 2 of Crome (1981) onto a grid.
[2]The entire gut, rather than just the stomach was examined, and hence latter would be inflated, compared to the other studies. The number of empty stomachs was not recorded so it isn't known whether all values of N were 100%, or whether empty stomachs were included in the calculation of the mean number of prey per gut. Were the former the case, the mean prey per gut would be comparable to the data from other species, except for the elevation caused by using the entire gut; were the latter the case, the effect would be to diminish the values of the mean prey per gut. For this reason, these values are not directly comparable to those of the other studies but are consistent within the study of Lunney *et al.* (1989).

Brown (1991) also examined the diet of *L. guichenoti* but the taxa were listed at the ordinal level for insects, by Class for other arthropods, and by Class, or Subphylum, for all other taxa. At those taxonomic levels Brown's (1983) data agreed with those summarised in Table 8.7.

INTENSITY OF FEEDING BY *LAMPROPHOLIS*

Proportion of digestive tracts with prey and mean incidence of prey

The intensity of predation can be expressed either as the percentage of stomachs/guts that contain prey or as the numbers of prey found within the digestive tract (Table 8.7). These values, of course, are lower earlier in the day before lizards have captured many prey, and Belmont (1977) noted that during cold seasons when lizards were not very active, fewer items were found in their stomachs. These sources of error probably are minimal, however, as samples were collected mainly during seasons when lizards are actively feeding and at times of day when stomachs were expected to be relatively full. In all studies of *L. delicata* and *L. guichenoti* for which data are available, 65–100% of the stomachs contained food, not only in healthy undisturbed woodland, but also in some areas with many trees dead from dieback, or where there was grazing by sheep and cattle (Table 8.7). There were no consistent differences in these figures between these different kinds of woodland. Similarly, the mean number of prey items per stomach with food, averaged 1.2–6.4 items per stomach in unburned areas (Table 8.7). For this measure, data were available from an area that had experienced a fire; there the mean level of 3.1 before the fire, had dropped to 0.5, post-fire (Lunney *et al.* 1989).

It seems that the ability to find food does not vary consistently among different kinds of woodlands inhabited by *Lampropholis*, except under extreme conditions such as fire or drought. Lizards in such woodlands tend to have fewer items in their stomachs (Lunney *et al.* 1989).

Brown (1991) also examined the diet of *L. guichenoti* but the taxa were listed at the ordinal level for insects, by Class for other arthropods, and by Class or Subphylum for all other taxa. At those taxonomic levels Brown's data agreed with those summarised in Table 8.7.

NON-FOOD ITEMS FOUND IN STOMACHS

A small amount of plant material was noted in the stomach of *Lampropholis* but was discounted by some of the authors in Appendix 4 as accidentally ingested while consuming prey. This topic is discussed in more detail in Chapter 11. Lunney *et al.* (1989) considered consumption of vegetation to be intentional and not merely ingested accidently when catching prey.

SIZES OF PREY

The size of prey could make an important difference in the trophic relations between different species or sexes of the same species of predatory lizards (see section *The trophic niche*, Chapter 11). If one species of lizard eats the smaller individuals of a given species or smaller species of prey within a given taxon, say Order, of insects and

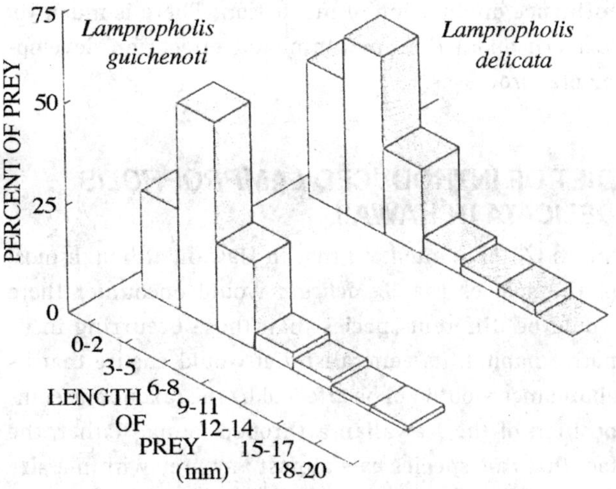

Fig. 8.4. Frequency of the lengths of prey consumed by *Lampropholis guichenoti* and *Lampropholis delicata*. From Lunney *et al.* (1989) with permission.

another species of lizards eats larger ones, then they could have identical taxonomic composition of diet at the ordinal level of identification but still occupy different trophic niches. For this reason it is important to identify the prey in stomach contents to the lowest taxonomic level possible. Even if such identification is not feasible, however, one can still detect such segregation of niches merely by assessing the size classes of prey eaten by paired species.

The ranges in classes of the length of prey in the guts of the two common species of *Lampropholis* were practically identical (2 to 18–20 mm) and both sets of histograms were skewed to the left with peaks at the 3–5 mm size range (Lunney *et al.* 1989) (Fig. 8.4). Belmont (1977) also found that the size of prey did not differ significantly between these two species nor between different sizes of adults in either species (her Fig. 2.13). The findings of Webb (1983) for *L. delicata* did not encompass that entire range of sizes of prey found by the other authors but did agree that the greatest bulk of prey (86%) was in the size-class centred on 5 mm; the next-highest (centred on 10 mm) included the remaining 14% of the sample.

RELATION OF DIET TO TYPE OF LIZARDS' HABITAT

Simbotwe (1985) studied the diet of *L. guichenoti* and found similar results to those of Belmont (1977), although his sample size was smaller and therefore probably missed more of the rarer dietary items than did Belmont's sample for this species. For example, Belmont (1977) recorded seven broad taxa (isopods, mites, centipedes, collembolans, earwigs, roaches, and several kinds of hymenopterans) that did not occur in Simbotwe's samples, whilst conversely, Simbotwe (1985) obtained only one taxon (1 individual scorpionfly) not present in Belmont's sample. The most striking difference between the studies was the large number of individual collembolans (12.2% of the total dietary items) in Belmont's (1977) samples in comparison to the total absence of this group in Simbotwe's (1985) and Crome's (1981) samples. Belmont's samples came from a healthy, relatively undisturbed eucalypt woodland except for occasional light grazing, whereas Simbotwe's were from a similar habitat except it was more open and was grazed by cattle and sheep. Crome conjectured that the reason why none of the *L. delicata* or *L. guichenoti* in her study had consumed collembolans was because collembolans in general prefer moist conditions and the habitat she studied was drier than the one studied by Simbotwe.

For an assessment of the ability of *Lampropholis* to surmount changes in its habitat because of its broad diet, see Chapter 12).

DEVELOPMENTAL INFLUENCES ON THE EFFICIENCY OF PROCESSING PREY

The species of *Lampropholis* are active foragers, rather than waiting in camouflage to ambush passing prey so presumably their effectiveness as predators probably is related to their sensory modes of detecting and/or tracking prey (see Chapter 7), the extent of their visual fields in different types of vegetation, and the behaviour and defensive mechanisms of potential prey. Also, the more ground they would cover during their foraging, the greater would be the number of potential prey they would encounter. In addition, their success in predation surprisingly also depends on the temperature to which the eggs from which they hatched were subjected. Bilcke *et al.* (2006) divided the finding and processing of prey by *L. delicata* into four categories that they termed 'mobility' (the amount of ground covered per unit time), 'capture time' (the time between finding the prey and its first successful seizure by the lizard), 'subdue time' (time between the first successful seizure of the prey and the beginning of swallowing it), and the 'handle time' (the time between the beginning of swallowing the prey and its complete disappearance into the lizard's mouth). They also recorded 'failures' (unsuccessful attempts to capture prey), 'drops' (subsequent releases of the prey after capture), and 'shakes' (jerky movements of the head and/or body of the lizard after the prey was captured and before swallowing it; see Chapter 4).

Lizards from eggs that had been incubated at 28°C had greater mobility, failed less often in attempts to catch prey, allowed fewer prey to escape after capture, caught prey items in less time, subdued them more quickly, and handled (swallowed) prey faster, when at an ambient temperature of 28°C than they did under cooler conditions (20°C). Lizards from eggs incubated at 20°C showed the inverse relationship and were more efficient in all the above phases of feeding under an ambient temperature of 20°C than when they were under 28°C, except for two (subduing and swallowing prey), activities that they performed equally well at high and low ambient temperatures. Shaking of the prey did not conform to either of these two patterns. Hot-incubated lizards shook their prey at a lower rate, rather than a higher one under warmer ambient conditions, whereas cold-incubated lizards shook prey at the same rate at both 20°C and 28°C.

It would seem from the above results that the temperature at which eggs incubate constitutes a major influence on the subsequent efficiency of feeding by the hatchlings. This conclusion is an oversimplification, however, as Bilcke *et al.* (2006) also showed that the body size of the hatchlings significantly influences their efficiency in all phases of feeding, except for the times to subdue and to swallow prey (no effect of temperature on efficiency of feeding in these two phases was identified). There seems to be at least three ways that the two operative stimuli may affect the efficiency of feeding: (1) with body size and temperature of incubation, each exerting a direct, independent influence on efficiency of feeding; or (2) with temperature of incubation influencing the body-size of hatchlings that, in turn affects the efficiency of feeding, i.e. effect of temperature being indirect through its influence on the size of the body, and (3) in an even more indirect way, with temperature directly affecting the duration of incubation, which, in turn, affects the size of hatchlings that finally has an impact on feeding efficiency. Shine (1983) and Downes and Shine (1999) found that low temperatures during incubation result in longer time of incubation and in larger sizes of the bodies of the lizards at the time of hatching. Even more complex scenarios are possible. For example, the temperature of incubation could affect both duration of incubation and size of hatchlings directly, as well as exerting an indirect effect on size of hatchlings via an influence on duration of incubation. There is much yet to learn about the environmental effects on developmental processes.

DIET OF INTRODUCED *LAMPROPHOLIS DELICATA* IN HAWAII

Smith (2016) pointed out that in Hawaii, although most of the species that *L. delicata* would encounter there would be different species than those occurring in its native habitat, its generalist diet would ensure that its abundance would not be affected by the taxonomic composition of the Hawaiian arthropod fauna. Rather, the fact that this species eats almost any prey within a size range it can manage means that it could be a competitive threat to other small lizards in Hawaii.

HYPOTHESES GENERATED AND MEANS OF TESTING THEM

There are two approaches to science, inductive and experimental, both of which are valid, but only the latter currently is in vogue. The inductive method is most effective when little or nothing is known about the taxon under consideration. It does not involve answering a specific question but rather of assembling data over a wide spectrum so as to be able to frame a meaningful hypothesis. This currently is perceived by many as mere 'data collecting' and considered as an inferior procedure not worthy of modern scientific study. Yet, many of the major advances in biology and other fields have been achieved by that means and it is a necessary first step for topics that are virtually unstudied. The best example is the formulation of the concept of natural selection as the mechanism driving evolution. Today, Darwin's *Origin of Species and Descent of Man* (1859) would be rejected as a PhD thesis because he 'didn't ask a question' or formulate a preformed hypothesis for testing. His exhaustive, careful observations without a particular plan in advance, and the subsequent interpretation of the massive amount of information thus gathered formed the basis of the most important contribution to biology ever made. In general, as the base of knowledge widens sufficiently to formulate an intelligent question, the focus can narrow and a transition made from inductive to experimental methods. For example, knowing what we now do about the diets of the species of *Lampropholis*, informed hypotheses can be made and tested.

Both authors of this book are retired and now concentrate on publishing their backlogs arising from previously collected data, rather than engaging in additional experimentation or supervising the programs of PhD students. Consequently, we leave the resolution of the questions we have raised in this chapter to the next generation of herpetologists and ecologists.

REFERENCES

Belmont CM (1977) Resource partitioning and habitat selection in ground skinks. BSc (Hons) thesis. University of New England, Armidale, NSW.

Bilcke J, Downes S, Büscher I (2006) Combined effect of incubation and ambient temperature on the feeding performance of a small ectotherm. *Austral Ecology* **31**(8), 937–947. doi:10.1111/j.1442-9993.2006.01663.x

Brown GW (1983) Comparative feeding ecology of South-Eastern Australian scincids. PhD Dissertation. La Trobe University, Melbourne.

Brown GW (1991) Ecological feeding analysis of south-eastern Australian scincids (Reptilia: Lacertilia). *Australian Journal of Zoology* **39**(1), 9–29. doi:10.1071/ZO9910009

Chapple DG, Slavenko A, Tingley R, Farquhar JE, Camaiti M, Roll U, Meiri S (2023) Built for success: distribution, morphology, ecology and life history of the world's skinks. *Ecology and Evolution* **13**(12), e10791. doi:10.1002/ece3.10791

Crome B (1981) The diet of some ground-layer lizards in three woodlands of the New England Tableland of Australia. *Herpetofauna* **13**(1), 4–11.

Darwin C (1859) *Origin of Species by Means of Natural Selection, or the Preservation of Favored Races in the Struggle for Life*. Undated reprint of the Sixth London Edition. A. L. Burt, Publisher, New York.

Downes SJ, Shine R (1999) Do incubation-induced changes in a lizard's phenotype influence its vulnerability to predators? *Oecologia* **120**(1), 9–18. doi:10.1007/s004420050827

Green RH (1965) Two skink lizards newly recorded from Tasmania. *Records of the Queen Victoria Museum* **19**, 1–4.

Greene HW, Jaksić FM (1983) Food-niche relationships among sympatric predators: effect of level of prey identification. *Oikos* **40**(1), 151–154. doi:10.2307/3544212

Greer AE (1979) A phylogenetic subdivision of Australian skinks. *Records of The Australian Museum* **32**(8), 339–371.

Lunney D, Ashby E, Grigg J, O'Connell JM (1989) Diets of scincid lizards *Lampropholis guichenoti* (Dumeril & Bibron) and *L. delicata* (De Vis) in Mumbulla State Forest on the South Coast of New South Wales. *Australian Wildlife Research* **16**(3), 307–312. doi:10.1071/WR9890307

Martin LJ (2015) Selective foraging behaviour in the scincid lizard *Lampropholis guichenoti*. *Australian Zoologist* **37**(4), 508–509. doi:10.7882/AZ.2014.026

Metcalfe DC (2013) *Lampropholis delicata* (Common Garden Skink): coprophagy. *Herpetological Bulletin* **124**, 23.

Miller KA, Duran A, Melville J, Thompson MB, Chapple DG (2017) Sex-specific shifts in morphology and colour pattern polymorphism during range expansion of an invasive lizard. *Journal of Biogeography* **44**(12), 2778–2788. doi:10.1111/jbi.13075

Rose AB (1974) Gut contents of some amphibians and reptiles. *Herpetofauna* **7**, 4–8.

Shine R (1983) Reptilian viviparity in cold climates: testing the assumptions of an evolutionary hypothesis. *Oecologia* **57**(3), 397–405. doi:10.1007/BF00377186

Simbotwe MP (1985) Comparative ecology and demography of two populations of Australian reptiles (Lacertilia, Scincidae). PhD Dissertation. University of New England, Armidale, NSW.

Smith AH (2016) Invasion ecology of the plague skink (*Lampropholis delicata*) in Hawaii. MSc thesis, University of Hawaii, Manoa.

Warner FJ (1995) A comparison of the dietary contents of two sympatric Scincid lizards, *Lampropholis delicata* (de Vis, 1888), and *Lampropholis guichenoti* (Duméril and Bibron, 1839) (Lacertilia, Scincidae). BAppSc (Hons) thesis. University of Western Sydney, Hawkesbury.

Webb GA (1983) Diet in a herpetofaunal community on the Hawkesbury Sandstone Formation in the Sydney area. *Herpetofauna* **14**(2), 87–91.

Webb GA (1995) Diet and food selection in a community of small ground-dwelling lizards and frogs in south-eastern Australia. *Herpetofauna* **25**, 36–44.

9
Predation on *Lampropholis*

INTRODUCTION

Numerous external factors influence a skink's ability to evade predation. An individual's attributes may combine in ways that either increase or decrease the chance of predation.

As with other reptiles, *Lampropholis* survive by constantly trading off the need to forage for resources against the risk of predation. Being heliotherms, at first emergence in the morning, they immediately seek an area of open sunlight to bask (see Chapter 6), and consequently are exposed to view. When they have warmed sufficiently to forage, they move about in search of food and their activity may attract attention. Social activities such as mating or fighting may also attract the attention of predators and at the same time may distract the participants in these activities from approaching dangers. Howard *et al.* (2003) suggested that the preference of *L. delicata* for a habitat with loose structure of the litter reduced the conflicting need of exposure for foraging and basking on the one hand to shelter from predators on the other.

One way of assessing the pressure of predation upon a population is to list the potential species of predators that are syntopic with the species of prey of interest and examine the dietary habits of those predators as a means of identifying their relative importance. The senior author and her students conducted such research on the Hawkesbury Campus of Western Sydney University (hereafter 'Hawkesbury Site'). Clearing of the native vegetation and the development of dams on the site commenced in 1891 in preparation for the foundation of the Hawkesbury Agricultural College (now the Hawkesbury Campus of Western Sydney University). The land where the studies were carried out was in remnants within the farmland matrix (Burgin 2010). The results of these, and other similar studies, will be recounted for the relevant available taxa in each appropriate section below.

PREDATION BY BIRDS

Seven species of birds (Table 9.1) have been observed to prey on the two common species of *Lampropholis*. At the Hawkesbury Site the small skinks were predominately *L. delicata* and *L. guichenoti* but two other species, *Cryptoblepharus virgatus* and *Saproscincus mustelinus*, also were included in the analyses. However, the latter two species were present in very low numbers (Anderson and Burgin 2002, 2008) and thus would have had an insignificant effect on the outcome of the results.

Data were collected from eight remnant woodlands ranging in size from 20 to 176 ha. Contemporary research revealed that the microhabitats of these skinks did not differ significantly along a gradient from the edge to the core of the study site (Baiada 2002). This result is in contrast with other localities, for example where Kremaster and Bunnell (1999) determined that the influence of ecotones is influential within 40 m of the edge. Eight sites, four on the perimeter and four in the core, were randomly sampled in mid-morning (when birds and these skinks are both active) in the autumn and spring of 2001 and the summer of 2002. The distribution of trees and the limited intrusion of weeds into these long-established remnants allowed for equivalent sight for point-of-transects at both edge and core, and yielded data that were more accurate than obtained from pitfall trapping (Anderson 2007; Anderson and Burgin 2008).

Table 9.1. Number of each observed species of avian predator on *Lampropholis* recorded at the edges and cores of woodland remnants on the Hawkesbury Campus of Western Sydney University in 2001–2002.

N = Number of samples; SD = standard deviation; ns = not significant; ** significance level = 0.01; *** significance level = 0.001. Table from Anderson (2007) with author's permission.

Species (common name)	Edge (N)	Edge mean (SD)	Core (N)	Core mean (SD)	Significance level
Corcorax melanorhamphos (white-winged chough)	56	14 (15.10)	57	14.3 (6.40)	0.82 ns
Corvus coronoides (raven)	58	14.5 (11.09)	30	7.5 (1.92)	48.91**
Cracticus torquatus (grey butcherbird)	34	8.5 (3.97)	5	1.3 (0.96)	21.57***
Gymnorhina tibicen (magpie)	43	10.8 (8.18)	8	5.3 (6.65)	32.32***
Manorina melanocephala (noisy miner)	101	25.3 (7.37)	21	5.3 (6.65)	49.24***
Dacelo novaeguineae (kookaburra)	17	4.3 (5.32)	12	3 (3.16)	0.86 ns
Sturnus vulgaris (European starling)	4	1 (2)	0	0	-

The numbers of *Lampropholis* were greater in the inner areas of remnants but, by contrast, the distribution, abundance, and diversity of their avian predators were greatest on the edges (Anderson 2007; Anderson and Burgin 2008). As observed elsewhere (Barrows *et al.* 2006; Chace and Walsh 2006), these predators cross-forage between the urban and peri-urban areas near the remnants. Individuals of one species, *Manorina melanocephala* (noisy miner), were quite common in the remnants sampled. This species is known as a 'honeyeater' but it supplements its diet with arthropods and small reptiles (Grey *et al.* 1998; Parsons *et al.* 2006) and it may occur in greater numbers where it cross-forages between local backyards where 'meat' is available than where only seeds and/or nectar are present. Thus, this species may be more carnivorous than previously recognised. For example, Anderson (2007) observed noisy miners successfully preying upon *Lampropholis*.

Other modifications by humans in the area (e.g. fences, power lines) were also advantageous for birds. They provided convenient perches for magpies, butcherbirds, and kookaburras from which to locate prey (Sewell and Catterall 1998; Anderson and Burgin 2008).

Four of the seven species of birds preyed on *Lampropholis* in significantly higher numbers at the edge than in the core of remnants; another species (*Sturnus vulgaris*, European starling) also occurred more commonly at the edge of remnants but was present only in small numbers. *Corcorax melanorhamphos* (white-winged chough) was encountered equally between the edge and core (Table 9.1). There were fewer *L. delicata* and *L. guichenoti* at the edge of remnants of woodland than in the core, whereas, collectively, predatory birds were in greater abundance and diversity at the edges of these remnants than within them (i.e. there was a strong negative correlation between the numbers of birds and numbers of skinks). These results suggest that birds supplement their diet by cross-foraging between the edge of natural areas and nearby urban areas (Anderson and Burgin 2002, 2008; Barrows *et al.* 2006; Chace and Walsh 2006). Certainly, fragmentation of habitats such as that occurring in north-western Sydney, provides opportunities for predators to cross-forage between human-modified areas (urban and/or peri-urban) and native habitats, typically resulting in increased numbers of avian predators at the edge of remnants. Using decoys that mimicked skinks, it was confirmed that the rate of strikes was significantly higher at the edges of remnants than in their cores. These data confirm previous observations of greater predation pressure at ecotonal borders (Caterall *et al.* 1991; Saunders *et al.* 1991; Berry 2001; Major *et al.* 2001).

When a species is introduced into a new location it not only faces a new assemblage of potential competitors and perhaps somewhat different physical environments, but also perhaps an unfamiliar set of predators. The latter aspect of the ecology of invasions has received less attention than other topics. Two species of birds, however, are known to prey on introduced *L. delicata* on Lord Howe Island. One of these is the buff-banded rail (*Gallirallus philippensis*) observed by Schulz (2009) to capture an adult *L. delicata* and feed it to its chick. This species of bird also occurs in *L. delicata*'s native home, so perhaps it was not a novel predator for the population on Lord Howe. The other predator was the Lord Howe Island currawong (*Strepera graculina crissalus*) a species closely related to the native Australian pied currawong (*Strepera graculina graculina*) so again probably not a species significantly different from the one occurring in the skink's native land.

PREDATION BY REPTILES

Snakes

Snakes are notorious for consuming a considerable range of sizes of prey, including some that are large in relation to the size of the snake's own body. Shine (1977) noted that the upper size limit of prey may increase with the increasing size of the predatory snake and Greer (1997) found that an adult, white-lipped snake (*Drysdalia coronoides*; weight 11.9 g) readily accepted hatchlings of *L. delicata* (~0.12 g) while also consuming an adult *Ctenotus robustus* (weight 5.48 g), a difference in size of prey between 1 and 45% of the snake's body weight. The eastern small-eyed snake (*Cryptophis nigrescens*) forages widely for inactive prey and does not discriminate between large and small *L. guichenoti*. Some other snakes appear to be more selective. The yellow-faced whip snake (*Demansia psammophis*) actively forages and most frequently consumes larger, rather than smaller *L. guichenoti*; the death adder (*Acanthophis antarcticus*) that employs a sit-and-wait strategy most frequently consumes smaller individuals. Although the sizes of prey consumed may be based directly on an active response to body size or on certain integumentary chemicals (Wall and Shine 2009), other choices may be artefacts. Downes (2002a) concluded that some preferences for size of prey were not based on active choice but rather upon species-specific predatory tactics and the behaviour of the prey (*L. guichenoti*), which varied between small and large individuals. For example, small individuals are much more active than are larger ones and may attract greater attention. The vulnerability of *L. guichenoti* to predation, therefore, may have been influenced by a combination of the behavioural characteristics of the prey and the foraging habits of the predator, rather than on size *per se*.

Most of the above studies of predation by snakes upon *Lampropholis* have been undertaken under controlled conditions, with numbers of predators and prey manipulated appropriate to the research hypothesis. The relative abundance of predator and prey after perturbation, for example fire – a common occurrence in extensive areas of the Australian environment – has received limited attention. In one such study, however, Foster *et al.* (2016) investigated the response of reptiles to a matrix of burnt versus unburnt habitat and observed that, despite *C. nigrescens* being a predator upon *Lampropholis*, both *L. delicata* and its predator were more common in burnt areas compared to unburnt ones. That *L. delicata* sought more exposed (i.e. post-fire) habitat appears incongruous with other behavioural observations of the species. For example, in open woodland at the Hawkesbury Site, there were fewer *L. delicata* and *L. guichenoti* in perimeter areas of open bushland than in the core (Jelbart 1998; Anderson and Burgin 2002; Anderson 2007) despite environmental conditions (e.g. temperature, humidity) being equivalent throughout the woodlands (Baiada 2002).

In the study by Foster *et al.* (2016), mentioned above, refuges in burnt areas likely had less shading from overhanging vegetation than did unburnt areas and it was suggested that the thermal characteristics of these retreats would have been more suited both to predator and to prey than those within unburnt areas. However, both prey species cope within a broad range of ambient temperatures. For example, Spellerberg (1972) determined that the critical thermal range was 4.6–41.3°C for *L. delicata* and 1.9–41.3°C for *L. guichenoti*. It is thus unlikely that ambient temperature within refuges was the influential factor in the selection of artificial retreats in open areas except under the most extreme conditions.

The co-location of both predator and prey under artificial refuges in burnt areas may have resulted from a lack of alternative refuges post-fire (Foster *et al.* 2016). For example, Croft *et al.* (2010) found that *L. delicata*, *L. guichenoti*, and *C. nigrescens* were less likely to use charred timber as a refuge than uncharred retreats. Foster *et al.* (2016) observed that these species – predator and prey – sought the unburnt artificial refuges over potential refuges that had been burnt. Although, intuitively, it could be assumed that avoidance of predation would be a major contributing factor in the selection of preferred retreats, apparently, that is not always the case.

Despite the observations above, Foster *et al.* (2016) noted that post-fire, there was a positive correlation between large numbers of skinks (*L. delicata* and *L. guichenoti*) and their predator, *C. nigrescens*, in artificial refuges in burnt areas. In addition to the direct impact of predation by *C. nigrescens* upon *Lampropholis*, the presence of the snake's odour may have longer-term impacts. Downes (2001) observed that for the first 6 months after hatching, *L. guichenoti* maintained in enclosures with scent levels that mimicked a high density of a natural snake predator, emerged and became active later in the day, were less mobile, and selected safer microhabitats (substrates) than those that were raised in areas without the predator's scent. Such behaviour reduced opportunities for foraging and basking. However, over time,

Table 9.2. The time before *Lampropholis guichenoti* initiated activity, the frequency with which they basked, and their level of activity.

Lizards were sampled from predator-scented and control-scented enclosures (see text) and transferred to cages with substrates either of fresh paper or as paper impregnated with the odour of the predatory snake.

		From heavily snake-scented enclosure: Mean ± Standard error		From control-scented enclosure: Mean ± Standard error	
Characteristic	**Type of paper**	**Hatchlings**	**Yearlings**	**Hatchlings**	**Yearlings**
Time before activity was initiated (minutes)	Fresh	41.3 ± 16.4	51.1 ± 17.6	37.4 ± 13.3	57.7 ± 16.0
	With snakes' scent	66.3 ± 15.1	58.1 ± 14.4	34.4 ± 11.7	56.5 ± 14.0
Number of moves per trial	Fresh	128.1 ± 16.8	135.7 ± 20.4	175.0 ± 18.7)	124.1 ± 23.6
	With snakes' scent	112.8 ± 13.7	112.6 ± 21.2	151.3 ± 10.3	90.4 ± 16.8
Times basked per trial	Fresh	15.0 ± 1.0	26.8 ± 2.1	15.1 ± 1.1	25.0 ± 2.3
	With snakes' scent	9.5 ± 1.3	28.3 ± 2.1	10.3 ± 1.4	19.8 ± 1.9

individuals became less sensitive to the snake's odour. After 1 year, activity patterns of the skinks that were exposed to the predator's odour showed the same behavioural patterns and use of substrate as those that had not been exposed. However, at the end of 6 months there had been differences in growth rates between the two treatments – *L. guichenoti* exposed to the predator's scent grew more slowly, weighed less, and were shorter at maturity. Such individuals, therefore, attained a smaller body size; they also produced fewer offspring. Such exposure to the scent of a predator clearly had reduced long-term fitness. These disadvantages would, presumably, render such individuals more susceptible to predation when predators did not discriminate between prey of different sizes.

When a diurnal lizard, such as *L. delicata*, retires to its shelter for the night, not only may it become more vulnerable to nocturnally foraging predators, such as *C. nigrescens*, that can gain access to the shelter, but it may also encounter its own prey under the same objects. Leung *et al.* (2006) conducted an experiment to find out whether the odour of either their potential prey and/or their potential predators could influence their choice of shelters. The lizards were subjected to three types of shelters, one scented with the odour of *C. nigrescens*, another with the odour of one of their prey, the cricket *Acheta domesticus*, and the third an unscented control. When offered only control shelters only 50% of the skinks sought shelter and then only after they explored the arena. By contrast, in response to the odour either of prey or predator 96% sought immediate cover but did not discriminate between scented and unscented cover. The authors suggested that the scent either of a predator or of prey prompts *L. delicata* to move under cover, but to go to any cover regardless of the location of the scent.

Lamprophlis guichenoti also was subjected to a study of its responses to the scent of predators. Downes (2001) reared hatchlings to maturity in outdoor enclosures in which the optimal sites for basking and foraging were covered with the scent of a species of snake (*Demansia psammophis*) whose diet consists exclusively of lizards (*Demansia psammophis*). There were two types of enclosures, one mimicking high densities of the snakes (heavily scented habitat) and the other lightly scented as a 'control-scented' habitat mimicking low densities of the snake. The refuges in which the lizards spent their periods of inactivity were not scented. Neonates that were hatched in the laboratory were tested early in life and again as sexually mature yearlings. Many aspects of these lizards' lives were affected by their exposure to high levels of the snakes' odours. (1) During the early months of their life fewer lizards in the heavily scented enclosure emerged from shelter during the day compared with their counterparts in the control-scented enclosure; however, over time, the percentage of lizards steadily increased in the heavily scented enclosure. (2) Of the lizards that were outside of shelters a consistently smaller percentage of those from predator-scented enclosures were mobile compared with lizards in control-scented enclosures, although the percentage that were mobile when abroad increased over time. (3) When put into cages on paper that was heavily scented by the snake's odour versus on unscented (fresh) paper, hatchlings from control-scented enclosures initiated activity at the same time as they did on unscented paper but those from heavily scented enclosures later became more active on scented paper than on fresh paper (Table 9.2). (4) Hatchlings from both types of enclosures had higher levels of activity (number of moves per trial) on unscented paper than those on papers scented with snake's odour but yearlings from both types of enclosures, having experienced prolonged exposure to their respective concentrations of odour, initiated activity similarly in both types of test cages (Table 9.2). (5) Hatchlings from both types of enclosures basked more frequently when on fresh papers than when on scented papers, and so did yearlings from control-scented enclosures, but those from heavily scented enclosures showed little difference in frequency of basking between fresh and scented papers (slightly higher on scented) (Table 9.2). (6) Neither a lizard's environment during rearing nor the age at which it had been exposed to the snake's odour affected its speed of sprinting. (7) Lizards raised in predator-scented enclosures and control-scented enclosures used different substrates but only during the initial few months of the experiment; afterwards they used the same microhabitats.

Another factor that may influence the susceptibility to predation is polymorphism in colouration of the skin. (see Chapter 5). Forsman and Shine (1995) investigated temporal and spatial variation in the dimorphic colour pattern (i.e. presence/absence of a distinct mid-lateral white stripe) in *L. delicata* from localities within 50 km of the coastline in eastern New South Wales. In both sexes a decrease in the relative frequency of the striped polymorph resulted in an increase in the incidence of regenerated tails (55–77%). Additionally, striped females appeared to be more likely to survive and they produced larger clutches than did unstriped ones. Experimentally, there was a difference between males and females in survival rate between the two morphs. Striped males were less likely to survive than were the non-striped phenotype. The authors suggested that gravid females with a white lateral stripe were less susceptible to predation than those without it. By contrast, males of this morph were more susceptible to predation.

Shine (1977) listed the prey of the elapid snakes of New England, NSW; five species from this area are known predators of *Lampropholis*: *Austrelaps ramsayi* [nomenclature updated]: highland copperhead, *Hemiaspis signata, Pseudechis porphyriacus, Pseudonaja textilis*, and *Unechis gouldii* [now *Suta dwyeri*]. Webb (1987) reported *L. guichenoti* from the stomach contents of *A. ramsayi*.

Five species of snakes have been recorded from the Hawkesbury Site where *Lampropholis* were studied and a further three species have been predicted to occur at this locality although they have not been recorded there (*Furina diadema, Hemiaspis signata*, and *Cryptophis nigrescens*) (Burgin 2010). In the past, some staff at that site chose to kill snakes despite protection of snakes under New South Wales legislation. Over recent decades, however, the focus of the institution on agricultural activities has declined together with numbers of outdoor staff and the campus has become a more secure haven for snakes. In addition, although not widely known, local snake catchers do release snakes removed from nearby peri-urban/urban areas into the bushlands of the campus or those nearby. Overall, however, snakes have remained uncommon (Burgin, pers. obs.).

Of the five species of snakes known from the area, Cogger (2018) listed only one of them, *Demansia psammophis* (yellow-faced whip snake) to have a diet consisting mainly of small skinks. Of the other four species one of them, *Morelia spilota* (diamond python), is an ambush predator and adults typically prey upon mammals, although Slip and Shine (1988) reported that reptiles made up 23% of the diet of juveniles. Species-level data are not available but since *Lampropholis* are the most common skinks across much of this python's range, juvenile snakes may ambush them. Fearn *et al.* (2002) found that although *M. spilota* consumes small skinks such predation was not common; of 258 pythons collected from the urban areas of Brisbane and Ipswich, only one small individual was found to have consumed a lizard despite *L. delicata* and/or *L. guichenoti* being common throughout the area.

Dendrelaphis punctulatus (green tree snake) tends to be diurnal (Shine 1991; Gray *et al.* 2020) and, as its common name implies, spends much of its time in the canopy of trees and shrubs although, to a lesser extent, it may forage at ground level where *Lampropholis* also forage. Cogger (2018) listed this snake's diet as predominately frogs and birds. Others, however (Shine 1991; Shea 2010; Fearn and Trembath 2011), reported that these snakes also consume lizards. Webb and Rose (1984) reported *L. delicata* from the stomach of an individual of this species.

Pseudonaja textilis (the eastern brown snake) preys on a diversity of vertebrates that include frogs, reptiles and their eggs, birds, and mammals, although smaller individuals (70 mm SVL) consume proportionately more ectothermic prey such as small skinks than do larger individuals (Beatson 2022).

Demansia psammophis consumes predominantly small diurnal lizards, their eggs, and frogs. Unlike sit-and-wait predators, this species may chase and capture fleeing lizards (Anonymous 2020). Some species of snakes that were not documented as present at the Hawkesbury Site (although they may have been there) also are known, or potential, predators of *Lampropholis*. One such species, *Cacophis squamulosus* (golden-crowned snake) is secretive (Cogger 2018) and nocturnal (Llewelyn *et al.* 2005), and this may account for the species not being observed at the Hawkesbury Site. Where present, however, it likely is a substantial predator of *Lampropholis*. For example, Shine (1980) found that more than 70% of its prey consisted of either *Lampropholis* or *Leiolopisma*. *Cacophis squamulosus* also consumes lizards' eggs.

Can *Lampropholis* detect snakes by odour?

Downes and Shine (1999) compared the response of naïve hatchlings of *L. delicata* to cotton swabs impregnated with integumentary chemicals from the predatory *Drysdalia coronoides* (white-lipped snake) to their reaction to a neutral control of distilled water, and to a pungency control of a commercial cologne. They found that lizards tested at the lower ambient temperatures of 18° and 23.5°C did not flee in response to the snake's odour and only a small proportion fled even at the higher test temperature of 29°C (Table 9.3). The lizards incubated at 27°C did not respond to the snake's odour at the lowest temperature, but the proportion of them fleeing increased with progressively increasing test temperatures. Thus, these lizards are more vulnerable to predation at low ambient temperatures than at higher ones but are somewhat better protected by previous exposure to higher temperature during incubation as embryos.

Naimo *et al.* (2021) tested the response of *L. delicata* to the odour of one of its natural ophidian predators, *Pseudechis porphyriacus* (Australian red-bellied black snake), from its original homeland, following 40 generations of isolation from it after introduction of the lizard onto Lord Howe Island. As a control they tested lizards' responses to a predator common to the original natural habitat in Australia and to the invaded one on Lord Howe Island (the huntsman spider, *Heteropoda* sp.). They recorded the time it took for the lizards to emerge from a shelter and begin to either bask or forage in the presence and absence of the odor of each predator's scent. Surprisingly, even after such an extended period of isolation from the predatory snake, there was no difference in the response between native and invasive skinks. In a different experiment, when the lizards were subjected to a simulated attack by a bird, with the option of fleeing to either a snake-scented or an unscented shelter they did not avoid the one with snake's scent. The only difference that was detected between the Australian skinks and the invasive ones on Lord Howe Island was that prior to exposure to cues was a difference in activity and exploratory behaviour between the two groups. Clearly, even after long isolation from predatory snakes there had been no change in response to the odour of *P. porphyriacus*. Since neither group avoided the odour of this species it may mean that *L. delicata* never did have a negative response to *P. porphyriacus* in its native homeland despite it being known to be preyed upon by this species. The species to which a negative response was given by *L. delicata* in the experiment by Downes and Shine (1999; see above) was not the same one that was used in the experiment by Naimo *et al.* (2021) and it may be that *Drysdalia coronoides* is a more common predator on *L. delicata* than the other, and that *L. delicata* may develop a negative response only to the odour of a common predator.

Table 9.3. The effect of incubation temperatures on the subsequent fleeing-response of *L. delicata* to the odour of the white-lipped snake, *Drysdalia coronoides*, as tested at different ambient temperatures.

No lizards fled from either the neutral or pungency controls. Data from Downes and Shine (1999).

Ambient temperataure during test (°C)	Per cent of lizards fleeing from swab containing odour	
	Incubated at 20°C	Incubated at 27°C
18°C	0	0
23.5°C	0	12
29°C	2	29

Because *Lampropholis* can detect the odour of snakes, one might expect them to engage in antipredatory behaviours in response to the scent, indicating a high risk of predation. Downes and Hoefer (2004) compared the antipredator responses of *L. delicata* in the absence of snake's odour (low risk) to its presence (high risk). At high risk, the time individuals spent in antipredator behaviour, such as responsivity, immobility, and hiding, decreased with increased number of lizards in a group, presumably because a snake can only process one prey at a time, thereby reducing the chances of a given individual lizard becoming a victim. Hence basking in groups rather than as individuals is adaptive.

Can snakes detect *Lampropholis* by odour?

The strength of a snake's response to potential prey can be gauged by the number of times it flicks its tongue (picking up scent) when presented with chemical (olfactory) cues of the lizard. Downes and Shine (1999) found that *Drysdalia coronoides* responded more strongly to some species of skinks than to others of similar size and ecological niche. Its response to *L. delicata* (32.9 ± 5.1 tongue-flicks/min) and *Saproscincus mustelina* (33 ± 4.9 tongue-flicks/min), two rather closely related species, was stronger than it was to *A. maccoyi* (26 ± 4.9 tongue-flicks/min) (for a discussion of the effect of social groupings of *Lampropholis* on the responses of individuals to threat from snakes, see Chapter 4).

Downes and Shine (1998) found the preferred prey of juvenile broad-headed snakes (*Hoplocephalus bungaroides*) to be the gecko *Amalosia lesueurii.* When offered the choice between that species and equally sized *Lampropholis* spp. the snakes consistently chose the former. Adults also prefer the gecko but will take mice as well. To test whether neonate snakes could distinguish various odours, Downes (1999) presented them with cotton swabs scented with a novel odour (cologne), the scent of mice, the scent of *Amalosia*, and the scent of *L. delicata*, compared to a control (cotton swab with distilled water). The number of tongue-flicks in response to the control was low and somewhat higher, but still low, for the odours of cologne and mice. By contrast, the number of flicks of the tongue in response to the scent of *Lampropholis* was more than double that for mice, and it almost doubled again in response to the scent of the gecko. Thus, the snakes not only could distinguish between the odours of prey and non-prey, but could distinguish between the species of prey, giving the strongest responses to the preferred species.

This species of snake retreats under objects at night and an experiment was performed by giving them a choice of retreats (slabs of artificial sandstone), with the slabs scented with the same scents used for the experiment for tongue-flicks. Once the snake had retreated under a slab (initial retreat), the slab was removed and the next morning its choice of its second slab was recorded (final retreat). As in the experiment with tongue-flicks, there was a graded response, both for low and final retreats, from low and practically equal choices for retreats smelling of cologne or mice, a higher number for those scented with the odour of *Lampropholis* and finally a still higher choice for retreats with the odour of the gecko. Thus, in both types of experiments responses were weak to a novel and a prey that was not yet taken by neonates but much stronger in ascending order of preference of prey.

Vulnerability of hatchling *Lampropholis delicata* to predation by snakes

Downes and Shine (1999) staged trials in which hatchling *L. delicata* of two different ages post-hatching, and incubated at different temperatures, were offered as prey to neonatal white-lipped snakes (*Drysdalia coronoides*) as a means of ascertaining the factors influencing the vulnerability of this species of lizard to a potential predator. The predator-prey interactions were complex as they involved the attributes both of predator and prey, as well as maturation of the prey over time, and the influence of different thermal histories. Finally, there was an interaction of various of these factors with each other.

Some aspects of the relationship were consistent between the different temperatures at which the prey had been incubated. For example, at both incubation temperatures a higher percentage of the younger hatchlings fled from the odour of the snakes than did older ones, but within each age-group more of those incubated at the higher temperature fled than did those incubated at the lower temperature (Table 9.4). From this analysis alone, one would expect that a greater proportion of hatchlings incubated at low temperature would be caught by snakes than would those incubated at a higher temperature, merely because fewer of them would flee from the snake's odour. It is uncertain, however, whether fleeing is an advantage or a disadvantage initially, as movement may attract the attention of a potential predator when remaining immobile may avoid detection – it depends on whether the predator was relying on vision or olfaction for detection of potential prey (Downes and Shine 1999).

The hypothesis of an advantage to being incubated at higher temperature was reinforced by the finding that for both ages of hatchlings, the snakes missed a greater proportion of their strikes against the lizards incubated at the higher temperature than they did against those incubated at the lower temperature and by the fact that for both incubation temperatures the time to subdue and eat 6-week-old lizards was greater than for 2-week-old ones, thereby giving the older ones greater opportunity to escape. These advantages to the older hatchlings, however, were offset by two other results: (1) a greater percentage of those that were 6 weeks old and incubated at 27°C a were captured on the snake's first attempt than was the case for younger skinks (2 weeks old) under the same conditions, and (2) the percentage of escapes from the initial strike by the snake was lower for the older skinks than for the younger ones (Table 9.4).

Several aspects of the predator-prey relationship did not give a consistent advantage to either age-group but rather the advantage depended more upon the temperature of incubation. For example, of the hatchlings incubated at 20°C, a greater proportion of the 6-week-old ones escaped as a consequence of autotomy than did the

Table 9.4. Effectiveness of neonatal *Drysdalia coronoides* in capturing hatchlings of *Lampropholis delicata* of different ages post-hatching and having been incubated at different temperatures.

Data from Downes and Shine (1999).

	Prey incubated at 20°C		Prey incubated at 27°C	
Trait	**Two weeks old**	**Six weeks old**	**Two weeks old**	**Six weeks old**
Prey captured on first attempt (%)	36	36	7	36
Prey fled from snake's odour (%)	57	14	60	50
Prey struck at and missed (%)	43	14	71	29
Prey escaped by autotomy (%)	21	43	36	21
Time spent handling prey (min)	311 ± 50	339 ± 53	372 ± 52	384 ± 43
Total prey captured (%)	36	50	50	29

2-week-old ones, whereas the values were approximately reversed for those incubated at 27°C (Table 9.4). Similarly, probably the most important measure of all, the total percentage of the potential prey that was captured, also showed this pattern; the values were very similar for the two temperatures of incubation, but reversed in terms of the two age-groups (the values for the 6-week-old individuals incubated at 20°C were the same as those for 6-week-olds incubated at 27°C, and each higher than the values for the opposite age-group. Another complicating factor mentioned by the authors was that the effectiveness of the initially naïve predators may have improved during the experiments. The fact that they consistently required more time to process the 6-week-old prey then the 2-week-old prey, however, suggests that this may not have been a critical factor (Downes and Shine 1999).

Downes and Shine (1999) conducted similar tests on two other sympatric species of small skinks with similar life histories, microhabitats, and thermal preferences (*Saproscincus* [formerly *Lampropholis*] *mustelina* and *Anepischetosia* [formerly *Nannnoscincus*] *maccoyi*) and found interspecific differences in the magnitude and direction of the above kinds of phenotypic shifts. Thus, *L. delicata* can serve as a model against which other taxa can be compared (see Chapter 15).

Lizards

Snakes are not the only reptilian predator of *Lampropholis*. Turner (2010) reported a chase that involved an adult brown-tailed bar-lipped skink (*Glaphyromorphus fuscicaudis*) in determined pursuit for at least 10 minutes of an adult *L. coggeri*. The chase passed through dense leaf litter and over low buttresses of fig trees. During the pursuit, the *G. fuscicaudis* seized the smaller skink in its mouth and violently shook it for several seconds but when it momentarily released its grip the *L. coggeri* fled, apparently unharmed, whereupon the larger skink continued pursuit of its intended prey. This observation appears to be the only record of attempted predation by *G. fuscicaudis* on a species of *Lampropholis*. However, along with numerous other species of Australian skinks, Cameron (1993) reported *G. fuscicaudis* to be 'rare or insufficiently known'. This classification does not appear to have changed in the intervening decades and there does not appear to be any other record of this species being a predator upon *Lampropholis*. There is, therefore, insufficient data to determine whether this genus is a substantial component of the diet of *G. fuscicaudis*. Even if it is, the apparent rarity of *G. fuscicaudis* and the effort expended in its attempt to capture the intended prey, suggests that it is unlikely to be a serious predator of *Lampropholis*.

Meek *et al.* (2001) confirmed that the Australian water dragon *Intellagama lesueurii* uses sit-and-wait tactics to capture and consume *L. delicata*. Over several days, repeated predation upon these small skinks was observed. This predator can detect potential prey from distances up to about 7 m.

Of the 11 species of lizards inhabiting the Hawkesbury Site, eight are sufficiently small that their gape presumably would preclude consumption of all lizards except hatchlings and thus it is unlikely that their predation would have much impact on *Lampropholis*. Only three species were sufficiently large to potentially prey on adult *Lampropholis*. *Tiliqua scincoides* (blue-tongued lizard) is one such species but there is no evidence that it actually does (Burgin 2010).

Another species of large lizard at the Hawkesbury Site is *Eulamprus quoyii* (Eastern water skink; Burgin 2010). Information on this species' diet is scant but as its common name suggests, it mainly forages in moist habitats such as wetlands, small creeks, and riverine vegetation

(Law and Bradley 1990; Cogger 2018). However, it is not restricted to such areas but also occurs in dry sclerophyll forests (Cogger 2018). It consumes fish, small lizards, native fruit (Anonymous 2020), and tadpoles and metamorphs of frogs (Pyke and Miehs 2001). This species would have a broader overlap in range with *L. delicata* than with *L. guichenoti* since the former prefers a moister habitat than does the latter.

The third species of large lizard that Burgin (2010) listed as being resident on-site is *Varanus varius* (lace monitor). It prefers semi-arboreal prey (Jessop *et al.* 2010); however, its diet covers a wide range of prey consisting of insects (including pupae), reptiles, small mammals, nestling birds, and carrion (Weavers 1989; Guarino 2002; Jessop *et al.* 2010; Cogger 2018). During more than 20 years of observation of this species foraging at the Hawkesbury Site (Burgin, pers. obs.), it was never seen to show even a cursory interest in reptiles as small as *Lampropholis*. Another hint that *V. varius* is not necessarily a major predator of skinks was provided by Lloyd *et al.* (2009) who tested the response of three litter-dwelling skinks (not *Lampropholis*) to the odour of two varanid species. None of the skinks responded to the odour of *V. varius* although two species did respond to that of *Varanus tristis*. Thus, the skinks could avoid predation by the latter species. We suggest, therefore, that *Lampropholis* seldom would be consumed by *V. varius*.

PREDATION BY MAMMALS

A range of alien predators, including cats (*Felis catus*), foxes (*Vulpes vulpes*) and domestic dogs (*Canis lupus familiaris*) and/or wild dogs (i.e. those that are cross bred between the domestic dog and the dingo, *Canis lupus dingo*) potentially prey on small skinks, including *Lampropholis* (Meek 1998; Shine and Tamayo 2016).

Slobo-Wilson *et al.* (2021) compared the respective roles of mammals and birds as predators on Australian reptiles and found that a total of 108 species of reptiles (20 of them listed as threatened) had been recorded as consumed by foxes, 263 species (15 threatened) consumed by cats, and 86 species by both. Data on individual species of reptiles were not reported but it was stated that squamate species known to be consumed by foxes were also those deemed likely to be undergoing declines in their populations. The likelihood of predation by foxes was higher for larger squamates than for smaller species, whereas cats tended to prey predominately on the smaller ones (size of prey peaked at 217 g). Thus, cats are more likely to prey upon *Lampropolis* than are foxes.

The domestic cat, *Felis catus*

Cats are the most commonly maintained domestic pet mammal in the developed world (Turner and Bateson 2000) and on a global scale, they are considered among the top 100 of the world's worst invasive alien species (Scanes 2018). Their ownership is deeply entrenched within Western society and generally, including in Australia, there is no limit on the number of cats an individual may own (Lloyd *et al.* 2013). Many are allowed to roam free (van Heezik 2010) and 'unowned' cats inhabit all urban and peri-urban areas of Australia (Swarbrick and Rand 2018). Cats are efficient and abundant predators (Lloyd *et al.* 2013). They are thus significant predators of wildlife in cities (Turner and Bateson 2000) and because many are subsidised predators, they place significant pressure on native populations. Additionally, due to the large number of domestic cats, and their large and flexible home range when feral, predation by cats may have an impact virtually everywhere (van Heezik 2010).

In Australia, cats are considered the reason for the decline, and even the extinction, of some species of native wildlife (Dickman 1996; Risbey *et al.* 2000; Letnic *et al.* 2009; Doherty *et al.* 2015, 2017; McGregor *et al.* 2015; Legge *et al.* 2017; Stokeld *et al.* 2018; Hu *et al.* 2019). Woinarski *et al.* (2018) estimated a loss of ~1.8 million individual native Australian reptiles per day to predation by the combination of feral cats and pet cats. They emphasised that such estimates are very rough and that the impact of cats differs among investigators and is subject to at least several sources of error: (1) the number of reptiles consumed by feral cats is higher in drier areas [beyond the range of any *Lampropholis*] and in drier years compared to wetter ones, (2) cats may kill snakes that also prey upon skinks, thereby reducing predation-pressure upon skinks from that source, and (3) the impact of pet cats on skinks may be underestimated because of their undetected consumption of skinks completely in the field without bringing partial remains home.

As devastating as cats are to reptiles generally, cats' dietary preferences are variable and in Australia, taxa other than skinks collectively make up a greater dietary component (Barratt 1997; Meek 1998; Murphy *et al.* 2004; Dickman 2009) than do skinks.

In an online survey that investigated the prey taken by pet dogs, Franklin *et al.* (2021) found that 41% (*n* = 1,346) of their respondents also were cat owners. Of these, 36% observed that their cat captured reptiles, a number similar to those whose cat(s) caught birds. The median number of individual reptiles caught was four (range 1–150). Almost all these captures of reptiles by cats (97%) over the 6 months of the study were 'small reptiles' compared to <10% in that category for pet dogs (see section *Dingoes, domestic dogs, and their crosses* below).

Unlike small skinks that often were not recorded even to generic level by Franklin *et al.* (2021), geckos and the larger reptiles were much more likely to be recognisable and thus were more frequently identified to species. However, species of *Lampropholis*, particularly *L. delicata* and/or *L. guichenoti* are by far the most common small reptiles in urban areas in most of the regions from which the responses were drawn. It is well known that cats consume *Lampropholis* (Griffiths 1987; Smith and Smith 2013; Dickman and Newsome 2015) and it is likely that some of the small skinks that Franklin *et al.* (2021) did not identify to species were *Lampropholis*.

Another survey of predation by cats was restricted to urban Sydney in an area where domestic cats potentially lived within foraging distance of urban bushland reserves. In this study, it was confirmed by owners that at least 55% (*n* = 113) of these cats hunted wildlife during the 7–13 months that owners reported on the prey their cats returned to them. Thirty-three of these cats returned between zero and 28 small lizards per cat spanning four species, including *L. delicata and L. guichenoti*). The prey items of 26 cats that returned 10 or more prey showed some specialisation in their dietary preference. Five cats showed a marked preference (87.5–94.4%) for lizards, although the largest group of cats (*n* = 8) returned between 66.7–100% small birds whereas a further three (70–90%) most often returned large birds; four cats returned between 81.8–97.1% rats. Six cats demonstrated no such specialisation. These cats returned four to six different prey taxa, some of which may have been *Lampropholis* (Dickman and Newsome 2015).

Predation by cats upon native Australian wildlife, including some *Lampropholis*, would certainly have been occurring since the arrival of Europeans in Australia in 1788 as cats were imported in large numbers over the next hundred years (Tidemann *et al.* 1994; Abbott 2002) and had spread into arid Central Australia at least several decades before permanent settlement there by Europeans (Finlayson 1961). The stomach contents of two cats from that region contained four lizards from one individual and one lizard and one marsupial mouse from the other (Heatwole 1970). Feral cats are now widespread throughout the Australian continent and some associated islands (Abbott *et al.* 2014; Spencer *et al.* 2016). Legge *et al.* (2017) indicated that there were more than 2 million feral cats in Australia with estimates varying between 1.4 and 5.6 million, dependent on prevailing climatic conditions. By contrast, on Lord Howe Island, an offshore island of Australia, cats cannot be implicated in predation upon *L. delicata* because domestic cats and their feral descendants, were eradicated there in the 1980s (Miller and Mullette 1985), approximately a decade before the recorded establishment of *L. delicata* on the island (Hutton 2003).

Lampropholis delicata and *L. guichenoti* are common in urban areas, including backyards, throughout much of their range. For example, in one of the two most populous of Australian cities, Sydney, these two species constitute the most abundant urban reptiles (Griffiths 1987) and consequently are in contact with feline populations. In other cities and towns within a broad geographic region one, or both, of these species is/are also common (Cogger 2018).

Due to their abundance and sympatric association with humans in the cities of the eastern coast of Australia, it is not surprising that Shine and Koenig (2001) suggested that *Lampropholis* were killed in greater numbers by domestic pets and motor vehicles than was any other reptilian taxon in the Sydney region. Despite this carnage, along with representatives of other small species and the juveniles of larger lizards, *Lampropholis* were absent from the records of injured animals submitted to WIRES (Wildlife Information, Rescue and Education Service) over the decade 1989–1998, when Shine and Koenig (2001) analysed data from wildlife rescues. Despite such absence of evidence, predation by cats on wildlife (including *Lampropholis*) has been considered an issue in Australia, at least since the 1990s (e.g. Paton 1990; Dickman 1996; Meek 1998; Abbott 2002).

Cats can find their preferred prey even when the prey has become scarce (Mendl and Harcourt 1988; Dickman 2009; Bradshaw 2013; Dickman and Newsome 2015) and, subsequently, as Smith and Smith (2013) suggested, cats – domestic and feral – may deplete prey populations. The carnage recorded to be inflicted on wildlife by cats, however, is underestimated. For example, based on the responses to their survey, Dickman and Newsome (2015) found that 43.1% (*n* = 51) of the respondents who did not supply evidence were uncertain whether their cats hunted. Some prey could be fully consumed on site and thus not observed by the owner of the pet (Lloyd *et al.* 2013). Some previous researchers (Kays and DeWan 2004; Loss *et al.* 2013) found that only 33–50% of all prey successfully hunted were returned to their owners and, that neither size of the prey nor habitat was found to influence whether a cat abandoned or completely consumed a kill as opposed to returning it home (Lloyd *et al.* 2013).

To determine whether cats preferred specific prey, Dickman and Newsome (2015) surveyed a sample of householders from the eastern suburbs of Sydney within 0.5 km radius of a bushland reserve. Of 362 responses to letter-drops, 44% of respondents owned cats that could potentially access bushland reserves. The replies of the 105 cat owners that participated in the study revealed that cats tended to exhibit 'between-phenotype' hunting behaviour. Overall, a minimum of 667 prey items were returned to cat owners (range 1–58 per individual). At least 51 cats did not return any prey and 26 returned ≥ 10 prey. Five of these showed a marked preference for lizards. These lizards formed between 87.5 and 94.4% of all prey returned, second only in number to the eight cats that preferred small birds. In suburban Adelaide, Paton (1990) reported that 34% (n = 700) of the domestic cats he studied delivered reptiles to their owners. In a similar study carried out in Canberra that included 214 domestic cats, 75% hunted prey in addition to the food provide by their owners; herpetofauna (reptiles and frogs) formed the smallest group of prey (<10%) that were captured and returned home (Barratt 1997). Bamford and Calver (2012) reported that a single pet cat exterminated a population of an estimated 40–50 *Ctenotus fallens* over two years in a single suburban garden in Perth. After the cat was removed it took six years for the species to begin recolonisation.

In Hawaii, cats have been present for hundreds of years (Tomich 1969; Lohr 2012). Many individual cats commonly occur in urban and rural areas, and many of those that are feral occur in colonies (Lohr 2012). *Lampropholis delicata* has also been introduced there; individuals first appeared in museum collections obtained from the island of Oahu in 1900. Subsequently *L. delicata* was introduced onto other islands of Hawaii (see Chapter 3). In a discussion of potential influences on the extinction of the Hawaiian endemic lizard, *Emoia impar*, Fisher and Ineich (2012) noted that introduced domestic cats and the mongoose, *Herpestes auropunctatus* are predators of small skinks elsewhere in the Pacific and that in Fiji, Samoa, and Tonga where cats and/or mongooses have been introduced and occur in large numbers, skinks remain extant, albeit in reduced numbers.

Lampropholis delicata was also introduced accidently into New Zealand where it continues to expand its range (Chapple *et al.* 2016; Tingley *et al.* 2016). Over a 1-year period, Gillies and Clout (2003) studied the diet of domestic cats in two of Auckland's suburbs – one suburb was imbedded within this metropolis whereas the other was at the urban/forest fringe. The mean number of lizards caught per cat was equivalent in the two suburbs. Since geckos were considered to have been lost from the suburban area, and *L. delicata* had expanded its range into the suburbs, *L. delicata* was assumed to be among the reptiles captured by cats but not among those from the peri-urban area. However, unlike the Australian studies reported above, rodents were apparently not preferred prey of the domestic cats, although their capture per cat was far greater in peri-urban areas. It is likely, therefore, that this species made up a greater proportion of the reptilian component of the diet at the urban site than on the urban fringe of the forest.

Whilst acknowledging that cats do consume reptiles, Braysher (1993) suggested that species of *Lampropholis* were not likely to be greatly impacted by cats or other predators, even in urban areas. Others have disputed this suggestion. For example, Woinarski *et al.* (2018) found that in Australia cats – feral and domestic – consumed a higher proportion of reptiles than on any other continent. Based on data from >80 studies and the diet of >10,000 cats in Australia's natural environment, it was estimated that a single feral cat killed up to 225 endemic reptiles annually. However, large sections of the Australian continent are beyond the range of any species of *Lampropholis* (see Chapter 2 and Cogger 2018), and those areas tend to be where the diet of feral cats has the highest frequency of reptiles. In highly modified landscapes and based on more limited data, although undoubtedly incorporating more extensive areas of the range of at least one species of *Lampropholis*, Woinarski *et al.* (2018) found that feral cats consumed 130 million reptiles annually while pet cats consumed 53 million. Doherty *et al.* (2015) estimated that reptiles constituted an average of 24% of the diet of feral cats although, as also demonstrated by Woinarski *et al.* (2018), the abundance was higher in hotter, drier regions of Australia beyond the range of any species of *Lampropholis*. However, while the prediction of Braysher (1993) may be reality for the common and widespread *L. delicata* and *L. guichenoti*, predation on species that have a more restricted range (e.g. *L. colossus*, *L. couperii*, *L. elongata*) would be under greater threat of extinction if similar predation rates by cats occurred. At a minimum, the impact of cats has the potential to interfere with the dynamics within native ecosystems (Dent 2016). Norbury (2001) suggested that, where they existed in high numbers, feral rabbits could be the primary prey of cats and, in turn, support a higher population of cats and thus potentially exacerbate predation upon lizard populations.

While both domestic cats (*Felis catus*) and dogs (*Canis lupus familiaris*) prey on small vertebrates, including reptiles (Shine and Koenig 2001; Shine and Tamayo 2016), predation on reptiles by pet cats has been estimated to be much higher (26.3%) than that by dogs (16.1%; Woinarski *et al.* 2018). Although there does not appear to be data for the loss of *Lampropholis* to predation by cats, Bamford and Calver (2012) reported that a single domestic cat may have exterminated a population of an estimated 40–50 *Ctenotus fallens* (a species of skink with a snout-to-vent length a little over twice the length of the largest individual of *Lampropholis*) in a 2-year period. The greatest impact occurred within the first few months. When the predation is by feral cats, and thus not provisioned by an owner, the impact may be even greater.

Overall, however, there are extremely limited data on the predation of cats upon *Lampropholis*. Braysher (1993, p. 36) suggested that *L. delicata* 'seem quite capable of withstanding substantial predation by cats and other predators within the urban environment'. In a single day's observation in an urban backyard in Pymble (Sydney), of the nine *L. guichenoti* captured, five had obviously recently suffered caudal autotomy. This indicated that they had escaped a previous attempt at predation. Although there is no certainty that these autotomised lizards were all escapees from attempted predation by cats, birds were seldom observed to pay attention to reptiles at that site; they appeared to be more interested in the provisioning provided by households in the area than in seeking small ground-dwelling skinks in heavily vegetated backyards (Burgin, pers. obs.).

The European red fox, *Vulpes vulpes*

The assessment of the effect of foxes on lizards varies among different studies.

Cogger *et al.* (1993) ranked the impact of predation on reptiles by foxes second only to destruction of habitat and it has been estimated that reptiles may represent as much as 55% of a fox's diet (Catling 1988; Molsher *et al.* 2000; Read and Bowen 2001; Paltridge 2002). By contrast other studies have indicated that foxes consume very few lizards, typically <5% of their diet (Triggs *et al.* 1984; Lunney *et al.* 1990; Risbey *et al.* 2000; Mitchell and Banks 2005; Urlus 2009). A possible explanation of these very different results may be that not all sizes of reptiles are equally impacted by foxes. For example, Hu *et al.* (2019) observed that with successful baiting to suppress foxes, there was no change in the species richness or abundance of the small reptiles although a large species (the lace monitor, *Varanus varius*) did increase in numbers following the reduction of the population of foxes. Cupples *et al.* (2011) found that the mature size of reptilian species identified in the scats of foxes were all larger than either *L. delicata* or *L. guichenoti*, albeit in an area beyond the range of any species of *Lampropholis*.

Another possible explanation of the discrepancy among different studies is that predation may be higher in some habitats than in others. Most of the high rates of predation were measured in the more arid areas of the country where *Lampropholis* do not exist. Thus, while foxes do consume small skinks, because the distribution of *Lampropholis* does not extend into arid regions where predation by foxes is more focused on small prey than elsewhere (perhaps because of a scarcity of larger prey), the impact of foxes is lessened. Paltridge (2002), Newsome *et al.* (2014), and Moore *et al.* (2018) only found species in the diet of foxes there that were larger than any species of *Lampropholis*.

Where foxes and *Lampropholis* do occur syntopically, the former might be expected to have an indirect effect on predation upon the latter. Braysher (1993) and Sutherland *et al.* (2011) suggested that control of mammalian predators upon mesopredators could result in a cascade of impacts on community structure.

Foxes and their lairs were present at the Hawkesbury Site (Burgin, pers. obs.). Based on the information presented above, we now believe that predation by foxes would be inconsequential for *L. delicata* and *L. guichenoti* at that locality despite the now apparently erroneous suggestion by Anderson and Burgin (2002) that 'feral species' predation (including ... foxes) may influence *Lampropholis* species numbers' on the Hawkesbury Site. The impact of such predation may, however, vary among localities and/or biotic communities.

Foxes may prey upon the eggs of *Lampropholis*. King (1986–1988) found a nest of *L. guichenoti* that had been dug up by a predator; scats nearby indicated the predator probably was a fox. Some of the remaining eggs were undamaged and were taken to the laboratory and hatched; they were *L. guichenoti*.

Dingoes, domestic dogs, and their crosses

There has been some controversy as to the distinctiveness of domestic dogs (*Canis lupus familiaris*) and dingoes (*Canis lupus dingo*) because of the extent of their interbreeding. Collectively, wild dogs of various extents of cross-breeding are widespread throughout Australia, including within much of the range of species within the genus *Lampropholis*.

The New South Wales governmental legislation has removed the word 'dingo' from the State's legislation and the term 'wild dog' has been expanded to encompass any remnant of either dingoes or domestic dogs. The justification for this is that it is assumed that the extent of hybridisation between dingoes and domestic dogs has resulted in 'dingoes' becoming functionally extinct, particularly in south-eastern Australia. However, recent studies refute this assertion. For example, Cairns *et al.* (2020) found that of a sample of 783 wild-caught canids from eastern New South Wales ancestry ranged from 9% to 23% 'pure dingo' whereas 76–88% had some domestic-dog heritage. However, Crowther *et al.* (2021) reported that, despite introgression from domestic dogs, there has been limited change to the species' functional morphology and ecological role. Herein we make the distinction between domestic dogs and dingoes, although recognising that a considerable, but geographically variable, degree of crossing between the two occurs and that the distinction between feral domestic dogs and wild dingoes may be slight in some localities.

Despite some predation by dingoes upon foxes, the dietary overlap between these two canines is high (≥85%) except that dingoes tend to take larger prey than do foxes. Both species preferentially hunt identical mammalian prey in greater proportion than the prey's relative abundance. This observation was consistent across three arid sites. Although beyond the range of any species of *Lampropholis*, the mature size of reptilian species identified in scats were all larger than either *L. delicata* or *L. guichenoti* (Cupples *et al.* 2011).

Within arid regions (Simpson and Strzelecki deserts and Nullarbor Plain), again outside the range of *Lampropholis*, Cupples *et al.* (2011) provided an example of the small, although similar, contribution that skinks may make to the diet of dingoes. The frequency of occurrence of skinks in the diet both of foxes (20.2%) and dingoes (17.6%) was modest by comparison. Unlike the three arid regions where Cupples *et al.* (2011) undertook their study, the abundance of mammalian prey was insufficient in the Tanami Desert for feral foxes, dingoes, and cats to persist (Paltridge 2002). Where all three of these predatory species do co-occur, however, their diet overlapped considerably. Dingoes were found to take greater numbers of larger prey than did foxes and cats although, at least seasonally, reptiles contributed to the diet of all three species (Paltridge 2002). Although the sites of these studies were all beyond the natural range of *Lampropholis*, these data do provide evidence that dingoes, and the dog-dingo interbreeds sometimes consume small reptiles.

Although in general, wild dogs seldom consume small reptiles (Mitchell and Banks 2005; Glen *et al.* 2006; Hu *et al.* 2019) they sometimes do prey on reptiles as small as *Lampropholis* (Shine and Koenig 2001; Martin and Murray 2013; Shine and Tamayo 2016; Franklin *et al.* 2021) but the extent of their overall impact on wildlife is largely unknown (Hughes and Macdonald 2013) or considered to be underestimated (Doherty *et al.* 2017). However, two studies, provide relevant quantitative data, one from a forested area and one from an urban situation. Lunney *et al.* (1990) found that reptiles represented 3% of the diet of dogs from coastal forests lying within the range of both *L. delicata* and *L. guichenoti* in New South Wales. Franklin *et al.* (2021) in an online, self-selected questionnaire, asked participants to record the number of prey animals each of their pet dogs and/or cats had caught in the previous 6 months. Responses were received from across Queensland, New South Wales, and Victoria – largely within the range of at least one species of the two common species of *Lampropholis*. Of the 1,529 valid responses received, 24% of the 261 dog owners observed that their dogs preyed on animals, of which 62% were native species. Whilst the greatest proportion of the prey consisted of skinks and other lizards, the dogs also successfully preyed on much larger species. Thirty-four per cent of the prey that were returned to the owner included at least one reptile, a similar percentage to the proportion of birds caught. The median range of reptiles observed as dogs' prey was 2 (range 1–40), although <10% of the taxa were small skinks and lizards. We assume these figures to be a substantial underestimate of the true number of small individuals since small prey are more likely to be completely consumed where they are captured than are larger prey.

Domestic dogs span the spectrum from complete dependence on their owner to free-ranging feral animals, some with an intermixture of dingo. All hunt and scavenge for food, even those that are restricted to home environs; thus, all have the potential to take a toll on *Lampropholis*. While they do prey upon small vertebrates, including reptiles (Shine and Koenig 2001; Shine and Tamayo 2016), specific reference to their predation upon *Lampropholis* has not been mentioned. Predation is more likely to occur in urbanised areas where dogs and *Lampropholis* coexist in habitats produced by humans but are much less likely to have an impact there than are cats (Burgin 2015).

Alien rodents

Globally, introduced predators typically have an average of twice the impact on prey than do their native predators (Smith *et al.* 2017) and rodents of the genus *Rattus* have been considered the most damaging and widespread of alien mammalian predators (Barney *et al.* 2021). There are three species of the genus established within Australia (*Rattus rattus*, black rat; *Rattus norvegicus*, brown rat; *Rattus exulans*, Pacific rat), together with one additional species of rodent (*Mus musculus*, house mouse). Of these species, *R. exulans* and *R. rattus* have been predicted to be a severe threat to native wildlife. However, in research that investigated the exclusion of introduced *R. rattus* from both local and landscape scales, Smith *et al.* (2017) found that the impact of this species on *L. delicata* was less than expected and more closely equivalent to that of native predators. The authors suggested this outcome to be due to the co-evolution of *Lampropholis* with native rodents.

Alien rodents have frequently been identified as a factor in the decline of reptiles (Pickard 1984; Newman 1994; Caut *et al.* 2008; Norbury *et al.* 2014; Smith *et al.* 2017), most notably associated with insular faunas, for example in New Zealand (Towns and Ferreira 2001; Towns *et al.* 2003; Harris 2009; Smith *et al.* 2017) and the Hawaiian Islands (Chapple *et al.* 2016; Smith 2016; Towns *et al.* 2016). Salo *et al.* (2007) suggested that introduced predatory rodents may have twice the impact on native species as do native predators. However, there appears to be limited research on predation by rodents upon *Lampropholis* (Thibault *et al.* 2017).

The available information sometimes is contradictory. Hoare *et al.* (2007) reported that the Pacific rat (*R. exulans*) impacted *Hoplodactylus duvaucelii* (Duvaucel's gecko), whereas Shiels *et al.* (2013) did not observe representatives of any reptile in the stomach contents of *R. exulans* (Polynesian rat), *R. rattus* (black rat), or *M. musculus* (house mouse) in the Hawaiian Islands. This was despite introduced *L. delicata* being present in sufficient numbers to be referred to as the 'plague skink'. In their study of predation by *R. rattus* on *L. delicata*, Smith *et al.* (2017) reported that – at the local and landscape scale – when rats were present, they suppressed the activity of *L. delicata*. In the absence of rats, significantly more *Lampropholis* were sighted although the authors concluded that the impact of the rats on *L. delicata* was moderate. They based this assessment on the finding that *R. rattus* only occasionally consumed *Lampropholis* and concluded that *Lampropholis* may respond to *R. rattus* as if it were a native predator (e.g. *Rattus fuscipes*) rather than an alien predator since, unlike many islands where rats were not historically present, in Australia native rodents occur.

Data are scant on the impact of rodents on small skinks such as *L. delicata* and *L. guichenoti* within these lizards' natural range. In the Hawaiian Islands where alien *L. delicata* have long been established (Baker 1979), Shiels *et al.* (2013) investigated the diet of the three alien species of rodents (*R. rattus*, *R. exulans*, *M. musculus*) that also are alien rodents within Australia. The diets of these rodents were found to be mainly arthropods and plant material. None of the stomachs sampled of any of the species contained lizards. However, this situation does not always prevail. For example, on Île Walpole (New Caledonia) where *R. exulans* was present in exceptionally large numbers (Sadlier *et al.* 2019), this rodent was found to be a significant predator of lizards (Jourdan *et al.* 2015; Thibault *et al.* 2017). These observations are reinforced by the observations of Monks *et al.* (2014), who found that when the numbers of prey are reduced substantially, *R. rattus* may opportunistically shift diet rapidly to an alternative source.

On a small, uninhabited island (Surprise Island) in New Caledonia, Caut *et al.* (2008) found evidence that *R. rattus*, did switch diet opportunistically. During much of the year, it consumed a wide range of food including a variety of plant material (Shiels *et al.* 2013). In the breeding season of seabirds, however, up to 24% of this rats' diet consisted of chicks and eggs. When this source of food was spent, the rats increased their intake of hatchling turtles; outside of the breeding season for seabirds and turtles, all rats consumed 100% plant material.

In Australia, the house mouse, *M. musculus*, also switches its diet when more favourable food becomes limited; if an alternative food source is not available, the mouse population crashes. This is a common phenomenon towards the end of major outbreaks of mice in Australia's broadscale cropping areas. For example, Hanifie (2018) reported that it is the inevitable end to a dramatic increase in mice to plague proportions after above-average crops. When the grain is harvested and removed the numbers of mice drop dramatically. There was no mention of predation by mice on *Lampropholis* although this very likely occurs.

By contrast, in New Zealand, *M. musculus* were found to prey on native lizards. For example, Pickard (1984) observed that skinks constituted 2–12% of the diet of mice with the quantity varying throughout the year. It was concluded that *M. musculus* were opportunistic predators upon small reptiles with higher levels of predation preceding, and shortly after, the hatching of skinks – a period when there would be a greater abundance of smaller naïve prey. Shiels and Pitt (2014) reviewed the diets of introduced rodents on Pacific Islands more generally and did not mention reptiles as a dietary component of any rat species on any of the islands they surveyed. More generally, however, there have been observations of predation by *M. musculus* on live lizards (e.g. Burt 1927; Pickard 1984; Norbury *et al.* 2014), together with some convincing indirect evidence (Newman 1994; Lettink and Cree 2006). Shiels *et al.* (2013) retrieved lizards' remains from the stomach contents of rodents very infrequently (*R. rattus* n = 95, 4.9 ± 0.7; *R. exulans* n = 12, 4.2 ± 1.4; *M. musculus* n = 47, 7.2 ± 1.5). However, these authors reported that no *Lampropholis* were among the skinks consumed. In New Caledonian rainforests where two exotic rat species (*R. rattus, R. exulans*) coexist, Quiterie *et al.* (2019) observed that two-thirds (n = 15) of the scincid and gekkonid taxa of these rainforests were consumed by rodents – nine by *R. rattus* (n = 374) and six by *R. exulans* (n = 87).

Despite the apparent inconsistencies and contradictions demonstrated above, there is no doubt that alien rodents (*R. rattus, R. norvegicus, R. exulans, M. musculus*) are predators upon skinks (Whitaker 1973; Newman 1994; Dent 2016). However, as has been previously observed (e.g. in Australia, Hawaii, and New Zealand) *Lampropholis* make up only a small percentage of the diet of alien rats. Nevertheless, some observations indicate that where rodent numbers are substantial, the lizard fauna is less diverse and lizard populations are smaller; the impact of alien rodents may be more substantial. The lizards also behave differently in environs free of predatory rats (McCallum 1986; Hoare 2006; Dent 2016) or, undoubtedly, when an eruptive episode dramatically increases the number of rodents in the population (Blackwell *et al.* 2001; Pavey *et al.* 2008). Such plagues occur intermittently, both in Australia and New Zealand (Singleton *et al.* 2005). In addition to applying predation pressure on skinks, rodents may compete directly for invertebrate prey that constitute a considerable component of skinks' diet (McCallum 1986; Jones *et al.* 2005). Rodents may also have an indirect impact by supporting populations of other predators (Dent 2016). In Australia, they may constitute supplementary options for predatory birds.

Dasyurids: Native Australian carnivorous mammals

Of the Australian carnivorous marsupials in the Family Dasyuridae more than half are restricted to the arid zone (Dickman 2003; Geiser and Pavey 2007; Stannard *et al.* 2014; Baker and Dickman 2018) and thus outside the range of any species of *Lampropholis* (see Cogger 2018). Most of these predators are sufficiently small (<500 g) to be referred to as 'micro-carnivores' (Dickman 2014; Jones *et al.* 2014; Contos and Letnic 2019). Although considered ubiquitous and often abundant, only 25 species have been studied; most are secretive and lead a solitary existence. Under such circumstances, it is not surprising that knowledge of their dietary regimen is limited (Stannard *et al.* 2014). An added difficulty is that some of the information was derived from investigation of dasyurids' scats (Lunney *et al.* 1989, 2001; Dickman 2014) and the prey could not be identified to species.

Typically, dasyurids are broadly insectivorous (Nagy *et al.* 1988; Green 1989; Lunney *et al.* 2001) and fewer than 10% of these species' prey include vertebrates, such as reptiles, alive or scavenged, in their diet (Dickman 2014; Baker and Dickman 2018). The impact of dasyurid predation on *Lampropholis* is, therefore, unlikely to be of great consequence. In addition, the geographic ranges of many dasyurid species do not overlap with any species of *Lampropholis* (compare Baker and Dickman 2018 [mammals] and Cogger 2018 [skinks]). Even where dasyurids and *Lampropholis* do overlap (Lindenmayer *et al.* 2001; Lunney *et al.* 2001; MacNalley *et al.* 2001) they may occupy different microhabitats and thus be ecologically isolated. For example, Foster *et al.* (2016) observed that, post-fire, there was a separation between *L. delicata* and *Antechinus stuartii.* Skinks were most common in sites with high densities of large herbivores, whereas the dasyurids preferred areas in which these herbivores were absent. Separation of niches was also observed in the forests of south-eastern New South Wales where Lunney *et al.* (2001) found that *Antechinus agilis*, but not it's congener, *Antechinus swainsonii*, occasionally consumed *L. delicata* and *L. guichenoti*. Such differences in diet may result from *A. swainsonii* extensively foraging at ground level in leaf litter and on logs where *Lampropholis* also forage, whereas *L. agilis* principally forage above ground on trees, typically above the level at which *Lampropholis* are found.

Other dasyurid taxa that consume reptiles include *Sarcophilus* (Tasmanian devil), *Dasyurus* (quolls), *Phascogale* (phascogales), and *Sminthopsis* (dunnarts) (Strahan 1983; Baker and Dickman 2018). Significantly larger than any other extant species of dasyurid, the Tasmanian devil weighs up to 13 kg and is restricted to Tasmania. Reptiles make up a minor component of its diet (Fleay 1952; Guiler 1970; Taylor 1986; Jones and Barmuta 1998). Whilst knowledge of its diet has relied on an investigation of scats, there is no direct record of its predation on *L. delicata*. However, the low diversity of skinks on Tasmania, the combination of the low level of predation of this species upon reptiles in general, the declining populations of *S. harrisii*, and the observation of Pemberton *et al.* (2008) that Tasmanian devils are predominately scavengers, there is no reason to assume that they would have a great impact on native Tasmanian *L. delicata*.

Quolls potentially prey on *Lampropholis* but are also unlikely to have a significant impact. For example, the diet of the spotted-tailed quoll, *Dasyurus maculatus*, consists predominantly of larger vertebrates (gliders, rabbits, bandicoots, and possums) and only infrequently do they consume insects, birds, or reptiles (Glen and Dickman 2006). If *Lampropholis* are eaten at all, such consumption likely occurs infrequently (Belcher 1995; Glen and Dickman 2006; Glen *et al.* 2006; Jarman *et al.* 2007).

Of the two species of bilby (*Macrotis*), one species does not overlap the distribution of any species of *Lampropholis*, and the other does so only marginally (Strahan 1983). Although species of other groups, including bandicoots such as, *Isoodon obesulus obesulus* (Quin 1988) and *Isoodon obesulus peninsulae* (Keiper and Johnson 2004), are omnivores they apparently do not include vertebrates in their diet. Likewise, although based on extremely limited data from outside the range of the species that coexist with *Lampropholis*, dunnarts (*Sminthopsis macroura* and *Sminthopsis crassicaudata*) have not been reported to consume vertebrates (Morton *et al.* 1983).

Only the range of one species of phascogale, *Phascogale tapoatafa* (brush-tailed phascogale), overlaps that of any species of *Lampropholis*. This threatened phascogale occurs in low density over most of its former range in eastern Australia, where it overlaps the ranges of *L. delicata* and *L. guichenoti*. Since Europeans arrived in Australia, much of the brush-tailed phascogale's range has been cleared, with loss of the hollows in trees used by *P. tapoatafa*, thereby reducing the quality and extent of its suitable habitat (Soderquist and Serena 1994; van der Ree *et al.* 2001, 2006).

There is limited information on the diet of the species in eastern and south-eastern Australia where phascogales and species of *Lampropholis* overlap in range. However, as with all other species of dasyurids, coexistence with phascogales would not be of major threat to any species of *Lampropholis*. For example, brush-tailed phascogales are largely arboreal (Rhind 2004), rare, and their deliberate pursuit of vertebrates negligible, although they may scavenge vertebrates occasionally (Scarff *et al.* 1998).

EVOLUTIONARY CHANGE IN ANTIPREDATOR RESPONSE TO MAMMALIAN PREDATORS

The lizards of New Zealand historically were subjected to predation by birds, but not by mammals prior to the evolutionarily recent introduction of predatory mammals to New Zealand. Hoare (2006) assessed the shift in antipredatory strategy arising from this transition by comparing the behavioural responses of native New Zealand lizards to predatory mammals with those of introduced *L. delicata*, which had co-evolved with mammals in Australia. She found that some native species shifted from 'freezing' into immobility, as a response facilitating avoidance of detection by visually hunting predatory birds, to one of fleeing at the approach of predatory mammals.

PREDATION BY AMPHIBIANS

Most Australian anurans are nocturnal (Ferraro and Burgin 1993; Schell 2002) and primarily feed on invertebrates (Tyler 1999; Madani 2021); thus they would not likely encounter *Lampropholis* or eat them. There are, however, occasional records of the Australian green tree frog (*Litoria caerulea*) consuming vertebrate prey, e.g. micro-bats and snakes (Fleay 1952; Mikula 2015).

The cane toad (*Bufo marinus* [*cf. Rhinella marina*, Cogger 2018]) has been reported to be a voracious predator (DEWHA 2010) that may dramatically reduce populations of native species of reptiles but most research undertaken on the impact of its diet has not confirmed this accusation (Phillips *et al.* 2003; Beckmann and Shine 2009; Radford *et al.* 2020).

PREDATION BY INVERTEBRATES

Native ants

Although evidence of the impact of native ants on *Lampropholis* species is scant, an indication that they may attack skinks is provided by Twigg and Fox (1991) who reported that some pitfall traps were treated with fly spray to minimise the mortality of *Lampropholis* caught in the traps. In undertaking research on the distribution of lizards (including *Lampropholis*), Bragg *et al.* (2005) also sprayed their pitfall traps twice daily to prevent predation upon captured skinks. The Australian Wildlife Conservancy Guidelines (AWC 2013) also instructs researchers on the use of repellents or insecticides when setting pitfall traps for reptiles in the event of activity by ants. Native ants inhabit all terrestrial ecosystems in Australia and are the most dominant of the terrestrial invertebrates (Shattuck 1999); however, based on extensive observation of *Lampropholis*, we report that predation upon them by native ants is not common.

Introduced ants

Fisher and Ineich (2012) reported that alien ants are known to cause the decline and maybe even extinction of reptiles. Lach and Thomas (2008) found that, in the 20 years prior to publication of their paper, >6,700 ants representing 105 species in 73 genera had been intercepted at Australia's entry ports; at least six of these have become established and are major threats to native biodiversity (Thomas and Holway 2005). One of these, the Argentine ant (*Linepithema humile*) has expanded its range to include the eastern coastal states, South Australia, and Tasmania. Other imported ants such as the African big-headed ant (*Pheidole megacephala*), the African yellow crazy ant *Anoplolepis gracilipes* (formerly *A. longipes*), the Central American tropical fire ant (*Solenopsis geminata*), the South American red fire ant (*Solenopsis invicta*), and Central and South American little fire ant (*Wasmannia auropunctata*) occur in Queensland (Moloney and Vanderwoude 2002). At least some of these species are likely, therefore, to become established within the range of *L. delicata* and *L. guichenoti*, and perhaps of some other species of *Lampropholis*. These six species of ants are among the world's most destructive (Lach and Thomas 2008). Their impact on small Australian skinks, however, is largely unknown, although the New South Wales Threatened Species Scientific Committee named *L. delicata* as one of the species that could become threatened by an invasion of *A. gracilipes* (OE&H 2021). This species constructs large nests that join to form supercolonies. These ants lack stings but attack by biting and spraying formic acid (O'Dowd *et al.* 2003; Lee and Yang 2022).

In a discussion of the history of *A. gracilipes* on the Australian Territory of Christmas Island, Lenancker *et al.* (2021) provided some hint of its potential impact elsewhere. It has been present on the Island since the early 1900s (Donisthorpe 1935; Abbott 2004) but remained largely unnoticed for decades. By late 2002, supercolonies of high density had expanded to cover >2,500 ha of the Island's rainforest with >2,000 ants/m^2 within these colonies. The boundaries of the colonies are not discrete and ants may be found up to 200 m from the main colony. The number of colonies may also expand rapidly. For example, Abbott (2004, 2006) reported that 34 discrete, high-density supercolonies spanning areas of 0.9–787 ha had developed between 1989 and 2003.

There has been some evidence available for at least two decades that skinks are adversely affected by *S. invicta* in south-eastern Queensland (Nattrass and Vanderwoude 2001). This species preys upon lizards that are hatching in the nest (Chalcraft and Andrews 1999; Moloney and Vanderwoude 2002) and thus *Lampropholis* would be susceptible, particularly where there is a large aggregation of their eggs (see Chapter 4).

Being more widespread in Australia than the other species of introduced ants, *L. humile* is likely to encroach on the habitat of multiple species of *Lampropholis*. Human *et al.* (1998) reported that 91.6% of their food consisted of liquid carried to the nest in their distended gasters and 5.5% consisted of insects and 0.4% of seeds. Only 2.5% of its diet was unidentified material that potentially could have included morsels of vertebrates such as reptiles. Also, Abril *et al.* (2007) reported that the solid component of its diet consisted mainly of insects, especially of aphids. Thus the extent to which this species preys or scavenges on *Lampropholis* is unknown but if such occurs, the impact must be low.

There is no indication in the literature that *P. megacephala* is a predator of vertebrates (Castiñeiras and Castellanos 1983; Dejean *et al.* 2007; Arakaki *et al.* 2009; Taniguchi 2011). However, at least within the areas where they coexist with *Lampropholis*, they have been shown to have a significant impact on native invertebrates (Heterick 1997; Vanderwoude *et al.* 2000; Dejean 2007; Hoffmann and Parr 2008). There is thus the potential for their presence to have a dramatic indirect impact due to competition with *Lampropholis* for invertebrate prey. In general, competition between species of invasive alien ants and native species is likely to result in the displacement of native ants with multiple consequences for local biota (Lach and Thomas 2008).

In summary, aggressive alien ants are known to cause the decline and even extinction of reptiles either directly by predation at some stage of the reptiles' life cycle (from eggs to adults) or indirectly through competition for resources or transmission of disease (Harris 2002; Abbott 2006; Lach and Thomas 2008; Kenis *et al.* 2009). The extent to which these processes apply to species of *Lampropholis* requires further investigation.

Spiders

Notable among the spiders that are predatory upon reptiles are those of the genus *Latrodectus* (red-backed spiders or black widows), known to prey upon seven species of lizards and two species of snakes in Europe, the Middle East, and the United States, and 13 species of lizards and five of snakes in Australia (O'Shea and Kelly 2014). *Latrodectus hasselti* is the species most often reported as preying upon Australian lizards, especially on lygosomine skinks, including *L. guichenoti* (Copland 1953) and *L. delicata* (see immediately below).

Metcalfe and Ridgeway (2013) observed an adult *L. delicata* in a natural impression on a rock platform where it was tightly bound by the silk strands of the web of a female *Latrodectus hasselti*; the spider remained on guard despite the authors' interference to encourage it to move so that they could determine whether the lizard was dead. Shine and Tamayo (2016) witnessed a red-backed spider encasing an adult *L. delicata* in silk threads in a suburban backyard in Sydney (Plate 6). The process of encasing the skink continued for 2 hours.

Other taxa of spiders also have been implicated as predators of *Lampropholis*. Shine and Tamayo (2016) observed a huntsman spider (*cf. Heteropoda longipes*) holding an adult *L. delicata* in its jaws and Clifford (1988) watched *L. delicata* being caught in webs of *Achaearanea* (now *Parasteatoda*) *tepidariorum* (Family Theridiidae). This spider trails its web along the ground where the lizards stumble into it and on one occasion the author witnessed the spider hoisting the lizard aloft. The spider hauled up the struggling lizard by alternately raising the head and abdomen at the rate of ~10 cm per hour.

Turner (2021) reported an attack by a Union Jack wolf spider (*Tasmanicosa* sp.; Family Lycosidae) on a subadult *L. delicata* (Plate 7) in the following terms:

> *'initially the skink was alive and was seen to breathe, mouth-gape, and wriggle several times; ... it also made several sudden violent movements in its attempts to break free. After approximately 15 mins. of observation the skink had ceased moving and appeared to be immobilised – breathing was infrequent and laboured. After a further 5 mins, there was no breathing and the skink was limp ... For much of the observation period the spider was exposed on the ground surface and several times moved short distances ... This movement was quite rapid despite the relatively large size of the skink which it dragged with apparently little effort.'*

Shine and Tamayo (2016) suggested that predation by spiders on skinks undoubtedly is commonplace in Sydney due to the high population densities of small skinks and the occurrence of numerous taxa of large spiders.

Possible predation by centipedes and scorpions

There do not appear to be any reports of predation by centipedes upon *Lampropholis*. It is likely, however, that such may occur as *Ethmostigmus rubripes*, a centipede found in the same habitat as *L. delicata* and *L. guichenoti* (Heatwole, pers. obs.) belongs to the Scolopendromorpha. This is an Order containing large species that prey not only on invertebrates but also on rodents, birds, toads, snakes, and geckos. Their defensive bites have been known to result in fatalities in humans and the venom of *E. rubripes* has been shown to be highly toxic to vertebrates (mice) (Ménez *et al.* 1990) although it is even more toxic to various insects.

Scorpions (*Urodacus* spp.) also occupy some of the same microhabitats as *L. delicata* and *L. guichenoti*

(Heatwole, pers. obs.) and being generalised predators could potentially take young *Lampropholis*. Quinlan *et al.* (1995) examined the hard parts of the remains of prey found in the burrows of two species in the genus *Urodacus* and found them to consist of beetles and ants but cautioned that more soft-bodied prey would not be detected by their methods and that these scorpions should be capable of consuming larger prey than indicated by the hard parts they found.

Accidents

Although not predation, various structures may accidently trap lizards and cause their demise. For example, Langkilde *et al.* (2002) rescued an adult *L. delicata* that had its head firmly wedged in the seed capsule of the introduced ornamental garden plant (*Tibouchina*), a species widespread in gardens in Sydney, and Shine and Shine (2016) released an adult male *L. delicata* from a sticky piece of toffee discarded on their front porch by a Halloween 'trick-or-treater'. The lizard was adhered to the sweet by its hind limbs and the underside of its tail and the posterior part of its abdomen. Neither of these animals seemed to be able to escape from their predicament unaided and serve as examples of the hazards faced by living in proximity to humans.

PREDATORS UPON *LAMPROPHOLIS*' EGGS

Eggs may be disrupted and/or consumed by scavengers including ants, bandicoots, snakes, and monitor lizards (Carter 1992; also see sections on snakes, lizards, and ants above). In the laboratory, Doody and Paull (2013) used vibration to mimic disturbance of *L. delicata* eggs by predators. They observed that the immediate response of embryos was 'explosive' hatching. Compared with clutch-mates that were left to emerge from their egg spontaneously 2 to 4 days later, those that were stimulated to hatch prematurely were significantly smaller, ran slower, and more residual yolk was retained in the shell post-hatching. The nutrients discarded with the shell at premature hatching otherwise would have facilitated further maturation of the embryo over the last few days of normal incubation.

In the field, Doody and Paull (2013) simulated attempted predation of *L. delicata* eggs by removing some of them and placing them on the ground nearby. The results in the field were the same as those in the laboratory. Eggs hatched within seconds after disturbance and immediately sprinted approximately 40 cm from the nest with most moving under leaf litter. Thus, mechanical disturbance releases the hatchling from the egg, thereby allowing it to run and attempt escape rather than remaining trapped within the egg's membranes. These observations were consistent with the outcome of studies of other oviparous species, for example the sand lizard, *Lacerta agilis* (Olsson *et al.* 1996); hatchlings that emerged from their shells 2 days early were significantly smaller and grew more slowly than full-term individuals.

Disturbance of the nest site by whatever means may initiate premature hatching of young, provided the embryos are sufficiently developed. A study by Humphreys (1976) focused on the wolf spider, *Geolycosa godeffroyi* (*Schizocosa leuckartii* was also present and it has similar habits to *G. godeffroyi*). He evacuated approximately 1,750 spider burrows that were ~10 cm wide × 18 cm deep. Approximately 18% of these were occupied by adult and/or subadult spiders. Eggs of *Lampropholis delicata* were recovered from the tailings produced by resident spiders in four additional burrows and eggs also were found in the tailings produced by the resident spider's diggings from a further 15 burrows. It was also determined that resident spiders ejected skinks' eggs that were placed in the burrow by the researcher.

Eggs that had escaped the spider's attention in the burrows were collected and were found to be still viable, whereas those collected from the tailings were desiccated, or would have become so, and thus would have failed to hatch. Overall, Humphreys (1976), predicted that due to the spider's behaviour, mortality of the eggs deposited in the spiders' burrows would have approximated 44%.

AUTOTOMY: PROTECTION FROM PREDATION

As reviewed above, predators of *Lampropholis* include a wide range of fauna spanning birds, snakes, lizards, cats, foxes, and dogs (dingoes, domestic dogs, and cross breeds), rodents, antechinus, and invertebrates such as ants and spiders and possibly other taxa. The outcome of any encounter between predator and prey will result in escape, death, or injury (loss of some body part, among *Lampropholis* typically the tail). Although such caudal autotomy may occur whenever a predator grasps a skink, it is frequently attributed to unsuccessful predatory attempts by birds (Anderson 2007; Anderson and Burgin 2008).

Autotomy also is a major tactic used by *Lampropholis* to escape attack by predatory birds (Bateman and Fleming 2009; Cromie and Chapple 2012; Gills and Higham

2016). Chapple *et al.* (2015) reported that on Lord Howe Island, where *L. delicata* were accidently introduced multiple times during the 1980s (see Chapter 3), caudal autotomy is common. They observed that 55% (n = 1,375) of adults had suffered loss of the tail. Females, however, were more likely to have lost at least a segment of their tail than were males. By contrast, Simbotwe (1985) found no significant difference between the sexes in the incidence of aututomised or regenerating tails in *L. guichenoti*. At least within some areas of the range of the various species of *Lampropholis* within Australia, loss of the tail does not appear to be quite so prevalent (Burgin, pers. obs.). The casual observations that led to this comment are supported to some extent by a study of predation by birds on *L. delicata* and *L. guichenoti* in the open woodlands of peri-urban north-western Sydney. In that study (Anderson 2007; Anderson and Burgin 2008), 32.3% (n = 2,640) of the recorded attacks on skink decoys were unambiguously attacks by birds. Effectively equal numbers of decoys of sizes representing juvenile, subadult, and adult *Lampropholis* were attacked by birds and there was no significant difference in the attack zone (head, body, or tail). Of these attacks, 28.5% of the birds' strikes were in the tail region which, presumably, would have resulted in caudal autotomy among those individuals that escaped while other strikes may well have resulted in successful predation by the bird.

The vulnerability of individuals of *L. guichenoti* that have suffered caudal autotomy may be, at least in part, explained by the observations of Downes and Shine (2001) that individuals that had lost their tails were more likely to succumb to predation by a diurnal snake (*Demansia psammophis*) than by a nocturnal one (*Cryptophis nigrescens*), which predominantly preys on small skinks (Cogger 2018). Since there was no difference in an individual snake's level of detection of tailless lizards compared to those with intact tails, and no difference in response that could be attributed to chemoreception, it was concluded that the higher rate of successful predation by the diurnal snake was probably due to *Lampropholis* having poorer locomotory performance post-autotomy, and a greater tendency to flee sooner when a snake approached, thus eliciting an attack by the snake. More generally, the detection of the chemical cues of a predatory species – and the subsequent responsiveness to such cues – may enhance a skink's ability to evade a predatory snake (Thoen *et al.* 1986; Webb *et al.* 2009). Indeed, Ortega *et al.* (2019) suggested that early detection of the threat of predation by prey was essential to avoid capture. Downes (2002b), however, observed that chemosensory response in *L. guichenoti* differed among individuals. Some responded by decreasing their activity and mobility. Although counter-intuitive, the individuals that did not flee were more likely to escape detection and thus suffered less predation than those that attempted to escape.

Overall, loss of the tail in *Lampropholis* results in poor locomotory performance, altered antipredator behaviour, and ultimately, survival (Chapple and Swain 2004; Bateman and Flemming 2009; Cromie and Chapple 2012; Chapple *et al.* 2015). Cromie and Chapple (2012) found that both *L. delicata* and *L. guichenoti* had substantially reduced the speed with which they sprinted after loss of the tail. Caudal autotomy also resulted in an overall decrease in activity although, more generally, these species had similar behavioural responses to loss of their tails; both were less active post-autotomy and a preference for more sheltered basking sites than otherwise observed. However, *L. delicata* marginally tended to seek more sheltered sites than did *L. guichenoti*.

Since attempts at predation upon a lizard may be cumulative over the life of a lizard the incidence of broken and regenerated tails might be expected in increase with body size (age). This expectation was realised for both sexes of *L. guichenoti* in the study by Simbotwe (1985) in Australia (Table 9.5) and by Chapple *et al.* (2014) on Lord Howe Island. It should be noted, however, that very small lizards would be likely to be taken whole, rather than just their tail, more often than would larger individuals.

Table 9.5. The percentage of different sizes and sexes of *Lampropholis guichenoti* with broken or regenerating tails.

Data from Simbotwe (1985).

Snout-to-vent length (mm)	Per cent with damaged tails	
	Males	**Females**
24.0–26.9	0	0
27.0–29.9	0	20.0
30.0–32.9	0	20.0
33.0–35.9	33.0	28.5
36.0–38.9	37.5	35.7
39.0–41.9	64.7	53.1
42.0–44.9	50.0	58.3
45.0–47.9	66.6	80.0

That caudal autotomy may be an impediment to movement by *Lampropholis* can be gleaned from the work of Cromie and Chapple (2012). They observed that the post-autotomy wriggling of the segment of tail detached during an encounter with a predator was equivalent in *L. guichenoti* and *L. delicata*. There was also a positive relationship between the length of the shed segment of tail and the distance it moved. Such movement of the detached tail may serve to distract the predator and thus enhance the chance of the skink's escape (Arnold 1988; Downes and Shine 2001). Following caudal autotomy, *L. delicata* and *L. guichenoti* both showed similar behavioural responses although *L. delicata* remained more active than *L. guichenoti*. There was, however, a substantial decline in their speed of sprinting (*L. delicata* 28%; *L. guichenoti* 39%) at the temperature at which they performed optimally (30°C). A decrease in activity, and a preference for more sheltered retreats also was observed (Downes and Shine 2001) – activities that may have longer-term impacts on growth and condition.

There is greater risk of predation among gravid females than among those that are non-gravid, although Shine (2003) did suggest that even at a moderate decrease in body temperature, caudal autotomy or a large meal may inhibit *L. guichenoti* more than does gravidity. However, based on another study (Qualls and Shine 1997), it was observed that even mature *L. guichenoti* from adjacent populations with equivalent morphological and reproductive characteristics (e.g. body size and shape; reproductive output) may have dramatically different locomotory performance in gravid females as compared to non-gravid ones. Gravid females from one population ran much faster than did those that were non-gravid, whereas individuals from an alternate population ran much slower. Presumably, individuals moving more slowly would be more vulnerable to predation than the faster-moving skinks.

Since larger – and typically older – *Lampropholis* females tend to have larger clutches (Baker 1979; Burgin 1989; Chapple *et al.* 2015), larger gravid females may more readily draw the attention of potential predators. However, as discussed above, based on the use of decoys, Anderson and Burgin (2008) found that birds appeared not to discriminate based on the size of the skink. Contrary to these observations, however, indirect evidence for larger individuals being exposed to predation more often than smaller ones was reported by Chapple *et al.* (2015) who found that attempted predation on *L. delicata* resulting in autotomy occurred most frequently in adult females (i.e. larger individuals). It would appear, therefore, that size alone, is not necessarily the factor involved in higher levels of predation on large female *Lampropholis*. Smaller individuals with less experience of predators and/or lesser speed across the same distance than larger individuals, may be less competent in evading predation than the larger ones.

Regeneration of autotomised tails

Reptiles are the highest taxon of vertebrates that have any significant power of regenerating lost appendages. In *Lampropholis* autotomised tails regenerate. Purvis (1979) removed the tails of *L. guichenoti* and described the histology of the tail at succeeding stages of regeneration. He recognised six stages: (1) formation of a scab over the wound and withdrawal of the spinal chord internally (by day 2 following autotomisation); (2) formation of new epidermis and the blastema beneath the scab, followed by loss of the scab (less than day 6); (3) growth of an ependymal tube into the blastema and formation of an ependymal sac (within 6 days); (4) early differentiation of the blastema into muscle and a cartilaginous tube. Appearance of blood vessels at the distal end of the blastema. Growth of longitudinal nerves into the blastema (6 to 11 days); (5) further differentiation of the blastema, including segmentation of muscle into blocks separated by septa of connective tissue; then (6) appearance of mature regenerated scales at the proximal end of the regenerating part (the regenerate). The proximal end of the regenerate becomes fully differentiated into a cartilaginous tube, band of fat, segmented muscles, and a covering of scales (19–30 days); see Purvis (1979) for photographs of histological preparations illustrating each stage.

Regeneration of a tail that is damaged but not completely lost may lead to a second tail growing from the site of damage, thereby resulting in a bifurcated tail (a 'two-tailed' lizard). Dissnayake and Clemann (2022) reported an individual of *L. guichenoti* with a bifurcated tail and noted that it was 'particularly slow-moving and easy to capture', suggesting that the double tail impeded its locomotion and made it more susceptible to a potential later attack.

EFFECT OF EMBRYOS' EXPERIENCE ON SUSCEPTIBILITY TO PREDATION

In this book we have repeatedly reported research showing that genetic determinants are influenced by, or even superseded by, their effect on *Lampropholis*' phenotype – morphology (e.g. body size; see Chapter 5), growth (see section on predation by snakes in this chapter), and physiological ability (Chapter 7). Here we relate the role of the embryos' environment on the effect of the temperature of incubation on juveniles' post-hatching ability to avoid predation at two intervals following post-hatching. Downes and Shine (1999) first examined the performance of hatchling *L. delicata* in activities that would be expected to facilitate escaping predators and found that within their normal range of operating temperatures the speed of sprinting over a short distance (0.25 m) not only increased with increasing ambient temperature (Table 9.6), but that those that had been incubated at a higher temperature (27°C) ran faster than did those that had been incubated at a lower one (20°C). The potential for escaping a pursuing predator increases with increasing temperature within the normal operating range unless the predator's speed also increases to the same extent. Since mammalian and avian predators keep a relatively constant body temperature the increase in the speed of sprinting would seem to be an advantage to the ectothermic lizard; if the potential predator is also an ectotherm, e.g. a snake, then relative advantage would be greatest either to the predator or prey, depending on which experienced the greatest thermal effect. However, regardless of the predators' speed or relative increase, the lizard prey is better off if its speed increases with ambient temperature than if it did not. A final attribute that may enhance a lizard's chance of escaping a predator is how far it will run before pausing, either because it is fatigued and unable to continue or merely decides to stop. Downes and Shine (1999) also examined this trait in *L. delicata* and found that relatively fewer of the lizards paused with increase in ambient temperature (Table 9.7). Thus, they not only ran faster at the higher ambient temperatures but they went farther without stopping. Again, the previously experienced temperature of incubation was an important influence. Lizards incubated at higher temperature within the normal range of active lizards stopped more often than did those incubated somewhat lower on the scale. Stopping would seem to be maladaptive in escaping predators as it would allow a predator to approach closer. It might be an advantage to the prey if it also involved a change in direction of movement (as the authors mentioned sometimes occurred). Confusion of a would-be predator by a change of direction may more than compensate for a momentary stop.

Several of the variables discussed above, had interactive effects with each other so that the situation is somewhat more complex than appears from a consideration only of individual single factors, one at a time.

Table 9.6. Comparison of the relationship of the speed of sprinting to ambient temperature of *Lampropholis delicata* incubated at two different temperatures.

Data from Downes and Shine (1999).

Ambient temperature	Speed of sprinting (m/sec)	
	Lizards incubated at 20°C	Lizards incubated at 27°C
18°C	0.27 ± 0.01	0.30 ± 0.02
23.5°C	0.32 ± 0.08	0.44 ± 0.02
29°C	0.45 ± 0.03	0.54 ± 0.0

Table 9.7. Effect of incubation temperature on the percentage of *L. delicata* that paused during their locomotory tests at different ambient temperatures.

Ambient temperature during test	Percentage of animals that paused during ther locomotary trials	
	Lizards incubated at 20°C	Lizards incubated at 27°C
18°C	12	54
23.5°C	15	20
29°C	2	5

CONCLUSION

As outlined in this chapter, birds, reptiles, rodents, carnivorous marsupials, canines, and some invertebrates from the most pristine environments to urban backyards threaten *Lampropholis* skinks with predation. Despite native and introduced predators, however, these skinks apparently are thriving in the habitats created by humans (Recher 2010) and, even within urban environments, are withstanding the impacts of substantial pressure from predation (Braysher 1993).

With such a diversity of predators, even when ≤10% of their diets consist of reptiles, predation must result in substantial impacts on the numbers of *Lampropholis*, particularly when the predators cross-forage between urban areas and local bushland as do numerous birds, dogs, cats, and snakes (Anderson and Burgin 2002).

Adding to the challenges of interpreting predators' impacts on prey is that the population's viability – and thus the degree of vulnerability and the need for conservation – of most Australian native species is unknown (Woinarski *et al.* 2018). Species of *Lampropholis* are no exception. While in recent decades research has increased substantially on *L. delicata* and *L. guichenoti*, most described species of the genus have had, at best, extremely limited attention and there are additional species yet to be described. For most other species of *Lampropholis* there is limited mention (if any) beyond their taxonomic description or comment based on little more than incidental observations and/or inclusion on a list of the species. Based on what is known of the biology of this genus, any species of *Lampropholis* would face similar ranges of predators with just one caveat – the extent to which a predatory species overlaps the range of a particular species of prey.

REFERENCES

Abbott I (2002) Origin and spread of the cat, *Felis catus*, on mainland Australia, with a discussion of the magnitude of its early impact on native fauna. *Wildlife Research* **29**(1), 51–74. doi:10.1071/WR01011

Abbott I, Peacock D, Short J (2014) The new guard: the arrival and impact of cats and foxes. In *Carnivores of Australia: Past, Present and Future*. (Eds AS Glen and CR Dickman) pp. 69–104. CSIRO Publishing, Collingwood.

Abbott KL (2004) Alien ant invasion on Christmas Island, Indian Ocean: the role of ant-scale associations in the dynamics of supercolonies of the yellow crazy ant, *Anoplolepis gracilipes*. PhD thesis. Monash University, Clayton, Vic.

Abbott KL (2006) Spatial dynamics of supercolonies of the invasive yellow crazy ant, *Anoplolepis gracilipes*, on Christmas Island, Indian Ocean. *Diversity and Distributions* **12**(1), 101–110. doi:10.1111/j.1366-9516.2006.00193.x

Abril S, Oliveras J, Gómez C (2007) Foraging activity and dietary spectrum of the Argentine ant (Hymenoptera: Formicidae) in invaded natural areas of the northeast Iberian Peninsula. *Environmental Entomology* **36**(5), 1166–1173.doi:10.1603/0046-225X(2007)36[1166:FAADSO]2.0.CO;2

Anderson L (2007) Edge effects on small skinks: their prey, predators and competitors in peri-urban remnants in north-western Sydney. PhD thesis. University of Western Sydney, Penrith.

Anderson L, Burgin S (2002) The influence of woodland remnant edges on small skinks (Richmond, New South Wales). *Austral Ecology* **27**(6), 630–637. doi:10.1046/j.1442-9993-2002.01224.x

Anderson L, Burgin S (2008) Patterns of bird predation on reptiles in small woodland remnant edges in peri-urban north-west Sydney, Australia. *Landscape Ecology* **23**(9), 1039–1047. doi:10.1007/s10980-008-9252-5

Anonymous (2020) *Yellow-faced Whip Snake*. Australian Museum. <https://australian.museum/learn/animals/reptiles/yellow-faced-whip-snake/>

Arakaki A, Taniguchi G, Arce R, Sugano I, Fukuda S (2009) Control of the big-headed ant, *Pheidole megacephala* (Hymenoptera: Formicidae), in coffee fields using bait stations containing Amdro (Hydramethylnon). *Sociobiology* **53**, 403–410.

Arnold EN (1988) Caudal autotomy as a defense. In *Biology of the Reptilia* (Volume 16). (Eds C Gans and RC Huey) pp. 235–274. AR Liss, New York.

AWC (2013) *Australian Wildlife Conservancy: Animal Trapping, Handling, Sampling and Photographing Guidelines*. Australian Wildlife Conservancy. <https://www.dcceew.gov.au/sites/default/files/env/pages/a8cac765-28ca-4da4-9195-9069626f04a1/files/e2016-0120-application-attachment-handling-guidelines.pdf>

Baiada A (2002) Determining biological indicators to define edge-zones within remnants of the Cumberland Plan. Bachelor of Land Management and Conservation (Hons). University of Western Sydney, Richmond.

Baker JK (1979) The rainbow skink, *Lampropholis delicata*, in Hawaii. *Pacific Science* **33**(2), 207–212.

Baker A, Dickman C (2018) *Secret Lives of Carnivorous Marsupials*. CSIRO Publishing, Clayton South.

Bamford M, Calver MC (2012) Cat predation and suburban lizards: a 22 year study at a suburban Australian property. *The Open Conservation Biology Journal* **6**(1), 25–29. doi:10.2174/1874839201206010025

Barney SK, Leopold DR, Francisco K, Flaspohler DJ, Fukami T, Giardina CP, Gruner DS, Knowlton JL, Pitt WC, Wilson Rankin EE (2021) Successful management of invasive rats across a fragmented landscape. *Environmental Conservation* **48**(3), 200–207. doi:10.1017/S0376892921000205

Barratt DG (1997) Predation by house cats, *Felis catus* (L.), in Canberra, Australia. I. Prey composition and preference. *Wildlife Research* **24**(3), 263–277. doi:10.1071/WR96020

Barrows CW, Allen MF, Rotenberry JT (2006) Boundary processes between a desert sand dune community and an encroaching suburban landscape. *Biological Conservation* **131**(4), 486–494. doi:10.1016/j.biocon.2005.12.024

Bateman PW, Flemming PA (2009) To cut a long tail short: a review of lizard caudal autotomy studies carried out over the last 20 years. *Journal of Zoology* **277**(1), 1–14. doi:10.1111/j.1469-7998-2008-00484.x

Beatson C (2022) *Eastern Brown Snake*. Australian Museum. <https://australian.museum/learn/animals/reptiles/eastern-brown-snake/>

Beckmann C, Shine R (2009) Impact of invasive cane toads on Australian birds. *Conservation Biology* **23**(6), 1544–1549. doi:10.1111/j.1523-1739.2009.01261.x

Belcher CA (1995) Diet of the tiger quoll (*Dasyurus maculatus*). *Wildlife Research* **22**(3), 341–357. doi:10.1071/WR9950341

Berry L (2001) Edge effects on the distribution and abundance of birds in a southern Victorian forest. *Wildlife Research* **28**(3), 239–245. doi:10.1071/WR00057

Blackwell GL, Potter MA, Minot EO (2001) Rodent and predator population dynamics in an eruptive system. *Ecological Modelling* **142**(3), 227–245. doi:10.1016/S0304-3800(01)00327-1

Bradshaw J (2013) *Cat Sense: The Feline Enigma Revealed*. Penguin UK, London.

Bragg JG, Taylor JE, Fox BJ (2005) Distributions of lizard species across edges delimiting open-forest and sand-mined areas. *Austral Ecology* **30**(2), 188–200. doi:10.1111/j.1442-9993.2005.01436.x

Braysher M (1993) The need for a critical review of the impact of pest animals on Australia's herpetofauna. In *Herpetology in Australia: A Diverse Discipline*. (Eds D Lunney and D Ayers) p. 36, Transactions of the Royal Zoological Society of New South Wales, Surrey Beatty & Sons, Chipping Norton, NSW.

Burgin S (1989) The taxonomic and phylogenetic relationships of some Australian scincid lizards (Scincidae: Lygosominae). PhD thesis. Macquarie University, Sydney.

Burgin S (2010) A natural history of the Ham Common, the lands sandwiched between Richmond and Windsor – two of Governor Macquarie's five towns. In *The Natural History of Sydney*. (Eds D Lunney, P Hutchings and D Hochuli) pp. 263–276. Royal Zoological Society of New South Wales, Sydney.

Burgin S (2015) Rainbow lorikeets, possums and pythons: the wildlife of the Gold Coast. In *The Gold Coast Transformed: From Wilderness to Urban Ecosystem*. (Eds T Hundloe, B McDougall and C Page) pp. 103–106. CSIRO Publishing, Clayton South.

Burt C (1927) A note on the mouse *Mus musculus* Linn. as an enemy to lizards. *Copeia* **162**, 15–16.

Cairns KM, Nesbitt BJ, Laffan SW, Letnic M, Crowther MS (2020) Geographic hot spots of dingo genetic ancestry in southeastern Australia despite hybridisation with domestic dogs. *Conservation Genetics* **21**(1), 71–90. doi:10.1007/s10592-019-01230-z

Cameron E (1993) The development of the action plan for Australian reptiles. In *Herpetology in Australia: A Diverse Discipline*. (Eds D Lunney and D Ayers) pp. 109–119. Transactions of the Royal Zoological Society of New South Wales, Surrey Beatty & Sons, Chipping Norton, NSW.

Carter DB (1992) Reproductive ecology of the lace monitor *Varanus varius* in South-eastern Australia. PhD thesis. Australian National University, Canberra.

Castiñeiras A, Castellanos JA (1983) Report of *Pheidole megacephala* (Hymenoptera: Formicidae: Myrmecinae) as a predator of *Diatraea saccharalis* in Cuba. *Ciencia y Técmoca en la Agricultura, Protección de Plantas* **6**(3), 7–10.

Catling PC (1988) Similarities and contrasts in the diets of foxes, *Vulpes vulpes*, and cats, *Felis catus*, relative to fluctuating prey populations and drought. *Wildlife Research* **25**, 307–317. doi:10.1071/WR9880307

Catterall CP, Green RJ, Jones DN (1991) Habitat use by birds across a forest-shrub interact in Brisbane: implications for corridors. In *Nature Conservation 2: The Roles of Corridors*. (Eds DA Saunders and RJ Hobbs) pp. 247–258. Surrey Beatty & Sons, Chipping Norton, NSW.

Caut S, Angulo E, Courchamp F (2008) Dietary shift of an invasive predator: rats, seabirds and sea turtles. *Journal of Applied Ecology* **45**(2), 428–437. doi:10.1111/j.1365-2664.2007.01438.x

Chace JF, Walsh JJ (2006) Urban effects on native avifauna: a review. *Landscape Urban Planning* **74**(1), 46–69. doi:10.1016/j.landurbplan.2004.08.007

Chalcraft DR, Andrews RM (1999) Predation on lizard eggs by ants: species interactions in a variable physical environment. *Oecologia* **119**(2), 285–292. doi:10.1007/s004420050788

Chapple DG, Swain R (2004) Inter-populational variation in the cost of autotomy in the metallic skink (*Niveoscincus metallicus*). *Journal of Zoology* **264**(4), 411–418. doi:10.1017/S0952836904005941

Chapple DG, Miller KA, Chaplin K, Barnett L, Thompson MB, Bray RD (2014) Biology of the invasive delicate skink (*Lampropholis delicata*) on Lord Howe Island. *Australian Journal of Zoology* **62**(6), 498–506. doi:10.1071/ZO14098

Chapple DG, Reardon JT, Peace JE (2016) Origin, spread and biology of the invasive plague skink (*Lampropholis delicata*) in New Zealand. In *New Zealand Skinks*. (Ed. DG Chapple) pp. 341–359. Springer, Cham. doi:10.1007/978-3-319-41674-8_13

Clifford HT (1988) Predation of skinks by *Achaearanea tepidariorum*. *Newsletter of the Australasian Arachnological Society* **32**, 2–3.

Cogger HG (2018) *Reptiles and Amphibians of Australia*. Updated 7th edn. CSIRO Publishing, Clayton South, Vic.

Cogger HG, Cameron E, Sadlier R, Eggler P (1993) *The Action Plan for Australian Reptiles*. Australian Nature Conservation Agency, Canberra.

Contos P, Letnic M (2019) Diet of the crest-trailed mulgara (*Dasycercus cristicauda*) in the Strzelecki Desert. *Australian Mammalogy* **42**(2), 211–215. doi:10.1071/AM19003

Copland SJ (1953) Recent Australian herpetology. *The Proceedings of the Linnean Society of New South Wales* **78**, v–xxxvii.

Croft P, Reid N, Hunter JT (2010) Experimental burning changes the quality of fallen timber as habitat for vertebrate and invertebrate fauna: implications for fire management. *Wildlife Research* **37**(7), 574–581. doi:10.1071/WR10053

Cromie GL, Chapple DG (2012) Impact of tail loss on the behaviour of locomotor performance of two sympatric *Lampropholis* skink species. *PLoS One* **7**(4), e34732. doi:10.1371/journal.pone.0034732

Crowther MS, Cairns KM, van Eeden LM, Letnic M (2021) Introgression does not influence the positive ecological and functional role of dingo populations. *Australian Zoologist* **41**(3), 338–346. doi:10.7882/AZ.2020.016

Cupples JB, Crowther MS, Story G, Letnic M (2011) Dietary overlap and prey selectivity among sympatric carnivores: could dingoes suppress foxes through competition for prey? *Journal of Mammalogy* **92**(3), 590–600. doi:10.1644/10-MAMM-A164.1

Dejean A (2007) Predatory abilities favour the success of the invasive ant *Pheidole megacephala* in an introduced area. *Journal of Applied Entomology* **131**(9–10), 625–629. doi:10.1111/j.1439-0418.2007.01223.x

Dejean A, Moreau CS, Uzac P, Le Breton J, Kenne M (2007) The predatory behavior of *Pheidole megacephala* [*Le comportement prédateur de Pheidole megacephala*]. *Comptes Rendus Biologies* **330**(9), 701–709. doi:10.1016/j.crvi.2007.06.005

Dent E (2016) The impacts of an introduced mammalian predator (*Mus musculus*) on tree weta (*Hemideina trewicki*) and skinks (*Oligosoma polychroma, Oligosoma infrapunctatum and Oligosoma lineoocellatum*) in Cape Sanctuary, Hawkes Bay. MSc thesis. Victoria University, Wellington.

DEWHA (2010) The cane toad (*Bufo marinus*). Fact sheet. Department of the Environment, Water, Heritage and the Arts. <https://www.dcceew.gov.au/environment/invasive-species/publications/factsheet-cane-toad-bufo-marinus>

Dickman CR (1996) *Overview of the Impacts of Feral Cats on Australian Native Fauna*. Australian Nature Conservation Agency, Canberra.

Dickman CR (2003) Distributional ecology of dasyurid marsupials. In *Predators with Pouches: The Biology of Carnivorous Marsupials*. pp. 318–331. CSIRO Publishing, Collingwood.

Dickman CR (2009) House cats as predators in the Australian environment: impacts and management. *Human–Wildlife Conflicts* **3**(1), 41–48. doi:10.26077/55nn-p702

Dickman CR (2014) Micro-carnivores: the ecological role of small dasyurid predators in Australia. In *Carnivores of Australia: Past, Present and Future*. (Eds S Glen and CR Dickman) pp. 241–263. CSIRO Publishing, Collingwood.

Dickman CR, Newsome TM (2015) Individual hunting behaviour and prey specialisation in the house cat *Felis catus*: implications for conservation and management. *Applied Animal Behaviour Science* **173**, 76–87. doi:10.1016/j.applanim.2014.09.021

Dissanayake DSB, Clemann N (2022) A common garden skink (*Lampropholis guichenoti*) with a bifurcated tail. *Reptiles & Amphibians* **29**(1), 139–140. doi:10.17161/randa.v29i1.16365

Doherty TS, Davis RA, Etten EJB, Algar D, Collier N, Dickman CR, Edwards G, Masters P, Palmer R, Robinson S (2015) A continental-scale analysis of feral cat diet in Australia. *Australian Mammalogy* **37**, 219–224. doi:10.1071/AM14038

Doherty TS, Dickman CR, Johnson CN, Legge SM, Ritchie EG, Woinarski CZ (2017) Impacts and management of feral cats *Felis catus* in Australia. *Mammal Review* **47**(5), 83–97. doi:10.1111/mam.12080

Donisthorpe H (1935) The ants of Christmas Island. *Annals and Magazine of National History* **15**(90), 629–635. doi:10.1080/00222933508655010

Doody JS, Paull P (2013) Hitting the ground running: environmentally cued hatching in a lizard. *Copeia* **2013**(1), 160–165. doi:10.1643/CE-12-111

Downes S (1999) Prey odor influences retreat-site selection by naïve broadheaded snakes (*Hoplocephalus bungaroides*). *Journal of Herpetology* **33**(1), 156–159. doi:10.2307/1565560

Downes S (2001) Trading heat and food for safety: costs of predator avoidance in a lizard. *Ecology* **82**(10), 2870–2881. doi:10.1890/0012-9658(2001)082[2870:THAFFS]2.0.CO;2

Downes SJ (2002a) Size-dependent predation by snakes: selective foraging or differential prey vulnerability? *Behavioral Ecology* **13**(4), 551–560. doi:10.1093/beheco/13.4.551

Downes SJ (2002b) Does responsiveness to predator scents affect lizard survivorship? *Behavioral Ecology and Sociobiology* **52**(1), 38–42. doi:10.1007/s00265-002-0477-z

Downes S, Hoefer AM (2004) Antipredatory behaviour in lizards: interactions between group size and predation risk. *Animal Behaviour* **67**, 485–492. doi:10.1016/j.anbehav.2003.05.010

Downes SJ, Shine R (1998) Sedentary snakes and gullible geckos: predator-prey coevolution in nocturnal rock-dwelling reptiles. *Animal Behaviour* **55**(5), 1371–1385. doi:10.1006/anbe.1997.0704

Downes SJ, Shine R (1999) Do incubation-induced changes in a lizard's phenotype influence its vulnerability to predators? *Oecologia* **120**(1), 9–18. doi:10.1007/s004420050827

Downes SJ, Shine R (2001) Why does tail loss increase a lizard's later vulnerability to snake predators? *Ecology* **82**(5), 1293–1303. doi:10.1890/0012-9658(2001)082[1293:WDTLIA]2.0.CO;2

Fearn D, Trembath DF (2011) Natural history of the common tree snake, *Dendrelaphis punctulatus* (Serpentes: Colubridae), in the wet-dry tropics of north Queensland. *Australian Journal of Zoology* **58**(6), 384–389. doi:10.1071/ZO10059

Fearn D, Robinson B, Sambono J, Shine R (2002) Pythons in the pergola: the ecology of 'nuisance' carpet pythons (*Morelia spilota*) from suburban habitats in south-eastern Queensland. *Wildlife Research* **28**(6), 573–579. doi:10.1071/WR00106

Ferraro TJ, Burgin S (1993) Review of environmental factors influencing the decline of Australian frogs. In *Herpetology in Australia: A Diverse Discipline*. (Eds D Lunney and D Ayers) Transactions of the Royal Zoological Society of New South Wales, pp. 205–218. Surrey Beatty & Sons, Chipping Norton, NSW.

Finlayson HH (1961) On Central Australian mammals. Part IV: the distribution and status of Central Australian species. *Records of the South Australian Museum* **14**, 141–191.

Fisher R, Ineich I (2012) Cryptic extinction of a common Pacific lizard *Emoia impar* (Squamata, Scincidae) from the Hawaiian Islands. *Oryx* **46**(2), 1–9. doi:10.1017/S0030605310001778

Fleay D (1952) The Tasmanian or marsupial devil: its habits and family life. *Australian Museum Magazine* **10**(9), 275–280.

Forsman A, Shine R (1995) The adaptive significance of colour pattern polymorphism in the Australian scincid lizard *Lampropholis delicata*. *Biological Journal of the Linnean Society* **55**(4), 273–291. doi:10.1111/j.1095-8312.1995.tb01066.x

Foster CN, Barton PS, Sato CF, Wood JT, MacGregor CI, Lindenmayer DB (2016) Herbivory and fire interact to affect forest understorey habits, but not its use by small vertebrates. *Animal Conservation* **19**(1), 15–25. doi:10.1111/acv.12210

Franklin M, Rand J, Marston L, Morton J (2021) Do pet cats deserve the disproportionate blame for wildlife predation compared to pet dogs? *Frontiers in Veterinary Science* **8**, 731689. doi:10.3389/fvets.2021.731689

Geiser F, Pavey CR (2007) Basking and torpor in a rock-dwelling desert marsupial: survival strategies in a resource-poor environment. *Journal of Comparative Physiology, B: Biochemical, Systemic and Environmental Physiology* **177**(8), 885–892. doi:10.1007/s00360-007-0186-z

Gillies C, Clout M (2003) The prey of domestic cats (*Felis catus*) in two suburbs of Auckland City, New Zealand. *Journal of Zoology (London)* **259**(3), 309–315. doi:10.1017/S095283690200328X

Gills G, Higham TE (2016) Consequences of lost endings: caudal autotomy as a lens for focusing attention on tail function during locomotion. *Journal of Experimental Biology* **219**(16), 2416–2422. doi:10.1242/jeb.124024

Glen AS, Dickman CR (2006) Diet of the spotted-tailed quoll (*Dasyurus maculatus*) in eastern Australia: effects of season, sex and size. *Journal of Zoology* **269**(2), 241–248. doi:10.1111/j.1469-7998.2006.00046.x

Glen AS, Fay AR, Dickman CR (2006) Diets of sympatric red foxes *Vulpes* and wild dogs *Canis lupus* in the Northern Rivers Region, New South Wales. *Australian Mammalogy* **28**(1), 101–104. doi:10.1971/AM06013

Gray C, Doody JS, Kimes K, Enloe C, Bartek B, Clulow S, Deering S, Soennichsen K, Fryer H (2020) *Dendrelaphis punctulatus* (common tree snake). *Herpetological Review* **51**(2), 348.

Green K (1989) Altitudinal and seasonal differences in the diets of *Antechinus swainsonii* and *Antechinus stuartii* (Marsupialia, Dasyuridae) in relation to the availability of prey in the Snowy Mountains. *Australian Wildlife Research* **16**, 581–592. doi:10.1071/WR9890581

Greer A (1997) *The Biology and Evolution of Australian Snakes*. Surrey Beatty & Sons, Chipping Norton, NSW.

Grey MJ, Clarke MF, Loyn RH (1998) Influence of the noisy miner *Manorina melanocephala* on avian diversity and abundance in remnant grey box. *Pacific Conservation* **4**, 55–69. doi:10.1071/PC980055

Griffiths K (1987) *Reptiles of the Sydney Region*. Three Sisters Productions, Winmalee.

Guarino F (2002) Spatial ecology of a large carnivorous lizard, *Varanus varius* (Squamata: Varanidae). *Journal of Zoology* **258**(4), 449–457. doi:10.1017/S0952836902001607

Guiler E (1970) Observations on the Tasmanian devil, *Sarcophilus harrisii* (Marsupialia: Dasyuridae) 1. Numbers, home range, movements and food in two populations. *Australian Journal of Zoology* **18**, 49–62. doi:10.1071/ZO9700063

Hanifie S (2018) *Cannibal Mice Eat Other as Grain Harvest Ends, Lowering the Rodent Population*. ABC Riverland, <https://www.abc.net.au/news/2018-04-27/hungry-mice-turn-to-cannibalism-lowering-mice-popultions/9700080>.

Harris DB (2009) Review of negative effects of introduced rodents on small mammals on islands. *Biological Invasions* **11**(7), 1611–1630. doi:10.1007/s10530-008-9393-0

Harris RJ (2002) *Potential Impacts of the Argentine Ant (Linepithema humile) in New Zealand and Options for Its Control*. Department of Conservation, Wellington, New Zealand.

Heatwole H (1970) Thermal ecology of the desert dragon *Amphibolurus inermis*. *Ecological Monographs* **40**(4), 425–457. doi:10.2307/1942339

Heterick B (1997) The interaction between the coastal brown ant, *Pheidole megacephala* (Fabricius), and other invertebrate fauna of Mt Coot-tha (Brisbane, Australia). *Australian Journal of Ecology* **22**(2), 218–221. doi:10.1111/j.1442-9993.1997/tb00661.x

Hoare JM (2006) Novel predators and naïve prey: how introduced mammals shape behaviours and populations of New Zealand lizards. PhD thesis. Victoria University of Wellington, Wellington.

Hoare JM, Pledger S, Nelson NJ, Daugherty CH (2007) Avoiding aliens: behavioural plasticity in habitat use enables large, nocturnal geckos to survive Pacific rat invasions. *Biological Conservation* **136**(4), 510–519. doi:10.1016/j.biocon.2006.12.022

Hoffmann BD, Parr CL (2008) An invasion revisited: the African big-headed ant (*Pheidole megacephala*) in northern Australia. *Biological Invasions* **10**, 1171–1181. doi:10.1007/s10530-007-9194-x

Howard R, Williamson I, Mather P (2003) Structural aspects of microhabitat selection by the skink *Lampropholis delicata*. *Journal of Herpetology* **37**(3), 613–617. doi:10.1670/222-99N

Hu Y, Gillespie G, Jessop TS (2019) Variable reptile responses to introduce predator control in southern Australia. *Wildlife Research* **46**(1), 64–75. doi:10.1071/WR18047

Hughes J, Macdonald DW (2013) A review of the interactions between free-roaming domestic dogs and wildlife. *Biological Conservation* **157**, 341–351. doi:10.1016/j.biocon.2012.07.005

Human KG, Weiss S, Weiss A, Sandler B, Gordon DM (1998) Effects of abiotic factors on the distribution and activity of the invasive Argentine ant (Hymenoptera, Formicidae). *Environmental Entomology* **27**(4), 822–833. doi:10.1093/ee/27.4.822

Humphreys WF (1976) Spider induced egg mortality in a skink population. *Copeia* **1947**, 404.

Hutton I (2003) *The Australian Geographic Book of Lord Howe Island*. Australian Geographic, Sydney.

Jarman PJ, Allen LR, Boschma DJ, Green SW (2007) Scat contents of the spotted-tailed quoll *Dasyurus maculatus* in the New England gorges, north-eastern New South Wales. *Australian Journal of Zoology* **55**(1), 63–72. doi:10.1071/ZO06014

Jelbart J (1998) An investigation of the influence of edge effects on the distribution and abundance of skinks in Remnant Cumberland Plain Woodland in Western Sydney. Master in Environmental Science, University of Western Sydney – Hawkesbury, Richmond.

Jessop T, Urlus J, Lockwood T, Gillespie G (2010) Preying possum: assessment of the diet of lack monitors (*Varanus varius*) from coastal forests in Southern Victoria. *Biawak* **4**, 59–63.

Jones C, Moss K, Sanders M (2005) Diet of hedgehogs (*Erinaceus europaeus*) in the upper Waitaki Basin, New Zealand: implications for conservation. *New Zealand Journal of Ecology* **29**(1), 29–35. <https://www.jstor.org/stable/24056190>

Jones ME, Barmuta LA (1998) Diet overlap and relative abundance of sympatric dasyurid carnivores: a hypothesis of competition. *Archaeologica Oceania* **25**, 102–113.

Jones ME, Burnett S, Claridge AW, Fancourt B, Körtner G, Morris K, Peacock D, Troy S, Woinarski J (2014) Australia's surviving marsupial carnivores: threats and conservation. In *Carnivores of Australia: Past, Present and Future*. (Eds AS Glen and CR Dickman) pp. 197–240. CSIRO Publishing, Collingwood.

Jourdan H, Brescia F, Vidal E (2015) Impact des espèces invasives sur les reptiles des massifs miniers. Rapport Scientifique CNRT/IRD.IAC.

Kays RW, DeWan AA (2004) Ecological impact of inside/outside house cats around a suburban nature preserve. *Animal Conservation* **7**(3), 273–283. doi:10.1017/S1367943004001489

Keiper P, Johnson CN (2004) Diet and habitat preference of the Cape York short-nosed bandicoot (*Isoodon obesulus peninsulae*) in north-east Queensland. *Wildlife Research* **31**(3), 259–265. doi:10.1071/WR02030

Kenis M, Auger-Rozenberg M-A, Rogues A, Timms L, Péré C, Cock MJW, Settele J, Augustin S, Lopez-Vaamonde C (2009) Ecological effects of invasive alien insects. *Biological Invasions* **11**, 21–45. doi:10.1007/s10530-008-9318-y

King DF (1966–1968) Eggs and hatching of the lizard *Leiolopisma guichenoti* (Scincidae). *Geelong Naturalist: Journal of the Geelong Field Naturalists Club* **23–24**, 47–48.

Kremaster L, Bunnell FL (1999) Edge effects: theory, evidence and implications to management of western Northern American forests. In *Forest Fragmentation: Wildlife and Management Implications*. (Eds J Wisniewski, JA Rochelle and L Lehmann) pp. 117–153. Brill, Leiden.

Lach L, Thomas ML (2008) Invasive ants in Australia: documented and potential ecological consequences. *Australian Journal of Entomology* **47**(4), 275–288. doi:10.1111/j.1440-6055.2008.00659.x

Langkilde T, Smith V, Phillips S, Barrott E, Shine R (2002) Ornamental plant traps lizard. *Herpetofauna* **32**, 131.

Law BS, Bradley RA (1990) Habitat use and basking site selection in the water skink, *Eulamprus quoyii*. *Journal of Herpetology* **24**(3), 235–240. doi:10.2307/1564388

Lee C-Y, Yang C-CS (2022) Biology, ecology, and management of the invasive long-legged ant, *Anoplolepis gracilipes*. *Annual Review of Entomology* **67**, 43–63. doi:10.1146/annurev-ento-033121-102332

Legge S, Murphy BP, McGregor H, Woinarski JCZ, Augusteyn J, Ballard G, Baseler M, Buckmaster T, Dickman CR, Doherty T, Edwards G, Eyre T, Fancourt B, Ferguson D, Forsyth DM, Geary WL, Gentle M, Gillespie G, Greenwood L, Hohnen R, Hume S, Johnson CN, Maxwell N, McDonald P, Morris K, Moseby K, Newsome T, Nimmo D, Paltridge R, Ramsay D, Read J, Rendall A, Rich M, Ritchie E, Rowland J, Short J, Stokeld D, Sutherland DR, Wayne AF, Woodford L, Zewe F (2017) Enumerating a continental-scale threat: how many feral cats are in Australia? *Biological Conservation* **206**, 293–303. doi:10.1016/j.biocon.2016.11.032

Lenancker P, Feldhaar H, Holzinger A, Greenfield M, Strain A, Yeeles P, Hoffman BD, Tay WT, Lach L (2021) Origin, behaviour, and genetics of reproductive workers in an invasive ant. *Frontier in Zoology* **18**(1), 13. doi:10.1186/s12983-021-00392-2

Letnic M, Koch F, Gordon C, Crowther MS, Dickman CR (2009) Keystone effects of an alien top predator stem extinctions of native mammals. *Proceedings of the Royal Society B: Biological Sciences* **276**(1671), 3249–3256. doi:10.1098/rspb.2009.0574

Lettink M, Cree A (2006) Predation by the feral house mouse *Mus musculus*, of McCann's skinks (*Oligosoma maccanni*) constrained in pitfall traps. *Herpetofauna* **36**, 61–62.

Leung AN, Phu TTL, Lister D, Poolice J, Caon L, Jones R, Chowdhary N, Warner DA (2006) Diurnal skinks (*Lampropholis delicata*) seek shelter in the presence of scent from a nocturnal predator and prey. *Herpetofauna* **36**(2), 117–121.

Lindenmayer DB, Cunningham RB, MacGregor C, Tribolet C, Donnelly CF (2001) A prospective longitudinal study of landscape matrix effect on fauna in woodland remnants: experimental design and baseline data. *Biological Conservation* **101**(2), 157–169. doi:10.1016/S0006-3207(01)00061-1

Llewelyn J, Shine R, Webb JK (2005) Thermal regimes and diel activity patterns of four species of small elapid snakes from south-eastern Australia. *Australian Journal of Zoology* **53**(1), 1–8. doi:10.1071/ZO04037

Lloyd R, Alford RA, Schwarzkopf L (2009) Chemical discrimination among predators by lizards: responses of three skink species to the odours of high- and low threat varanid predators. *Austral Ecology* **34**(1), 50–54. doi:10.1111/j.1442-9993.2008.01881.x

Lloyd KA, Hernandez SM, Carroll JP, Abernathy KJ, Marshall GJ (2013) Quantifying free-roaming domestic cat predation using animal-borne video cameras. *Biological Conservation* **160**, 183–189. doi:10.1016/j.biocon.2013.01.008

Lohr CA (2012) Human dimensions of introduced terrestrial vertebrates in the Hawaiian Islands. PhD thesis. University of Hawaii, Manoa.

Loss SR, Will T, Marra PP (2013) The impact of free-ranging domestic cats on wildlife of the United States. *Nature Communications* **4**, 1396. doi:10.1038/ncomms2380

Lunney D, Ashby E, Grigg J, O'Connell M (1989) Diets of scincid lizards *Lampropholis guichenoti* (Dumeril and Bibron) and *Lampropholis delicata* in Mumbulla State Forest on the South Coast of New South Wales. *Australia Wildlife Research* **16**(3), 307–312. doi:10.1071/WR9890307

Lunney D, Triggs B, Eby P, Ashby E (1990) Analysis of scats of dogs *Canis familiaris* and foxes *Vulpes* (Canidae: Carnivora) in coastal forests near Bega, New South Wales. *Australian Wildlife Research* **17**(1), 61–68. doi:10.1071/WR9900061

Lunney D, Mathews A, Grigg J (2001) The diet of *Antechinus agilis* and *A. swainsonii* in unlogged and regenerating sites in Mumbulla State Forest, south-eastern New South Wales. *Wildlife Research* **28**(5), 459–464. doi:10.1071/WT00015

MacNalley R, Parkinson A, Horrocks G, Conole L, Tzaros C (2001) Relationships between terrestrial vertebrate diversity, abundance and availability of coarse woody debris on south-eastern Australian floodplains. *Biological Conservation* **99**(2), 191–205. doi:10.1016/S0006-3207(00)00180-4

Madani G (2021) Notes on the diet and foraging habits of the giant barred frog *Mixophyes iteratus*. *Australian Zoologist* **41**(2), 200–204. doi:10.7882/AZ.2020.028

Major RE, Christie FJ, Gowing G (2001) Influence of remnant and landscape attributes on Australian woodland bird communities. *Biological Conservation* **102**(1), 47–66. doi:10.1016/S0006-3207(01)00090-8

Martin LJ, Murray BR (2013) A preliminary assessment of the response of a native reptile assemblage to spot-spraying invasive bitou bush with glyphosate herbicide. *Ecological Management and Restoration* **14**(1), 59–62. doi:10.1111/emr.12019

McCallum J (1986) Evidence of predation by kiore upon lizards from the Mokohinau islands. *New Zealand Journal of Ecology* **9**, 83–87. <https://www.jstor.org/stable/24052769>

McGregor H, Legge S, Jones ME, Johnson CN (2015) Feral cats are better killers in open habitats, revealed by animal-borne video. *PLoS One* **10**(8), e0133915. doi:10.1371/journal.pone.0133915

Meek PD (1998) Food items brought home by domestic cats *Felis catus* (L) living in Booderee National Park, Jervis Bay. *Proceedings of The Linnean Society of New South Wales* **120**, 43–47. <http://biostor.org/reference/68324>

Meek R, Avery R, Weir E (2001) *Physignathus lesueurii* (Australian water dragon): predation on a skink (*Lampropholis delicata*). *Herpetological Bulletin* **76**, 31–32.

Mendl M, Harcourt R (1988) Individuality in the domestic cat (Paper). *American Society of Mechanical Engineers*, 41–54.

Ménez A, Zimmerman K, Zimmerman S, Heatwole H (1990) Venom apparatus and toxicity of the centipede *Ethmostigmus rubripes* (Chilopoda, Scolopendridae). *Journal of Morphology* **206**(3), 303–312. doi:10.1002/jmor.1052060307

Metcalfe DC, Ridgeway PA (2013) A case of web entanglement and apparent predation of the skink *Lampropholis delicata* (De Vis, 1888) (Sauria: Scincidae: Lygosominae) by the red-back spider *Latrodectus hasseltii* Thorell, 1870 (Aranea: Araneomorpha: Theridiidae) in an autochthonous mesic habitat in coastal southeast Australia. *Herpetology Notes* **6**(1), 375–377.

Mikula P (2015) Fish and amphibians as bat predators. *European Journal of Ecology* **1**(1), 71–80. doi:10.1515/eje-2015-0010

Miller B, Mullette KJ (1985) Rehabilitation of an endangered Australian bird: the Lord Howe Island woodhen *Tricholimnas sylvestris* (Sclater). *Biological Conservation* **34**(1), 55–95. doi:10.1016/0006-3207(85)90057-6

Mitchell BD, Banks PB (2005) Do wild dogs exclude foxes? Evidence for competition from dietary and spatial overlaps. *Austral Ecology* **30**(5), 581–591. doi:10.1111/j.1442-9993.2005.01473.x

Moloney S, Vanderwoude C (2002) Red imported fire ants: a threat to eastern Australia's wildlife? *Ecological Management & Restoration* **3**, 167–175. doi:10.1046/j.1442-8903.2002.t01-1-00109.x

Molsher RL, Gifford EJ, McIlroy JC (2000) Temporal, spatial and individual variation in the diet of red foxes (*Vulpes vulpes*) in central New South Wales. *Wildlife Research* **27**(6), 593–601. doi:10.1071/WR99015

Monks JM, Monks A, Towns DR (2014) Correlated recovery of five lizard populations following eradication of invasive mammals. *Biological Invasions* **16**(1), 167–175. doi:10.1007/s10530-013-0511-2

Moore D, Kearney MR, Paltridge R, McAlpin S, Stow A (2018) Feeling the pressure at home: predator activity at the burrow entrance of an endangered arid-zone skink. *Austral Ecology* **43**(1), 102–109. doi:10.1111/aec.12547

Morton SR, Denny MJS, Read DG (1983) Habitat preferences and diet of sympatric *Sminthopsis crassicaudata* and *S. macroura* (Marsupialia: Dasyuridae). *Australian Mammalogy* **6**(1), 29–34.

Murphy RC, Keedwell RJ, Brown KP, Westbrooke I (2004) Diet of mammalian predators in braided river beds in the central South Island, New Zealand. *Wildlife Research* **31**(6), 631–638. doi:10.1071/WR03033

Nagy KA, Lee AK, Martin RW, Fleming MR (1988) Field metabolic-rate and food requirement of a small dasyurid marsupial, Sminthopsis-Crassicaudata. *Australian Journal of Zoology* **36**(3), 293–299. doi:10.1071/ZO9880293

Naimo AC, Jones C, Chapple DG, Wong BBM (2021) Has an invasive lizard lost it antipredator behaviours following 40 generations of isolation from snake predators? *Behavioral Ecology and Sociobiology* **75**(9), 131. doi:10.1007/s00265-0212-03064-7

Nattrass R, Vanderwoude C (2001) A preliminary investigation of the ecological effects of fire ants (*Solenopsis invicta*) in Brisbane. *Ecological Management and Restoration* **2**(3), 220–223.

Newman SG (1994) Effects of a mouse *Mus musculus* eradication programme and habitat change on lizard population of Mana Island with special reference to McGregor's skink, *Cyclodina macgregori*. *New Zealand Journal of Zoology* **21**(4), 443–456. doi:10.1080/03014223.1994.9518015

Newsome TM, Crowther MS, Dickman CR (2014) Rapid recolonisation by the European red fox: how effective are uncoordinated and isolated control programs? *European Journal of Wildlife Research* **60**(5), 749–757. doi:10.1007/s10344-014-0844-x

Norbury G (2001) Conserving dryland lizards by reducing predator-mediated apparent competition and direct competition with introduced rabbits. *Journal of Applied Ecology* **38**(6), 1350–1361. doi:10.1046/j.0021-8901.2001.00685.x

Norbury G, van der Munckhof M, Neitzel S, Hutcheon A, Reardon JT, Ludwig K (2014) Impacts of invasive house mice on post-release survival of translocated lizards. *New Zealand Journal of Ecology* **38**(2), 322–327. <https://www.jstor.org/stable/24060810>

O'Dowd DJ, Green PT, Lake PS (2003) Invasional 'meltdown' on an oceanic island. *Ecological Letters* **6**(9), 812–817. doi:10.1046/j.1461-0248.2003.00512.x

OE&H (2021) *Invasion of the yellow crazy ant,* Anoplolepis gracilipes *(Fr. Smith) into NSW – profile.* Office of Environment and Heritage. <https://threatenedspecies.bionet.nsw.gov.au/profile?id=20010>

Olsson M, Gullberg A, Shine R, Madsen T, Tegelström H (1996) Paternal genotype influences incubation period, offspring size, and offspring shape in an oviparous reptile. *Evolution* **50**(3), 1328–1333. doi:10.1111/j.1558-5646.1996.tb02372.x

Ortega Z, Mencía A, Velentín P-M (2019) Antipredatory behaviour of a mountain lizard towards the chemical cues of its predatory snakes. *Behaviour* **155**(10), 817–840. doi:10.1163/1568539X-00003504

O'Shea M, Kelly K (2014) Predation on a weasel skink (*Saproscincus mustelinus*) (Squamata: Scincidae: Lygosominae) by a redback spider (*Latrodectus hasselti*) (Araneae: Araneomorpha: Theridiidae), with a review of other *Latrodectus* predation events involving squamates. *Herpetofauna* **44**(1&2), 49–53.

Paltridge R (2002) The diets of cats, foxes and dingoes in relation to prey availability in the Tanami Desert, Northern Territory. *Wildlife Research* **29**(4), 389–403. doi:10.1071/WR00010

Parsons H, Major RE, French K (2006) Species interactions and habitat associations of birds inhabiting urban areas of Sydney, Australia. *Austral Ecology* **31**(2), 212–217. doi:10.1111/j.1442-9993.2006.01584.x

Paton DC (1990) Domestic cats and wildlife. *Bird Observer* **696**, 34–35.

Pavey CR, Eldridge SR, Heywood M (2008) Population dynamics and prey selection of native and introduced predators during a rodent outbreak in arid Australia. *Journal of Mammalogy* **89**(3), 674–683. doi:10.1644/07-MAMM-A-168R.1

Pemberton D, Gales S, Bauer B, Gales R, Lazenby B, Medlock K (2008) The diet of the Tasmanian devil, *Sarcophilus harrisii*, as determined from analysis of scat and stomach contents. *Papers and Proceedings of the Royal Society of Tasmania*, **142**, 13–21. doi:10.26749/rstpp.142.2.13

Phillips BL, Brown GP, Shine R (2003) Assessing the potential impact of cane toads on Australian snakes. *Conservation Biology* **17**(6), 1738–1747. doi:10.1111/j.1523-1739.2003.00353.x

Pickard CR (1984) The population ecology of the house mouse (*Mus musculus*) on Man Island. MSc thesis. Victoria University, Wellington.

Purvis MD (1979) Early stages of tail regeneration in *Lampropholis guichenoti* (Lacertilia: Scincidae). *Australian Zoologist* **20**(2), 289–296.

Pyke GH and Miehs A (2001) Predation by water skinks (*Eulamprus quoyii*) on tadpoles and metamorphs of the green and golden bell frog (*Litoria aurea*). *Herpetofauna* **31**, 90–101.

Qualls F, Shine R (1997) Geographic variation in 'costs of reproduction' in the scincid lizard *Lampropholis guichenoti*. *Functional Ecology* **11**(6), 757–763. doi:10.1046/j.1365-2435.1997.00150.x

Quin DG (1988) Observations on the diet of the southern brown bandicoot, *Isoodon obsulus* (Marsupialia: Peramelidae), in Southern Tasmania. *Australian Mammalogy* **11**, 15–25.

Quinlan TG, Smith GT, Calver MC (1995) Relationships between morphology and feeding behaviour in the syntopic scorpions *Urodacus armatus* Pocock and Urodacus *novaehollandiae* Peters (Scorpiones: Scorpionidae). *Journal of the Australian Entomological Society* **34**(4), 277–279. doi:10.1111/j.1440-6055.1995.tb01339.x

Quiterie D, Bourguet E, Martin T, Sarah S, Gouyet R, Mathilde M, Vidal E (2019) Sympatric invasive rats show different diets in a tropical rainforest of an island biodiversity hotspot. *Pacific Science* **73**(2), 199–214. doi:10.2984/73.2.2

Radford IJ, Woolley L-A, Dickman CR, Corey B, Trembath D, Fairman R (2020) Invasive anuran driven tropical cascade: an alternative hypothesis for recent critical weight range mammal collapses across northern Australia. *Biological Invasions* **22**(6), 1967–1982. doi:10.1007/s10530-020-02226-4

Read J, Bowen Z (2001) Population dynamics, diet and aspects of the biology of feral cats and foxes in arid South Australia. *Wildlife Research* **28**(2), 195–2003. doi:10.1071/WR99065

Recher HF (2010) A not so natural history: the vertebrate fauna of Sydney. In *The Natural History of Sydney*. (Eds D Lunney, P Hutchings and D Hochuli) pp. 125–142. Royal Zoological Society of New South Wales, Sydney.

Rhind SG (2004) Direct impacts of logging and forest management on the brush-tailed phascogale *tapoatafa* and other arboreal marsupials in a jarrah forest in Western Australia. In *Conservation of Australia's Forest Fauna*. 2nd edn. (Ed. D Lunney) pp. 639–655. Royal Zoological Society of New South Wales, Sydney.

Risbey DA, Calver MC, Short J, Bradley JS, Wright IW (2000) The impact of cats and foxes on the small vertebrate fauna of Heirisson Prong, Western Australia (II) A field experiment. *Wildlife Research* **27**(3), 223–235. doi:10.1071/WR98092

Sadlier RA, Debar L, Chavis M, Bauer AM, Jourdan H, Jackman TR (2019) *Epibator insularis*, a new species of scincid lizard from l'Île Walpole, New Caledonia. *Pacific Science* **73**(1), 143–161. doi:10.2984/73.1.7

Salo P, Korpimäki E, Banks PB, Nordström M, Dickman CR (2007) Alien predators are more dangerous than native predators to prey populations. *Proceedings of the Royal*

Society B: Biological Sciences **274**(1615), 1237–1243. doi:10.1098/rspb.2006.0444

Saunders DA, Hobbs RJ, Margules CR (1991) Biological consequences of ecosystem fragmentation: a review. *Conservation Biology* **5**(1), 18–32. doi:10.1111/j.1523-1739.1991.tb00384.x

Scanes CG (2018) Invasive species. In *Animals and Human Society*. (Eds CG Scanes and SR Toukhsati) pp. 413–426. Elsevier, Amsterdam. doi:10.1016/B978-0-12-805247-1.00024-1

Scarff FR, Rhind SG, Bradley JS (1998) Diet and foraging behaviour of brush-tailed phascogales (*Phascogale tapoatafa*) in the jarrah forest of south-western Australia. *Wildlife Research* **25**(5), 511–526. doi:10.1071/WR98007

Schell CB (2002) Ecology and life-history variation within a population of the frog *Limnodynastes tasmaniensis* (Anura: Myobatrachidae) from a remnant woodland of the Cumberland Plain in north-western Sydney. PhD thesis. University of Western Sydney, Sydney. <http://handle.uws.edu.au:8081/1959.7/692>

Schulz M (2009) Notes on the delicate skink *Lampropholis delicata* occurring on Lord Howe Island. *Herpetofauna* **39**, 48–52.

Sewell SR, Catterall CP (1998) Bushland modification and styles of urban development: their effects on birds in south-eastern Queensland. *Wildlife Research* **25**, 41–63. doi:10.1071/WR96078

Shattuck SO (1999) *Australian Ants: Their Biology and Identification*. CSIRO Publishing, Collingwood.

Shea GM (2010) The suburban terrestrial reptile fauna of Sydney – winners and losers. In *The Natural History of Sydney*. (Eds D Lunney, P Hutchings and D Hochuli) pp. 154–197. Royal Zoological Society of New South Wales, Sydney. doi:10.7882/FS.2010.015

Shiels AB, Pitt EC (2014) A review of invasive rodent (*Rattus* spp. and *Mus musculus*) diets on Pacific Islands. *Proceedings of the 26th Vertebrate Pest Conference*. doi:10.5070/V426110336

Shiels AB, Flores CA, Khamsing A, Krushelnycky PD, Mosher SM, Drake DR (2013) Dietary niche differentiation among three species of invasive rodents (*Rattus rattus, R. exulans, Mus musculus*). *Biological Invasions* **15**, 1037–1048. doi:10.1007/s10530-012-0348-0

Shine R (1977) Habitats, diets, and sympatry in snakes: a study from Australia. *Canadian Journal of Zoology* **55**(7), 1118–1128. doi:10.1139/z77-144

Shine R (1980) Comparative ecology of three Australian snake species of the genus *Cacophis* (Serpentes: Elapidae). *Copeia* **1980**(4), 831–838. doi:10.2307/1444462

Shine R (1991) Strangers in a strange land: ecology of the Australian colubrid snakes. *Copeia* **1991**(1), 120–131. doi:10.2307/1446254

Shine R (2003) Locomotor speeds of gravid lizards: placing 'costs of reproduction' within an ecological context. *Functional Ecology* **17**(4), 526–533. doi:10.1046/j.1365-2435.2003.00756.x

Shine R, Koenig J (2001) Snakes in the garden: an analysis of reptiles 'rescued' by community-based wildlife carers. *Biological Conservation* **102**(3), 271–283. doi:10.1016/S0006-3207(01)00102-1

Shine R, Tamayo B (2016) When predators become prey: the lizard-eating spiders of suburbia. *Australian Zoologist* **38**(2), 212–213. doi:10.7882/AZ.2016.021

Shine T, Shine R (2016) Trick or treat? The lingering danger of Halloween for suburban fauna. *Australian Zoologist* **38**(1), 16. doi:10.7882/AZ.2016.002

Simbotwe MP (1985) Comparative ecology and demography of two populations of Australian reptiles (Lacertilia, Scincidae). PhD Dissertation. University of New England, Armidale, NSW.

Singleton GR, Brown PR, Pech RP, Jacob J, Mutze GJ, Krebs CJ (2005) One hundred years of eruptions of house mice in Australia: a natural biological curio. *Biological Journal of the Linnean Society* **84**(3), 617–627. doi:10.1111/j.1095-8312.2005.00458.x

Slip DJ, Shine R (1988) Feeding habits of the diamond python, *Morelia s. spilota*: ambush predation by a boid snake. *Journal of Herpetology* **22**(3), 323–330. doi:10.2307/1564156

Slobo-Wilson AM, Murphy BP, Legge SM, Chapple DG, Crawford HM, Dawson SJ, Dickmzn CR, Doherty TS, Fleming PA, Gentle M, Newsome TM, Palmer R, Rees MW, Ritchie EG, Speed J, Stuart J-M, Thompson E, Turpin J, Woinarski JCZ (2021) Reptiles as food: predation of Australian reptiles by introduced foxes compounds and complements predation by cats. *Wildlife Research* **48**(5), 470–480. doi:10.1071/WR20194

Smith AH (2016) Invasion ecology of the plague skink (*Lampropholis delicata*) in Hawaii. MSc thesis. University of Hawaii, Manoa.

Smith P, Smith J (2013) Flora and Fauna Survey of Pied Piper Playground and Melody Gardens Reserves, Seven Hills. Report prepared for Blacktown City Council. P & J Smith Ecological Consultants, Blaxland.

Smith HM, Dickman CR, Banks PB (2017) Using effect size benchmarks to assess when alien impacts are actually alien. *Scientific Reports* **7**, 38627. doi:10.1038/srep38627

Soderquist TR, Serena M (1994) An experimental reintroduction programme for brush-tailed phascogales (*Phacogale tapoatafa*): the interface between captivity and the wild. In *Creative Conservation: Interactive Management of Wild and Captive Animals*. (Eds PJS Olney, GM Mace and ATC Feistner) pp. 431–438. Springer, Dordrecht. doi:10.1007/978.94-011-0721-1_25

Spellerberg IF (1972) Temperature tolerances of southeast Australian reptiles examined in relation to reptile thermoregulatory behaviour and distribution. *Oecologia* **9**(1), 23–46. doi:10.1007/BF00345241

Spencer PBS, Yurchenko AA, David VA, Scott R, Koepfli K-P, Driscoll C, O'Brien SJ, Menotti-Raymond M (2016) The population origins and expansion of feral cats in Australia. *Journal of Heredity* **107**(2), 104–114. doi:10.1093/jhered/esv095

Stannard HJ, McAllan BM, Old JM (2014) Dietary composition and nutritional outcomes in two marsupials, *Sminthopsis macroura* and *S. crassicaudata*. *Journal of Mammalogy* **95**(3), 503–515. doi:10.1644/13-MAMM-A-071

Stokeld D, Fisher A, Gentles T, Hill B, Triggs B, Woinarski, Gillespie GR (2018) What do predator diets tell us about mammal declines in Kakadu National Park? *Wildlife Research* **45**(1), 92–101. doi:10.1071.WR17101

Strahan R (1983) *The Complete Book of Australian Mammals*. Angus & Robertson Publishers, Sydney.

Sutherland DR, Glen AS, de Torres PJ (2011) Could controlling mammalian carnivores lead to mesopredator release of carnivorous reptiles? *Proceedings of the Royal Society B: Biological Sciences* **278**(1706), 641–648. doi:10.1098/rspb.2010.2103

Swarbrick H, Rand J (2018) Application of a protocol based on trap-neuter-return (TNR) to manage unowned urban cats on an Australian University Campus. *Animals* **8**(5), 77. doi:10.3390/an/8050077

Taniguchi G (2011) Clumped versus broadcast applications of bait to control big-headed ants, *Pheidole megacephala* (F.), (Hymenoptera: Formicidae) in pineapple fields. *Sociobiology* **57**(2), 285–290.

Taylor RJ (1986) Notes on the diet of the carnivorous mammals of the upper Henry River region, western Tasmania. *Papers and Proceedings of the Royal Society of Tasmania* **120**, 7–10. doi:10.26749/rstpp.120.7

Thibault M, Brescia F, Jourdan H, Vidal E (2017) Invasive rodents, an overlooked threat for skinks in a tropical island hotspot of biodiversity. *New Zealand Journal of Ecology* **41**(1), 74–83. doi:10.20417/nzjecol.41.9

Thoen C, Bauwens D, Verheyen RF (1986) Chemoreceptive and behavioural responses of the common lizard *Lacerta vivipara* to snake chemical deposits. *Animal Behaviour* **34**(6), 1805–1813. doi:10.1016/S0003-3472(86)80266-4

Thomas M, Holway DA (2005) Condition-specific completion between invasive Argentine ants and Australian *Iridomyrmex*. *Journal of Animal Ecology* **74**(3), 532–542. doi:10.1111/j.1365-2656.2005.00952.x

Tidemann CR, Yorkston HD, Russack AJ (1994) The diet of cats, *Felis catus*, on Christmas Island, Indian Ocean. *Wildlife Research* **21**(3), 279–286. doi:10.1071/WR9940279

Tingley R, Thompson MB, Hartley S, Chapple DG (2016) Patterns of niche filling and expansion across the invaded ranges of an Australian lizard. *Ecography* **39**(3), 270–280. doi:10.1111/ecog.01576

Tomich PQ (1969) *Mammals in Hawaii, Honolulu*. Bishop Museum Press, Hawaiʻi.

Towns DR, Ferreira SM (2001) Conservation of New Zealand lizards (Lacertilia: Scincidae) by translocation of small populations. *Biological Conservation* **98**(2), 211–222. doi:10.1016/S0006-3207(00)00156-7

Towns DR, Hitchmough RA, Perrott J (2016) Conservation of New Zealand lizards: a fauna not forgotten but undervalued? In *New Zealand Lizards*. (Ed. D Chapple) pp. 293–320. Springer, Cham.

Towns DR, Parrish R, Westbrooke I (2003) Inferring vulnerability to introduced predators without experimental demonstration: case study of Suter's skink in New Zealand. *Conservation Biology* **17**(5), 1361–1371. doi:10.1046/j.1523-1739.2003.02175.x

Triggs BH, Brunner H, Cullen JM (1984) The food of the fox, dog and cat in Croajingalong National Park, southeastern Victoria. *Australian Wildlife Research* **11**(3), 491–499. doi:10.1071/WR9840491

Turner G (2010) Some predatory-prey interactions involving reptiles and frogs from the Wet Tropics Region, North Queensland. *Queensland Naturalist* **48**(6), 78–83.

Turner G (2021) Predation of a garden skink *Lampropholis delicata* by a union-jack wolf spider *Tasmanicosa* sp. (Lycosidae). *Queensland Naturalist* **59**(1–3), 24–27.

Turner DC, Bateson P (2000) Why the cat? In *The Domestic Cat: The Biology of Its Behaviour*. (Eds DC Turner, P Bateson and PPG Bateson) pp. 3–8. Cambridge University Press, Cambridge.

Twigg LE, Fox BJ (1991) Recolonization of regenerating open forest by terrestrial lizards following sand mining. *Australian Journal of Ecology* **16**(2), 137–148. doi:10.1111/j.1442-9993.1991.tb01041.x

Tyler M (1999) *Australian Frogs: A Natural History*. New Holland, Frenchs Forest.

Underwood H, Calaban M (1987) Pineal melatonin rhythms in the lizard *Anolis carolinensis*: I. Response to light and temperature cycles. *Journal of Biological Rhythms* **2**, 179–193.

Urlus J (2009) The influence of vegetation structure and land management on East Gippsland reptile communities. (Hons) thesis. Deakin University, Melbourne.

van der Ree R, Soderquist TR, Bennett AF (2001) Home-range use by the brush-tailed phascogale (*Phascogale tapoatafa*) (Marsupialia) in high-quality, spatially limited habitat. *Wildlife Research* **28**(5), 517–525. doi:10.1071/WR00051

van der Ree R, Bennett AF, Soderquist TR (2006) Nest-tree selection by the threatened brush-tailed phascogale (*Phascogale tapoatafa*) (Marsupialia: Dasyuridae) in a highly fragmented agricultural landscape. *Wildlife Research* **33**(2), 113–119. doi:10.1071/WR05007

Vanderwoude C, De Bruyn LA, House APN (2000) Response of an open-forest ant community to invasion by the introduced ant, *Pheidole megacephala*. *Austral Ecology* **25**(3), 253–259. doi:10.1046/j.1442-9993.2000.01021.x

van Heezik Y (2010) Pussyfooting around the issue of cat predation in urban areas. *Oryx* **44**(2), 153–154. doi:10.1017/S003060531000027X

Wall M, Shine R (2009) The relationship between foraging ecology and lizard chemoreception: can a snake analogue (Burton's legless lizard, *Lialis burtonis*) detect prey scent? *Ethology* **115**(3), 264–272. doi:10.1111/j.1439-0310.2008.01595.x

Weavers BW (1989) Diet of the lace monitor (*Varanus varius*) in south-eastern Queensland. *Australian Zoologist* **25**(3), 83–85. doi:10.7882/AZ.1989.007

Webb GA (1987) Stomach contents of some reptiles killed on roads in the Bombala area, New South Wales. *Herpetofauna* **17**, 7–8.

Webb GA, Rose AB (1984) The food of some Australian snakes. *Herpetofauna* **16**, 21–26.
Webb JK, Du W-G, Pike DA, Shine R (2009) Chemical cues from both dangerous and nondangerous snakes elicit antipredator behaviours from a nocturnal lizard. *Animal Behaviour* **77**(6), 1471–1478. doi:10.1016/j.anbehav.2009.02.020
Whitaker AH (1973) Lizard populations on islands with and without Polynesian rats, *Rattus exulans* (Peale). *Proceedings of the New Zealand Ecological Society* **20**(1973), 121–130. <https://www.jstor.org/stable/24061519>
Woinarski JCZ, Murphy BP, Palmer R, Legge SM, Dickman CR, Doherty TS, Edwards G, Nankivell A, Read JL, Stokeld D (2018) How many reptiles are killed by cats in Australia? *Wildlife Research* **45**(3), 247–266. doi:10.1071/WR17160

10
Parasites and pathogens

INTRODUCTION

Reptiles are infested by a variety of parasites that include viruses, bacteria, and eukaryotic organisms (Schall 1982; Bower *et al.* 2019). A literature search in Google Scholar and other academic research tools, using the key words '*Lampropholis parasite*', '*reptile parasite*' and related terms found widely scattered reports on this topic.

The information on the parasites and diseases of reptiles in the wild is limited and focuses on the taxonomy of the pathogens rather than on the pathogenic outcomes of hosts (Gartrell 2016). Bower *et al.* (2019) suggested that the reasons for such limited study included challenges caused by difficulties in visual identification and classification of parasites, small sample sizes with consequent limited statistical power, and inadequate preservation of sufficient specimens of hosts to enable subsequent examination of their parasites. These issues are exacerbated when the host is small, innocuous, inconspicuous, and not perceived to be of any commercial value.

Most mention of parasites of skinks worldwide consist of passing comments on their potential role as vectors of disease, description of new species, infestation of new species of host, and/or description of the morphology of the parasite. Less frequently, studies have dealt with the impact of such organisms on the health and wellbeing of the host species concerned. For example, despite being common throughout urban Sydney, no sick or injured *Lampropholis* skinks were recorded by the Wildlife Information and Rescue Service (WIRES) database over a 10-year period (1989–1998), even though more than 11,000 lizards were examined during that time (Shine and Koenig 2001). Although, *L. delicata* and *L. guichenoti* are common species in the region of reporting, they are small, harmless, and draw limited attention. Whereas reptiles often are presented to wildlife-care facilities in Australia, they are almost always admitted because of traumatic injury and seldom with clinically apparent parasitic diseases (Rose 2005). It is not surprising, therefore, that Shine and Koenig (2001) did not find a record of *Lampropholis* presented to WIRES. Because of the small size of even fully grown *Lampropholis*, traumatic injuries, other than caudal autotomy, would commonly result in death, rather than in sub-lethal injuries. Free-ranging reptiles may, however, become infected with a range of parasites although, typically, the load is likely to be lower than in captivity (Telford 1971; Papini *et al.* 2011; Okulewicz *et al.* 2015), although Ko-Huan Lee (pers. comm.) found that in captivity, *L. guichenoti* can shed its load of parasites within 2 weeks.

This paucity of research occurs despite the recognition that studies of species-level competition may be complicated by, among other factors, an individual's number of parasites and/or its disease-load. In recent decades, however, Simberloff *et al.* (1997), Mack *et al.* (2000), and Resasco *et al.* (2019) have noted that, increasingly, parasites and disease have become recognised as potentially influencing the functioning of food webs and the disruption of ecosystems. Diseases may also endanger entire species.

There is a growing concern that the introduction of exotic organisms will facilitate the spread of their associated diseases (Bram and George 2000; Walter and Shaw 2002; Okulewicz *et al.* 2015). External globalisation and free trade and within-country failures in biosecurity; disruption of existing pests by more recent introductions; climatic change; intensification of land-use; withdrawal or reduction of the efficacy of pesticides; and on-going reduction in the acceptance of solutions for managing pests likely accelerates the rate at which exotic species are introduced and are spread (Goldson *et al.* 2015; Gartrell 2016). Specifically, the repeated introduction of *L. delicata* into New Zealand has the potential to introduce alien parasites and pathogens to native reptiles (Gill *et al.* 2001; Peace 2004; Chapple *et al.* 2015, 2016; Wairepo 2015; also see Chapter 3).

The prevalence of parasites in an area and the relative population sizes of host and parasite affect the degree of stress on the host and the nature of its immune responses (McCoy *et al.* 2012). Elevation may also play a role in host-parasite dynamics. For example, Hamilton *et al.* (2021) reported that the diversity and abundance both of ectoparasites and internal parasites of skink assemblages decreased with elevation.

Argaez *et al.* (2018) found that in several species of phrynosomatid lizards, individuals that had lost their tails were more susceptible to a higher parasite load than were those that had not previously experienced autotomy. Caudal autotomy is common in *Lampropholis* (see Chapter 9) and it would be of interest whether it is affected by elevation in the same way.

The etiology of parasitism may differ between wild populations of hosts and those in captivity or in the pet trade. Okulewicz *et al.* (2015) found parasitic infection to be widespread, with a diversity of endoparasitic 'Protozoa', Cestoda, Trematoda, and pseudoparasites among exotic lizards in private captive-breeding collections and in the trade network, the latter probably including wild-caught animals. The total prevalence of endoparasites was 42.35% ($n = 168$) but was higher (59.5%) in reptiles that had been bred in captivity than in those from the trade network. Whereas a wide range of species of lizards were present in the trade network of pet shops and wholesalers, diversity of parasites in private collections was rather low and consisted mostly of Oxyuridae and 'Protozoa'. The diversity was higher within the trade-network cohort, with some of the parasitic species regarded as rare.

Introduced species of lizards may have an elevated diversity of parasites as they would not only have the complement from their place of origin but also new ones picked up in their new habitat from native species (Goldberg and Bursey 2000).

In this chapter the ectoparasites, endoparasites, and diseases of *Lampropholis* are addressed. However, despite seeking to cover all species of the genus, as with other chapters, in this case it was not possible as *L. delicata* and *L. guichenoti* appear to be the only species in their genus to have received any attention.

ECTOPARASITES

Ectoparasites are parasites that live on the surface of animals, rather than inside them. The most common ones associated with reptiles are mites (Acari: Trombiculidae) and ticks (Acari: Ixodidae) (Rose 2005). Many mites and ticks are reported to be associated with leaf litter (e.g. Norris and Conroy 1999; York 1999), the habitat that *Lampropholis* species have been shown to prefer (Torr and Shine 1996; Taylor and Fox 2001; Howard *et al.* 2003; Anderson 2007). Ticks attack only the larger species of lizards, whereas mites parasitise species of all sizes of hosts (see review by Heatwole and Pianka 1993).

Mites

Although mites effectively have a worldwide distribution and parasitise a wide range of reptiles (Divers 2020), data on their infestation of Australian skinks is limited. Arnold (1986, 1993) suggested that Australian endemic reptiles may host a range of species of mites. Domrow and Lester (1985) indicated that at the time of their publication, 38 species of squamates were known to host them, but that assessment probably was incomplete. An indication of the prevalence of mites was reported by Heath and Whitaker (2015), who observed that most species of skinks, including *L. delicata*, intercepted on entry into New Zealand, carried a range of ectoparasites, mostly mites. When such infestation is heavy, the host may experience reduced vitality. Death may subsequently occur due to anaemia and/or concurrent bacterial and viral infections (see below). Perhaps the heaviest infestation by mites for any Australian reptile was reported by Domrow (1981) who found a moribund *L. delicata* that had its whole dorsum, sides, limbs, and tail infested by hundreds of nymphs of the mite *Urosactinia* sp.

In captivity, parasitic mites, predominantly the snake mite, *Ophionyssus natricis*, may severely debilitate reptiles, including skinks (Domrow 1987; Walter and Shaw 2002). On occasion, they may even cause morbidity due to induced anaemia and/or due to the transfer of haemoparasites, mainly to snakes but also to skinks (Pasmans *et al.* 2008). Other species of the genus (e.g. *O. scincorum*) have been observed to be harmful vectors of haemoparasites in some skinks (Allison and Desser 1981). The impact of infection on reproductive success is, however, not well understood (Hare *et al.* 2010). In addition to these two species of *Ophionyssus*, a diversity of endemic mites, particularly Trombiculidae (chiggers) may infect free-ranging lizards (Arnold 1986, 1993).

Stapley and Keogh (2006) reported that the loads of mites on male *Pseudemoia entrecasteauxii* were positively related to body size, condition, and 'head depth' (dimensions of the head differed between the sexes in this species). In another reptilian species (*Quedenfeldtia trachyblepharus*) it was observed that the number of mites was not a function of sex of the host but that large individuals carried more mites than smaller animals (Blouin-Demers *et al.* 2013). Since Hamilton *et al.* (2021) observed that ectoparasites did not discriminate among species and *P. entrecasteauxii* was one of the three taxa encountered in their study, this also may occur in *L. guichenoti* when parasitised by the same species of mite. Lee *et al.* (2023) found that 30% of the *L. guichenoti* from a locality in Sydney harboured trombiculid mites (chiggers). The intensity of infection ranged from 5 to 40 mites per individual (Ko-Huan Lee, pers. comm.).

Red mites are more easily detected than drably coloured ones because they contrast with the colour of skinks' scales, and accordingly they commonly have been observed in clusters in the axillae of *Lampropholis*, particularly in the more tropical climes of Queensland (Burgin, pers. obs.). Burgin also observed 'red mites' as parasites of *Lampropholis* in the region of Sydney. For example, in a modest sample collected on a single day in February (1985) 22.2% (*n* = 18) of *L. delicata* and 14.3% (*n* = 14) of *L. guichenoti* were observed to be parasitised in the axillary region with red mites at very low (1–7) infestations. In the description of the type series of *Saproscincus saltus* (Cape Melville shade skink), a species within the size range of *Lampropholis*, Hoskin (2013) described obvious clusters of red mites in the same body region as previously observed in northern populations of *Lampropholis* by Burgin. Red mites are also obvious in the photograph of *S. saltus* in Cogger (2018). Additionally, in northern Queensland, Turner (2007) reported that he observed the adults of another small skink (*Leiolopisma* [now *Techmarscincus*] *jigurru*), also within the size range of *Lampropholis*, to have clusters of red mites around the axillae.

An indication of the potentially detrimental effects of mites on the reproduction of skinks was provided by Hare *et al.* (2010). Predominantly concerned with the effects of palpation on the failure of pregnancy in captive New Zealand skinks, *Oligosoma maccanni* (McCann's skink), they observed that the removal of the mite, *Ophionyssus natricis* – the most common ectoparasite among captive reptiles (Tomkins and Begon 1999; Pasmans *et al.* 2008) – resulted in an 80% increase in reproductive success in females. The level of parasitism, however, did not affect morbidity. By contrast, despite the observation that three species of *Carlia* frequently are host to large numbers of conspicuous larval trombiculid mites (a taxon also observed on *Lampropholis*, Burgin, pers. obs.), Manicom (2010) found no evidence to suggest that the burden of mites had affected body condition, volume of stomach contents, growth, reproductive success, or avoidance of predators (i.e. there were no apparent 'fitness costs'). Of these three species, only *C. rostralis* is substantially longer than any of the described *Lampropholis* species, and most (81.0%, *n* = 11) are within the same range as the smaller species of *Carlia*. Additionally, Hamilton *et al.* (2021) observed that only 37.5% (*n* = 16) of the skink assemblage in the south-eastern range of her study site carried ectoparasitic mites and ticks and that these mites were not host-specific.

Ticks

Ticks are significant pests and act as vectors of pathogens in reptiles worldwide (Nicholson *et al.* 2019). In a review, Heatwole and Pianka (1993) noted that of the eight genera and 58 species of ticks known from Australia at that time only 11 species in two genera (*Amblyomma* and *Aponnomma*, now combined into the single genus *Amblyomma*) had been reliably reported from squamates and two additional genera (one species each) doubtfully recorded. There is limited information on most species, including their host-specificity and their impact on their host (Baker *et al.* 2014; Irwin *et al.* 2018; Natusch *et al.* 2018). It is known, however, that the Australian tick fauna is unique and distinct to this continent and that they may harbour a diversity of microorganisms including viruses, bacteria, and eukaryotes with zoonotic potential (Irwin *et al.* 2018). Debilitating disease and death may also occur. For example, argasid ticks may cause paralysis that results in the degeneration of muscle at the site of the bite whilst transmission of green-lizard papilloma-associated virus, several hemogregarines, and the filarid worm *Macdonaldius oschei* have all been observed in association with ticks (Divers 2020). However, as with research on the parasitism of *Lampropholis* more generally, the information on the effects of ticks on these skinks is, at best, limited.

In Australia, research on ticks has been largely restricted to species that infest humans and domesticated animals (see e.g. Baker *et al.* 2014; Greay *et al.* 2016; Harvey *et al.* 2019). The significant exception is the work of Professor Mike Bull. He co-authored more than 80 papers on the parasitism of *Tiliqua rugosa* (sleepy lizard) by the ticks *Amblyomma limbatum* and *Bothriocroton hydrosauri* (Godfrey and Gardner 2017), together with host-parasite interactions of various other reptiles (Godfrey *et al.* 2009; Fenner *et al.* 2011). *Tiliqua rugosa* is a substantially larger lizard (250 mm snout-to-vent length) than any *Lampropholis* species (50 mm, snout-to-vent length). Also, *Tiliqua* typically pair-bonds for the long-term and lives in social groups (Cogger 2018). There is no evidence of pair-bonding in any species of *Lampropholis*; males appear to mate with females when a chance encounter occurs, and such activity does not involve prolonged contact, thereby reducing the likelihood of transferring ticks between them (Burgin, pers. obs.).

Environmental effects on infestation by mites and ticks

Burned sites versus unburned sites

Ticks and mites both have been observed to be most common in 'long unburned' sites compared to more recently burned ones (Ash 1988; Norris and Conroy 1999; York 1999). The same is true for *L. delicata* and *L. guichenoti* (Watson 2001) and thus the habitat preferences of host and parasite match, or more likely, the presence of the host influences the presence of the parasite.

Effect of season and differential survival of hosts

Argaez *et al.* (2020) studied the survival over a 2-year period of three species of American lizards of the genus *Sceloporus* parasitised by mites and ticks. Two of the species appeared to be unaffected whereas there was an increase in survivability with an increased load of parasites in one species. The researchers speculated that such unexpected results may have been due to mites discriminating among hosts and more frequently parasitising healthy individuals. Alternatively, they suggested that robust individuals may have a greater chance of becoming infected because they were more likely to be in contact with parasites as they moved through their environment than were more sedentary unhealthy individuals. Another suggestion was that the former likely had more social contacts than did less healthy individuals and hence more chances of transmission of parasites to them. Comparable studies need to be carried out on *Lampropholis*, as model organisms, to ascertain whether a comparable situation is more broadly applicable or merely a special condition in the species studied by Argaez *et al.* (2020).

Effect of elevation

Elevation also may play a role in host-parasite dynamics. For example, Hamilton *et al.* (2021) reported that there was a negative correlation between ectoparasites' abundance and increased elevation. Whilst only a single species of skink was observed with ectoparasites in the alpine zone, in the subalpine and montane zones (i.e. across the range of *L. guichenoti* in the area), ectoparasites were prevalent in both males and females. The number of mites did not vary across the elevational range of *L. guichenoti* or among any of the six parasitised taxa of skinks; however, the burden of parasites did increase across the active season (November–January).

Clinal variation of parasites

With the dearth of information on the ectoparasites of *Lampropholis* more generally, it is not surprising that the data are particularly scant on the effects of clinal variation of their parasites. Smyth (1973), Chilton *et al.* (2000), and Hamilton *et al.* (2021) found the ectoparasitic ticks and mites of reptiles to be sensitive to a range of environmental conditions including temperature, moisture, vegetation, and characteristics of the soil. The study by Hamilton *et al.* (2021) on the prevalence of skinks and their ectoparasites in the Southern Alps of south-eastern Australia revealed that there was clinal variation in their ectoparasites. Species of ticks (Acari: Ixodidae) and one undescribed species of mite (Trombiculidae) were common and there was clinal as well as temporal variation in their abundance.

Effect of behaviour on transfer of mites and ticks

Any direct contact between individuals such as during mating or fighting would facilitate the transfer of ectoparasites from one individual to another. Accordingly, the communal nesting and communal use of sites for brumation by *Lampropholis* (see Chapter 4), may also increase incidence of infestation by ticks or mites. Sharing of refuges may not have to be synchronous to affect transfer. In an estimation of the risk of cross-infection by ticks (and other parasites) due to asynchronous use of refuges, Leu *et al.* (2010) concluded that *Tiliqua rugosa* self-infects during repeated use of the same refuges. and it also has much higher loads of ticks when multiple individuals share refuges. The same is probably true of *Lampropholis*.

Secondary infections facilitated by mites and ticks

The prevalence of parasites in an area and the relative population sizes of host and parasite affect the degree of stress on the host and on the nature of its immune responses (McCoy *et al.* 2012). In the process of an ectoparasite gaining access through the skin to feed on the blood and/or other tissues of their host, the parasite necessarily creates lesions that may include excoriation, papules, scaling, and inflammation. In addition to the immediate, short-term, external injury that they inflict, ectoparasites may facilitate the entry of viruses and bacteria (e.g. *Rickettsia* [Gartrell 2016] and *Aeromonas hydrophila* [Divers 2020]) that cause secondary infections.

ENDOPARASITES

Introduction

The most significant multicellular endoparasitic animals that infect *Lampropholis* are the helminth worms. 'Helminth' is a general, collective term for parasitic worms of the phyla Nematoda (roundworms or nematodes), Cestoda (tapeworms or cestodes), and Trematoda (flukes and flatworms). Both the Nematoda and Trematoda contain free-living forms as well as endoparasites.

Numerous species of helminths may be soil-transmitted and they often infect the intestinal tract of reptiles and other vertebrates. Other endoparasites, for example schistosomes (a group of trematodes), invade the blood vessels (Divers 2020). Helminths are taxonomically defined by their external characteristics, the host organ they inhabit, and the internal and external morphology of their eggs, larvae, and adult stages (Castro 1996). The most comprehensive checklist of the helminth parasites of Australian reptiles was provided by Pichelin *et al.* (1999).

The 'Protozoans' (single-celled eukaryotic microorganisms; see below) are also common endoparasites (Gartrell 2016).

Many taxa of endoparasites attacking reptiles can be dispersed into new localities within the bodies of their introduced hosts and are subsequently transmitted to previously uninfected species in their new environment. Also, introduced reptiles may newly acquire parasites transmitted by local native hosts. For example, Goldberg and Bursey (2000) found that 89% (n = 62) of the lizard, *Anolis sagrei*, introduced into Hawaii presented with two species of trematode and three species of nematode as well as one acanthocephalan and one pentastomid. All helminth species had been previously reported from other hosts and all of the parasites had been introduced to Hawaii.

It is likely that transfer of endoparasites is facilitated by direct contact during mating or fighting, or by the sharing of sites for oviposition or brumation, either synchronously or asynchronously.

Helminths

The helminths are a highly diverse group of worms that include Platyhelminthes, a phylum consisting not only of free-living flatworms (planarians) but also parasitic tapeworms (cestodes) and flukes (trematodes). They are common parasites of reptiles, including skinks. Goldberg *et al.* (2008) in a study of 446 skinks, spanning 11 species of *Emoia* from Papua New Guinea, found that, among infected skinks, the mean species richness of helminths was 6.4 ± 4.7 and individual lizards were infested by 1–14 species.

Cestodes (tapeworms)

The cestodes that parasitise reptiles may have complex and varied life cycles. For example, a species may have a variety of intermediate hosts, or be a definitive or a paratenic parasite in a single species of host. Cestodes are not, however, very often pathogenic but, in the extreme, their presence may induce loss of weight and ultimately death of the host (Divers 2020). There are over 5,000 known species of these ribbon-like parasitic tapeworms. In the adult form, cestodes have been reported to occur in the digestive tract of all Orders of reptiles (Divers 2020), birds (Denegri *et al.* 1998) and mammals (Denegri *et al.* 1998; Sanchis-Monsonís *et al.* 2020).

Jones (1985) described the cestode *Baerietta hickmani* (later reassigned to the genus *Cylindrotaenia*; see Jones 1987) known from several species of skinks including *L. delicata* and *L. guichenoti*, but mainly the former (Korneva and Kornienko 2013). Other publications by Jones (1988, 1989, 1994) focused on the morphology and ultrastructure of *C. hickmani*.

Trematodes (flukes)

The limited number of publications that address trematodes parasitising *Lampropholis* are either descriptions of new species (Jones 1985, 1989, 1994; Goldberg and Bursey 2012) or brief comments on the role of particular species as primary hosts. The list of hosts, however, includes substantial numbers of other reptilian species (Jones and Cappello 2004), often infected due to the consumption of encysted metacercariae in intermediate hosts (O'Rourke and Lertpiriyapong 2015).

Gartrell (2016) reported that, compared to nematodes and cestodes, trematodes were uncommon parasites of *L. delicata* in New Zealand. In the reptiles of Papua New Guinea, however, Goldberg *et al.* (2008) found that the species diversity of trematodes (5 species) was more similar to that of cestodes (2) than to that of nematodes (14). Overall, however, any conclusions from any species of *Lampropholis* should be considered with caution due to the lack of research undertaken to date, none of which appears to have been focused directly on *Lampropholis* within its natural range.

Nematodes (roundworms)

Parasitic nematodes attack the mesentery, coelomic cavity, and the blood vessels of all Orders of reptiles. Divers (2020) suggested that such parasitism is overwhelmingly common in the presence of poor hygiene resulting from highly contaminated living conditions.

Jones and Resasco (2016) noted that although many museum specimens of *L. guichenoti* had gastric nematodes, the diversity and level of infection by these parasites was limited. They described *Hedruris wogwogensis* from *L. guichenoti* as its type host. Nineteen per cent ($N = 182$) of the nematodes recovered were of this species. The only other nematode observed was a single individual of the genus *Skrjabinodon* in one specimen and *Maxvachonia chabaudi* parasitising three females.

Resasco *et al.* (2019) observed that *H. wogwogensis* has intermediate hosts that include the amphipod *Arcitalitrus sylvaticus* (note that in Chapter 8 it was reported that *Lampropholis* eat amphipods, albeit rarely). Rasasco *et al.* (2019) also determined that habitat fragmentation has the potential to negatively impact the balance of host–parasite dynamics between *H. wogwogensis* and its hosts, *A. sylvaticus* and *L. guichenoti,* and may, in addition, influence the distribution and abundance of these parasites. *Arcitalitrus sylvaticus* was not present in cleared habitat and *H. wogwogensis* did not occur in the gut of *L. guichenoti* from pitfall traps in such areas. In fragmented forest, *H. wogwogensis* did infect *L. guichenoti*. The incidence of these nematodes in cleared forest was ~25% of that found in continuous forest. Although fragmentation of the habitat was shown to disrupt the dynamics between hosts and parasites due to alteration of the distribution and abundance of the intermediate host, Resasco *et al.* (2019) suggested that such changes in key players' life cycles may result in either greater infection or in other forms of disruption of their interaction.

Brown (1991) reported on the gut contents of skinks in south-eastern Australia and of the 40 *L. guichenoti* investigated, only one individual was found to carry nematodes. Most other mentions of nematodes parasitising *Lampropholis* are merely minor comments, often referring to extralimital introduced populations of *L. delicata* in Hawaii (Goldberg and Bursey 2000; Fisher and Ineich 2012) and New Zealand (Bell 2016; Mockett *et al.* 2017).

Lee *et al.* (2022) investigated the seasonality of parasitism by nematodes in *L. guichenoti* by collecting faecal samples of the lizards and comparing the parasite loads (as measured by number of nematode eggs per gram of faeces) between the season at which the lizards were active and the season of inactivity. Two kinds of nematodes were found. The eggs of trichinelloid nematodes were too few for meaningful statistical treatment and were found only during the season of inactivity; hence they were excluded from further analysis. In ascarid nematodes, however, the yield from infected ascarids was 1,923 eggs per gram of faeces (range 125–6,539) during the season lizards were inactive to 21,245 (range 111–74,692) eggs per gram during the season of activity, a result that was highly statistically significant ($P < 0.001$). The prevalence of parasites increased from 15.9% in the non-activity season to 72.5% in the activity season. The authors hypothesised that the reason for the greater incidence of nematodes during the season of the lizards' activity was that warmer conditions favoured: (1) increased activity by the lizards than in cooler periods, and (2) improved survival and development of the nematodes.

Acanthocephalans (thorny-headed worms or spiny-headed worms)

Species in the phylum Acanthocephala have an indirect life cycle with arthropods as their intermediate host. Mature individuals parasitise reptiles although they more commonly occur in fish, birds, mammals, and amphibians. They attach to their hosts intestinal mucosa by hooks and spines located on their proboscises (Margulis and Chapman 2009).

Sphaerechinorhynchus rotundocapitatus is the only acathocephalan known to be a parasite of *Lampropholis*. It occurs in *L. guichenoti* (Daniels and Simbotwe 1984; Pichelin *et al.* 1999; Goldberg and Bursey 2001). It was reported as cystacanths from the intestine of *L. guichenoti* by Daniels and Simbotwe (1984) of which only 2% (*n* = 254) were found to be infected, a smaller percentage than in either of two other species of skink (*Eulamprus quoyii* 34%, *n* = 53%; *Hemiergis decresiensis* 6%, *n* = 49). Daniels and Simbotwe (1984) suggested that the incidence of infection decreased with increasing aridity of the habitat. *Lampropholis delicata*, in particular, overlaps the range of *L. guichenoti* more than any other *Lampropholis* species (see Cogger 2018) and at least this species may be exposed to *S. rotundocapitatus*.

Pentastomida (tongue worms)

Pentastomes are wormlike arthropods that live exclusively in the respiratory tract of vertebrates (Waloszek *et al.* 2005; Paré 2008) including various reptiles (Spratt and Beveridge 2019). The overwhelming majority of these parasites infect the lungs of their host but may also be present in other tissues (Paré 2008; Divers 2020). They are obligate parasites with a life cycle that typically involves at least one intermediate host (Paré 2008). Although pathogenicity may vary as the pathogen migrates though the host's tissues, occasionally the host may present with pneumonic symptoms (Divers 2020).

In a review of pentastomes of Australian reptiles, Riley *et al.* (1985) mentioned the presence of these parasites in some snakes and larger skinks but does not mention smaller skinks such as *Lampropholis*. There also does not appear to be any mention of their infection of any *Lampropholis* species elsewhere in the literature, although in view of the overall dearth of research on the parasites of these small skinks such parasitism may have gone unnoticed.

'Protozoa'

'Protozoa' is an informal name of convenience for various phyla of unicellular eukaryotes in the Kingdom Protista that no longer are considered to constitute a monophyletic or holophyletic entity. Included are such diverse taxa as flagellates, opalinids (phylum Chromista), amoebae, apicomplexans, microsporans, haplosporidians, paramyxeans (O'Donoghue and Adlard 2000). The group includes both non-pathogenic and pathogenic taxa (Santín 2013; Issa 2014), the latter of which infect a wide range of invertebrates and vertebrates, including reptiles (Brice and Thomas undated).

Richter *et al.* (2008) reported that numerous species, across the spectrum from harmless commensals to pathogenic taxa, have been found to inhabit the gastrointestinal tract of reptiles; at least one taxon, *Cryptosporidium*, may be lethal (Santín 2013). Most studies involving reptiles have been restricted to parasites in captive animals (Xiao *et al.* 2004; Pedraza-Díaz *et al.* 2009; Santín 2013). Scullion and Scullion (2009) reported that the principal gastrointestinal diseases of pet reptiles were caused by 'protozoans'. They observed nine genera of flagellates/amoebae, one ciliate genus, and five coccidian genera of apicomplexans in the reptiles they sampled.

Data on Australian reptiles are limited (Brice and Thomas undated). However, the genus *Eimeria* (enteric coccidians [phylum Apicomplexa] causing diarrhoeal coccidiosis) has been implicated in infections of Australian lizards (Cannon 1967a, b; O'Donoghue and Adlard 2000). Cannon (1967a) recorded one of these species, *Eimeria lampropholidus* from *L. guichenoti*.

FUNGAL INFECTIONS

Fungal infections are caused by emerging obligate pathogenic fungi of the families Clavicipitaceae and Onygenaceae (Schmidt 2015) and veterinarians routinely observe lizards with such skin diseases (Hellebuyck *et al.* 2012). These may be exacerbated by a range of environmental stressors that include low temperature, extreme humidity, and/or other concurrent disease (Divers 2020).

Other, commonly identified fungi found in lesions of reptiles include *Aspergillus, Muco, Paecilomyces, Penicillium, Fusarium*, and *Geotrichium*, whereas agents of histoplasmosis, coccidiodomycosis, and cryptococcosis were only rarely observed (Jacobson *et al.* 2000). In addition to the first three taxa listed above as being common, Gartrell (2016) reported that *Beauveria, Candida, Cladosporium, Parananizziopis*, and *Rhizopus* had been cultured from reptiles in New Zealand.

The culture and identification of fungi that infect reptiles are problematic (Schmidt 2015; Gartrell 2016; Paré and Conley 2020), at least in part because determination of the pathogenicity of a fungal infection requires isolation, differentiation, and evaluation of the pathogen (Schmidt 2015). When discussing investigation into the cause of death of reptiles (including *L. delicata*) in New Zealand, Gartrell (2016) noted that many diseases routinely are not analysed due, in part, to lack of funding. He also commented that often it is unrealistic to conduct a post-mortem on small reptiles because they rapidly decompose and desiccate upon death. Live examination of small reptiles is also problematic. For example, Jacobson *et al.* (2000) recommended that identification of a fungal infection in reptiles requires obtaining a good biopsy; however, without euthanasia it may be unrealistic to biopsy for fungal infections of even the largest *Lampropholis*.

The fungal disease chytridiomycosis that has caused the decline, extirpation, and extinction of amphibians worldwide (Berger *et al.* 2009; Heatwole 2013) has been reported to infect individuals of introduced *L. delicata* in New Zealand, although without causing serious disease (Goldson *et al.* 2015). In general, mention of a detrimental impact of any fungal species on reptiles is rare and only appears to have been previously mentioned in the context of the potential death of embryos due to fungal infection of eggs when the nest site is too moist (Brown and Shine 2004).

Although potentially amenable to study, fungal infection of eggs has received only limited attention (see Doody 2011; Doody and Paul 2013). When clutches of several females are in contact as a result of communal nesting (see Chapter 4), infertile eggs or non-viable fertilised ones may detrimentally impact the viability of other fertile eggs in the nest. Among oviparous reptiles, fungi growing on dead or defective eggs may spread to adjacent healthy ones (Moreira and Barata 2005; Carvalho *et al.* 2020). For example, *L. delicata* have a clutch size of between one and seven (mean 3.4), which may contribute to a communal nest of, perhaps more than 200 eggs (Doody and Paul 2013; Chapple *et al.* 2015). Because of communal nesting, the infection of eggs of one species of *Lampropholis* may spread to the eggs of another species within the nest. In such situations, the spread of fungal infection, or other diseases, could cause the loss of entire clutches of multiple females of more than one species.

In the absence of data for *Lampropholis*, we provide examples of the fate of eggs from other species with fungal infections. In the laboratory, Moreira and Barata (2005) observed that fungal infections colonised infertile eggs of the Iberian rock lizard (*Lacerta monticola*) and resulted in the death of embryos during incubation; a following field experiment provided less inconclusive results than the laboratory-based study. Pike *et al.* (2016) did, however, note that fungal infections were visible on the eggs of *Eutropis longicaudata* that failed to hatch. These authors were able to exclude temperature and relative humidity as contributing factors. Moreira and Barata (2005), Huang (2006), Warner and Chapman (2011), and Carvalho *et al.* (2020) have been more definite. For example, in their study of the tegu (*Salvator merianae*), Carvalho *et al.* (2020) initially observed that fungi (including yeasts) infected infertile eggs and, subsequently, attacked fertile ones, thereby causing damage and ultimately the death of the infected embryos. It is very likely that this same sequence of events would occur in a communal aggregation of healthy *Lampropholis* eggs in the presence of infertile or damaged eggs.

BACTERIAL INFECTIONS

Alibardi (2010) noted that even when a terrarium in which lizards (including *L. delicata*) are maintained is neglected, rarely do bacterial infections develop. A search for information about those bacteria known to infect *Lampropholis* netted passing reference to two taxa: *Borrelia* (Resasco *et al.* 2019), *Ehrlichia* (Kraus 2009), and *Rickettsia* (Kraus 2009; Barry *et al.* 2011). In general, bacterial infections probably are more easily transferred by contact during mating, fighting, or when sharing sites for oviposition or winter refugia, than are ectoparasites or multicellular endoparasites (Bull *et al.* 2012).

Escherichia coli

Escherichia coli is a taxon that is widely distributed across a broad taxonomic spectrum of hosts and ranges from harmless commensal strains to pathogenic variants that may cause both intestinal and extra-intestinal disease in animals including reptiles (Leimbach *et al.* 2013). The chance of unique *E. coli* serovars becoming established within a naïve native population would depend on a range of factors but primarily on: (1) the frequency of the new host's exposure to *E coli*; (2) the probability of unique *E. coli* serovars becoming established in the naïve population; and (3) the persistence of the *E. coli* population within the naïve population or, conversely, its rate of loss from the newly acquired population (Gordon and Cowling 2003).

In Australia, Gordon and Cowling (2003) surveyed 2,300 non-domesticated vertebrate hosts. They often encountered *E. coli* in mammals, whereas it was less prevalent in birds, and uncommon in fish, frogs, and reptiles. They observed that 10% (n = 314) of endemic species of lizards carried *E. coli* and that it was relatively common only in four species. Of those four, two were *Lampropholis* (*L. delicata* 12%, *n* = 8; *L. guichenoti* 22%, *n* = 23). Presumably, other species of the genus also carry *E. coli*. Apparently, as for humans, most strains are non-pathogenic for *Lampropholis*.

Salmonella

Although not always resulting in disease, *Salmonella* occurs in humans and other animals more than any other potential pathogen. It is commonly associated with reptiles (Woodward *et al.* 1997; Kikillus *et al.* 2011) but only occasionally causes disease and/or death in them (Le Souëf *et al.* 2015; Gartrell 2016). In Australia, Scheelings *et al.* (2011) examined 504 reptiles from 57 endemic species in four families. Although no *Lampropholis* were among the samples taken, their study does indicate widespread occurrence of *Salmonella* among Australian reptiles. Whilst 47% (*n* = 210) of the captive reptiles sampled were found to be shedding *Salmonella*, somewhat fewer (41%, *n* = 294) of the wild-caught ones were infected.

Bull *et al.* (2012) found evidence that in sleepy lizards (*Tiliqua rugosa*) pairs of lizards that shared the same genotypes of *Salmonella* were more closely connected in the same social network than pairs that did not and suggested that the bacteria were transmitted from host to host around the network, rather than lizards each independently contracting the same bacterial genotype. Given that multiple individuals of *Lampropholis* share nesting sites and shelters (see Chapter 4), it is likely that their mode of transmission is similar.

In New Zealand, where *L. delicata* is an introduced species, Middleton (2008) studied the prevalence and spatial distribution of *Salmonella* serovars collected from faecal excretion of native lizards on offshore islands.[1] Most islands surveyed were found to host between one and three unique serovars and skinks (8.5%) were more likely to be carriers of *Salmonella* than were the resident geckos (1.6%). Some of the identified serovars were unique to specific islands, probably because of lack of dispersal of the hosts. Bell (2016) also reported that all known *Salmonella* serovars identified in New Zealand were previously known. However, this may not be the case for those of *L. delicata*; as it expands its range (Moreno *et al.* 2011; Goldson *et al.* 2015) potentially it could spread additional *Salmonella* serovars to native reptiles. Gartrell (2016) commented that of the only bacterial and fungal infections that could cause death among New Zealand's native geckos and the tuatara (*Sphenodon punctatus*) would be those transmitted by exotic species. To date, however, there does not appear to have been any direct proof of *L. delicata* transmitting unique serovars of *Salmonella* to native species in New Zealand or other islands onto which it has been introduced.

Lampropholis delicata is the only introduced reptilian species established in New Zealand and thus presently it is the most likely carrier of foreign *Salmonella* serovars. Some support for this proposal was demonstrated by Middleton *et al.* (2011) who undertook an in-depth spatial and temporal immunological study of *Salmonella* in endemic reptiles on Stephens Island with a focus on the tuatara. This species was not found to carry intestinal *Salmonella* although this bacterium was retrieved from 19% of the *Oligosoma* skinks tested from the same area. *Salmonella* was also retrieved from soil samples. Further investigation identified that both captive and free-ranging *S. punctatus* possessed antibodies to *Salmonella*; based on investigation of cross-reactivity, these antibodies were shown not to be closely related either to those of *E. coli* or those of *Citrobacter*. This research indicates that *S. punctatus* is able to develop an immune response to *Salmonella*.

INTERACTION OF INFESTATION BY DIFFERENT COMBINATIONS OF PARASITES INFECTING SPECIES OF *LAMPROPHOLIS*

Lee *et al.* (2023) studied the separate and combined effects of parasitism by endoparasitic nematodes and ectoparasitic mites on various behavioural and physiological attributes of *L. guichenoti*. Individuals that were free of mites maintained high maximum speed of sprinting whether or not they were infested by nematodes; however, individuals that *were* infested by mites had slower maximum speeds of sprinting if they also were infested by nematodes. By contrast, neither endurance (time to exhaustion of lizards forced to run in a treadmill), foraging efficiency (an index based on number of crickets attacked, number of crickets caught and number of crickets eaten in 15 minutes under controlled conditions after having fasted for 2 days), and time to handle prey did not differ significantly between

1 *Lampropholis delicata* was not established within the study islands.

individuals that were parasite-free, were infested by either category of parasite, or host to both. Nor did the lizards' preferred temperature differ among these degrees of parasitism (see Chapter 6).

Slowing of the maximum speed of sprinting caused by the combined parasitism by nematodes and mites would be expected to reduce the effectiveness of *Lampropholis* in escaping predators, thereby increasing the mortality of doubly infested lizards. Similarly one would expect that reduction in speed would also reduce the host's ability to capture fast-moving prey. The lack of parasitism having such an effect on foraging is surprising. Perhaps the spatial limitation of the experimental conditions relative to the more open space in the field restricted the ability of prey to avoid capture by the lizards.

Although acknowledging that native animals frequently are concurrently infected by multiple parasites, Lee *et al.* 2022 reported that there was growing evidence that endoparasitic and ectoparasitic load, have only negligible effect on the 'whole-organism performance' of reptiles. In their study of the cumulative effects of both endoparasites and ectoparasites on *L. guichenoti*, they found that there was not a significant relationship between body condition or parasitism. In contrast, they did observe significantly reduced sprint speed when skinks were infected with mites. There was, however, no significant relationship in the endurance of *L. guichenoti*'s speed of sprinting or foraging efficiency when individuals were not infected with nematodes. Based on these results the authors suggested that 'short-lived lizards my simply endure parasitic infections'.

REFERENCES

Alibardi L (2010) Tail regeneration: ultrastructural and cytological aspects. In *Morphological and Cellular Aspects of Tail and Limb Regeneration in Lizards*. Volume 207. Springer Berlin, Heidelberg.

Allison B, Desser SS (1981) Developmental stages of *Hepatozoon lygosomarum* (Doré 1919) comb. N. (Protozoa: Haemogregarinidae), a parasite of a New Zealand skink, *Leiolopisma nigriplantare*. *Journal of Parasitology* **67**(6), 852–858. doi:10.2307.3280710

Anderson L (2007) Edge effects on small skinks: their prey, predators and competitors in peri-urban remnants in north-western Sydney. PhD thesis. University of Western Sydney, Sydney.

Argaez V, Solano-Zavaleta I, Zúñiga-Vega JJ (2018) Another potential cost of tail autotomy: tail loss may result in high ectoparasite loads in *Sceloporus* lizards. *Amphibia-Reptilia* **39**(2), 191–202. doi:10.1163/15685381-17000156

Argaez V, Solano-Zavaleta I, Zúñiga-Vega JJ (2020) Do ectoparasites affect survival of three species of lizards of the genus *Sceloporus*? *Zoology* **138**, 125723. doi:10.1016/j.zool.2019.125723

Arnold EN (1986) Mite pockets of lizards, a possible means of reducing damage by ectoparasites. *Biological Journal of the Linnean Society* **29**(1), 862–864. doi:10.1111/j.1095-8312.1986.tb01767.x

Arnold EN (1993) Function of the mite pockets of lizards: an assessment of a recent attempted test. *Canadian Journal of Zoology* **71**(4), 862–864. doi:10.1139/z93-111

Ash J (1988) The location and stability of rainforest boundaries in north-eastern Queensland, Australia. *Journal of Biogeography* **15**(4), 619–630. doi:10.2307/2845440

Baker SC, Walker AR, Campelo D (2014) A list of the 70 species of Australian ticks: diagnostic guides to the species accounts of *Ixodes holocyclus* (paralysis tick), *Ixodes cornuatus* (southern paralysis tick) and *Rhipicephalus australis* (Australian cattle tick); and consideration of the place of Australia in the evolution of ticks with comments on four controversial ideas. *International Journal of Parasitology* **44**(12), 941–953. doi:10.1016/j.ijpara.2014.08.008

Barry M, Peirce MA, Heath CG (2011) A new blood parasite within the relict endemic New Zealand gecko *Hoplodactylus duvaucelii*. *Veterinary Parasitology* **179**(1–3), 199–202. doi:10.1016/j.vetpar.2011.02.004

Bell BD (2016) A review of potential alpine newt (*Ichthyosaura alpestris*) impacts on native frogs in New Zealand. *Journal of the Royal Society of New Zealand* **46**(3–4), 214–231. doi:10.1080/03036758.2016.1216455

Berger L, Longcore JE, Speare R, Hyatt A, Skerratt LF (2009) Fungal diseases of amphibians. In *Amphibian Decline: Diseases, Parasites, Maladies and Pollution*. (Eds H Heatwole and JW Wilkinson) pp. 2986–3050, Surrey Beatty & Sons, Chipping Norton, NSW.

Blouin-Demers G, Lourdais O, Bouazza A, Verreault C, El Mouden H, Slimani T (2013) Patterns of throat colour variation in *Quedenfeldtia trachyblepharus*, a high-altitude gecko endemic to the High Atlas Mountains of Morocco. *Amphibia-Reptilia* **34**, 567–572. doi:10.1163/15685381-00002900

Bower DS, Brannelly LA, McDonald CA, Webb RJ, Greenspan SE, Vickers M, Gardener MG, Greenees MJ (2019) A review of the role of parasites in the ecology of reptiles and amphibians. *Austral Ecology* **44**(3), 433–448. doi:10.1111/aec.12695

Bram RA, George JE (2000) Introduction of nonindigenous arthropod pests of animals. *Journal of Medical Entomology* **37**(1), 1–8. doi:10.1603/0022-2585-37.1.1

Brice B, Thomas G (undated) *Coccidian Tales*. In *Proceedings of the Australian Wildlife Rehabilitation Conference*, Sydney 2018. Kanyana Wildlife Rehabilitation Centre. <https://www.awrc.org.au/uploads/5/8/6/6/5866843/07_coccidian_tales_april_final.pdf>

Brown GW (1991) Ecological feeding analysis of south-eastern Australian scincids (Reptilia, Lacertilia). *Australian Journal of Zoology* **39**(1), 9–29. doi:10.1071/ZO9910009

Brown GP, Shine R (2004) Maternal nest-site choice and offspring fitness in a tropical snake (*Tropidonophis mairii*, Columbridae). *Ecology* **85**(6), 1627–1634. doi:10.1890/03-0107

Bull CM, Godfrey SS, Gordon DM (2012) Social networks and the spread of *Salmonella* in a sleepy lizard population. *Molecular Ecology* **21**(17), 4386–4392. doi:10.1111/j.1365-294X.2012.05653.x

Cannon LRG (1967a) New coccidian from Australian lizards. II. *Eimeria*. *Parasitology* **57**, 237–250. doi:10.1017/S0031182000072061

Cannon LRG (1967b) New coccidian from Australian lizards. I. *Isospora*. *Parasitology* **57**, 227–235. doi:10.1017/S003118200007205X

Carvalho AM, Souza LKH, Ataides FS, Péres Junior AK (2020) Eggs of tegu lizard, *Salvator merianae* (Duméril & Bibron, 1839) (Squamata: Teiidae). *Brazilian Journal of Biology* **80**(1), 112–114. doi:10.1590/1519-6984.193388

Castro GA (1996) Helminths: structure, classification, growth, and development. In *Medical Microbiology*. 4th edn. (Ed. S Baron). University of Texas Medical Branch, Galveston.

Chapple DG, Miller KA, Chaplin K, Barnett L, Thompson MB, Bray RD (2015) Biology of the invasive delicate skink (*Lampropholis delicata*) on Lord Howe Island. *Australian Journal of Zoology* **62**(6), 498–506. doi:10.1071/ZO14098

Chapple DG, Knegtmans J, Kikillus H, van Winkel D (2016) Biosecurity of exotic reptiles and amphibians in New Zealand: building up Tony Whitaker's legacy. *Journal of the Royal Society of New Zealand* **46**(1), 66–84. doi:10.1080/03036758.2015.1108344

Chilton NB, Andrews RH, Bull CM (2000) Influence of temperature and relative humidity on the mounting success of *Amblyomma limbatum* and *Aponomma hydrosauri* (Acari: Ixodidae) larvae and nymphs. *International Journal of Parasitology* **30**(9), 973–979. doi:10.1016/s00200085-0

Cogger HG (2018) *Reptiles and Amphibians of Australia*. Updated 7th edn. CSIRO Publishing, Clayton South, Vic.

Cole NR (2005) The ecological impact of the invasive House Gecko *Hemidactylus frenatus* upon endemic Mauritian Geckos. PhD thesis. University of Bristol, Bristol.

Daniels CB, Simbotwe MP (1984) The biology of acanthocephalan parasites of Australian skinks. *Journal of Herpetology* **18**(2), 211–213.

Denegri G, Bernadina W, Perez-Serrano J, Rodriguez-Caabeiro F (1998) Anoplocephalid cestodes of veterinary and medical significance: a review. *Folia Parasitologica* **45**, 1–8.

Divers SJ (2020) *Parasitic Diseases of Reptiles*. MSD Manual: Veterinary Manual. <https://www.msdvetmanual.com/exotic-and-laboratory-animals/reptiles/parasitic-diseases-of-reptiles>

Domrow R (1981) A small lizard stifled by phoretic deutonymphal mites (*Uropodina*). *Acarologia* **22**(3), 247–252.

Domrow R (1987) Acari Mesostigmata parasitic on Australian vertebrates: an annotated checklist, keys and bibliography. *Invertebrate Taxonomy* **1**(7), 817–948. doi:10.1071/IT9870817

Domrow R, Lester LN (1985) Chiggers of Australia (Acari: Trombiculidae): an annotated checklist, keys and bibliography. *Australian Journal of Zoology Supplementary Series* **33**(114), 1–11. doi:10.1071/AJZS114

Doody JS (2011) Environmentally cued hatching in reptiles. *Integrative and Comparative Biology* **51**(1), 49–61. doi:10.1093/icb.icr043

Doody JS, Paull P (2013) Hitting the ground running: environmentally cued hatching in a lizard. *Copeia* **1**, 160–165. doi:10.1643/CE-12-111

Fenner AL, Godfrey SS, Bull CM (2011) Using social networks to deduce whether residents or dispersers spread parasites in a lizard population. *Journal of Animal Ecology* **80**(4), 835–843. doi:10.1111/j.1365-2656.2011.01825.x

Fisher R, Ineich I (2012) Cryptic extinction of a common Pacific lizard *Emoia impar* (Squamata, Scincidae) from the Hawaiian Islands. *Oryx* **46**(2), 187–195. doi:10.1017/S0030605310001778

Gartrell B (2016) Diseases of New Zealand reptiles. In *New Zealand Lizards*. (Ed. DG Chapple) pp. 207–238. Springer, Cham. doi:10.1007/978-3-319-41674-8_8

Gill BJ, Bejakovtch D, Whitaker AH (2001) Records of foreign reptiles and amphibians accidently imported to New Zealand. *New Zealand Journal of Zoology* **28**(3), 351–359. doi:10.1080/03014223.2001.9518274

Godfrey SS, Gardner MG (2017) Lizards, ticks and contributions to Australian parasitology: C. Michael Bull (1947–2016). *International Journal for Parasitology: Parasites and Wildlife* **6**(3), 295–298. doi:10.1016/j.ijppaw.2017.08.010

Godfrey SS, Bull CM, James R, Murray K (2009) Network structure and parasite transmission in a group living lizard, the gidgee skink, *Egernia stokesii*. *Behavioral Ecology and Sociobiology* **63**(7), 1045–1056. doi:10.1007/d00265.009.0730-9

Goldberg SR, Bursey CR (2000) Transport of helminths to Hawaii via the brown anole, *Anolis sagrei* (Polychrotidae). *Journal of Parasitology* **86**(4), 750–755. doi:10.1645/0022-3395(2000)086[0750:TOHTHV]2.0.CO;2

Goldberg SR, Bursey CR (2001) Intestinal helminths of seven species of gekkonid lizards (Sauria: Gekkonidae) from Western Australia. *Journal of the Royal Society of Western Australia* **84**(1), 23–27.

Goldberg SR, Bursey CR (2012) Intestinal helminths in nine species of endemic Australian lizards, *Lophognathus longirostris* (Agamidae), *Heteronotia binoei* and *Lucasium stenodactylum* (Gekkonidae), *Ctenotus grandis, Ctenotus helenae, Cyclodomorphus branchialis, Egernia depressa, Eremiascincus richardsonii, Morethia butleri*, and *Morethia lineoocellata* (Scincidae), with a review of Australian lizard helminths. *Comparative Parasitology* **79**(2), 247–268. doi:10.1654/4570.1

Goldberg SR, Bursey CR, Kraus F (2008) Gastrointestinal helminths of eleven species of *Emoia* (Squamata: Scincidae) from Papua New Guinea. *Journal of Natural History* **42**, 29–30. doi:10.1080/00222930802254789

Goldson SL, Bourdôt GW, Brockerhoff EG, Byron AE, Clout MN, McGlone MS (2015) New Zealand pest management: current and future challenges. *Journal of the Royal Society of New Zealand* **45**(1), 31–58, doi:10.1080/03036758.2014.1000343

Gordon DM, Cowling A (2003) The distribution and genetic structure of *Escherichia coli* in Australian vertebrates: host and geographic effects. *Microbiology* **149**(Pt 2), 3575–3586. doi:10.1099/mic.0.26486-0

Greay TL, Oskam CL, Gofton AW, Rees RL, Ryan UM, Irwin PJ (2016) A survey of ticks (Acari: Ixodidae) of companion animals in Australia. *Parasites and Vectors* **9**(1), 207. doi:10.1186/s13071-016-1480-y

Hamilton K, Goulet CT, Drummond EM, Senior AF, Schroder M, Gardner MG, White GM, Chapple DG (2021) Decline in lizard species diversity, abundance and ectoparasite load across an elevational gradient in the Australian alps. *Austral Ecology* **46**(1), 8–19. doi:10.1111/aec.12951

Hare KM, Hare JR, Cree A (2010) Parasites, but not palpation, are associated with pregnancy failure in a captive viviparous lizard. *Herpetological Conservation and Biology* **5**, 563–570.

Harvey E, Rose K, Eden J-S, Lo N, Abeyasuriya T, Shi M, Doggett SL, Holmes EC (2019) Extensive diversity of RNA viruses in Australian ticks. *Journal of Virology* **93**(3), e01358-18. doi:10.1128/JVI.01358-18

Heath ACG, Whitaker AH (2015) Mites (Acari: Pterygosomatidae, Macronyssidae) taken from lizards intercepted at the New Zealand border. *Systematic and Applied Acarology* **20**(7), 739–756. doi:10.11158/saa.20.7.3

Heatwole H (2013) Worldwide decline and extinction of amphibians. In *The Balance of Nature and Human Impact*. (Ed. K Rohde) pp. 259–278. Cambridge University Press, Cambridge.

Heatwole H, Pianka ER (1993) Natural history of the Squamata. In *Amphibia and Reptilia, Vol. 2A, Fauna of Australia*. (Eds CJ Glasby, GJB Ross, and PL Beesley) pp. 197–209. Australian Biological Resources Studies, Canberra.

Hellebuyck T, Pasmans F, Haesebrouck F, Martel A (2012) Dermatological diseases in lizards. *The Veterinary Journal* **193**(1), 38–45. doi:10.1016/j.tvjl.2012.02.001

Hoskin CJ (2013) A new skink (Scincidae: *Saproscincus*) from rocky rainforest habitat on Cape Melville, north-east Australia. *Zootaxa* **3722**(3), 385–395. doi:10.11646/zootaxa.3722.3.7

Howard R, Williamson I, Mather P (2003) Structural aspects of microhabitat selection by the *Lampropholis delicata*. *Journal of Herpetology* **37**(3), 613–617. <https://www.jstor.org/stable/1566074>

Huang W-S (2006) Parental care in the long-tailed skink, *Mabuya longicaudata*, on a tropical Asian island. *Animal Behaviour* **72**(4), 791–795. doi:10.1016/j.anbehav.2005.12.011

Irwin P, Egan S, Greay T, Oskam C (2018) Bacterial tick-associated infections in Australia: current studies and future directions. *Microbiology Australia* **39**, 200–202. doi:10.1071/MA18063

Issa R (2014) Non-pathogenic protozoa [Review Article]. *International Journal of Pharmacy and Pharmaceutical Science* **6**, 30–40.

Jacobson ER, Cheatwood JL, Maxwell LK (2000) Mycotic diseases of reptiles. *Seminars in Avian and Exotic Pet Medicine* **9**(2), 94–101. doi:10.1053/AX.2000.4621

Jones BF, Cappello M (2004) Hookworm infection: molecular mechanisms of disease and targets for control. *Drug Discovery Today: Disease Mechanisms* **1**(2), 217–222. doi:10.1016/j.ddmec.2004.10.004

Jones HI, Resasco J (2016) A new species of *Hedruris* (Nematoda: Hedruridae) from the Australian skink *Lampropholis guichenoti* (Reptilia: Scincidae). *Comparative Parasitology* **83**(2), 173–177. doi:10.1654/4806i.1

Jones MK (1985) Morphology of *Baerietta hickmani* n. sp. (Cestoda, Nematotaeniidae) from Australian Scincid lizards. *The Journal of Parasitology* **71**(1), 4–6. doi:10.2307/3281969

Jones MK (1987) A taxonomic revision of the Nematotaeniidae Lühe, 1910 (Cestoda: Cyclophyllidea). *Systematic Parasitology* **10**, 165–245. doi:10.1007/BF00145034

Jones MK (1988) Formation of the paruterine capsules and embryonic envelopes in *Cylindrotaenia hickmani* (Jones, 1985) (Cestoda, Nematotaeniidae). *Australian Journal of Zoology* **36**(5), 545–563. doi:10.1071/ZO9880545

Jones MK (1989) Ultrastructure of the cirrus pouch of *Cylindrotaenia hickmani* (Jones, 1985) (Cestoda, Nematotaeniidae). *International Journal of Parasitology* **19**(8), 919–930. doi:10.1016/0020-7519(89)90120-3

Jones MK (1994) Ultrastructure of the male accessory glands and sperm ducts of *Cylindrotaenia hickmani* (Cestoda, Cyclophyllidae). *Acta Zoologica* **75**(3), 269–275. doi:10.1111/j.1463-6395.1994.tb01214.x

Kikillus KH, Gartrell BD, Motion E (2011) Prevalence of *Salmonella* spp., and serovars isolated from captive exotic reptiles in New Zealand. *New Zealand Veterinary Journal* **59**(4), 174–178. doi:10.1080/00480169.2011.579246

Korneva ZV, Kornienko SA (2013) Morphology and ultrastructure of the uterus of *Lineolepis scutigera* (Dujardin, 1845) Karpenko, 1985 (Cestoda, Cyclophyllidae, Hymenolepididae) in the formation of uterine capsules. *Inland Water Biology* **6**(4), 259–267. doi:10.1134/S1995082913040111

Kraus F (2009) Alien reptiles and amphibians In *Alien Reptiles and Amphibians: A Scientific Compendium and Analysis, Volume 4*. (Ed. F Kraus) pp. 57–93. Springer Netherlands, Dordrecht. doi:10.1007/978-1-4020-8946-6

Lee K-H, Whiting MJ, Leu ST (2022) Seasonal differences in parasite load in a short-lived lizard. *Australian Journal of Zoology* **70**(1), 36–41.

Lee K-H, Whiting MJ, Leu ST (2023) Are whole-organism performances and thermal preference linked to endo- and ectoparasites in a short-lived lizard? *Ethology* **129**(8), 432–443. doi:10.1111/eth.13379

Leimbach A, Hacker J, Dobrindt U (2013) *E. coli* as an all rounder: the thin line between commensalism and pathogenicity. In *Between Pathogenicity and Commensalism. Current Topics in Microbiology and Immunology*. (Eds U Dobrindt, J Hacker and C Svanborg) **358**, pp. 3–32. Springer, Heidelberg. doi:10.1007/82_2012_303

Le Souëf AT, Barry M, Brunton DH, Jakob-Hoff R, Jackson B (2015) Ovariectomy as treatment for ovarian bacteria granulomas in a Duvaucel's gecko (*Hoplodactylus duvaucelii*). *New Zealand Veterinary Journal* **63**(6), 340–344. doi:10.1080/00480169.2015.1063468

Leu ST, Kappeler PM, Bull CM (2010) Refuge sharing network predicts ectoparasite load in a lizard. *Behavioral Ecology and Sociobiology* **64**(9), 1495–1503. doi:10.1007/s00265-010-0964-6

Mack RN, Simberloff D, Lonsdate WM, Evans H, Clout M, Bazzaz FA (2000) Biotic invasions: cause, epidemiology, global consequences, and control. *Ecological Applications* **10**(3), 689–710. doi:10.1890/1051-0761(2000)010[0689.BICEGC]2/0.CO;2

Manicom C (2010) Beyond abundance: the direct and indirect effects of predation in a Terrestrial Trophic Web. PhD thesis. James Cook University, Townsville.

Margulis L, Chapman MJ (2009) *Acanthocephala*. In *Kingdoms and Domains: An Illustrated Guide to the Phyla of Life on Earth*. 4th edn. Academic Press, Elsevier, London.

McCoy ED, Styga JM, Rizkalla CE, Mushinsky HR (2012) Time since fire affects ectoparasite prevalence on lizards in the Florida scrub ecosystem. *Fire Ecology* **8**(3), 32–40 doi:10.4996/fireecology.0803032

Middleton DMRL (2008) *The Prevalence of* Salmonella *and the Spatial Distribution of Its Serovars amongst New Zealand' Native Lizards*. Master of Science in Zoology, Massey University, Palmerston North.

Middleton DM, La Flamme AC, Gartrell BD, Nelson NJ (2011) The epidemiology of Salmonella in New Zealand island fauna. p. 273. In *Abstracts of Papers Presented at the 14th Biennial Conference of the Society for Research on Amphibians and Reptiles in New Zealand, Tautuku, Otago, New Zealand, 11–13 February 2011*. **38**, 269–279. doi:10.1080/03014223.2011.577789

Mockett S, Bell T, Jorge F (2017) The diversity and evolution of nematodes (Pharyngodonidae) infecting New Zealand lizards. *Parasitology*, **144**(5), 680–691. doi:10.1017/S0031182016002365

Moreira PL, Barata MST (2005) Egg mortality and early embryo hatching caused by fungal infection of Iberian rock lizard (*Lacerta monticola*) clutches. *The Herpetological Journal* **15**, 265–272.

Moreno V, Aguayo CA, Brunton DH (2011) A survey for the amphibian chytrid fungus *Batrachochytrium dendrobatidis* in New Zealand's endemic Hochstetter's frog (*Leiopelma hochstetteri*). *New Zealand Journal of Zoology* **38**(2), 181–184. doi:10.1080/03014223.2010.548073

Natusch DJN, Lyons JA, Dubey S, Shine R (2018) Ticks on snakes: the ecological correlates of ectoparasite infection in free-ranging snakes in tropical Australia. *Austral Ecology* **43**, 534–546. doi:10.1111/aec.12590

Nicholson WL, Sonenshine DE, Noden BH, Brown RN (2019) Ticks. In *Medical and Veterinary Entomology*. 3rd edn. (Eds Mullen GR and Durden LA) pp. 603–672. Academic Press, San Diego, CA. doi:10.1016/B978-0-12-814043-7.00027-3

Norris P, Conroy B (1999) Fire ecology of soil and leaf litter invertebrates. *Australian Bushfire Conference*, Albury 1999.

O'Donoghue PJ, Adlard RD (2000) Catalogue of protozoan parasites recorded in Australia. *Memoirs of the Queensland Museum* **45**(1), 1–164.

Okulewicz A, Kazmierczak M, Hildebrand J, Adamczyk M (2015) Endoparasites of lizards (Lacertilia) from captive breeding and trade networks. *Helminthologia* **52**(1), 34–40. doi:10.1515/helmin-2015-0008

O'Rourke DP, Lertpiriyapong K (2015) Biology and diseases of reptiles. In *Laboratory Animal Medicine*. 3rd edn. (Eds JG Fox, LC Anderson, GM Otto, KR Pritchett-Corning, MT Whary) pp. 967–1013. American College of Laboratory Animal Medicine. Academic Press, New York. doi:10.1016/B978-0-12-409527-4.00019-5

Papini R, Manetti C, Manciani F (2011) Coprological survey in pet reptiles of Italy. *Veterinary Record* **169**(8), 207–210. doi:10.1136/vr.d4398

Paré JA (2008) An overview of pentastomiasis in reptiles and other vertebrates. *Journal of Exotic Pet Medicine* **17**(4), 285–294. doi:10.1053/j.jepm.2008.07.005

Paré JA, Conley KJ (2020) Mycotic diseases of reptiles. In *Infectious Mycotic and Pathology of Reptiles: Color Atlas and Text, Diseases and Pathology of Reptiles Volume 1*. 2nd edn. Chapter 12. (Eds ER Jacobson and M Garner). CRC Press, Boca Raton, FL.

Pasmans F, Blahak S, Martel A, Pantchev N (2008) Introducing reptiles into a captive collection: the role of the veterinarian. *The Veterinary Journal* **175**(1), 53–68.

Peace JE (2004) Distribution, habitat use, breeding and behavioural ecology of the rainbow skinks (*Lampropholis delicata*) in New Zealand. MSc thesis, University of Auckland, Auckland.

Pedraza-Díaz S, Ortega-Mora LM, Carrió BA, Navarro V, Gómez-Bautista M (2009) Molecular characterisation of Cryptosporidium isolates from pet animals. *Veterinary Parasitology* **160**(3–4), 204–210. doi:10.1016/j.vetpar.2008.11.003

Pichelin S, Thomas PM, Hutchinson MN (1999) A checklist of helminth parasites of Australian reptiles. *Records of the South Australian Museum, Monograph Series Number* **5**, 1–16.

Pike DA, Clark RW, Manica A, Tseng H-Y, Hsu J-Y, Huang W-S (2016) Surf and turf: predation by egg-eating snakes has led to the evolution of parental care in a terrestrial lizard. *Scientific Reports* **6**, 22207. doi:10.1038/srep22207

Resasco J, Bitters ME, Cunningham SA, Jones HI, McKenzie VJ, Davies KF (2019) Experimental habitat fragmentation disrupts nematode infections in Australian skinks. *Ecology* **100**(1), e02547. doi:e02547.10.1002/ecy.2547

Richter B, Kübber-Heiss A, Weissenböck H, Schmidt P (2008) Detection of *Cryptosporidium* spp., *Entamoeba* spp. and *Monocercomonas* spp. in the gastrointestinal tract of snakes by in-situ hybridization. *Journal of Comparative Pathology* **138**(2–3), 63–71. doi:10.1016/j.jcpa.2007.11.001

Riley L, Spratt DM, Presidente PJA (1985) Pentastomids (Arthropoda) parasitic in Australian reptiles and mammals. *Australian Journal of Zoology* **33**(1), 39–53. doi:10.1071/ZO9850039

Rose K (2005) *Common Diseases of Urban Wildlife: Reptiles.* The Australian Registry of Wildlife Health, Zoological Parks Board of New South Wales.

Sanchis-Monsonís G, Fanelli A, Martinez-Carrasco C, Tizzani P (2020) The typical cestodes of the red fox in eastern areas of the Iberian Peninsula have a grouped distribution. *Veterinary Parasitology* **283**, 109168. doi:10.1016/j.vetpar.2020.109168

Santin M (2013) Clinical and subclinical infections with *Cryptosporidium* in animals. *New Zealand Veterinary Journal* **61**, 1–10. doi:10.1080/00480169.2012.731681

Schall JJ (1982) Lizards infected with malaria: physiological and behavioural consequences. *Science* **217**(4564), 1057–1059. doi:10.1126/science.7112113

Scheelings TF, Lightfoot D, Holz P (2011) Prevalence of *Salmonella* in Australian reptiles. *Journal of Wildlife Diseases* **47**(1), 1–11. doi:10.7589.0090-3558-47.1.1

Schmidt V (2015) Fungal infections in reptiles – an emerging problem. *Journal of Exotic Pet Medicine* **24**(3), 267–275. doi:10.1053/j.jepm.2015.06.014

Scullion FT, Scullion MG (2009) Gastrointestinal protozoal diseases in reptiles. *Journal of Exotic Pet Medicine* **18**(4), 266–278. doi:10.1053/j.epm.2009.09.004

Shine R, Koenig J (2001) Snakes in the garden: an analysis of reptiles 'rescued' by community-based wildlife carers. *Biological Conservation* **102**(3), 271–283. doi:10.1016/S0006-3207(01)00102-1

Simberloff D, Schmitz DC, Brown TC (Eds) (1997) *Strangers in Paradise.* Island Press, Washington, DC.

Smyth M (1973) The distribution of three species of reptile ticks, *Aponomma hydrosauri* (Denny), *Amblyomma abloimbatus* Neumann, and *Amb. limbatum* 1. Distribution and hosts. *Australian Journal of Zoology* **21**(1), 91–101. doi:10.1071/ZO9730091

Spratt DM, Beveridge I (2019) Wildlife parasitology in Australia: past, present and future. *Australian Journal of Zoology* **66**(4), 286–305. doi:10.1071/ZO19017

Stapley J, Keogh JS (2006) Experimental and molecular evidence that body size and ventral colour interact to influence male reproductive success in a lizard. *Ethology Ecology and Evolution* **18**(4), 275–288. doi:10.1080/08927014.2006.9522696

Taylor JE, Fox BJ (2001) Disturbance effects from fire and mining produce different lizard communities in eastern Australian forests. *Austral Ecology* **26**(2), 193–204. doi:10.1046/j.1442-9993.2001.01105.x

Telford S Jr (1971) A malaria parasite, *Plasmodium aurulentum* sp. nov. from the neotropical forest gecko *Thecadactylus rapicaudus*. *The Journal of Protozoology* **18**(2), 308–311. doi:10.1111/j.1550-7408.1071.tn03322.x

Tomkins DM, Begon M (1999) Parasites can regulate wildlife populations. *Parasitology Today* **15**(8), 311–313. doi:10.1016/s0169-4758(99)01484-2

Torr GA, Shine R (1996) Patterns of dominance in the small skink lizard *Lampropholis guichenoti*. *Journal of Herpetology* **30**(2), 230–237. doi:10.2307/1565514

Turner G (2007) Notes on the habits of three skinks from the Mt Bartle Frere Summit, North Queensland. *Herpetofauna* **37**(1), 2–7.

Wairepo J (2015) *Developing Biosecurity Strategies for an Invasive Reptile, the Plague Skink (*Lampropholis delicata*) on Great Barrier Island*. Master of Science in Conservation Biology, Massey University, Albany.

Waloszek D, Repetski JE, Maas A (2005) A new Late Cambrian pentastomid and a review of the relationships of this parasitic group. *Earth and Environmental Science Transactions of the Royal Society of Edinburgh* **92**(2), 163–176. doi:10.1017/S0263593300001280

Walter DE, Shaw M (2002) First record of the mite *Hirstiella diolii* Baker (Prostigmata: Pterygosomatidae) from Australia, with a review of mites found on Australian lizards. *Australian Journal of Entomology* **41**(1), 30–34. doi:10.1046/j.1440-6055.2002.00272.x

Warner DA, Chapman MN (2011) Does solitary incubation enhance egg with uptake and offspring quality in a lizard that produces single-egg clutches? *Journal of Experimental Zoology* **315A**(3), 149–155. doi:10.1002/jez.660

Watson P (2001) The role and use of fire for biodiversity conservation in south-east Queensland: fire management guidelines derived from ecological research. SEQ Fire and Biodiversity Consortium. <https://fireandbiodiversity.org.au/images/publications/The_role_and_use_of_fire_for_biodiversity_conservation_in_south-east_Queensland_-_Fire_management_guidelines_derived_from_ecological_research.pdf>

Woodward D, Khakhria R, Johnson W (1997) Human salmonellosis associated with exotic pets. *Journal of Clinical Microbiology* **35**(11), 2786–2090. doi:10.1128/jcm.35.11.2786-2790.1997

Xiao L, Ryan UM, Graczyk TK, Limor J, Lixia L, Kombert M, Junge R, Sulaiman IM, Zhou L, Arrowood MJ, Koudela B, Modrý D, Lal AA (2004) Genetic diversity of *Cryptosporidium* spp. in captive reptiles. *Applied and Environmental Biology* **70**(2), 891–899. doi:10.1128/AEM.70.2.891-899.2004

York A (1999) Long-term effects of repeated prescribed burning on forest invertebrates: management implications for the conservation of biodiversity. In *Australia's Biodiversity: Responses to Fire. Plants, Birds and Invertebrates.* (Eds AM Gill and JCZ Woinarski) pp. 186–266. Department of the Environment and Heritage, Canberra.

11
Population biology and the ecological niche

POPULATION DENSITY AND BIOMASS

Multiple factors interact in influencing the population density of a species and seldom are all of them identified for a particular species at any time or place. However, Mulhall *et al.* (2024) identified the amount of cover of the canopy as a key predictor of the abundance of *L. guichenoti*. Simbotwe (1985) estimated the population density of *L. guichenoti* by counting those he saw while walking belt transects during a dry year when vegetative cover was sparse and during a wet one when the vegetation was denser. Some lizards may have been inactive under cover or visually obstructed by leaf litter or low vegetation, or sought cover before he saw them, so the population densities probably are underestimated. His estimate for the wet year was almost double that for the dry one and biomass accordingly was also higher then (Table 11.1). The exceptionally high value of 3–4/m^2 found by Baker (1979) referred to the number beneath a single mango tree in Hawaii and likely was not typical of the entire general area.

A striking feature of the data in Table 11.1 is the wide variation in the population density of *L. delicata*. Is this extreme variation because this widespread species occurs in many different habitats of varying suitability to its ecological tolerances or to the fact that at some localities there are competitors, whereas at others there are no species with sufficiently similar ecological requirements to compete with it? Webb (1985) censused transects across eight different microhabitats (litter, grass, rock, bare ground, logs, stumps, trees, and shrubs) in a eucalypt forest. Six species of lizards were resident but no *L. delicata* were there; only three individuals of *L. guichenoti* were found, and those only on litter. Two other species were similarly rare on litter and either missing or rare on all other substrates; a further two species not only occupied most of the microhabitats but were on litter in greater abundance than *L. guichenoti*. Finally, one species (*Pseudemoia spenceri*) greatly outnumbered (238) all other species on litter and also occurred on three other substrates (Hutchinson and Donnellan 1992; Hutchinson 1993). One wonders how the dynamics of this assemblage would have changed if *P. spenceri* would not have occurred there. For example would *L. guichenoti* have expanded into other microhabitats in the absence of the dominant species in the area? This discussion is now verging on the topic of the section immediately following. Also see the sections *The dilemma of* Lampropholis, *Ecological segregation*, and *The structural niche* below.

BASIC CONCEPTS OF THE NICHE

Biotic communities and the multifarious roles that individual species play in their structure and dynamics are exceedingly complex. A further complication is the inconsistent and often ambiguous terminology used to describe the processes and the organising principles associated with them. Accordingly, this chapter first reviews fundamental concepts and precisely defines basic terms and then addresses the structuring of the assemblage of small skinks of which *Lampropholis* are members.

The niche describes a broad spectrum of the biology of a species. There are two main categories: (1) the *ecological niche* refers to the role an organism plays in its environment – its activities and its relationships to its biotic and abiotic environment, and (2) the *habitat niche* (Hanson 1962). These categories can be further broken

Table 11.1. Population density and biomass of *Lampropholis guichenoti* and *Lampropholis delicata* at various locations and elevations.

Species **Locality** **Season**	**Elevation (m)**	**Lizards per hectare (Biomass g/ha)**	**Method**	**Authority**
Lamproholis guichenoti				
Newholme near Armidale, NSW				
Dry year (1981–1982)	1,058	287 (292)	Belt transect[1]	Simbotwe (1985)
Wet year (1983–1984)		560 (609)	10 × 100 m; 5 × 100 m	Simbotwe (1985)
Toohey Forest near Brisbane	–	85–90	Toe-clipping; mark and recapture	Milton (1980)
Toolangi State Forest and Maroondah Catchment, Vic.	560–1,050	Numbers so low that author excluded data from detailed analysis	Census of plots of 30 m^2 with varying visibility	Brown and Nelson (1993)
Lampropholis delicata				
Under a tree	–	30,000–40,000	Counted the number under a tree	Baker (1979)
Various open forests and woodlands	At least up to 800 m	1,075	Pitfall trapping; line transects; quadrats	Webb (1991)
Drypetes deplanchei and *Cryptocarya triplinervis* forest	–	1,213[2]	Counts along transects 2 m wide	Schulz (2009)
Syzygium fullargarii and *Chionanthus quadristamineus* forest	–	180		
Palm forest	–	200[3]		
Pandanus forest	–	330		
Tussock grass	–	400		
Various disturbed areas	–	30–3,900		

[1]Area corrected from miscalculation in the original thesis. [2]Mean of five sites. [3]Mean of two sites.

down into more specific subcategories, e.g. trophic niche, spatial niche, structural niche, and temporal niche (see below).

A *biotic community* is any naturally occurring group of organisms contemporaneously inhabiting a common environment and interacting with each other relatively independently of other such groups. It consists of all such organisms, with emphasis upon the word ALL as it is inclusive of all taxa: viruses, bacteria, fungi, algae, protists, invertebrates, vertebrates, and flowering and non-flowering plants, to the exclusion of none. Botanists and ornithologists are the worst offenders and one often reads of 'plant communities' and 'bird communities'. Such monolithic entities do not exist. There are no biotic communities consisting either only of plants or solely of birds. The main confusion in the literature is failure to distinguish between communities (all species) and assemblages (species of a particular taxon). Thus, 'plant communities' and 'bird communities' are oxymorons. The correct designations are 'plant assemblages' and 'bird assemblages'. There are, however, some biotic communities that lack plants altogether, such as some bare sand cays (no vegetation) on the Great Barrier Reef of Australia, which have food webs of invertebrates (including predators) of up to eight species, based entirely on scavengers feeding on the wash-up of marine carrion or on the faeces, cadavers, and food scraps of 'transfer' species such as nesting seabirds and intertidal crabs (Heatwole 1971). The trophic and energetic base of these terrestrial insular biotic communities thus is derived from the adjacent marine community.

A biotic community has no spatial scale; it can be as small as a clump of moss with its contained biota or as large as a coral reef. Often the unit studied is the biotic community encompassed by a particular type of vegetation, such as a tropical rainforest, or a tall-grass prairie.

Communities may be of varying sizes, and larger ones may contain smaller ones (Abercrombie *et al.* 1964). The components of communities can be categorised in various ways. The simplest is by species composition or biodiversity, which merely consists of listing the species contained therein. More structural detail can be added by indicating the distribution-abundance relationship.

The structure of a community can be viewed as based on taxonomically defined subunits known as assemblages. An assemblage contains all the members of a specified taxon that occur within a given community and can either be narrowly defined (e.g. the assemblage of species of *Lampropholis*), or expanded to include all small skinks, all lizards, all reptiles, and so on (see mention of 'plant assemblages' and 'bird assemblages' above).

Alternatively, a community may be considered as consisting of a group of 'guilds'. A *guild* consists of all the taxa of a biotic community that play similar roles in the community, without reference to taxonomic affiliation. For example, one could examine the guild of insectivorous species in the biotic community of the forest floor of a tropical rainforest. Small lizards would be a part of that, as would other taxa that live in the forest floor and prey upon insects (e.g. spiders, scorpions, centipedes, and predatory insects).

Note also that the species may not always be the fundamental unit in the above definitions, but rather sometimes it is the econe (Heatwole 1989; see below). Thus, in the above definitions, the word 'species' could be replaced by 'taxon or 'econe').

Finally, not all species interact equally at all times throughout the entire physical extent of the community of which they are members, but rather the number of interactants at any particular time and specific location will be a subset of the entire assemblage or guild, with the result that point-diversity is much lower than overall diversity (Heatwole *et al.* 2013).

The *ecological niche* of a species is its role in its environment, including the totality of its interactions both with its biotic environment (other species) and its abiotic (physical) environment (see Hanson 1962; Child and Child 1971). Another way of thinking of niches is that species or econes with similar niches are the subunits that constitute a particular guild.

Often the niche is divided into various functional components; for example, the spatial (habitat or microhabitat) niche, trophic niche, and temporal niche, these often having a thermal component. The extent to which niches overlap, or fail to do so, among species or the extent to which the niche can be altered in ways facilitating coexistence with other species often has a critical effect on the structure and dynamics of biotic assemblages and communities.

Econes

The intraspecific discrepancies between the niches of males compared to females and/or among those of different developmental or life-history stages largely have been ignored by ecologists and yet it pervades the very fabric of biological interactions within assemblages and communities of organisms. Heatwole (1989) introduced the concept of the econe, the fundamental ecological unit. He defined the econe as 'a species or some component of a species (such as life-history stage, age class, morph, or sex), whose members share common patterns of resource utilisation and niche characteristics but differ from other such components or species'. A species haves only one econe if both sexes as well as immatures live in the same places, eat similar foods, and interact in similar ways to the matrix of their biotic and physical environment; in short, they share the same ecological niche. In other species, however, the sexes and/or different life stages play different roles in their assemblages and communities. An extreme example is Amphibia, in which the young (tadpoles) may be diurnal, aquatic algivores whilst the adults may be nocturnal, terrestrial insectivores. The two stages share neither diet, activity period, nor the same space, and do not even interact with any other econe of their same species. They live in entirely separate biotic communities. Holometabolous insects constitute another extreme example, in which a diurnal folivorous larva undergoes a quiescent pupal period and emerges as a nocturnal nectarivorous moth. To treat such a species as a single ecological entity borders on the absurd; they have different econes. Econes are not restricted to such extreme examples. There are other, more subtle, but just as real, examples in lizards (see review by Heatwole 1977). Ruibal and Philibosian (1974) described a three-way econe with the juveniles foraging on the ground, the females low on tree trunks and the males higher up. Michelangeli *et al.* (2018), although not using the term, described two econes in *L. delicada* based on different thermoregulatory behaviours (see Chapter 4); some individuals selected hotter microhabitats than did the others, thereby segregating the population into separate spatial niches and forming two econes.

THE DILEMMA OF *LAMPROPHOLIS*

There are several ways that species can adapt to such a kaleidoscopic landscape. They can either become eurytopic and have broad physiological tolerances of, and behavioural predilections for, a wide range of physical environments and biological contexts, as *L. delicata* and *L. guichenoti* have done. Such animals are generalists with wide geographic ranges, extensive repertories of behavioural responses, and varied diets.

The opposite evolutionary strategy is specialisation into a narrower, and otherwise unoccupied, ecological niche, i.e. to become stenotopic in habitat and geographic range, a dietary specialist, and/or operate at times that allow avoidance of species with which interactions may not be favourable. In an evolutionary sequence, often the generalists are the enduring stem that repeatedly buds off taxa that become increasingly specialised, and although successful for a time in narrow, ephemeral niches, become extinct when conditions change and their restricted way of life is no longer tenable. The generalists persist because their broad ecological tolerances encompass a variety of situations, at least some of which are likely to outlast even drastic environmental perturbations. The genus *Lampropholis* may be a microcosm of both modes of coping with evolutionary change. *Lampropholis delicata* and *L. guichenoti*, are widespread and clearly eurytopic, and probably the stem group that gave rise to a swarm of satellite species with tiny geographic ranges (Chapter 2). Less research has been conducted on the latter group, but a vast literature is available for the former. For eurytopic species, wide tolerances may not be manifest at all levels; it is possible for the species to be eurytopic by virtue of individual populations adapting more narrowly to a series of local conditions that translates into a broad range of adaption for the species collectively. *Lampropholis delicata* and *L. guichenoti* have an exceedingly complex mode of adaptation that seems to have elements of both of these strategies.

To understand how assemblages function and how they are structured it is important to know how component species interact with each other and with the physical environment in different ecological contexts. How are these attributes altered by changes in the taxonomic composition of assemblages, or across environmental gradients?

Species of *Lampropholis* are denizens of the ground layer of woodlands, open forests, and urbanised habitats such as gardens (see Chapter 2). Within that general designation, they are eurytopic and occupy a wide range of habitats, such as different kinds of open woodlands and forests. These habitats have different assemblages of species, so that the taxonomic matrix in which these species find themselves may differ from one type of woodland to another, as may the physical environment, for example temperature, moisture, available cover, type of vegetation and openness to sunlight. The composition of the guild of predators upon them also may differ from habitat to habitat.

How do the species of *Lampropholis* meet these various challenges and conform to the differences in the structure of the assemblages in which they find themselves? Do their niches widen or compress in different physical and biological contexts? Are there biotic influences from other guilds in the community that interact peripherally with them. Eurytopic species might be expected to have a greater variety of interactions of lower intensity with a variety of other species compared to fewer but more intense interactions among similar stenotopic species.

The fact that two widely distributed species of *Lampropholis* are broadly syntopic (Chapter 2) raises questions as to the extent their niches overlap. On casual inspection, they may appear to be practically identical in habitat and general biology. Do they compete, and if so, why doesn't one eliminate the other? A welter of research on interspecific interactions of many taxa have suggested that coexisting species-pairs often either have subtle differences along one or more of four major axes of the ecological niche - diet, time, space, and thermal environment. For example, often species that occupy the same habitat, either have different activity periods, or eat different foods. The fact that *L. delicata* and *L. guichenoti* seem to be an apparent exception to ecological segregation of closely related species makes them of especial interest to evolutionists and community ecologists. Accordingly, it is important to take a detailed look at the niches of these two species in order to detect any nuances of difference that may illuminate the basis of their syntopy. Such information also may explain the slight differences in the geographic distribution of these congeners (see Chapter 2). For example, why does *L. delicata* extend farther north along the eastern Australian coast than does *L. guichenoti*, and why are there slight differences in the borders of their respective ranges in the south-western part of their distributions?

There have been a number of herpetological surveys of the occupants of particular regions or habitats that included some measure of the relative abundances of the

Table 11.2. The per cent of specimens of different species of *Lampropolis* caught in different habitats.

ad = *L. adonis*; am = *L. amicula*; bel = *L. bellendenkerensis*; cal = *L. caligula*; cog = *L. coggeri*; col = *L. colossus*; cou = *L. couperi*; del = *L. delicata*; ell = *L. elliotensis*; elo = *L. elongate*; gui = *L. guichenoti*; isl = *L. isla*; mir = *L. mirabilis*; rob = *L. robertsi*; sim = *L. similis*. N = number of individual specimens collected in each habitat; Per cent = percentage of the total number of specimens collected. Data from which these entries were calculated were downloaded from the *Atlas of Living Australia* at https://www.ala.org.au/ and trimmed to remove outliers as per Chapple *et al.* (2013); accessed 22 March 2024. Table prepared by Susan King. Data for the recently described *L. isla* was an exception; those were calculated from Hoskin (2025).

Habitat	ad	am	bel	cal	cog	col	cou	del	ell	elo	gui	isl	mir	rob	sim	N	%
Acacia Forests and Woodlands	0.36	0	0	0	0	0	0	0.04	0	0	0.17	0	9.79	0	0	74	0.14
Acacia Open Woodlands	0	0	0	0	0	0	0	0	0	0	0.02	0	0	0	0	5	0.01
Acacia Shrublands	0	0	0	0	0	0	0	0.11	0	0	0.16	0	0	0	0	69	0.13
Callitris Forests and Woodlands	0	0.14	0	0	0	0	0	0.11	0	0	0.04	0	0	0	0	37	0.07
Casuarina Forests and Woodlands	0	1.22	0	0	0	0	0	0.52	0	0	0.10	0	0	0	0	155	0.29
Chenopod Shrublands, Samphire Shrublands and Forblands	0	0	0	0	0	0	0	0.04	0	0	0.01	0	0	0	0	10	0.02
Cleared, non-native vegetation, buildings	24.46	18.32	2.27	6.19	18.34	4.65	13.61	36.10	0	48.68	37.65	0	5.59	0	19.46	19,233	36.03
Eucalypt Low Open Forests	0	0.14	0	0.88	0	0	0	0.50	0	0.00	0.24	0	0	0	0	185	0.35
Eucalypt Open Forests	18.64	35.01	0	39.82	1.04	0	39.27	19.73	0	14.47	27.59	0	4.90	0	2.70	12,744	23.87
Eucalypt Open Woodlands	0.12	0.81	0	0	1.38	0	0	0.29	0	0	1.21	0	0	0	1.62	418	0.78
Eucalypt Tall Open Forests	1.45	27.41	0	33.63	10.03	0	16.23	10.82	20.00	36.84	1.34	0	6.29	0	7.57	3,208	6.01
Eucalypt Woodlands	5.08	7.06	0	9.73	2.42	16.28	4.19	17.75	65.00	0	23.03	0	46.15	0	4.32	10,682	20.01
Heathlands	2.91	0.41	0	0.88	0	0	0	3.22	0	0	1.42	0	4.20	0	2.16	1,168	2.19
Hummock Grasslands	0	0	0	0	0	0	0	0.01	0	0	0.00	0	0	0	0	3	0.01
Shores of inland aquatic – freshwater, salt lakes, lagoons	0.85	0.68	0	0	3.46	0	0	1.03	0	0	0.79	0	0	0	0.54	476	0.89
Low Closed Forests and Tall Closed Shrublands	0	0.68	0	0	0	0	0	0.35	0	0	0.21	0	0	0	0	144	0.27
Mallee Open Woodlands and Sparse Mallee Shrublands	0	0	0	0	0	0	0	0.00	0	0	0.27	0	0	0	0	77	0.14

Habitat	ad	am	bel	cal	cog	col	cou	del	ell	elo	gui	isl	mir	rob	sim	N	%
Mallee Woodlands and Shrublands	0.12	0	0	0	0	0	0	0.74	0	0	1.81	0	0	0	0	673	1.26
Mangroves	0	0	0	0	1.04	0	0	0.42	0	0	0.13	0	0	0	1.62	137	0.26
Melaleuca Forests and Woodlands	0	0.54	0	0	0	0	0	0.91	0	0	0.12	0	7.69	0	1.08	258	0.48
Naturally bare – sand, rock, claypan, mudflat	0	0	0	0	0.35	0	0	0.11	0	0	0.09	0	0	0	0	49	0.09
Other Forests and Woodlands	0	0	0	0	0	0	1.57	0.67	0	0	1.13	0	0	0	3.78	477	0.89
Other Grasslands, Herblands, Sedgelands and Rushlands	0	0.14	6.82	2.65	0.35	0	0.52	0.50	0	0	0.53	0	0	0	0.54	271	0.51
Other Open Woodlands	0	0	0	0	0	0	0	0	0	0	0.03	0	0	0	0	10	0.02
Other Shrublands	0	0	0	0	0	0	0	0.51	0	0	0.84	0	0	0	1.62	353	0.66
Rainforests and Vine Thickets, or Moist Sclerophyll Forests	46.00	7.46	90.91	4.42	61.59	60.47	24.61	4.50	15.00	0	0.61	100	7.69	100	52.97	2,080	3.90
Regrowth, modified native vegetation	0	0	0	0	0	0	0	0.65	0	0	0.32	0	0	0	0	239	0.45
Sea and estuaries	0	0	0	0	0	0	0	0.24	0	0	0.02	0	0	0	0	61	0.12
Tropical Eucalypt Woodlands/ Grasslands	0	0	0	0	0	0	0	0	0	0	0	0	0	0	0	1	0.002
Tussock Grasslands	0	0	0	0	0	18.60	0	0.07	0	0	0.10	0	7.69	0	0	62	0.12
Unclassified native vegetation	0	0	0	1.77	0	0	0	0.06	0	0	0.03	0	0	0	0	24	0.05
TOTAL	**100**	**100**	**100**	**100**	**100**	**100**	**100**	**100**	**100**	**100**	**100**	**100**	**100**	**100**	**100**	**53,383**	**100**

various species native to the area. Often the methods were designed merely to list the species present and hence quantitatively were preliminary; some relied on small numbers of observations, inexact methods of sampling, or grouped several habitats together rather than distinguishing differences among habitats. Nevertheless they form a background for more-detailed studies and provide a basis for some speculative hypotheses and a few generalisations drawn from the metadata. The results of some exemplary surveys are portrayed in Table 11.2. The reader should keep in mind that proportional abundances based on small samples are much less reliable than those involving larger numbers of individuals. For example if only one animal of a species is found and only four for another, those data are unreliably indicative that one species is four times as abundant at the other.

It may be very difficult to unravel the complexity of species' interactions. A widespread species that eats a broad spectrum of prey, for example, may be successful because of its breadth of diet. Two dietary-specialist (stenotrophic) species that occupy the same microhabitat and eat the same species and size of prey may not be able to coexist; only the one with the most efficient method of finding and exploiting that prey will win (competitive exclusion). However, a dietary-generalist (eurytrophic) species usually can persist unless there are a number of other specialists that collectively cover its entire spectrum of prey. Also, it may persist if it feeds on the same species of prey as its competitor does, but does so in a different microhabitat, say, in grass rather than in leaf litter. Widespread species, like *L. delicata* and *L. guichenoti*, meet different assemblages of potential competitors over their geographic ranges so the vacant niches available to them may vary from place to place; the result is that over their vast geographic ranges they may occupy slightly different niches, depending on which other species are present.

The diets of *L. delicata* and *L. guichenoti* were compared in Chapter 8 but in some cases only one of these species was studied at a given locality. Clearly, comparison of the trophic niche of these two species from different areas, probably with different groups of potential competitors, would not be valid. Another technical difficulty, even when comparisons are made of both species at the same locality, is that one may be more abundant than the other, for whatever reason, resulting in smaller sample sizes of the less abundant one. Small samples fail to reveal the entire range of the species of prey consumed and introduce error into the similarity between two species of unequal abundance, thereby giving a false appearance of one species having a more limited diet than the other.

The above difficulties in analysis have not been completely resolved and much more research is required. The few available relevant data are reviewed here in hope it will provide an incentive for further research leading to a fuller understanding of *Lampropholis* and the assemblages and communities throughout their entire geographic ranges and array of biotic communities. Attacking these complex issues can proceed in an hierarchical way, the first stage being to discover which species coexist and under what conditions. A second stage could examine the ecological interactions of particular pairs of prominent species that appear to have similar niches and to assess their impacts on each other within the context of different assemblages of species. Third, comparative ecological studies could be expanded to include other members of the assemblage that overall seem less likely to interact with the primary constituents in general, but who may exert critical effect in one specific area of ecological overlap. Finally, the study of how the various assemblages and guilds are melded intricately into a dynamic, interacting community can be undertaken. In Chapter 8 we compared the prey consumed by the two common species of *Lampropholis* and the variables affecting their diets. In the present chapter we compare the trophic and other niches of *Lampropholis* with small skinks of some other genera that occur syntopically with *Lampropholis*. At times it will appear that different authors studying *L. delicata* and/or *L. guichenoti* in different contexts will disagree as to the relative niches occupied by these two species; such differences may be real and reflect the flexibility of these species in adjusting their behaviour and the type of microhabitat they occupy in ways that maximise their adaptation to local conditions. This topic is revisited in Chapter 16 with a fuller description of ways of approaching the topic.

THE SPATIAL NICHE

Habitats and microhabitats of *Lampropholis*: Studies in the field

The species of skinks differ substantially in the types of habitats they occupy; all of those in the Subfamily Acontinae are either completely fossorial or semi-fossorial and so are about two-thirds of those in the Family

Scincinae. By contrast the Lygosominae (containing the genus *Lampropholis*) are more varied, with 45% of the species being terrestrial, 22% being fossorial or semi-fossorial, 15% scansorial, 4% semi-aquatic and 14% occupying mixed habitats (Chapple *et al.* 2023). Although most *Lampropholis* are terrestrial some of the shorter-legged species perhaps spend much of their time either under cover or in leaf litter or grass.

The spatial niche has a geographic component. Much attention has been paid to mapping the distributional ranges of species, and this is a valuable exercise (see Chapter 2). Within its mapped geographic range, however, a given species is not continuously distributed (but see Chapter 16). To further complicate the matter, not all occupied habitats are equally favourable for a given species, and consequently, it may be abundant in one, and relatively rare in others (see previous section). Finally, there may be edge effects at the boundary with unsuitable habitats, or ecotonal effects where two differentially suitable habitats abut (Chapple *et al.* 2023). It is likely that the occupancy of particular habitats is influenced by the other species that also occur in the habitat (see Chapter 16 and section *Structure of assemblages* below).

Clearing of areas may dissect the original forested area into patches of remnant forest and thereby decrease the area of pristine habits and increase the length of ecotonal edges between the forest and the cleared areas. Anderson and Burgin (2002) compared the abundances of *L. delicata* and *L. guichenoti* in uncleared remnant woodland near Sydney with that of it edges with cleared land and found that both species were much more abundant in the interior of the woodland than on its edges (Fig. 11.1). In both situations *L. guichenoti* was the more abundant of the two species, with the disparity being greatest in the core of the uncleared woodland.

A vexing problem is that lists of component species of some assemblages may be incomplete. However, usually it would be the rarer species that would be overlooked and those would be expected to have a lesser impact on the assemblage at large.

Pengilley (1972) made a start in assessing the relative abundance of the various species by his study of the skinks at Coree Flats (Table 11.3). He found that two different methods of measuring that attribute largely agreed in the ranking of species, including *L. guichenoti* and *L. delicata*. By both measures *L. guichenoti* ranked the third most abundant species (10.4 and 9.9% of total skinks), whereas *L. delicata* ranked 8th (0.6%) by one measure and 9th (0.3%) by the other. Recall that in the total array of

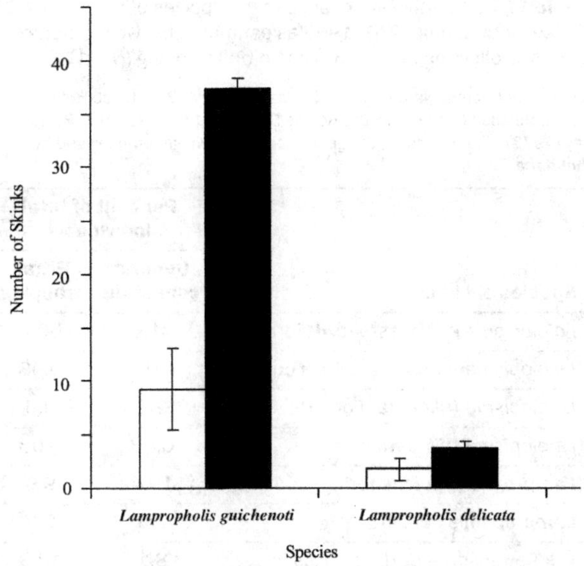

Fig. 11.1. Relative abundances of *Lampropholis guichenoti* and *Lampropholis delicata* caught in 15 pitfall traps in the core of remnant woodlands (black histograms) and 15 at its ecotonal edges (white histograms). From Anderson and Burgin (2002).

habitats, and even of microhabitats, these two species occupy similar environments. The question arises: why do two species with very similar ecologies have such great difference in abundance in the same area?

The methods of assessment can affect the results and thus care needs to be exercised in comparing the biodiversity and relative abundance of different areas. The Queensland CRA/RFA Steering Committee (1997) compared several ways of assessing these variables in 61 small species of lizards, including *L. adonis, L. amicula, L. couperi, L. delicata,* and *L. guichenoti*, and found that diurnal searches recorded the greatest number of species, followed by nocturnal searches, and finally by pitfall trapping. No single method ensured all species would be recorded so protocols should probably include a combination of methods.

Pengilley's (1972) photographs and vegetation maps of Coree Flats show considerable variation in kind and cover of vegetation and litter. Had he separately analysed each of these microhabitats individually, additional insights into subtle differences between these two species might have come to light. Nevertheless his study was a pioneering one leading to a better understanding of the complexity of the topic. He, in fact, did note that in savannah woodland in the lowlands *L. guichenoti* and *L. delicata* were the two most abundant species of small skink. Does this difference in their rank of abundance between the lowlands and the higher elevation of his

Table 11.3. Relative abundance of the species of skinks at Coree Flats during 1967–1968 as estimated by two methods, general collecting (N = 1,733) and pitfall trapping (N = 1,301).

For present nomenclature of the species labelled A, B, or C see Hutchinson and Donnellan (1992) and The Reptile Database (nd). Data from Pengilley (1972). The entries for the genus *Lampropholis* are emphasised by **boldface**.

	Per cent of total individuals	
Species of skink	**General collecting**	**Pitfall trapping**
Leiolopisma entrecasteauxii (Form B)	41.6	60.6
Leiolopisma entrecasteauxii (Form A)	0	0.08
Leiolopisma trilineata (Form B)	2.1	3.6
Leiolopisma trilineata (Form C)	0.06	0.08
Lampropholis guichenoti	**10.4**	**9.9**
Lampropholis delicata	**0.6**	**0.3**
L. x Rawlinson = *Carinascincus coventryi* ?	8.1	13.3
Pseudemoia spenceri	5.5	5.6
Sphenomorphus tympanum	23.7	5.0
Ctenotus taeniolatus	0.06	0
Hemiergis decresiensis	0.1	0
Anepischetosia maccoyi (formerly *Anotis maccoyi*)	7.2	0.9
Tiliqua nigrolutea	0.1	0.08
Egernia saxatilis	0.2	0
Egernia whitii	0.2	0.6

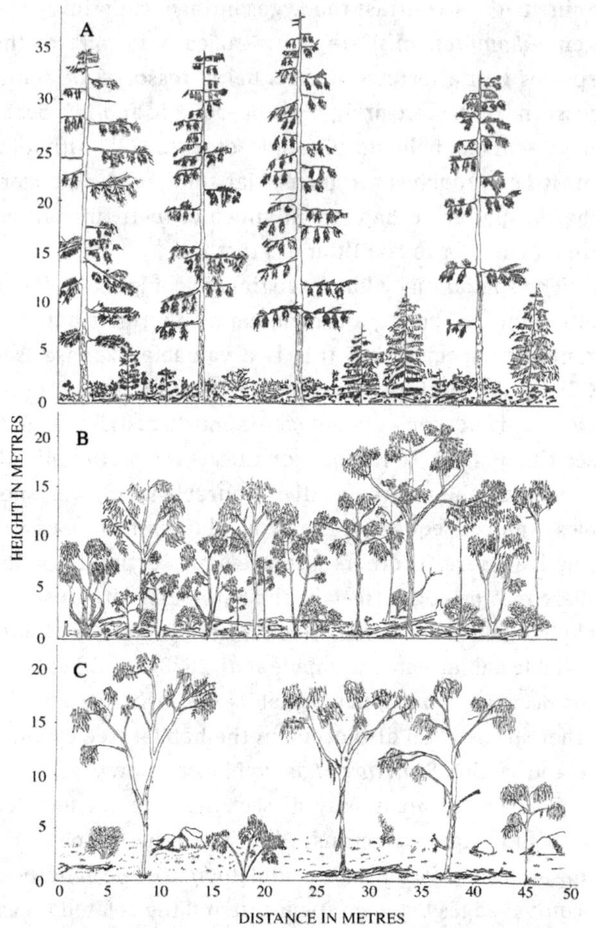

Fig. 11.2. Profiles of transects (50 m × 10 m) through different types of forest and woodland near Armidale, NSW. **(A)** Pine Forest. **(B)** *Eucalyptus* woodland at Puddledock. **(C)** *Eucalyptus* woodland at Blue Hole. From Belmont (1977) but note that in the original work the labels were reversed for Puddledock and Blue Hole (confirmed by Belmont, pers. comm.).

Coree Flats study area reflect an effect of the climatic differences between elevations, or is it a consequence of the different composition of other species in the assemblages at the two elevations – or a combination of both? This is the crux of the dilemma of *Lampropholis* (see section above).

Belmont (1977) recorded the respective locations of individuals of *L. delicata* and *L. guichenoti* when first sighted in three habitats near Armidale, NSW, a Pine Forest and two mixed eucalypt woodlands, one at Blue Hole, and the other at Puddledock (Fig. 11.2). Although the canopies of all three habitats were relatively open, the one with the most trees per hectare and the most complete closure of the canopy was Puddledock and the most open was Blue Hole (Fig. 11.2; Table 11.4). The density of shrubs was not greatly different in the three areas. The greatest cover of fallen logs was in Pine Forest and least at Blue Hole. All three areas had ground-level microhabitats of branches, grass, herbs, leaf litter, rocks, and a minor collective category of 'other', with the greatest cover in Pine Forest and least at Blue Hole (Fig. 11.2). The design of the research was to assess differences between these species' use of these structures that might provide a clue to attributes differentially influencing their microhabitat selection.

Lampropholis delicata occurred in Pine Forest and one of the eucalypt woodlands (Puddledock) but was absent from the other woodland (Blue Hole). *Lampropholis guichenoti* was present at all three locations. Puddledock and Pine Forest had a greater number of trees (Table 11.5; Fig. 11.2) and hence provided more shade than did Blue Hole. The more open canopy at Blue Hole would be expected to allow greater penetration of light to the forest floor and result in higher temperatures and/or drier conditions there than in either of the other two sites. The study by Anderson *et al.* (2023) found that *L. delicata* is

Table 11.4. Type of vegetative cover occupied by individuals of different species of skink when first observed in the Brindabella Range (mainly Coree Flats).

Note that the total number of observations exceeds 100% because some individuals when first observed were located at the border of two types of cover and were counted for both. Data from Pengilley (1972). **Boldface** indicates the highest score for that type of cover. Identity of *L.* x. not known. Taxonomic names brought up to date.

Habitat	Per cent of observed occupancy of different types of vegetative cover							
Type of cover	% of total area	*Pseudemoia entrecasteauxii* Males	*Pseudemoia entrecasteauxii* Females	*Lampropholis guichenoti*	*L.* x.	*Acritoscincus trilineata*	*Pseudemoia spenceri*	*Sphenomorphus tympanum*
Snowgrass (SG)	16.1	19.2	22.0	5.7	11.2	**27.0**	19.6	18.6
SG and Shrubs (SH)	21.5	21.2	23.0	21.5	17.1	**35.0**	19.6	30.2
SG and Leaf Litter (LL)	2.2	3.0	2.7	0	3.9	**4.2**	2.1	0
SG, SH, and LL	9.1	11.8	7.0	4.4	**15.1**	4.2	3.1	4.7
Dense SG and SH	0.7	3.9	3.7	0	0.7	0	**4.1**	0
Dense SG, SH, and LL	0.4	**0.5**	**1.1**	0	0	0	0	0
SH	15.4	10.3	9.7	20.9	11.8	17.0	7.2	**23.3**
SH and LL	15.8	15.8	18.5	22.8	17.8	6.3	**23.7**	9.3
Dense SH and LL	3.1	4.4	5.9	1.9	**7.2**	4.2	5.2	4.7
Dense SH	0.5	1.0	1.1	**1.9**	0.7	0	0	0
LL	2.2	1.5	1.1	3.2	**5.3**	0	4.1	0
Sparse*	13.2	7.4	5.4	**17.7**	9.2	2.1	11.3	9.3

*Little or no vegetation of litter.

Table 11.5. The density of woody vegetation and logs (number per hectare) in three study areas, as determined by the point-quarter method of Cottam and Curtis (1956).

Re-formatted from Belmont (1977) with her permission.

Locality Structure	Pine Forest	Eucalypt woodland Puddledock	Eucalypt woodland Blue Hole
Density (number per hectare)			
Trees	166	228	101
Shrubs	184	192	208
Logs	429	245	55

more abundant in woodlands with a denser canopy than in more open woodland or grassland and perhaps this explains its absence from Blue Hole. Another difference between these study areas is that both the Pine Forest and Puddledock had a much higher cover of logs that did Blue Hole (Table 11.5); *L. delicata* did not occur at Blue Hole. Both species tended to be found more often close to logs, rather than farther away, but *L. delicata* had a higher proportion of sightings actually on logs (zero distance; Table 11.2).

Taking all these facts into account, three hypotheses were entertained: (1) solid cover, such as fallen logs, are more important to *L. delicata* than to *L. guichenoti*, and hence habitats with few logs (such as Blue Hole) are not suitable for the former; (2) *L. guichenoti* has greater tolerance than does *L. delicata* to open, warmer, and/or drier habitats; and (3) *L. delicata* selects habitats with greater cover of vegetation in the ground layer. These are not necessarily mutually exclusive and could operate in combination. The study by Anderson *et al.* (2023) confirmed that *L. delicata* is more abundant in woodlands with a denser canopy (and hence greater shade and lower temperatures) than in more open woodlands or open grasslands.

The first hypothesis was addressed in several ways. One was to investigate whether in Pine Forest and at Puddledock *L. delicata* is more often associated with logs than is *L. guichenoti*, which turned out to be the case (Fig. 11.3). Levels of occupancy were assessed during the time of day that animals were foraging. *Lampropholis delicata* had its highest occupancy at logs in Pine Forest, whereas *L. guichenoti* was less often associated with logs

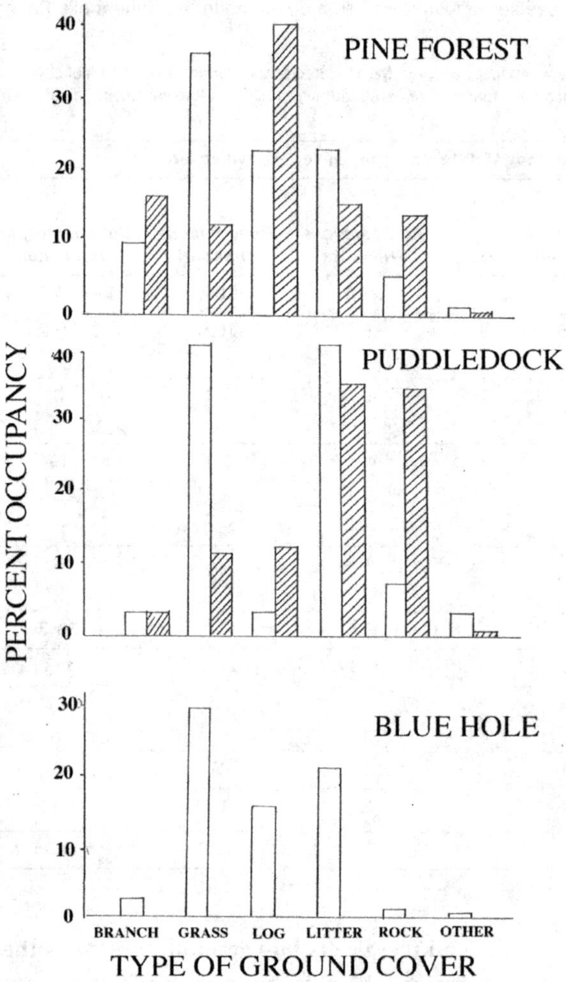

Fig. 11.3. Frequency of active *Lampropholis delicata* (hatched histograms) and *Lampropholis guichenoti* (white histograms) occurring at first sighting on various substrates at the three study sites. Note that *Lampropholis delicata* is not shown for Blue Hole because it did not occur there and therefore not subject to analysis. From Belmont (1977).

there and had its highest occupancy on grass. At Puddledock *L. delicata*'s highest occupancy was greatest equally on litter and rocks, and *L. guichenoti*'s equally divided between grass and litter (Fig. 11.3). This suggests that these two species may tend to forage in slightly different microhabitats and that the favoured habitat may shift depending on other characteristics of the forest, such as openness of the canopy and/or the relative availability or favoured substrates. This hypothesis requires rigorous experimental testing. The habitat selection of these two species is equivocal and seems to vary with geographic location and depending on the species composition of the assemblage (Anderson *et al.* 2023) (see Chapters 11 for further analysis of this topic).

Further, stronger support was provided by a more-detailed analysis of microhabitat selection. Where these species occurred together both were associated with leaf litter at similar, rather high, frequencies, but otherwise they diverged in their microhabitats. *Lampropholis delicata* had a much higher incidence of association with two solid substrates, logs and rocks, than did *L. guichenoti*, whereas the reverse was true for grass (Fig. 11.3).

The second hypothesis was tested by recording the frequency of occurrence of these two species under different conditions of exposure to the sun when first sighted. Recall that because of their relatively large surface area to volume ratio, small lizards heat and cool more rapidly than larger ones under the same conditions (see Chapter 6). A very small lizard walking over the dappled floor of a rather open woodland, alternately would pass through patches of shade and patches of direct sunlight and its body temperature would respond to these changing conditions almost immediately and fluctuate accordingly. Accurately recording body temperatures of such a small lizard was nearly impossible because attempt to capture it could cause its temperature to change while it fled through a thermally different patch of forest floor. Also, its temperature would change during the time it was restrained in a different thermal environment while its cloacal temperature was being registered. Conduction of heat through the investigator's hand could change the lizard's temperature significantly, even during the brief time required for reading the cloacal thermometer. Thus, the conventional means of measuring lizards' body temperatures at the time of Belmont's study was inapplicable for such small lizards. Alternatively, she recorded lizards' locations relative to patches of sun and shade, at two times of year, summer and autumn.

In summer, the first sightings of lizards in the sun were not significantly different from the first sightings in shade (Table 11.6); they spent little time in semi-shade (dappled sunlight in patches, such as would occur from the sun shining through a cluster of leaves, leaving small patches of sunlight and small patches of shade). It appears that the lizards were shuttling between patches of sun and shade while foraging over the forest floor, with about equal time spent under the two conditions. In autumn, the pattern changed in ways reflecting the lower ambient temperatures at that time of year. Then, both species of lizards were seen significantly more often in sunlight than in shade (Table 11.6).

Table 11.6. First-sightings of *Lampropholis* under different conditions of sun versus shade.

N = number of individuals. Chi-square (X^2) calculated with Yates correction from Belmont (1977).

Time of year Condition	*Lampropholis guichenoti*		*Lampropholis delicata*	
	N	%	N	%
January–February (Summer)				
In direct sun	35	42	31	50
In semi-shade	9	11	6	10
In shade	39	47	25	40
March–April (Autumn)				
In direct sun	40	76	32	76
In semi-shade	6	11	4	10
In shade	7	13	6	14
Sun versus shade[1] (Summer)				
X^2	0.216		0.64	
P	$0.70 > P > 0.50$		$0.50 > P > 0.30$	
(Autumn)				
X^2	21.78		16.44	
P	<0.001		<0.001	

[1]Excluding semi-shade.

The relative use of sun and shade by the two species was practically identical in autumn (76% of the sightings in the sun and 13–14% in the shade) (Table 11.6). In summer, *L. guichenoti* was seen slightly more often in shade than in sun, whereas the reverse was true for *L. delicata*, although the differences between the two species was not significant for either season (summer: $X^2 = 0.10 > P > 0.05$; autumn: $X^2 = 1.04 \pm 0.50 > P > 0.30$).

Thus, hypothesis 2 was not supported. This conclusion is reinforced by measurements in the laboratory of the upper thermal tolerances of these two species. Five independent studies of *L. guichenoti* gave values ranging from slightly below, to slightly above, that of a single one for *L. delicata* (see review by Heatwole and Taylor 1987). Given the fact that variables such as acclimation can affect the Critical Thermal Maximum, it appears that the difference, if any, between these two species is slight (see Chapter 6).

The validity of the third hypothesis is difficult to assess from present data. It would appear to be counter-indicated by the fact that *L. delicata* is less frequently associated with grass, than is *L. delicata*, but that observation ignores the fact that in the more closed habitats, other types of low vegetation are present, for example, mosses and forbs, that may be more attractive than grass to *L. delicata*. In a later study, Bilcke *et al.* (2006) found the structure of vegetation to be a likely influence on the ecology of *L. delicata* and suggested that the removal of a leafy exotic weed from its habitat would have important implications for its foraging and fitness.

The affinity for grass by *L. delicata* may not be consistent among all habitats but perhaps occurs only where grass occurs in conjunction with other specific features of the habitat. For example, Howland *et al.* (2014) found that the proportion of sites occupied by this species was very low in open grassland, higher in open woodland and still higher in denser woodland (Fig. 11.4).

Farquhar *et al.* (2023) investigated the microhabitats used by *L. delicata* in the Liverpool Range and analysed the importance of structure of the habitat, thermal physiology, microclimate, and macroclimate as predictors of the suitability of sites for occupancy by this species. The most important climatic predictor of this species' distribution was the seasonality of temperature, with deeper (and therefore cooler) litter being an important refuge from excessive heat.

Taking all the above studies in conjunction suggests that the optimal habitat for *L. guichenoti* includes thick litter, grass, or other low vegetation with shade from an overhead canopy and access to cover such as logs, litter, or rocks. All these attributes may be instrumental in facilitating the animals maintaining their body temperatures within favourable limits. Milton (1980) indicated logs and litter as favourable microhabitat for *L. guichenoti.*

Even for species that are syntopic (occupy the same habitat within the same geographic area) utilisation of different microhabitats by different species could lead to nuances of ecological segregation. The floor of a habitat, such as woodland, consists of a mosaic of microhabitats, in the present example consisting of branches; patches of grass, forbs, moss, and leaf litter; logs; and rocks. We have seen already that *L. delicata* uses logs more often than *L. guichenoti*, but the separation is incomplete and both species do avail themselves of that microhabitat. Are there others that also provide some ecological segregation?

Figure 11.5 shows the microhabitats being used by individual lizards at their initial sightings. It confirms that *L. delicata* uses logs more often than does *L. guichenoti* in the habitats in which both occur; in Pine Forest, the former species was first sighted on logs more than in any other microhabitat. At Puddledock, the

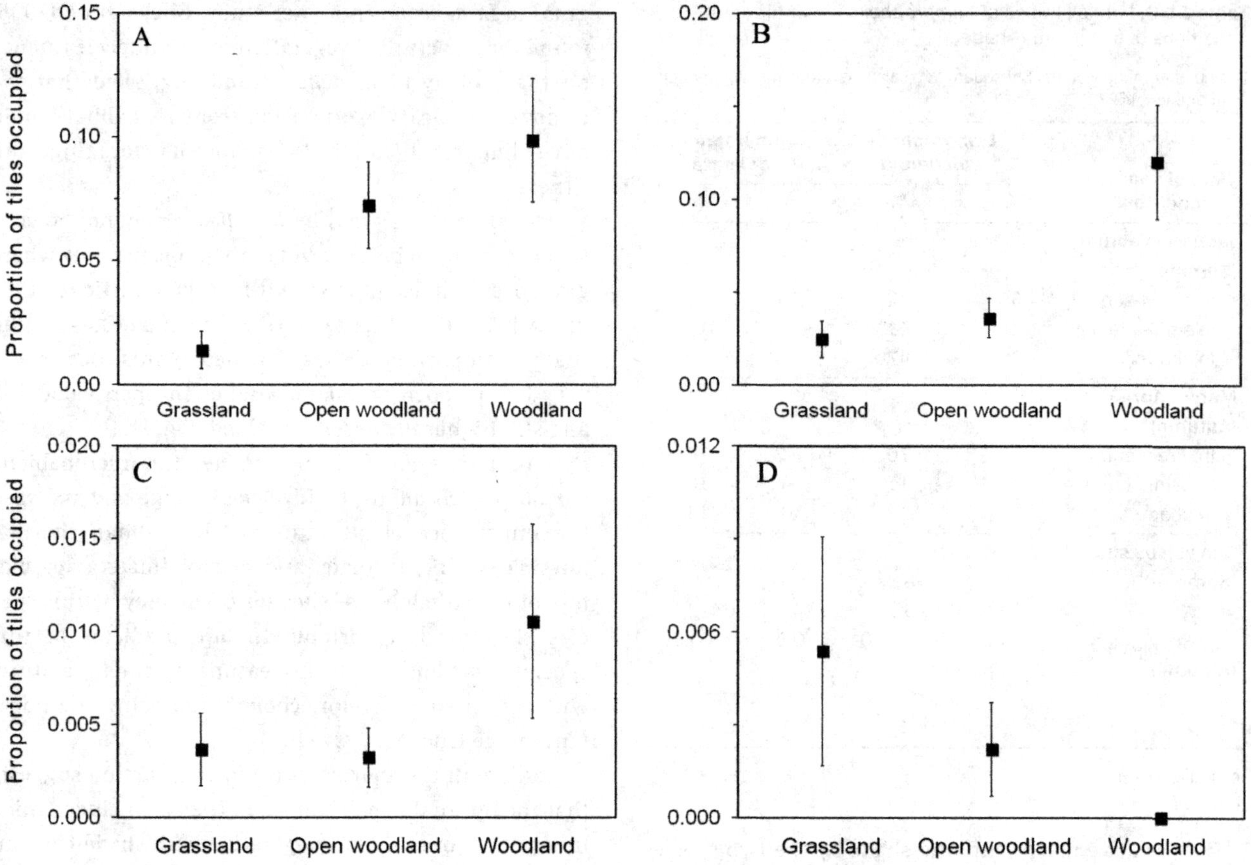

Fig. 11.4. Mean occupancy of sites in three types of habitat for *Morethia boulengeri*: (**A**), *Lampropholis delicata* (**B**), *Hemiergis talbingoensis* (**C**) and *Delma* spp. (**D**). Vertical lines indicate standard errors. From Howland *et al.* (2014).

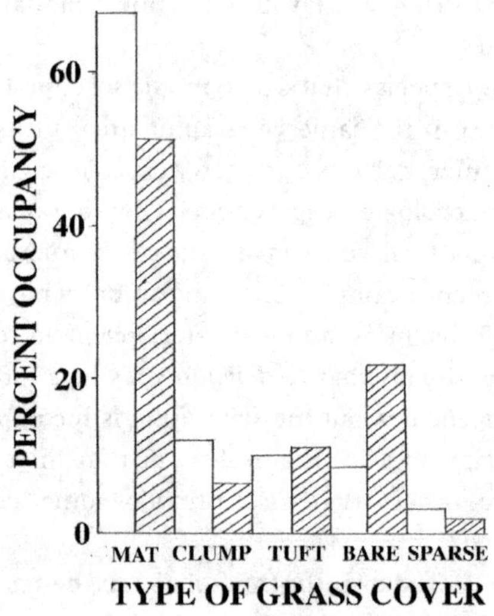

Fig. 11.5. Frequency of first sightings of active *Lampropholis delicata* (white histograms) and *Lampropholis guichenoti* (hatched histograms) on different structural forms of grass in the Pine Forest. Re-formatted from Belmont (1977) and published with her permission.

density of logs was lower by nearly half that in the Pine Forest (429 as opposed to 245 per hectare) and accordingly the first sightings of both species in that microhabitat were reduced, but with *L. delicata* still using logs proportionately more often than *L. guichenoti*. Thus, where logs were most abundant, first sightings on logs were high for both species, but where logs were less abundant, first sightings on them were reduced for both species, and where logs were scarce, the species that depended most on logs (*L. delicata*) was absent. In the absence of *L. delicata* at Blue Hole, *L. guichenoti* increased its relative use of this resource, thereby raising the possibility that in other habitats, perhaps the presence of *L. delicata* on logs may have inhibited their use by the opposite species.

An interesting phenomenon is that with a reduction in the density of available logs between Pine Forest and Puddledock, there was a shift from logs as the primary microhabitat used by *L. delicata* to rocks being equal-first with leaf litter. Logs and rocks have one feature in common: they both are slightly elevated above the other substrates and provide potential vantage points from which the lizards can scan the surrounding litter and grass for prey. If logs and rocks are equally effective as scanning posts, then the proportional use of such perches might depend on the relative number of each that is available. Unfortunately, the density of rocks in these habitats was not measured so a test of that hypothesis was not made. However, at Blue Hole, sightings of *L. guichenoti* on rocks was low but sightings on logs was rather high, suggesting that rock may not have been plentiful. If so, then perches of both types (rocks and logs) were both in low numbers and since both seem to be important for *L. delicata* in other areas, lack of suitable sites for scanning for prey might be the reason for lack of this species there. An alternative hypothesis is that paucity of rocks and logs may have limited the ability of this species to overwinter. Belmont (1977) searched under rocks and logs for brumating *Lampropholis* in winter and found *L. delicata* and *L. guichenoti*. If *L. delicata* depends on rocks and logs for overwintering, more than does *L. guichenoti*, it might explain the absence of the former species at Blue Hole where logs were relatively scarce and perhaps rocks also were.

Simbotwe (1985) also addressed the microhabitat selection of *L. guichenoti* in a eucalypt woodland (Newholme) near Armidale, NSW, albeit not at the same sites as those studied by Belmont (1977), and his comparison was between that species and *Hemiergis decresiensis*, not *L. delicata*. He included lizards in the open and found that category to be the largest of all. He lumped grass and litter into a single category and collectively those had the second-largest frequency of occupancy.

In the Pine Forest, *L. guichenoti* was first sighted most often on grass, whereas *L. delicata* had low representation there. The former species also was well represented on logs (see above) and litter. At Puddledock, *L. guichenoti* had high and nearly equal representation on grass and litter, almost to the exclusion of other categories. *Lampropholis delicata*'s two main representations there were on rock (see above) and litter. Clearly, where these two species occur together they respond differently to the array of microhabitats. Where *L. guichenoti* occurs alone (Blue Hole), its representations are high on grass, as in the other two areas, and litter and logs (see above); other categories are all low. Thus, for *L. guichenoti* grass and litter both rank high in all three habitats.

Given the importance of logs as microhabitat for both species (see above), a more-detailed analysis was conducted to see whether the state of decay of logs was related to microhabitat selection. Table 11.7 shows the relative use logs in different stages of decay. For both species firm logs were used more often than those either listed as 'soft' or 'rotten'. However, since the relative abundance of these stages in the environment was not reported, it is not possible to ascertain whether there was active selection on the part of either species for a particular stage, or whether they used logs in the proportion at which they occurred in the environment. Whether the species differed in their use of logs in the various stages of decay could be tested, however, and it was found that there was no significant difference between the two species ($X^2 = 0.31$; df = 4; $0.99 > P > 0.98$). It was not possible to test whether hollow logs versus solid ones were actively selected or merely used in the proportion they occurred in the environment, but the proportional use of these two categories of log did not differ between the species ($X^2 = 0.31$; df = 1; $0.70 > P > 0.50$).

Table 11.7. Comparison between the number of logs in different stages of decay used by *Lampropholis delicata* and *Lampropholis guichenoti* in the Pine Forest (Belmont 1977).

	Stage of decay		
Species	**Firm**	**Soft**	**Rotten**
Lampropholis delicata	45	16	12
Lampropholis guichenoti	55	16	15
	Solid		**Hollow**
Lampropholis delicata	47		26
Lampropholis guichenoti	48		32

Given the high incidence of occurrence of *L. guichenoti* on grass in all three habitats (Fig. 11.5), that relationship was examined in more detail. Grass occurred in several structural forms: mats, clumps, tufts, sparse growth, and was contrasted with bare ground. In all habitats examined, both species were observed foraging in all these microhabitats and it can be concluded, therefore, that with respect to type of grass both species are eurytopic. In the Pine Forest, both species displayed similar patterns in that their predominant proportional representation was, by far, on mats. There were minor differences, however; representation by *L. delicata* on mats was somewhat higher than it was for *L. guichenoti* (Fig. 11.5). Also, *L. guichenoti* had a relatively higher representation on bare ground, than did *L. delicata*. All other representations were relatively low. At Puddledock, *L. delicata* occurred on clumps of grass at low frequency compared with all other types of grass or with bare

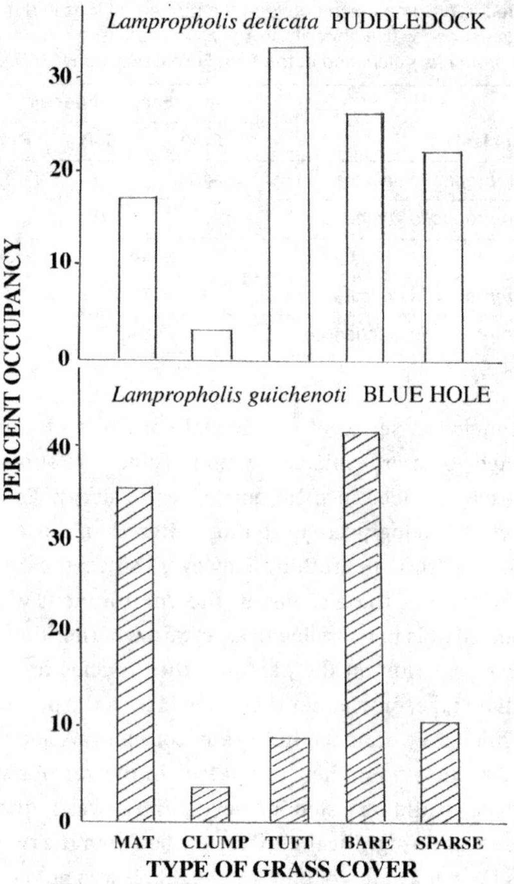

Fig. 11.6. Frequency of first sightings of active *Lampropholis delicata* (white histograms) at Puddledock and *Lampropholis guichenoti* (hatched histograms) at Blue Hole. Re-formatted from Belmont (1977) and published with her permission.

ground, and at Blue Hole *L. guichenoti* foraged mainly on bare ground or on mats of grass, with all other categories having low occupancy (Fig. 11.5). Thus, it seems that when the two species occur together in the same habitat (Pine Forest), their proportional representations on different structural types of grass are similar (Fig. 11.5), whereas each species displays a different distribution of frequencies in different habitats, for example compare *L. delicata* between Pine Forest and Puddledock, and *L. guichenoti* between Pine Forest and Blue Hole (Figs 11.5 and 11.6). This pattern could arise either from each species having different behavioural preferences for structural type of grass in different habitats, or it may merely reflect differences in the relative cover of different types of grass in different habitats, with the species foraging in the various structural types of grass in the same proportion as the relative cover of those types. The latter seems the more parsimonious of the two hypotheses and is reinforced by the fact that when both species are exposed to a similar array of structural types, they display similar frequencies of occupancy; that is, both species are eurytopic with respective to structural type of grass and are randomly distributed over those structural types. This issue could be clarified either by: (1) experiments in which lizards are given choices between structural types in the laboratory or in field enclosures, or by (2) observing lizards and recording the time spent foraging on different structural types and relating those results to the frequency of occurrence of those types.

Leaf litter ranked high as a microhabitat occupied by both species, especially at Puddledock (Fig. 11.3), and its percentage cover was measured in a quadrat of one square metre, centred on the locations at which *Lampropholis* were sighted. Both species were represented relatively evenly across the spectrum of percentages of coverage except for two higher values for *L. delicata* (at coverages of 20 and 80% litter cover). The correlation between occupancy and per cent litter cover was not significant (Spearman Rank correlation test: *L. delicata* $r_s = -0.38$; $P << 0.05$; *L guichenoti* $r_s = 0.051$; $P << 0.05$). Thus, it seems that these eurytopic lizards forage over leaf litter, but that their use of it is not related to the extent of its cover. It should be noted that the cover in the square-metre plot surrounding the first sighting of a lizard does not necessarily represent the total size of the patch of litter over which the lizard was foraging. For example, if a large patch of litter had only a small proportion of its area within the boundaries of the square-metre sample, the lizard would appear to be foraging in a small patch, when, in reality, it was foraging near the periphery of a large patch. This effect could confound the correlation between per cent cover and occupancy of litter by the lizards.

The depth of litter may be a more important feature than the size of its patches. Farquhar *et al.* (2023) found that *L. delicata* relied heavily upon deep litter as a refuge when high temperatures at midday, both in sun and shade, exceeded its preferred temperature (also see Chapters 2 and 12).

In order to summarise the results of these various field observations from different sites and types of forest, a metadata analysis was carried out (Table 11.8). Where *L. delicata* and *L. guichenoti* co-occur, they both use the complete array of microhabitats available, but show somewhat different frequencies; the former spent less time on grass than on any other type of substrate, and the latter spend more time there than on any other

Table 11.8. Order of rank of occurrence of *Lampropholis delicata* and *Lampropholis guichenoti* on different substrates at different localities.

1 to 5 indicates greatest to least extent of occupancy. Dashes indicate the substrate was not included in the study. Data from Belmont (1977) and Simbotwe (1985). For ease of comparison data were entered in pale type for *L. delicata* and in **boldface** for *L. guichenoti*. Data from the category 'other' excluded. Note that at Newholme grass and litter were combined, so tied ranks were given for each.

Habitat or locality	Type of substrate				
Species	**Bare**	**Grass**	**Litter**	**Logs**	**Rocks**
PINE FOREST					
Lampropholis delicata	–	4	2	1	3
Lampropholis guichenoti	–	**1**	**2.5**	**2.5**	**4**
EUCALYPT WOODLAND					
Lampropholis delicata	–	4	1	3	2
Lampropholis guichenoti	–	**1**	**2**	**4**	**3**
BLUE HOLE					
Lampropholis delicata	–	Absent	Absent	Absent	Absent
Lampropholis guichenoti	–	**1**	**2**	**3**	**4**
NEWHOLME					
Lampropholis guichenoti	**1**	**3.5**	**3.5**	**2**	**5**

type of cover. Since both species are occupying the same site, they are exposed to the same frequencies of occurrence of the different patches of habitat, which suggests they differ in their choice of microhabitat. There were no clear differences in choice for other habitats. In addition to the above observations, Simbotwe (1985) (Table 11.8) recorded the occurrence of *L. guichenoti* on bare patches and found that occupancy in the open was more frequent than for any type of cover. Since he did not study *L. delicata*, it was not clear whether the two species differed in their predilection for bare ground at the site he studied. Howard *et al.* (2003), however, did study *L. delicata* at a different location and found just the opposite. It did not favour open areas and this difference between the two species might be the secret to their co-occurrence in some localities. On the other hand, it may be that both species are flexible in regard to the type of substrate on which they forage and adjust their use of different microhabitats depending on the constitution of the local assemblage in which they are embedded (see Chapter 16). If so, this would account for some of the regional differences in their use of microhabitats and the divergent results obtained by different investigators (Table 11.8).

In assessing the microhabitats used by lizards, it is important to keep two considerations in mind: (1) Different types of substrates may resemble each other structurally more than different examples of the same type of substrate. For example leaf litter and some growth-forms of grass are both loosely structured with lots of interstices allowing even large-limbed lizards to move freely on or through it. By contrast, the open type of grass may differ greatly in structure from grass that is tightly bunched into tussocks and that inhibits fully legged lizards to move through it but allows easy passage to slender lizards with limbs of reduced size that can slither through it by sinuous 'grass-swimming' (see section *Relation of reduction of limbs to habitat*). (2) A lizard's requirements may encompass more than one type of structure in its habitat. For example, both logs and either litter or grass are important both for *L. delicata* and *L. guichenoti*, litter or grass being used in conjunction with bare ground for their diurnal foraging but they need logs or rocks or other solid objects under which to shelter during either nocturnal or seasonal periods of inactivity. Having either of these structural features missing may render a habitat unsuitable; both are needed.

Shoo *et al.* (2014) conducted two experiments to assess whether altering the structure of the forest floor would accelerate the restoration of abundance and diversity of reptiles in remnant forests and damaged forests. In one experiment they removed the existing cover of coarse woody debris and in the other they added either old fence posts or logs. They found that in stands where such debris was added the collective abundance of individuals of all reptiles increased in comparison to control plots. Some of the seven resident species of reptile, however, showed no change but *L. coggeri* was the only species whose numbers slightly decreased compared to numbers on control plots (see their Fig. 1). In the experiment in which the existing cover was removed on experimental plots, there was an overall decline in the small numbers of reptiles in general although some, including *L. delicata*, showed no change; only one of the seven resident species, *L. coggeri*, showed an increase. The authors cautioned that their results were based only on small numbers obtained from a preliminary study of only one year, and that more confident conclusions could be expected after monitoring over a longer period.

Disturbances such as logging, fire, and drought can modify habitats in ways that alters their suitability for species of *Lampropholis*. Lunney *et al.* (1991) found that *L. guichenoti* occurred in about equal numbers in unlogged and recently logged forest but after 10–15 years of regrowth its numbers had decreased, ostensibly because of the reduced amount of sunlight reaching the ground in the more heavily shaded regrowth; *L. delicata* exhibited a similar pattern. Fire reduced the number of *L. guichenoti*, at least on ridges, but had no effect on *L. delicata* or other species of skinks present. The numbers of all species were reduced by drought but *L. guichenoti* recovered more quickly than did *L. delicata*. Topography modified the response of *L. guichenoti*; in gullies this species was much more abundant than on ridges. Subtle differences, such as these, between the two species need to be factored into interpretation of their differences in distribution and in their relative abundance in different habitats, as well as in designing strategies for their conservation.

Habitat and microhabitats of *Lampropholis*: Studies in the laboratory

Given that *Lampropholis* often were associated with grass, rocks, and litter in the field and that the two species differed somewhat in the extent to which they were found in such associations, Belmont (1977) conducted experiments in the laboratory in plastic tanks 1.3 m × 0.6 m × 0.6 m high, equally divided between the two substrates being offered for choice in a particular experiment. The substrates tested were grass, litter, and rock, offered in each of the three possible combinations of choice: grass versus rocks; grass versus litter, and litter versus rocks. All choices were placed on 4 cm of soil; grass consisted of freshly cut short turf; rocks were flat, 4 cm in diameter, and adjacent but not overlapping; litter was a mixture of eucalypt leaves, twigs, and bark. A 100-watt incandescent light bulb supplied heat. Equal areas of each substrate type were offered. Ten animals of a given species were tested together in each experiment and the two species were tested separately. The animals were allowed to acclimatise to the apparatus for a day before readings were initiated. Then at half-hourly intervals during the day when temperatures were within the ambient range foraging animals experienced in nature, the number of animals foraging on each substrate was counted. Foraging was defined as active searching movements during which the head was moved from side to side. The results were that for *L. delicata* the numbers of individuals foraging did not differ significantly between any of the alternate choices offered; that is, this species foraged equally on all three substrates ($P > 0.05$ in all cases). The results were similar for *L. guichenoti* for two choices; there was no significant choice for grass versus litter ($X^2 = 3.17$; $P > 0.05$) or for rocks versus litter ($X^2 = 0.07$; $P > 0.05$). *Lampropholis guichenoti* did, however, forage significantly more frequently on grass than on rocks ($X^2 = 16.05$; $P < 0.05$). This agrees with the field data in which *L. guichenoti* had high frequency of occurrence on grass, although it often also occurred in the open and on litter (Fig. 11.3).

Choice of cover during periods of inactivity at night

Lampropholis overwinter under rocks and logs (see Chapter 6 section *Body temperatures in winter*) and it is likely that similar places are used as nocturnal retreats. Accordingly, Belmont (1977) conducted experiments to identify the attributes of cover important to lizards when selecting sheltered sites for periods of nocturnal inactivity. Seven different structural configurations were tested (Fig. 11.7A–G). For a given experiment, two different structures were centred on opposite halves of a glass terrarium covered with black paper; the structures rested on a bed of gravel 2 cm deep. The terrarium was covered by a lid of pine board and illuminated by a 60-watt light bulb above a hole cut in the lid directly above each structure. Lights were on from 07:00 to 18:30 in summer and from 08:30 to 17:30 in winter. Temperatures in the experimental apparatus ranged from 17.5°C to 28°C in summer and from 16°C to 26°C in winter. The locations of the cover types were reversed each day.

Six individuals of each species being tested were placed into the apparatus at noon of the day experiments were to begin and on the following five consecutive nights at 21:00 hours the number and locations of the lizards associated with each type of cover was recorded. All but one test involved both *L. delicata* and *L. guichenoti*. The sole exception to these procedures was one test involving 12 *L. delicata* but not *L. guichenoti*.

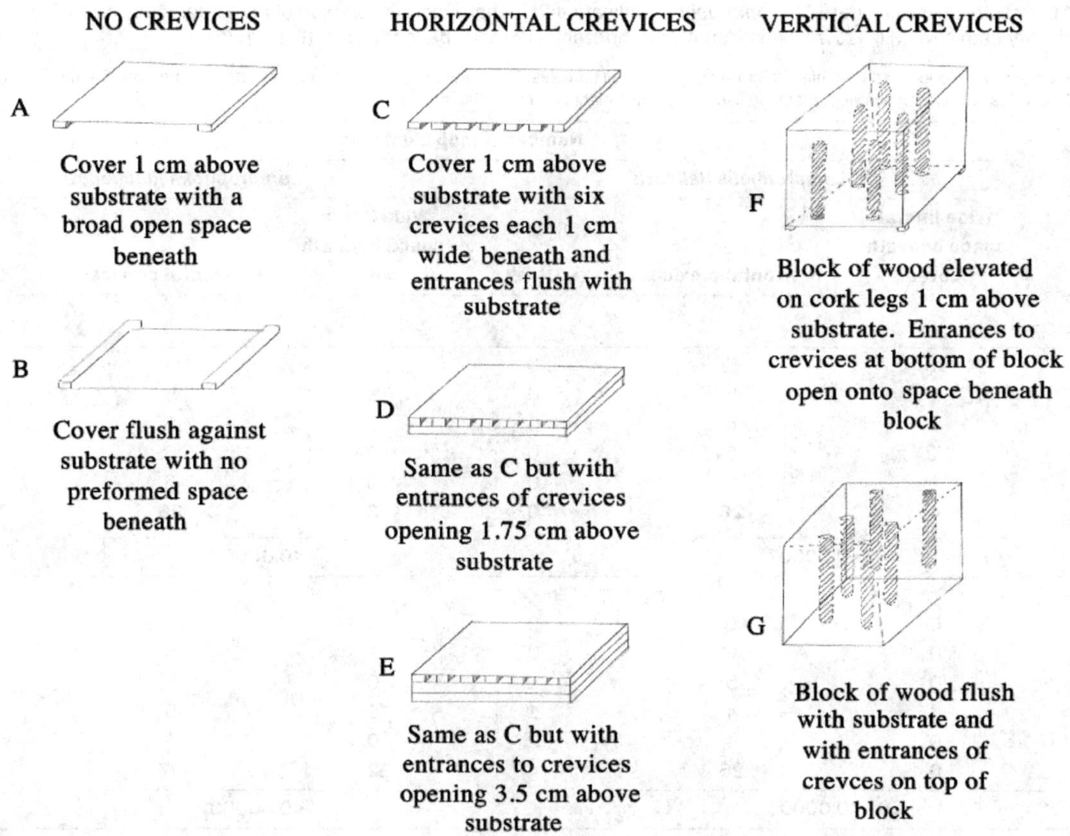

Fig. 11.7. Structures used to test preferences of *Lampropholis* for choice of sites for periods of nocturnal inactivity. These structures were made of wood; A–E provided cover of 20 cm × 20 cm. F–G were 19 cm × 12 cm × 11 cm thick. Redrawn from Belmont (1977) and published with her permission.

Table 11.9. Results of choice offered to *Lampropholis delicata* and *Lampropholis guichenoti* of nocturnal refuges consisting of cover in direct contact with the substrate (Fig. 11.7B) versus horizontal crevices under the cover and with the entrances at ground level (Fig. 11.7C).

Results were tested by the one-tailed binomial probability test for small samples. There were six individuals of each species. Data from Belmont (1977). Observations were made on five consecutive nights (n = 1–5) for each trial.

	Lampropholis delicata			*Lampropholis guichenoti*		
Day	**Cover in direct contact with substrate**	**Horizontal crevices**	**Other**	**Cover in direct contact with substrate**	**Horizontal crevices**	**Other**
1	3	3	0	1	5	0
2	1	5	0	2	4	0
3	1	5	0	1	5	0
4	2	4	0	0	6	0
5	1	5	0	3	3	0
Total	8	22	0	7	23	0
P		0.009			0.003	

When given the choice between forcibly sheltering under cover that was flat against the substrate, with no preformed space available (Fig. 11.7B), versus using horizontal crevices beneath the cover (Fig. 11.7C), both species chose crevices. The results were highly significant for both cases (Table 11.9). In this experiment all lizards sought cover during nights.

An experiment giving lizards a choice between a wide lateral space under cover, providing dorsal contact but not much opportunity for lateral contact (see Fig. 11.7A),

Table 11.10. Results of choice offered to *Lampropholis delicata* and *Lampropholis guichenoti* of nocturnal refuges consisting of a wide lateral space beneath cover (Fig. 11.7A) versus six narrow horizontal crevices beneath cover (Fig. 11.7C).

Results were tested by the one-tailed binomial probability test for small samples. There were six individuals for each species per trial. Each trial involved a different set of individuals. Observations were made on five consecutive nights for each trial (n = 1–5).

	Number of individuals of:					
	Lampropholis delicata			***Lampropholis guichenoti***		
Trial Day	**Wide lateral space beneath cover**	**Horizontal crevices**	**Other**	**Wide lateral space beneath cover**	**Horizontal crevices**	**Other**
Trial 1						
1	1	5	0	2	4	0
2	1	5	0	1	5	0
3	0	6	0	0	5	1
4	1	5	0	0	6	0
5	1	5	0	0	6	0
Total	**4**	**26**	**0**	**3**	**26**	**1**
P		<0.00007			<0.00003	
Trial 2						
1	0	6	0	0	6	0
2	0	6	0	1	5	0
3	0	5	1	0	5	1
4	0	4	1	1	5	0
5	0	4	2	0	6	0
Total	**0**	**26**	**4**	**2**	**27**	**1**
P		<0.00003			<0.00003	
Trial 3						
1	2	4	0	0	6	0
2	1	5	0	0	6	0
3	2	4	0	0	6	0
4	1	5	0	0	6	0
5	3	3	0	1	3	0
Total	**9**	**21**	**0**	**1**	**29**	**0**
P		0.022			<0.00003	
Trial 4						
1	1	5	0	0	6	0
2	0	6	0	0	6	0
3	1	5	0	0	6	0
4	1	5	0	0	6	0
5	0	6	0	0	6	0
Total	**3**	**27**	**0**	**0**	**30**	**0**
P		<0.0002			<0.00003	
GRAND TOTAL	**16**	**100**	**4**	**6**	**112**	**2**

versus a series of narrow horizontal crevices (Fig. 11.7C) was repeated four times, each time with a different set of lizards. The result was similar each time. During the night, most lizards sought shelter under one or other of structures A or C; on average only 3.3% of the *L. delicata* and 1.7% of the *L. guichenoti*, remained at large elsewhere in the terrarium, without seeking cover. For both species, there was an overwhelming, highly significant choice for the narrow horizontal crevices, rather than for the wider, more open space (Table 11.10).

The species of *Lampropholis* are not scansorial, although they do climb onto logs and rocks to bask and/

Table 11.11. Results of choice offered to *Lampropholis delicata* and *L. guichenoti* of nocturnal refuges consisting of a series of horizontal crevices either opening directly at ground level (Fig. 11.7C) or with entrances elevated 1.75 cm above the substrate (Fig. 11.7D).

Results were tested by the one-tailed binomial probability test for small samples. Different sets of six individuals were used for the two trials. Re-formatted from Belmont (1977) with her permission.

Trial Day	Entrances of horizontal crevices at ground level	Entrances of horizontal crevices elevated 1.75 cm above ground	Other
	Number of *Lampropholis delicata*		
Trial 1			
1	0	6	0
2	2	4	0
3	2	4	0
4	2	4	0
5	2	4	0
Total	**8**	**22**	**0**
P	0.009		
Trial 2			
1	2	3	0
2	2	4	0
3	1	5	0
4	3	3	0
5	3	3	0
Total	**11**	**18**	**0**
P	0.147		
	Number of *Lampropholis guichenoti*		
Trial 1			
1	2	4	0
2	2	4	0
3	4	2	0
4	3	3	0
5	2	4	0
Total	**13**	**17**	**0**
P	0.464		

or to scan the litter for prey (see above). The next experiment investigated whether the lizards responded differently to the crevices whose entrance were elevated slightly above the level of the substrate (1.75 cm; Fig. 11.7D) as opposed to those at ground level (Fig. 11.7C). All individuals of both species in this experiment entered crevices of one sort or another at night and more often used crevices elevated to 1.75 cm than they did those at ground level. The results, however, varied between the two species and were equivocal.

In one of two separate trials for *L. delicata* (Table 11.11) the statistical differences between the choices were not significant but in the other one they were. In the trial for *L. guichenoti* the differences were not significant.

A second experiment offered the choice between two heights of elevated entrances, one at 1.75 cm and the other at 3.5 cm (but none were available at ground level) (Table 11.12). Neither species showed significant preferences between the two heights. Thus, the results from these two experiments on choice of height of entrances to crevices are somewhat equivocal.

Finally, an experiment was conducted to see whether more clear-cut results would be obtained if the disparity between the two heights of entrances were increased. There were two trials for each species and the height compared were ground level and 3.5 cm above the substrate (Table 11.13). The greater use of elevated crevices was statistically significant in one of the two trials and at the boarder-line of significance in the other for *L. delicata*. Neither trial for *L. guichenoti* proved to be statistically significant, and indeed, on one of them more crevices were used at ground level than were those elevated at 3.5 cm. Thus, it appears that these two species may differ in their predilection

Table 11.12. Results of choice offered to *Lampropholis delicata* and *Lampropholis guichenoti* of nocturnal refuges consisting of horizontal crevices with entrances elevated above the substrate by 1.75 cm (Fig. 11.7D) versus those with entrances elevated above the substrate by 3.5 cm (Fig. 11.7E).

Results were tested by the one-tailed binomial probability test for small samples. There were six individuals for each species per trial. Each trial involved a different set of individuals. Re-formatted from Belmont (1977) with her permission.

	Number of individuals of:					
	Lampropholis delicata			***Lampropholis guichenoti***		
Day	**Entrances of horizontal crevices elevated 1.75 cm**	**Entrances of horizontal crevices elevated 3.5 cm**	**Other**	**Entrances of horizontal crevices elevated 1.75 cm**	**Entrances of horizontal crevices elevated 3.5 cm**	**Other**
1	4	2	0	4	2	0
2	4	2	0	3	2	1
3	4	2	0	2	4	0
4	2	4	0	1	5	0
5	2	4	0	5	1	0
Total	**16**	**14**	**0**	**15**	**14**	**1**
P	0.291			0.356		

Table 11.13. Results of choice offered to *Lampropholis delicata* and *Lampropholis guichenoti* of nocturnal refuges with entrances at ground level (Fig. 11.7C) versus those with entrances elevated 3.5 cm above the substrate (Fig. 11.7E).

Each trial involved a different set of individuals. Re-formatted from Belmont (1977) with her permission.

	Number of individuals of:					
	Lampropholis delicata			***Lampropholis guichenoti***		
Trial Day	**Entrances of horizontal crevices at ground level**	**Entrances of horizontal crevices elevated 3.5 cm**	**Other**	**Entrances of horizontal crevices at ground level**	**Entrances of horizontal crevices elevated 3.5 cm**	**Other**
Trial 1						
1	3	2	1	4	2	0
2	1	3	2	3	3	0
3	1*	4	0	1	5	0
4	0	5	0	1	5	0
5	2	3	0	2	4	0
Total	**7**	**17**	**3**	**11**	**19**	**0**
P	<0.001			0.100		
Trial 2						
1	2	4	0	4	2	0
2	2	4	0	4	2	0
3	2	4	0	3	3	0
4	2	4	0	5	1	0
5	2	4	0	0	5	1
Total	**10**	**20**	**0**	**16**	**13**	**1**
P	0.051			0.356		

*One lizard died between day 2 and day 3 from unknown causes.

for height above the substrate of entrances to crevices used during nocturnal sheltering, with *L. delicata* tending toward the more elevated ones, but *L. guichenoti* perhaps being less discriminatory, at least within the range of heights tested.

Crevices in logs and rocks or burrows in the substrate can occur at various angles. Belmont (1977) ran a test to see whether *L. delicata* and *L. guichenoti* prefer horizontal versus vertical crevices. In this test, one block of wood was raised slightly above the substrate on cork supports at each corner to provide ready access to vertical burrows from beneath the block (bottom entrance) (Fig. 11.7F). Neither species used the bottom-opening to vertical crevices at all but rather elected for the horizontal crevices at ground level (Table 11.14).

Table 11.14. Results of choice offered to *Lampropholis delicata* and *Lampropholis guichenoti* for nocturnal refuges consisting of either horizontal crevices opening at ground level (Fig. 11.7C) or vertical crevices with entrances opening upward into the covering block forming the cover (bottom entrance) (Fig. 11.7F).

Results were tested by the one-tailed binomial probability test for small samples. There were six individuals of each species. Re-formatted from Belmont (1977) with her permission.

	Number of individuals of:					
	Lampropholis delicata			*Lampropholis guichenoti*		
Day	Vertical crevices (opening at bottom)	Horizontal crevices (ground level)	Other	Vertical crevices (opening at bottom)	Horizontal crevices (ground level)	Other
1	0	2	4	0	1	5
2	0	5	1	0	0	6
3	0	5	1	0	0	6
4	0	5	1	0	4	2
5	0	5	1	0	3	3
Total	**0**	**20**	**8**	**0**	**8**	**22**

Table 11.15. Results of choice offered to *Lampropholis delicata* and *Lampropholis guichenoti* of nocturnal refuges consisting of vertical crevices opening at ground level (Fig. 11.7F) versus vertical crevices opening downward from the top of the block (Fig. 11.7G).

Results were tested by the one-tailed binomial probability test for small samples. There were six individuals of each species. Re-formatted from Belmont (1977) with her permission.

	Number of individuals of:					
	Lampropholis delicata			*Lampropholis guichenoti*		
Day	Vertical entrances opening at bottom	Vertical entrances opening at top	Other	Bottom-facing entrances	Top-facing entrances	Other
1	1	2	3	0	2	4
2	2	2	2	0	3	3
3	1	3	2	0	3	3
4	0	4	2	0	3	3
5	0	5	1	0	3	3
Total	**4**	**16**	**10**	**0**	**14**	**16**
P		0.013	0.0003			

The final trial was between vertical crevices with entrances from the bottom of the block forming the cover (as in the previous experiment: Fig. 11.7F), as opposed to entrances entering from the top of the block (Fig. 11.7G). As before, *L. guichenoti* did not use the vertical crevices with the bottom opening, but *L. delicata* did, although they significantly favoured the alternative of those with an entrance at the top (Table 11.15). This result is consistent with previous experiments in that *L. delicata* selected the highest available entrances of horizontal crevices in preference to those at ground level (Table 11.12) as well as entrances located at the top in the upper parts of its microhabitat in preference to those at lower levels. The results were enigmatic, however, for *L. guichenoti* showed no preference for the higher entrances to horizontal crevices, at least up to 3.5 cm above the substrate, yet it exclusively used the vertical crevices that opened high in the microhabitat (top-opening entrances in favour of those opening at ground level) (Table 11.15).

It is likely lizards as small as *Lampropholis* can use tunnels excavated by some of the larger wood-boring insects. Metcalfe and Annable (2016) reported a case of two *L. delicata* that died by getting stuck when both tried to pass through a hole bored by a larval cerambycid beetle.

THE CONSEQUENCES OF OVIPARITY

Oviparity may pose a restriction to the range of habitats that *Lampropholis* can use. Reptilian eggs do not survive under water-logged conditions. Adaptation of reptiles to fully aquatic habitats, or to very wet terrestrial situations, have followed the alternatives of viviparity, for example, true sea snakes, file snakes (Heatwole 1999) or of seasonal movement to terrestrial environments to oviposit, for example, sea kraits (Heatwole 1999) and sea turtles. A case in point for *Lampropholis* is their distribution relative to other lizards across moisture gradient patches in a swamp ecosystem in the Blue Mountains and Newnes Plateau of south-eastern Australia. Gorissen *et al.* (2017a) noted that this ecosystem could be divided into three zones: (1) the swamp proper with surface water or water-logged soil, densely shaded by sedges (*Gymnoschoenus sphaerocephalus*, *Xyris ustulata*, *Empodisma minus*, and *Lepidosperma limicola*), (2) a narrow, drier transitional zone of shrubs (*Leptospermum* spp.) and ferns (*Gleichenia* spp.), and (3) woodland with relatively dry soil. *Lampropholis delicata* and *L. guichenoti* as well as a number of other oviparous species and some viviparous ones, were present in the drier zones, but not in the swamp. *Eulamprus leuraensis*, an endangered species, rarely found in the narrow, transitional zone and not at all in the woodland, was abundant in the swamp and in fact was the most abundant vertebrate species there (Gorissen *et al.* 2017a, b). Perhaps the extraordinary abundance of this viviparous species in the swamp was because the wet conditions excluded potential oviparous competitors like *Lampropholis*.

ECOLOGICAL SEGREGATION

The composition of assemblages varies from place to place as the boundaries of some sympatric species are outdistanced and the geographic confines of a different set of species are overlapped. Consequently, not only do the climatic challenges change from one locality to another but so do the numbers and kinds of biotic interactions. A different consortium of selective forces emanates from both of these sources and one might expect such widespread, eurytopic species as *L. delicata* and *L. guichenoti* to have responded to these challenges by divergently adapting to them in ways leading to geographic differences in their attributes in a clinal fashion.

Pengilley (1972) investigated microhabitats occupied by skinks at Coree Flats in the alpine region of the Southern Highlands, NSW. He found species to be segregated both vertically and horizontally, with some species being mainly arboreal and others restricted to the ground. In general species differed in the time they spent on trees or fallen wood versus on the ground with at least one species being primarily arboreal and others mainly associated on the ground or associated with fallen wood (Fig. 11.8). Species at ground level differentially spent much of their time on leaf litter, grass, or wood. The two species that seemed ecologically most similar, in that they both lived on the ground and were frequently associated with fallen wood, were *L. guichenoti* and *Pseudemoia entrecasteauxii* (Fig. 11.8). They both were common species in the area although *P. entrecasteauxii* was the more abundant one. They both occupy coastal NSW and Victoria, differing only in that *L. guichenoti* extends somewhat farther inland in NSW and Victoria than does *P. entrecasteauxii*; on the other hand *P. entrecasteauxii* extends into Tasmania and some of the Bass Strait islands and *L. guichenoti* does not (Cogger 2018). It would seem that these two species are so similar ecologically that one would expect them to compete and that one should exclude the other.

Does the slightly larger *P. entrecasteauxii* feed on prey that are too large for the gape of *L. guichenoti* to accommodate? If so, is there competition between adult *L. guichenoti* and juvenile *P. entrecasteauxii*? Are there areas where *P. entrecasteauxii* does not occur but otherwise the assemblage is the same? If so, is *L. guichenoti* there a more dominant species than at Coree Flats? When the places where these two species were found in the plot at Coree Flats were superimposed on a map of the dead wood and the types of vegetation (Fig. 11.9), both were found at some time or other in all the types of habitats represented. The main difference was that *L. guichenoti* was less abundant than *P. entrecasteauxii* throughout, but especially in the lower end of the plot.

Rawlinson (1974) examined the composition of some reptilian assemblages (snakes and lizards) in several climatic regions of south-eastern Australia, Tasmania, and the islands of Bass Strait (Table 11.16). Here we consider only that portion of the overall terrestrial fauna that consisted of 10 species of lygosomine lizards. Activity is restricted mainly to dense layers especially of leaf litter; during periods of inactivity microenvironments under the litter or fallen logs are utilised.

Ecological segregation can be very subtle and can occur intraspecifically. For example, Michelangeli *et al.* (2018) found that some individuals of *L. delicata* selected hotter microhabitats than those occupied by other conspecific individuals during the same time and at the same general locality. This microsegregation may have exposed different econes of this species to different assemblages of competitors and predators that were similarly segregated into corresponding microhabitats.

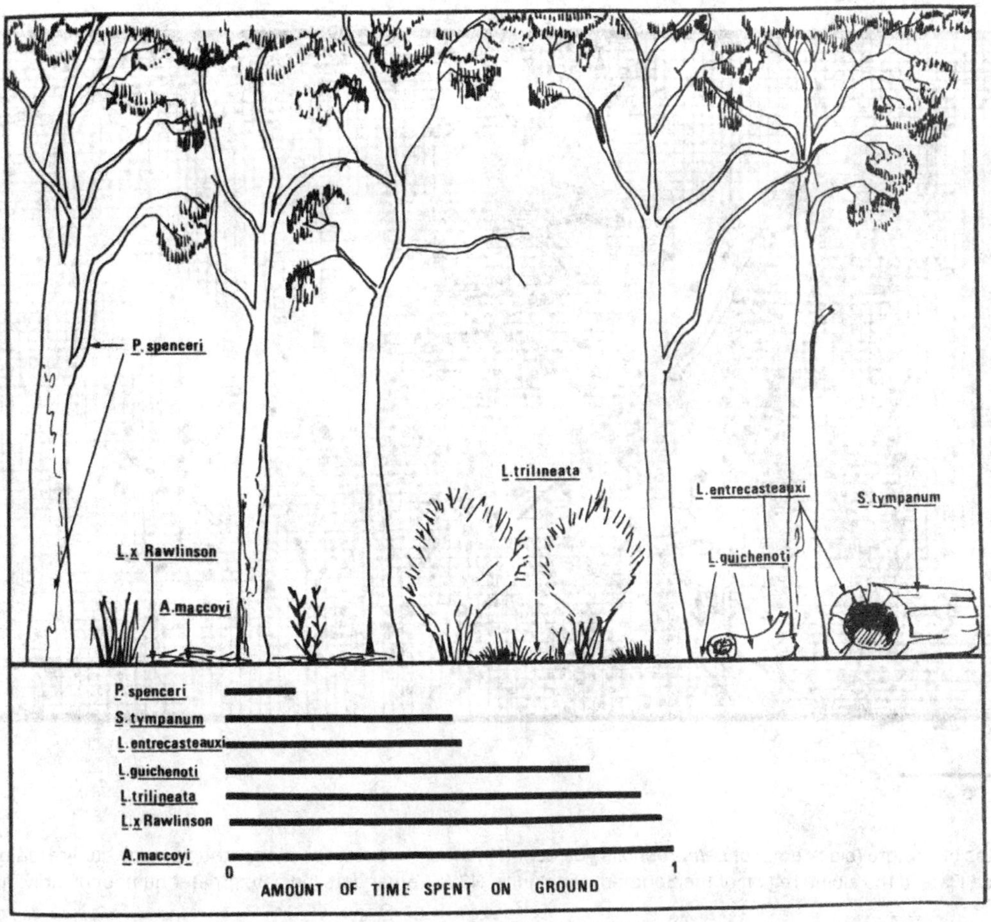

Fig. 11.8. Diagrammatic depiction of the spatial niches of the skinks at Coree Flats, alpine region of the Southern Highlands. Note that some of the species in this diagram have now been found to be members of different genera than originally allocated. The correct names from left to right are: *Psedemoia spenceri*, [*L. x.* Rawlinson: status uncertain], *Anepischetosia maccoyi, Acritoscincus trilineatus, Lampropholis guichenoti, Pseudemoia entrecasteauxii,* and *Eulamprus tympanum.* From Pengilley (1972).

Lampropholis delicata versus *Lampropholis guichenoti*

Distortions of perceived diet

Each method of sampling insects has its own bias, and perhaps the best way to assess the actual taxa of insects at a given place is quantitative sweeps in combination with either pitfall traps or excavation of enclosed quadrats (see Heatwole *et al.* 1981; Warner 1995). Even an accurate quantitative assessment of the species of potential prey present may not reflect the availability of that pool of prey to foraging lizards. Most skinks, including *Lampropholis,* use a combination of sight for detecting active prey and olfaction for finding inactive stages (Hutchinson 1993) (see section *Sensory perception* in Chapter 7). Nevertheless, prey that are secretive or camouflaged may not be encountered or detected in proportion to their abundances. It may be relevant in this context that the large number of termites that Belmont (1977) found in the stomachs of *L. guichenoti* were all reproductives, not workers.

A common life cycle of many ants is that during favourable weather, winged fertile males (drones) and females (queens) emerge, disperse, and mate during nuptial flights, often in large numbers; the male dies after mating and the inseminated female bites off her wings and sequesters herself in a secluded place and founds a new colony there; only her offspring of sterile worker and soldier castes regularly venture out of the nest where they are exposed to marauding predators. Eventually she produces new reproductives and the cycle repeats (Shattuck 1999). The result is that for a brief period there are swarms of defenseless numbers of reproductive ants.

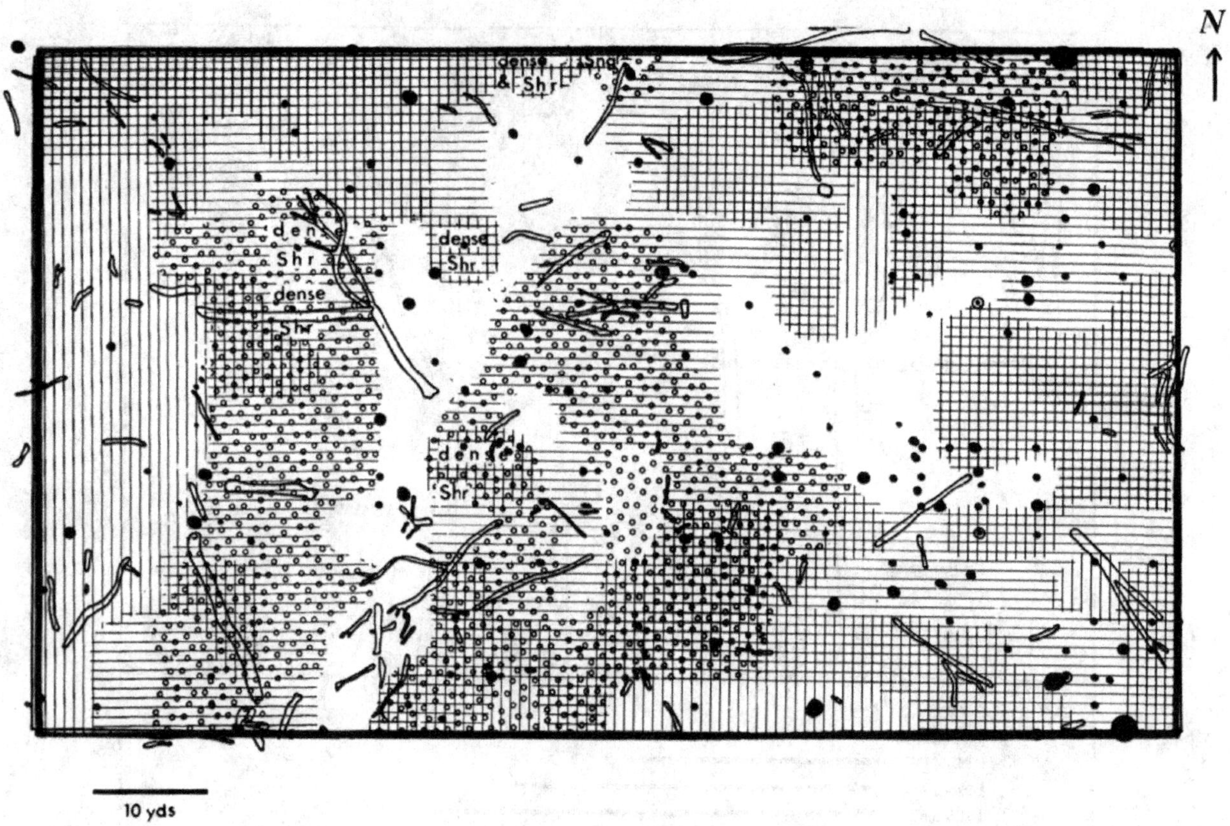

Fig. 11.9. Sites of capture (black dots) of *Lampropholis guichenoti* in the different types of vegetation in the study area of Pengilley (1972) at Coree Flats in the alpine region of the Southern Highlands, NSW. Larger dots indicate greater number of individuals than do smaller dots.

Table 11.16. Species of lygosomine skinks inhabiting the major terrestrial habitats in Tasmania.

Nomenclature of species brought up to date. Ld = *Lampropholis delicata*; Pe = *Pseudemoia entrecasteauxii*; Cm = *Carinascincus metallicus*; Co = *Carinascincus ocellatus*; Cp = *Carinascincus pretiosus*; Lb = *Lerista bougainvillii*; Ps = *Pseudemoia* sp.; Sp 1–3 = three species whose current names are not assignable. – means not present; + means present. *Lampropholis guichenoti* doesn't occur in Tasmania. Data from Rawlinson (1974).

	Species									
Habitat	**Ld**	**Pe**	**Cm**	**Co**	**Cp**	**Lb**	**Ps**	**Sp 1**	**Sp 2**	**Sp 3**
Low-elevation heath	+	–	+	+	+	+	–	+	–	–
Savannah woodland	–	+	+	+	+	+	–	+	–	–
Dry sclerophyll forest	+	+	+	+	+	–	–	+	–	+
Wet sclerophyll forest	–	–	+	–	+	–	–	–	–	+
Temperate rainforest	–	–	+	–	–	–	–	–	–	–
Subalpine forest	–	–	+	+	+	–	–	–	–	–
Dwarf montane forest and shrubland	–	–	+	+	+	–	–	–	+	–
High moors	–	+	–	–	–	–	–	–	+	–
Wet tussock sedgelands	–	+	+	–	–	–	–	–	–	–
Rock outcrops	–	–	+	–	+	–	+	–	–	–
Cleared and cultivated land	–	–	+	–	+	–	–	+	–	–
Number (%) of habitats occupied	2 (18)	4 (36)	10 (91)	5 (46)	8 (73)	2 (18)	1 (9)	4 (36)	2 (18)	2 (18)

Warner (1995) reported that most ants recovered from the stomachs of *Lampropholis* were winged reproductives and noted that such animals occurred in clusters and appeared 'weak, showing little or no retaliation when they land' and thus are easy prey in comparison to subsequent workers and soldiers who can defend themselves.

An example of lack of congruence between abundance and consumption of prey are species that sting or bite (e.g. worker and/or soldier ants just mentioned) or that are protected by noxious smells or tastes. Often the latter are aposematically (warningly) coloured and vertebrate predators learn to reject them as prey (Cott 1957). How relevant such colouration may be for *Lampropholis* requires experimental investigation.

To date, studies undertaken on the diet of any species of *Lampropholis* have been isolated in terms of duration, location, time of day, season, and the species investigated. For this reason alone, it must be assumed that the data are incomplete. For example, Appendix 3 is misleading in at least two respects. It appears that *L. guichenoti* has a much broader diet (minimum of 115 taxa) than *L. delicata* (minimum of 40 taxa). This is probably an artefact. The most complete identification of stomach contents of *Lampropholis*, by far, was that carried to the familial level by Brown (1983), but only *L. guichenoti* was available to him. Had it been possible to treat *L. delicata* as intensively as *L. guichenoti*, the differences between them may have disappeared or at least not have been so great. Even so, the length of the list of prey consumed, even by *L. delicata*, is impressive. Belmont (1977) also found a greater number of taxa in the stomachs of *L. guichenoti* than in those of *L. delicata*, but again, the two species were not equally represented because *L. delicata* was not present at one of the sites occupied by *L. guichenoti*. Since the data from all sites were lumped together for the stomach analyses, and the raw data are no longer available, it is not possible to pare the dataset to eliminate the results from *L. guichenoti* where it occurred alone. By contrast, Lunney *et al.* (1989) and Warner (1995) found about equal numbers of taxa in these two species (see section *Interspecific differences in diet* for a further treatment of this topic). The discrepancy arising from greater taxonomic detail devoted to one species in the absence of the other, is masked when identification is taken only to the ordinal level.

Another aspect that remains cryptic in the bald list of prey is identification of which life-history stages are consumed. Some authors did distinguish between life-history stages and, for Coleoptera, Hymenoptera, Lepidoptera, and Neuroptera, larvae and adults were listed separately. Belmont (1977) noted that most of the taxa of large spiders were represented in stomachs by immature individuals. It is likely that adults of some spiders, for example Lycosidae or Sparassidae, would be too massive for such small lizards to subdue and consume. Crome (1981b) recorded that the sizes of spiders in the stomachs of the two common species of *Lampropholis* were all small (2–8 mm in length; see also section *Size of prey*). Although Orthoptera frequently were found in the stomachs of *Lampropholis*, whenever specific details were given, they almost always referred to crickets. The few references to grasshoppers were in the context of being unusual and a result of habitats being disturbed and having a high cover of grass (e.g. Crome 1981b). Similarly, when details of Lepidoptera were made available, they involved only moths; there is no evidence of butterflies ever being eaten by *Lampropholis*. Most butterflies have large wings and it may not be physically possible for a *Lampropholis* to discard the wings of such prey and consume the digestible part.

It is obvious that the smaller the sample, the less reliable it is for assessing the variety of items in the diet. Different investigators have used different sample sizes to characterise the diets of the species of *Lampropholis*. Although some samples are so small that one can be certain they do not encompass the entire range of prey eaten by the lizards, even these data are of value because they provide metadata that collectively permit assessment of the relationship of sample size to the diversity of known prey. Such a relationship is presented in Fig. 8.1. Two points are clear from this figure. First, it is obvious that for *L. guichenoti* even large samples do not encompass the complete diet because results from slightly more than 100 stomachs still have not reached an asymptote, even for the number of Orders of invertebrates that are consumed. The number of families of invertebrates included in the diet are even less thoroughly sampled and the curve for those is still rising steeply even for large samples. For *L. delicata*, the curve perhaps may have been nearing an asymptote.

Brown (1983) presented data on *L. guichenoti* that also support the above view, although they are not included in Fig. 8.1 because five different habitats as diverse as grassland, mallee scrub, dry sclerophyll woodland, and wet sclerophyll forest, from sites as far apart as several hundred kilometres, were lumped together, and consequently would have a combined faunal diversity higher than those of any component site. The sample consisted of 40 stomachs representing 18 taxa at the level of Order or above and 53 at the familial level. These data show the same trend as seen in Fig. 8.1 but translated upward in actual values.

As a practical measure, Fig. 8.1 demonstrates that a portrayal of the complete diet of these lizards would necessitate sacrificing prohibitively large numbers of individuals. The alternate method of flushing the stomachs of live animals and releasing them is not tenable for this genus because of their small body size (see Warner 1995).

Lampropholis delicata versus *Lampropholis elongata*

These two species are not syntopic as they live in different structural kinds of grass (Graham *et al.* 2023). Whereas *L. delicata* has legs of normal size and forages in grassy areas under the shade of a canopy, *L. elongata* does not occur in such places but rather occupies a unique structural type of grass – tussocks of *Poa sieberiana* growing in open areas surrounded by the forest but not in the forest itself. *Lampropholis elongata* has reduced limbs and is capable of 'grass-swimming' (see section *Relation of reduction of limbs to habitat* below) because of its long, slender body and whose locomotion within the interstices of the tussocks is not impeded by large legs. This results in a mosaic pattern of mutually exclusive microdistributions with *L. delicata* within the forest under a canopy and *L. elongata* in more open areas. They are ecologically segregated by two structural features – their own adaptive morphology (shape of body and size of legs) and the structural configuration of the grass through which they are capable of manoeuvring (open mats of grass versus dense tussocks). Despite the difference in shade between these two habitats it is not likely that microclimate factors strongly in the ecological segregation of these two species, given that *L. delicata* lives in open areas at other locations. Thus these two species are sympatric in that they coexist at high elevations at the edge of the New England Tableland in NSW in an area of remnant old-growth eucalypt forest and pine plantations, but they are not syntopic because in that area they occupy different types of grassy microhabitats, each restricted to its own microhabitat by the restrictions imposed by its anatomy.

Lampropholis guichenoti versus *Hemiergis decresiensis*

Lampropholis guichenoti and *Hemiergis decresiensis* are two taxa of skinks that are broadly sympatric (Table 11.17), and occupy some of the same general natural and urban habitats. Both have the same pattern of diurnal activity centring on midday (see Fig. 11.11).

Table 11.17. Locations of the small skinks *Lampropholis guichenoti* and *Hemiergis decresiensis* in various structural microhabitats in eucalypt woodlands at Newholme, NSW, at the time the lizards were first sighted.

Modified from Simbotwe (1985a).

	Lampropholis guichenoti		***Hemiergis decresiensis***	
Microhabitat	**N**	**Per cent of animals**	**N**	**Per cent of animals**
Open ground	81	39.5	0	0
Grasses and litter	80	39.0	10	18.5
Fallen sticks and bark	41	20.0	30	55.6
Rock piles	3	1.5	14	25.9
Total	**205**	**100**	**54**	**100**
Index of diversity	**0.624**		**0.481**	
Index of niche overlap (Ojk)* = 0.492				

*See Pianka (1973).

Superficially, they seem to be similar to the *L. delicata*–*L. elongata* pair of species, *L. guichenoti* being an animal with well-developed legs and active on the surface (like *L. delicata*) and *H. decresiensis* being one with small vestigial legs (Huey and Bennett 1987) and a slithering serpentine mode of locomotion like *L. elongata*. There is, however, one important difference between these two pairs. Whereas *L. delicata* and *L. elongata* are ecologically segregated by their morphological differences sequestering them into two non-overlapping habitats, such is not the case with *L. guichenoti*–*H. decresiensis*. In fact the opposite occurs, and the latter species-pair overlap at least to some extent, and sometimes even occupy the same habitat. They have been known to shelter communally (see section *Communal brumation* in Chapter 4) and otherwise have been found beneath the same cover when not brumating (Heatwole, pers. obs.). Simbotwe (1985) compared their co-occurrences at Newholme, NSW, and found that they overlapped in four of the five microhabitats he identified at ground level although at significantly different frequencies. *Lampropholis guichenoti* predominated in the more open areas, including grasses and litter, whereas *H. decresiensis* was more abundant near complex structures such as fallen wood and rock piles. Accordingly, their segregation during foraging was not by habitat, but rather by microhabitat. This may seem to be a trivial distinction, but the essential difference is whether the two structural configurations are spatially separated (and the species isolated from each other) or intermixed (and the species coexisting closely).

The tiny vestigial legs of *H. decresiensis* (Huey and Bennett 1987) would not impede rapid movement in crevices within rock piles, among fallen wood, or through dense grass. The larger legs of *L. guichenoti* would permit rapid running over open surfaces. A habitat consisting of a structural mixture of grassy microhabitat as well as a cover of wood and rocks thus provides both species with favourable conditions. Tiny legs, or complete lack thereof, also is characteristic of fossorial lizards that seldom are active above ground (see section *Relation of reduction of limbs to habitat* below) and consequently one might think that *H. decresiensis* is such a species and therefore be spatially segregated into epigeal and subterranean microhabitats during their activity periods, even where they occupy the same locations. Such is not the case, however, as both rocks/wood and grass are combined in various habitats.

Simbotwe (1985) found both *L. guichenoti* and *H. decresiensis* to be diurnal with activity period beginning ~08:00 hours and lasting until ~18:00–19:00 hours, with the peak centring about midday (see section *Activity period in the field* below).

For a discussion of the interaction of introduced *L. delicata* with native species on some Pacific islands, see Chapter 3.

THE STRUCTURAL NICHE

The physical structure of the environment may be an important ingredient in microhabitat selection by animals (Sexton *et al.* 1964; Heatwole 1966). In the present context the importance of crevices and their orientation (vertical or horizontal) for *L. delicata* and *L. guichenoti* and the affinity of *L. mirabilis* for granite rocks and of *L. elongata* for tussocks of grass serve as examples for the genus *Lampropholis*. For example, even if everything else, such as favourable microclimates and abundance of suitable prey were available, even optimal, for *L. delicata* but appropriate crevices were lacking, that species might be excluded. Without such approaches as the experimental study by Belmont (1977) such critical nuances would be inexplicable. Perhaps it is time to recognise the reality of the structural niche.

Animals must select habitats that maximise benefits and minimise costs associated with avoiding predation; reducing competition with other species; obtaining food, shelter, and mates; and maintaining an optimal body temperature. Accordingly, Howard *et al.* (2003) assessed the role of key structural variables relevant to these vital requirements. In outdoor enclosures the lizards were provided pair-wise choices between either different types of substrates mimicking the respective depths and structure of litter in the field, or between two different examples of a given type of litter. The types of litter were bare ground, native (*Eucalyptus-Acacia*) litter, pine-needle litter, *Casuarina* litter, compact artificial litter of shredded paper, and loose artificial litter of ripped paper. Pitfall traps were installed in each type of litter but kept closed except for specified test-periods of 06:00–07:00, 08:00–09:00, 10:00–11:00, 12:00–13:00, 14:00–15:00, and 16:00–17:00 (over three consecutive days).

They expressed the results of their counts of lizards in the enclosures as 'the mean number of lizards caught per hour' in their pitfall traps. We have generalised this characterisation of their results by the term 'active occupancy' meaning an estimate of the 'number of active lizards in a given type of microhabitat'. Active occupancy can be viewed from two different perspectives. High values indicate that many individual lizards occur in the specific litter-type being tested and that many of them were active. However, low values of active occupancy may also be obtained at high population densities if the lizards are less frequently active and thus not being captured as often by falling into pitfall traps.

The most striking result of the study by Howard *et al.* (2003) is that lizards seem to avoid bare ground during their period of activity as he never found any there (Fig. 11.10). Thus, frequent and/or large patches of bare

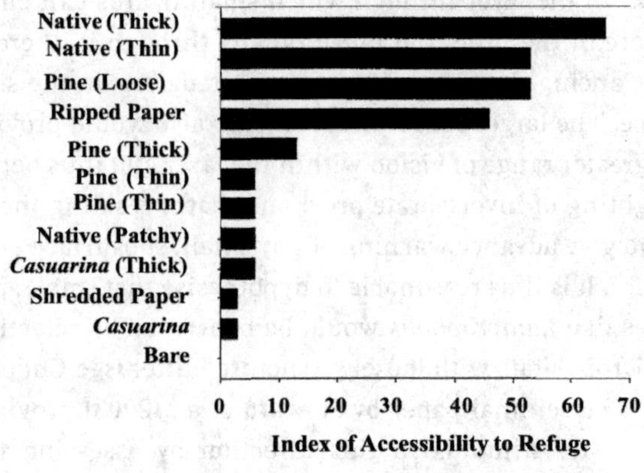

Fig. 11.10. The index of accessibility of different types of natural and artificial (experimental) litters as refuges for *Lampropholis delicata*. The higher the index the better protection if offers for animals living within it, as indicated by the visibility from above of glass marbles within the liter. Modified slightly from Howard *et al.* (2003). Printed with permission from the Society for the Study of Amphibians and Reptiles, USA.

Table 11.18. Active occupancy by *Lampropholis delicata* in different thicknesses of various of types of litter.

Data from Howard *et al.* (2003). See text for definition of 'active occupancy'.

Type of substrate Depth of litter	Number of trials	Mean active occupancy (range)
Bare soil (zero cm)	2	0
Eucalyptus-Acacia litter		
2.3 cm	7	(1.16-5.12)
5.3 cm	3	(4.46-5.27)
Casuarina litter		
2.2 cm	5	(1.38-2.55)
4.8 cm	3	(3.80-4.38)
Pine needles (non-native)		
2.4 cm	2	(1.77-3.02)
5.5 cm	7	(1.05-5.13)
9.8 cm	1	2.66

ground in the forest floor would be less favourable for *L. delicata* than those with at least some cover of litter, regardless of the type of litter on the rest of the forest floor. Most of the data for different depths of litter were obtained from the pair-wise trials and for *Casuarina* litter there seemed to be greater active occupancy in the deeper litter. For the more prevalent type of litter in Australia (*Eucalyptus-Acacia*), however, the results were more variable and active occupancy did not seem to vary much between the two depths of litter except for one low value in the shallower litter (2.3 cm; Table 11.18).

Compared to compact litter, litter composed of loosely spaced components contains larger, more-open spaces within the layer through which small lizards can enter more of the litter and move rapidly through it, thereby distancing themselves from large predators on the surface. The larger spaces in loose litter also would provide a greater range of vision within the layer and thus better sighting of invertebrate prey[1] and mates residing there, and give advance warning of any small, subsurface enemies. It is thus reasonable to hypothesise that small species like *Lampropholis* would be benefitted by selecting microhabitats with loosely structured litter (see Chapter 4). The seminal paper by Howard *et al.* (2003) provides heuristic stimulus in this direction by assessing the impedance of vision in different types of litter. They dropped glass marbles into litter and subsequent data on the percentage of those marbles that were not visible to investigators from above. They called this value the 'Index of accessibility to refuge'.

Figure 11.10 shows values of this index for the different types and thicknesses of the litters mentioned above. Bare ground of course offers no obstruction of vision, or routes of escape from predators on the surface, but all other substrates did. The types of litter that provides the best refuge from predators on the surface were thick layers of *Eucalyptus-Acacia* (native) litter, followed by thin layers of the same, loose pine needles, and an artificial litter of loose (ripped) paper; all of these had 50% or more of the glass marbles in the litter hidden from view above. All other types of litter had indices of less than 20%. Even small patches of *Eucalyptus-Acacia* litter had a low index (<10%) as did even thick layers of *Casuarina* litter (<5%). *Casuarina* litter had a low index in general but thick litter of this type had an index about twice as high as in thinner patches).

Howard *et al.* (2003) also considered that the availability of food in the types of litter might influence the index. Their results would apply to protection from predators that use vision to detect prey, but not to those relying on olfaction.

MORPHOLOGICAL AND HISTORICAL CORRELATES OF THE ECOLOGICAL NICHE

Sometimes the morphology of a lizard is a good indication of its lifestyle and the following examples have long been well-known. The type of an animal's dentition reveals the kind of food it eats and large size, especially of the head, can provide wider gape and allow manipulation of larger prey.

Clarke (1965) compared the sizes and bodily proportions of *L. guichenoti* and *L. delicata* to their nearest non-congeneric relatives, *Leiolopisma metallica* (now *Carinascincus metallicus*) and *Leiolopisma trilineatum* (now *Acritoscincus trilineatus*) both of which are larger than the two species of *Lampropholis* that Clarke (1965) studied. She found that even slight changes in the coefficients of growth were similar in all species (see Table 5.5 in Chapter 5) and that bodily proportions of an individual depended on its species, size, and sex. Consequently,

1 Most of the prey of *Lampropholis* are invertebrates that are active on the surface and and hence the range of vision within the interstices of the literature may not be highly relevant. However, a possibility not treated by Howard *et al.* (2003) is that some epigean predators may be able to detect subsurface prey by smell and dig them out.

individuals of all species, when of the same absolute size, also are practically identical in bodily proportions but since the other species grow larger, when *Lampropholis* reach sexual maturity their bodily proportions resemble those of immature individuals of the larger species and undergo no further changes. In short, *Lampropholis* are neotenic which means that they maintain juvenile bodily proportions throughout life (they 'never grow up'). An alternate explanation is that *Lampropholis*' condition is the primitive one and that the larger species have added additional growth with its continued concomitant allometric alteration of bodily proportions. Most other genera related to *Lampropholis* also are larger than *Lampropholis* and hence we consider neoteny as the more parsimonious of these two hypotheses.

Relation of reduction of limbs to habitat

About 24% of the species of skinks and 31.1% of the genera exhibit some reduction or loss of limbs as opposed to only 3% in non-scincid lizards; some skinks have lost only the forelimbs, others only the hind limbs (Camaiti *et al.* 2022; Chapple *et al.* 2023). Camaiti *et al.* (2023) in a review of the reduction of limbs in skinks concluded that:

> *On poor, sandy substrates of arid environments, body shapes with longer hindlimbs may be more efficient for locomotion in a granular fluid (i.e. sand) and exploit the air–substrate interface than complete limblessness. On richer, more humid substrates, such morphology is rare, indicating that navigating cluttered substrates selects for more equal and shorter limb lengths.*

Some species of *Lampropholis* have legs of normal size, e.g. *L. delicata* and *L. guicnenoti*, but others (*L. elongata, L adonis, L. amicula, L coggeri, L. couperi, L. robertsi* and *L. similis*) show some reduction of their legs relative to the size of their bodies; no species of this genus is completely legless.

Reduction of limbs, are associated with two different kinds of lizards: (1) fossorial species that remain underground most of their lives either in tunnels or burrowing through loose, sandy substrates (these species seldom are exposed to light and some have vestigial eyes); and (2) species that are active on the surface of the ground where they forage through dense vegetation or other substances where legs would be an impediment, rather than an aid, to movement through small interstices. Such lizards, adapted by their long slender bodies and reduction or loss of limbs have adopted a sinuous, serpentine style of locomotion dubbed as 'grass-swiming' (Carranza *et al.* 2008). The well-named *L. elongata* conforms to this assessment and accordingly its microhabitat consists mainly of the tussock form of grass (see section *Lampropholis delicata* versus *Lampropholis elongata* above).

THE TROPHIC NICHE

The diets of *L. delicata* and *L. guichenoti* have been treated in detail in Chapter 8 and found to be extremely similar, although not identical. These two species sometimes were found to differ but sometimes not, depending on time and place. A lot may depend on sample sizes and on the sensitivity of the analysis.

Although not using this terminology, Brown (1983) demonstrated three econes in *L. guichenoti*, a rather unexpected result for such a eurytopic, euryphagic lizard. Even after lumping the lower taxonomic categories of prey into only 22 larger categories of food, he found three econes, there being significant difference between the diets of males and females, and between adults and juveniles. He repeated the process, the second time using volumes of ingested prey instead of the number of individuals eaten, and again obtained a significant sexual dimorphism as well as an ontogenetic trophic difference. Even more importantly, he discovered that the same was true for all but three of the other 14 species of skinks he subjected to this analysis. The implication of this is enormous in terms of community dynamics. One is not dealing with a few syntopic species overlapping in the use of a common resource, but rather with about three different units per species. It may well be that for euryphagic species, the size of prey eaten may be of more significance than the actual taxonomic identity of the prey, and that adults of small species may be trophically separated from their parental generation but overlap trophically with the young of larger species. Thus, each econe of an individual species may need to be assessed and their interactions with a variety of different-sized species evaluated before the dynamics of the assemblages of which they are members are fully understood (also see section *Sizes of prey*).

Further expansion of the analysis to include comparison of *Lampropholis* with syntopic species of other small, terrestrial skinks, gave surprising results (Table 11.19). Comparison of *L. delicata* both with *Hemiergis decresiensis* and *Morethia boulengeri* produced similarities of 81%, that is, only slightly lower than the mean for those between *L. delicata* and *L. guichenoti* in syntopy (compare

Table 11.19. Comparison of the similarity and dissimilarity of the prey in the stomach/guts of *Lampropholis delicata* and *Lampropholis guichenoti* with other species of small skinks from the same study.

	Stomachs/Guts	
Author/*Species*	**Number**	**Per cent**
Crome: *Lampropholis delicata* vs *Hemiergis decresiensis*		
Similar prey	21	81
Dissimilar prey	5	19
Crome: *Lampropholis delicata* vs *Morethia boulengeri*		
Similar prey	21	81
Dissimilar prey	5	19
Crome: *Lampropholis guichenoti* vs *Hemiergis decresiensis*		
Similar prey	22	85
Dissimilar prey	3	15
Simbotwe: *Lampropholis guichenoti* vs *Hemiergis decresiensis*		
Similar prey	19	73
Dissimilar prey	7	27
Crome: *Lampropholis guichenoti* vs *Morethia boulengeri*		
Similar prey	18	69
Dissimilar prey	8	31
Similarity range	**18–22**	**69–85**
Dissimilarity range	**3–8**	**15–31**

Table 11.19 with Table 8.7 from Chapter 8). Comparison of *L. guichenoti* with the same two non-congeners also gave high values, ranging from about the level of the mean for *Lampropholis* at different sites and times to the mean for *Lampropholis* from the same site and at the same time of year. The number of studies of species other than *Lampropholis* were not as large as those for *Lampropholis*, and the variability in the latter was high, but the data suggest that similarities in the taxonomic composition of the gut contents of the non-*Lampropholis* skinks may have approximately the same similarities at the ordinal level of taxonomic composition of their prey, or perhaps even higher, than do *L. delicata* and *L. guichenoti*.

Hemiergis descresiensis is more fossorial and secretive than are *Lampropholis* and often are found under cover by day, whereas *Lampropholis* predominantly are animals that actively capture their prey on the surface of the ground. Thus, the high similarity in gut contents at the ordinal level between these taxa might be eroded if expressed at lower taxonomic levels. Although the ordinal composition of diet is similar in *Hemiergis* and the *Lampropholis*, perhaps *Hemiergis* tends to feed upon species belonging to Orders that are found under objects, whereas *Lampropholis* predominately consume different, but co-ordinal, species that are active on the surface. This hypothesis remains to be tested.

Another hypothesis worthy of testing is that euryphagic[2] species co-occur more easily than do stenophagic ones. If two species are restricted to a common narrow range of prey (in extreme cases a single species) and unable to switch to an alternate source of food should their species of prey decline due to some inclement condition, then competition between the two predators might become acute. If, however, two species have a wide range of prey, decline of a single species of prey would have a minimal effect as there would still be an array of alternate sources of food for both predators. The limited data from *L. delicata* and *L. guichenoti* and other syntopic small skinks are consistent with this interpretation but further studies are required for verification of this explanation for small scincids.

THE TEMPORAL AND THERMAL NICHE

Activity period in the field

Lizards can thermoregulate by shuttling between sun and shade, basking, altering their posture in relation to the angle of incidence toward the sun's rays, and

2 Referring to the variety of the diet: stenophagic means consumption of a narrow range of foods (See Hanson 1962) and euryphagic refers to species with of a broad diet.

climbing higher above a hot substrate (Chapter 6), thereby maintaining body temperature within a relatively narrow range in a thermally variable environment; nevertheless, seasonal differences in their thermal environment may necessitate altering the times of day that they can remain active (Díaz and Cabezas-Díaz 2004). Also, one could speculate that different diurnal lizards and nocturnal ones might feed on different species of insects and thereby escape competition through their occupancy of different thermal niches.

Lizards that are cryptic by virtue of living underground, beneath stones, or in or under logs are difficult to observe and to determine whether or not they are active, except under controlled conditions in the laboratory. However, lizards that are active on the surface can be identified readily as to whether they are active by day (diurnal) or by night (nocturnal). *Lampropholis delicata* and *L. guichenoti* are conspicuously diurnal and often can be seen foraging over the surface of ground during daylight hours during appropriate weather. However, the timing and duration of diurnal activity can vary daily, seasonally, and between different habitats.

Simbotwe (1985) studied the activity of *L. guichenoti* and *Hemiergis decresiensis* at Newholme, a habitat near Armidale, NSW, with sparse, open woodland subject to grazing by sheep. A composite graph for *L. guichenoti* from late spring to early autumn (November) gave a unimodal activity period during the warmest part of the day, centring on 12:00 noon for adults and an hour earlier for juveniles (Fig. 11.11 upper). *H. decresiensis* (Fig. 11.11 lower) had an activity period almost identical to that of *L. guichenoti.*

Belmont (1977) provided a more precise analysis for *L. guichenoti* and *L. delicata*, by separating the data from autumn and summer and comparing three different types of forest, also near Armidale. The two species showed broad overlap in their activity periods and in all three areas maintained a bimodal activity in summer with one peak in the morning and the second one in the afternoon, thereby avoiding the heat of midday. Under the cooler conditions of autumn, the pattern changed from bimodal to unimodal except for *L. delicata* at Pine Forest where the summer bimodal pattern was retained in autumn. Clearly, a diurnal internal rhythm, probably set by daylength, can be influenced by seasonal changes in local environmental conditions, likely temperature, and in winter even obliterated. The important point to note is that the similarities in the activity periods of these two species seem greater than the differences and do not fulfil the criterion for the ecological segregation of the temporal niche between these two species.

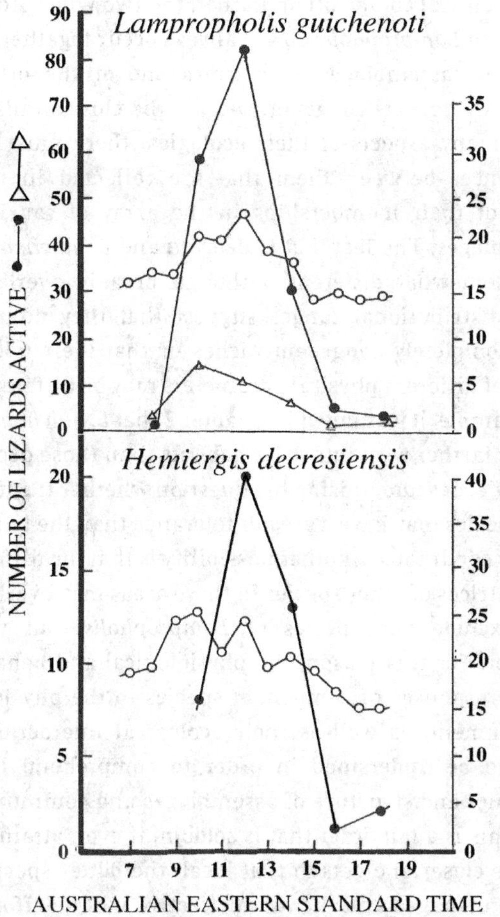

Fig. 11.11. The number of individuals of *Lampropholis guichenoti* and *Hemiergis decresiensis* active at different times of day at Newholme, NSW. Dots = adults; triangles = juveniles. Modified from Simbotwe (1985).

The size of their legs relative to that of their body differs between *Lampropholis* and *Hemiergis*, that despite their similar activity cycles, results in divergence in other axes of their niches and in their thermal and metabolic characteristics (see Chapters 4 and 7).

TAXONOMIC COMPOSITION OF ASSEMBLAGES

A prerequisite for assessing the type and intensity of interactions a species has with other species is to elucidate the taxonomic composition of the assemblages of which it is a member. Scarcely any two assemblages have identical composition. Rather, the two widespread species of *Lampropholis* do not always occur together in the same assemblages; sometimes one or the other occurs without its congener. Despite the close similarity of many aspects of their ecologies, there must be differences between them that are reflected in the scope of their memberships in the array of saurian assemblages. The fact that *L. delicata* and *L. guichenoti* have somewhat different, although broadly overlapping, distributional ranges suggest that they do not have completely congruent niches or that their tolerances of different physical parameters may be different. For example, it was noted in Chapter 2 that *L. guichenoti* ranges farther west into drier habitats than those occupied by *L. delicata*, raising the question whether the former species may have a greater tolerance than the latter for dry conditions. Another possibility is that the different matrices of other species in these areas may overlap and exclude one species of *Lampropholis*, but not another. For this reason, the physiological and behavioural responses of component species to the physical environment, as well as their ecological interactions need to be understood in order to comprehend the dynamics and structure of assemblages and communities. This is a tall order that is seldom, if ever, attained but the closer one gets to that ideal, the better species can be managed and the more effective will be efforts to conserve them.

Comparative studies of the niches of individual pairs or groups of species in the assemblage can aid in separating the more important attributes from those that are of lesser significance.

THE NICHES OF *LAMPROPHOLIS*

Character displacement is the adaptive divergence of sympatric, closely related species in some important aspect of their niches in ways that segregates them ecologically, thereby permitting coexistence. Invariably in studying any aspect of the biology of *Lampropholis*, topics are covered that relate directly to the niches these species occupy and bear upon how similar, or different, congeneric species are. They have been subjected to a number of studies on aspects of their niches, and so far none of these has uncovered any significant character displacement between them. For example, Chapter 8 compared the stomach contents of *L. delicata* and *L. guichenoti* and those results are not repeated here except to summarise that both species eat a similar taxonomic array of prey of about the same size. Indeed, the slight differences between these two species of *Lampropholis* at the same site is less than those between the populations of a given species at two different sites (and may be within sampling error). Thus, it seems that the trophic niches of these two species are practically identical and that they will consume any available prey of a size that they can manipulate and swallow (see Chapter 8). The differences in the lizards' diets between sites may merely reflect the variation in proportional representation of prey species in different areas, although this has yet to be empirically demonstrated in a conclusive way.

These two species of *Lampropholis* also live in practically identical microhabitats, have daily and seasonal activity patterns that are quite similar, and thus their trophic, temporal, and microspatial niches are nearly the same. They have broadly overlapping geographic ranges but with *L. guichenoti* extending farther westward than *L. delicata* into hotter, drier inland regions (Chapter 2), and in some of the more open, hotter and drier localities *L. delicata* is either absent or occurs at lower densities. These findings merge into suggesting that the physical attributes of the environment, such as moisture, temperature, and/or amount of solar radiation, play a greater role in determining the presence and abundance of these two congeners than do their biotic interactions. Is it possible that in many habitats some other physical or biotic factor keeps populations sufficiently low that food is not limiting to either species?

Overlap of *Lampropholis* with other taxa in their assemblages

Species that are genetically more distant from *Lampropholis* would be expected to have diverged farther from them than between congeners and thus would be expected to share fewer attributes. As a result by chance they would likely fit into the same assemblages more readily than would congeners. Nevertheless, overlap may be possible in one or more axes of the niche and character displacement of overlapping attributes could develop in areas of syntopy over time, just as it often does between congeners. For this reason, it is also important to compare the niches of species of *Lampropholis* to other components of their assemblages, especially if those species are abundant and likely to have a significant impact. A species that is rare, for whatever reason, would be expected to have less influence on an abundant species, than it would if the former were more common. There is some circularity in this reasoning, however, as the rarity of a species might be a direct consequence of competition between it and a more abundant species.

A species may be excluded from its favoured habitat, or its numbers reduced there, in the face of strong competition with one or more other species of lizard in that assemblage, whereas in another area with a different composition of the resident assemblage, one or more of those competitors may be lacking and our designated species could expand to become the major, or sole, occupant of its preferred habitat, rather than occupying the secondary habitat it was forced into in our previous example. Thus, the study of only one locality could give a distorted impression of the favoured habitat of a species. Only by studying a species' relative occupancy of habitats and microhabitats in different regions containing varying combinations of species and subject to different physical environments could one begin to unravel the intricacy and the complexity of the factors impinging upon the structuring of assemblages and understand the intrinsic preference for habitats by its components.

A further complication is that perhaps lizards view the structure of the environment in different ways than we do. For example, in the section *Habitat and microhabitats of* Lampropholis - *Studies in the field* above we noted that there were different opinions of the importance of grass as a favoured microhabitat of *L. delicata*. The conflicting conclusions that were reached by different investigators may have been because 'grass' was considered as a single kind of structure, when in fact, it may embrace a complex of structures to lizards. In fact, there is evidence that such is the case. Recall that Graham *et al.* (2023) found that although *L. delicata* and *L. elongata* both favoured grassy habitats, it was precisely the type of grass that spatially and ecologically separated them, the former species favouring the more open kind of grass and the latter restricted to a more densely structured grass (base of tussocks). Finally, Howard *et al.* (2003) found that *L. delicata* favoured open leaf litter in the habitat he studied and suggested that this preference was associated with the ability of the lizard to trade-off different activities, such as accommodating the balance of the need to bask and forage, rather than obtaining safety of shelter from predators. None of these three papers alone gave complete insight into the habitat selection of L. *delicata*. All three were required to explain that the preference was dependent, not on grass but on a diffuse structure of the substrate that allowed maximum flexibility for the lizard to carry out its daily activities. The outcome relative to other members of the assemblage varied depending on the preference of the interacting species. This conclusion illustrated the necessity of taking all physical and biotic influences, and their geographic distributions, into account in assessing the interacting forces structuring assemblages. The habitat or microhabitat can be influenced by all of these factors and lead to different outcomes. Consider how different the interpretation of *L. delicata* would have been if based only on one study. Depending on which one, it could have been concluded that presence of leaf litter was the essential feature governing this species' habitat selection or, if selection were based on a different feature, the conclusion could have been that grass was important, whereas neither structure is; the important attraction is a loosely structured substrate, regardless of whether that structure is provided by grass or by litter.

A few studies have addressed some aspects of the niches of a species of *Lampropholis* with those of a similarly abundant, but not congeneric, species. For example, Simbotwe (1985) compared the diet and activity cycle of *L. guichenoti* with that of *H. decresiensis*. *Hemiergis decresiensis* is slightly larger (50 mm SVL; Cogger 2018) than *L. guichenoti* (40 mm SVL). Cogger (2018) gives the habitat of *H. decresiensis* as 'a burrowing species usually found under stones and fallen timber' compared to that for *L. guichenoti* '... seen basking or foraging ... in leaf-litter, grass and rock piles in wet and dry sclerophyll forest, moister woodlands and in coastal mountain heaths. An abundant species in many suburban gardens in eastern Australia'. Although these two species take a similar taxonomic array of prey, there are proportional differences in the ordinal representation of those prey in the lizards' stomachs. Only four Orders individually accounted for more than 10% of the gut contents of *H. decresiensis*: Hymenoptera, Isoptera, Coleoptera, and Lepidoptera. Only Isoptera (termites), contain species restricted in the main to activity within the litter or found under cover, the putative habitat of *H. decresiensis*. The others all contain large numbers of species that fly in the adult stage. Thus, it would seem that a major part of the diet of *H. decresiensis* consists of volant species. This impression is misleading because it is not true at all. All the Hymenoptera in the stomach of *H. decresiensis* were ants, which are non-flying except very briefly when reproductives swarm and disperse. Ants constituted 28% of the total dietary items in this species, whereas for stomachs of *L. guichenoti* from the same area, the corresponding value was only 10%.

Coleoptera were represented in the stomachs of both species both by adults (volant) and larvae (non-volant). The proportions of these life stages differed between the two species of predatory lizards. In stomachs of *H. decresiensis* larvae constituted 17% of the coleopteran prey, whereas in *L. guichenoti* the corresponding value was less than half that amount (8%). A similar result was obtained for the Lepidoptera. When merely classifying the gut contents by Order, it appears that *H. decresiensis* proportionately eats nearly three times the number of this highly volant taxon (moths and butterflies) (13%) as does *L. guichenoti* (5%). However, all the lepidopterans eaten by *H. decresiensis* were caterpillars (100%), whilst only 78% of those consumed by *L. guichenoti* were. Clearly, the diets of these two species of lizards differ far more when life stages and lower taxonomic designations are considered. Taking all these factors into consideration, the result for the total stomach contents for *H. decresiensis* is that 78% of the dietary items were of non-flying individuals and 22% of volant ones. All Hemiptera and Orthoptera were considered volant, when in fact the nymphs of these Orders are not, so it is likely that *H. decresiensis* feeds on a slightly higher percentage of non-volant prey than the above figures reveal. In the stomachs of *L. guichenoti* the situation was reversed; 33% of the prey items were of non-volant species and 67% were of volant ones.

There is a lesson to be learned from this exercise: merely identifying prey to Order, as has been the procedure for most studies, may obscure subtle differences in the use of trophic resources and give a false impression of the degree of dietary overlap between species. By mining the more-detailed information presented by Simbotwe (1985) and Crome (1981a), it became clear that these two species were trophically separated, at least in part, and although overlapping in diet, they clearly tap somewhat different trophic resources, with *H. decresiensis* taking mostly non-volant species and *L. guichenoti* consuming primarily volant ones. Add to this separation the differences in size between the two species and the fact that at least for *Lampropholis* and seemingly true for lizards generally because larger individuals eat larger prey than do smaller ones, the trophic gap between the two species widens. It seems likely that *H. decresiensis* eats larger, more sedentary non-flying prey, crawling through the leaf litter whilst *L. guichenoti* concentrates on smaller, but volant, prey that lands on the surface. With these findings, one can begin to understand how these species can coexist in the same habitat, despite congruity of the temporal niche. However, current data are only highly suggestive, but not definitive, and should be viewed as guides for planning empirical research leading to conclusive understanding.

Proportion of digestive tracts with prey and mean incidence of prey

The intensity of predation can be expressed either as the percentage of stomachs/guts that contain prey or as the numbers of prey found within the digestive tract (Table 11.20). These values, of course, are lower earlier in the day before lizards have captured many prey, and Belmont (1977) noted that during cold seasons when lizards were not very active, fewer items were found in the stomachs. These sources of error probably are minimised, however, as samples are collected mainly during seasons when lizards are actively feeding and at times of day when stomachs were expected to be relatively full. In all studies of *L. delicata* and *L. guichenoti* for which data are available, 65–100% of the stomachs contained food, not only in healthy undisturbed woodland, but also in some with many dead trees from dieback, or where there was grazing by sheep and cattle (Table 11.20). There were no consistent differences between the different kinds of woodland. Similarly, the mean number of prey items per stomach with food averaged 1.2–5.3 for *L. delicata* and 0.4–6.4 items for *L. guichenoti* (see Table 11.20 For this measure, data were available from an area that had experienced a fire; there the mean level of 3.1 before the fire, had dropped to 0.5, post-fire (Lunney *et al.* 1989).

It seems that the ability to find food does not vary consistently among different kinds of woodlands inhabited by *Lampropholis*, except under extreme conditions such as fire or drought. Lizards in such woodlands tend to have fewer items in their stomachs.

INTRASPECIFIC FLEXIBILITY IN ADAPTING TO NEW ENVIRONMENTS

Species introduced into a novel environment may encounter a different array of niches than experienced in their land of origin. There likely are assemblages of native species in the new land that occupy niches in different ways than those the invaders have previously encountered, and there may be vacant niches similar to those they experienced before being translocated that are suitable, at least in part, for their unopposed occupancy. Furthermore in new areas, the physical environment may be different and perhaps more extreme. The versatility of the immigrant in successfully

Table 11.20. Percentage of stomachs with food and the number of prey per stomach in *Lampropholis delicata* and *Lampropholis guichenoti* compared to some other small syntopic Australian terrestrial skinks from various habitats.

N = number of stomachs examined.

Species	Habitat	N	Per cent of stomachs		Mean prey per stomach with food	Authority
			Empty	With food		
Lampropholis delicata	Hawkesbury Sandstone, Sydney	8	25	75	1.2	Webb (1983)
	Three habitats combined[1]	99	21	79	2.1	Belmont (1977)
	Angophora-Eucalyptus woodland	49	22	78	2.2	Warner (1995)
	Eucalypt woodland, Eastwood State Forest; healthy, abundant leaf litter	5	0	100	2.8	Crome (1981b)[2]
	Eucalypt woodland at Newholme					Crome (1981b)[2]
	Moderate dieback	10	0	100	7.6	
	Severe dieback	2	0	100	2.0	
	Mumbulla State Forest; healthy forest, 1980, pre-fire, pre-drought	20	–	–	3.1	Lunney *et al.* (1989)[3]
	Mumbulla State Forest; 1980, post-fire, pre-drought	11	–	–	0.5	Lunney *et al.* (1989)[3]
	Mumbulla State Forest	16	–	–	4.2	Lunney *et al.* (1989)[3]
	Mumbulla State Forest	17	–	–	5.3	Lunney *et al.* (1989)[3]
	Mumbulla State Forest	7	–	–	3.7	Lunney *et al.* (1989)[3]
	Eucalypt woodland at Newholme Site 2; mostly dead trees open, grassy, sparse litter	2	0	100	2.0	Crome (1981b)[2]
Lampropholis guichenoti	Three habitats combined[1]	104	9.6	90.4	2.8	Belmont (1977)
	Angophora-Eucalyptus woodland	73	22	78	1.7	Warner (1995)
	Eucalypt woodland, Eastwood State Forest; healthy, abundant leaf litter	12	35	65	2.3	Crome (1981b)[2]
	Eucalypt woodland at Newholme					
	Moderate dieback	20	30	70	1.5	Crome (1981b)[2]
	Severe dieback	24	8	92	3.4	
	Eucalypt woodland at Newholme Site 3. Grazed by cattle and sheep	–	18	82	1.2	Simbotwe (1985)
	Mumbulla State Forest					Lunney *et al.* (1989)[3]
	Pre-fire, pre-drought, 1980	17	–	–	4.5	
	Post-fire, pre-drought, 1980	5	–	–	0.4	
	Post-fire, during drought					
	1981	25	–	–	4.3	
	1983	26	–	–	6.4	
	Post-fire, post-drought					
	1984	17	–	–	3.6	

continued

Table 11.20. continued

Species	Habitat	N	Per cent of stomachs: Empty	Per cent of stomachs: With food	Mean prey per stomach with food	Authority
Hemiergis decresiensis	Eucalypt woodland, Eastwood State Forest; healthy, abundant leaf litter	2	0	100	9.5	Crome (1981b)[2]
	Eucalypt woodland at Newholme Site 1; moderate dieback	10	0	100	7.6	Crome (1981b)[2]
	Eucalypt woodland at Newholme Site 2; mostly dead trees	20	5	95	4.6	Crome (1981b)[2]
	Eucalypt woodland at Newholme Site 3	?	66	34	2.1	Simbotwe (1985)
Morethia boulengeri	Eucalypt woodland at Newholme Site 2; mostly dead trees	3	33	67	1.5	Crome (1981b)

[1]Pine Forest, eucalypt woodland at Puddledock, and eucalypt woodland at Blue Hole. Data for individual habitats are not provided; it is uncertain whether the means are based only on stomachs that contained some food, as was the case for the rest of the table, or whether empty stomachs were excluded.

[2]Values of percentages for each taxonomic category were obtained by projecting the histograms of figures 1 and 2 of Crome (1981b) onto a grid.

[3]The entire gut, rather than just the stomach, was examined and hence the latter would be inflated, compared to the other studies. The number of empty stomachs was not recorded so it isn't known whether all Ns were 100%, or whether empty stomachs were included in the calculation of the mean number of prey per gut. Were the former the case, the mean prey per gut would be comparable to the data from other species except for the increase caused by using the entire gut; were the latter the case, the effect would be to diminish the values of the mean prey per gut. For this reason these values are not comparable to those of the other studies but are consistent within the study of Lunney *et al.* (1989).

locking into the structure of the new assemblage and new environment is paramount int determining whether it persists in its new location. A species with rigid requirements would be less likely to become established than one with more malleable prerequisites. Tingley *et al.* (2016) examined the intraspecific flexibility of *L. delicata* in adjusting to the structure of the niches encountered in the novel physical and biotic environments it found upon being introduced into New Zealand, Lord Howe Island, and Hawaii. They recognised three outcomes of responses to the new conditions: (1) 'niche stasis' (the finding and filling of niches familiar to those experienced in their native Australia), (2) 'niche unfilling', a euphemism for failing to colonise a similar niche in their new location as the one they occupied in their native range, i.e. the proportion of their occupied native niche that did not overlap the occupied invaded niche, and (3) 'niche expansion' (extending their breadth of niche beyond that occupied in their native range).

They found that the equivalence of the niche realised by *L. delicata* in its native Australia with that in either New Zealand or Hawaii was low. However, the observed overlap between the niche realised in Australia and on Lord Howe Island did not depart significantly from expectations and was completely stable. Thus, invasion by this species onto Lord Howe Island required less of a shift in its niche than did invasion of either New Zealand or Hawaii. Also reflecting equivalency of niches between Australia and Lord Howe Island was the fact that when introduced onto Lord Howe island, this species showed little expansion of its native niche. Comparisons between New Zealand and Hawaii revealed that, although *L. delicata* expanded it native niche into wetter habitats in both places, the expansion was more pronounced in Hawaii.

The above regional differences indicate that invading species do not require finding a niche that is precisely equivalent to the one they occupy in their native land, but that a eurytopic species such as *L. delicata* can fashion new ones by expanding its original niche into new ones, or by emphasising parts of its previous niche while abandoning other parts, depending on the differences in the new physical environments and the overlap with novel guilds of prey, competitors, and predators in its acquired habitat.

An important lesson learned from the study by Tingley *et al.* (2016) is that multiple examples of invasions by a species must be investigated to unravel the full capacity of a species to adapt to new homes. As they pointed out, had they restricted their analyses only to Australia and New Zealand they would not have discovered the extent that *L. delicata* was capable of persisting in climates beyond the limits of its native realised niche (as in Hawaii). Indeed their analysis revealed that there are many places worldwide that *L. delicata* could colonise were it to reach those locations.

DEFINING THE ASSEMBLAGE: CASE STUDIES

Very few species, if any, of any taxon, have had the assemblages to which they belong completely defined in all the habitats they occupy. This is especially true of species such as *L. delicata* and *L. guichenoti*, that have broad niches, are highly eurytopic with respect to habitat and eat a wide variety of prey (see Chapter 8). Nor is it likely that all the different types of habitats in which they occur have been identified.

One can take the study by Homan (2008) of the Grampians National Park in Victoria, which, although incomplete, can serve as an example of the kind of preliminary information that is needed. He identified eight different habitats: damp forest, lowland forest, wet heathland, heathy woodland, heathy dry forest, sand heathland, shrubby foothill forest, and montane rocky shrubland and surveyed the vertebrate fauna of this park and found 115 species: 30 mammals, 65 birds, 17 reptiles (of which 12 were lizards, including *L. delicata* and *L. guichenoti*), and three amphibians; a general indication of abundance was also provided. He provided more specific notes on where a few particular species were found, but otherwise provided only the list for the park as whole, not individually for the individual habitats within the park. The only exception was reptiles, and then only for one type of vegetation: heathy dry forest 'amongst rocky outcrops with many fallen logs'. The assemblage of reptiles in that habitat consisted of seven of the 12 species of lizards known from the park: *Eulamprus tympanum, Egernia* (now *Liopholis*) *whitii, L. delicata, L. guichenoti, Morethia boulengeri, Niveoscincus* (now *Carinascincus*) *coventryi*, and *Tiliqua rugosa* (all skinks), and two of the five species of snakes: the elapids *Parasuta flagellum* and *Pseudonaja textilis*. In this relatively dry habitat, *L. guichenoti* was the second-most abundant reptile but *L. delicata* was in the lowest category of abundance.

Let's use the available information to gain an entry into the structure of this assemblage. Beginning with *L. guichenoti* we know that it eats practically any small arthropod within a certain size range, but that the principal taxa are ants, spiders, beetles, and (in some habitats) termites, probably because these are the most abundant taxa on the floor of woodlands; other taxa such as small snails, although rare, are relatively consistent dietary items (see Chapter 8). *Morethia boulengeri* also primarily feeds on arthropods. These two species are about the same size: 40–45 mm SVL (Cogger 2018) and thus probably overlap broadly in the size of their prey. They do seem to target somewhat different segments of the arthropod assemblage, however, with *M. boulengeri* taking mainly prey that is wholly, or partly, terrestrial or arboreal, but few aerial (flying) forms, and *L. guichenoti* preying upon all three categories nearly equally. In this regard, experiments to resolve whether this is because *L. guichenoti* has greater celerity and thus is better able to catch flying insects before they take flight, or because of overt preferences for particular prey species, would be enlightening. The higher incidence of wholly, or partly, arboreal species in the diet of *M. boulengeri* than in that of *L. guichenoti* may be related to the observation of Cogger (2018; and see below) that the former often shelters under the bark of standing ring-barked trees; thus, it may be more arboreal in its habits than is *L. guichenoti*.

The habitats of the four small species in this assemblage have been summarised by Cogger (2018) as follows:

Lampropholis delicata: 'Found in a wide variety of habitats from rainforest edges, wet and dry sclerophyll forests, coastal heaths and woodlands, where it forages among leaf-litter and other ground debris. Builds up large populations in suburban gardens'.

Lampropholis guichenoti: 'A diurnal lizard usually seen basking or foraging for ants and other small insects in leaf-litter, grass and rock piles in wet and dry sclerophyll forests, moister woodlands and in coastal and mountain heaths. An abundant species in many suburban gardens in eastern Australia'.

Morethia boulengeri: 'Found among ground litter and fallen timber in a wide variety of habitats: dry sclerophyll forest, woodlands, mallee, low shrublands and arid scrublands. Often shelters under the bark of fallen or standing dead, ring-barked trees'.

Carinascincus coventryi: 'Usually associated with decaying stumps and logs in wet and dry sclerophyll forests'.

Thus, all four of these species are diurnal inhabitants of the floors of forests, woodlands, and shrubby areas, forage at least partly at ground level for small arthropods and occur together in the same habitat, in the present case in heathy dry forest.

There are insufficient data for *C. coventryi* to permit detailed analysis; however, it seems to be more closely associated than the others with decaying logs and stumps. In some localities, some *Lampropholis* also tend to locate near logs, and more often were seen on firm ones as opposed to decaying ones (Fig. 11.3; Tables 11.7 and 11.8). Thus, there may be subtle, partial separation of the microhabitat of *C. coventryi* from the two common species of *Lampropholis*.

Lampropholis and *M. boulengeri* occupy similar temporal, spatial, and dietary axes of their niches, and are potential competitors, therefore warranting more-detailed investigation of their ecological interactions.

Liopholis whitii is even more disparate in its microhabitat in comparison to the above species. It is a 'swift-moving diurnal lizard found in a variety of habitats from coastal heaths and grasslands through dry sclerophyll forests to open woodland and often closely associated with rocky situations where it shelters under exfoliating rock slabs and in crevices, but also found in and around fallen timber and any other ground cover' (Cogger 2018). Its predilection for rock slabs as a microhabitat partly isolates it spatially from the trio of small skinks that ecologically are otherwise practically identical. *Liopholis whitii*, however, also uses shelter under fallen timber that does overlap with that of the *Lampropholis*. Note that *L. whitii* is larger (95 mm snout-to-vent) than the small species of skinks and thus would require larger crevices and even in shared type of cover, would still be ecologically segregated at the level of microhabitat.

Eulamprus tympanum is relatively distant from the above species. It resembles them by being diurnal (Cogger 2018) and by preying upon arthropods; it differs from them in size, however; it is larger than the species previously discussed. It is 80 mm SVL (Cogger 2018) and thus larger than any of the other skinks in the heathy dry forest and hence likely to eat prey beyond the size manageable for the smaller species. Its ecological segregation from the smaller species is further enhanced by marked difference in habitat. Although, it occurs in heathy dry forest, it lives in a particular habitat within that type of vegetation. As its common name of Water Skink suggests, it is most commonly encountered along small creeks, where it usually is seen basking on rocks, or logs. It readily takes to the water when alarmed. Thus, this species lives in a completely different habitat than the other lizards in the heathy dry forest in that it is restricted to the vicinity of streams and has aquatic tendencies. There may be some slight dietary overlap with *L. guichenoti* when individuals of the latter have home ranges bordering streams.

THE STRUCTURE OF ASSEMBLAGES

Now that the assemblage in heathy dry forest has been defined, we can begin to delve into the interactions that structure it. A long-held dictum of ecology was that closely related species could not coexist in the same habitat, without diverging into separate axes of their ecological niche. For example, species that share the same space and activity, may have different diets; those that are trophically similar and occupy the same microhabitat, usually are active at different times in the diel cycle, and so on. Consequently, it is important to study the interactions of species that are phylogenetically closely related, or ecologically similar, and to understand how they partition resources within their common habitats.

Extending such studies from the confines of the assemblage to the biotic community at large adds a higher level of complexity but certain steps can be taken in that direction. Even with simple lists of species, and knowledge of the general dietary habits of component species in assemblages or communities, one can assemble tentative food webs to serve as hypotheses for testing. Although seldom achieved, assembling the potential range of prey of individual predators and then relating that to the known grouping of species into assemblages or communities can be heuristic. For example, are avian predators that take a heavy toll on snakes that prey upon *Lampropholis*, and thus indirectly benefit the lizards, present in one habitat but not in others, and if so, is this reflected in the population densities of those lizards in different places? Do guilds of species that are directly predatory upon *Lampropholis* vary among habitats and affect the lizards' populations to lesser or greater extents in different habitats. Does the lack of some species of competitor, or conversely, does an abundance of several predators, differentially affect the density of *Lampropholis* in different habitats? There are myriads of such questions that need to be answered before the role of the various species of *Lampropholis* in different biotic communities can be completely understood. In Chapter 9 we address predation. In the present chapter we focus on the interaction of the two widespread species of *Lampropholis* with each other and with other syntopic taxa of lizards in an attempt to assess their role in these saurian assemblages.

The most comprehensive study of the niches of the species of *Lampropholis* with restricted geographic ranges was carried out by Goodman (2007). He studied three axes of the ecological niche, the spatial or structural niche (microhabitat), temporal niche (time of activity) and body size (which can be a surrogate for the trophic niche as different sizes of animals eat different sizes of prey) of 21 species of small skinks (1,679 specimens) from 24 populations in five assemblages in Queensland. The habitats were: (1) dry open forest at Alligator Creek, (2) a mosaic of closed tropical rainforest and coastal monsoonal forest at Trinity Beach, Cairns, and (3) cloud forest and monsoonal vine forest at Mt Bartle Frere. The structural features of microhabitats included type of cover on the ground (rocks, leaf litter, undergrowth), proximity to vegetation, extent of cover by the canopy. Two of his study areas did not contain any *Lampropholis* and they are not discussed here. The other three areas each contained a different species of *Lampropholis*, with *L. mirabilis* at Alligator Creek (Fig. 11.12a), *L. coggeri* at Cairns (Fig. 11.12b), and *L. robertsi* at Mt Bartle Frere (Fig. 11.12e).

At Alligator Creek there were eight species of skinks, of which *Lampropholis mirabilis* was one (Fig. 11.12a). It had a bimodal activity cycle, with a major peak at 11:00–13:00 hours and a minor peak at 13:00–17:00 hours. This cycle also occurred in two other species, *Eulamprus quoyii* and *Morethia taeniopleura. Eulamprus brachysoma* (name later changed to *Concinnia brachysoma*) had a unimodal activity cycle and shared, *L. mirabilis*' earlier major peak but not its later minor one and conversely *Carlia rhomboidalis* and *Carlia pectoralis*, both also unimodal, peaked in activity during *L. mirabilis*' minor peak in the afternoon. *Carlia jarnoldae* did not show any peaks but was active relatively uniformly throughout the day. No data on the activity of *Carlia vivax* was presented. In summary, activity of all species occurred throughout most of the day but with differences in the times of both peaks of *L. mirabilis* overlapping with some species and overlapping only in one peak with others.

At Cairns, *Lampropholis coggeri* had very low activity through the day except for a very strong early peak at 11:00–13:00 hours (Fig. 11.12b). No other species showed such a narrowly defined peak, although *Carlia storri* peaked in activity at the same time but also had

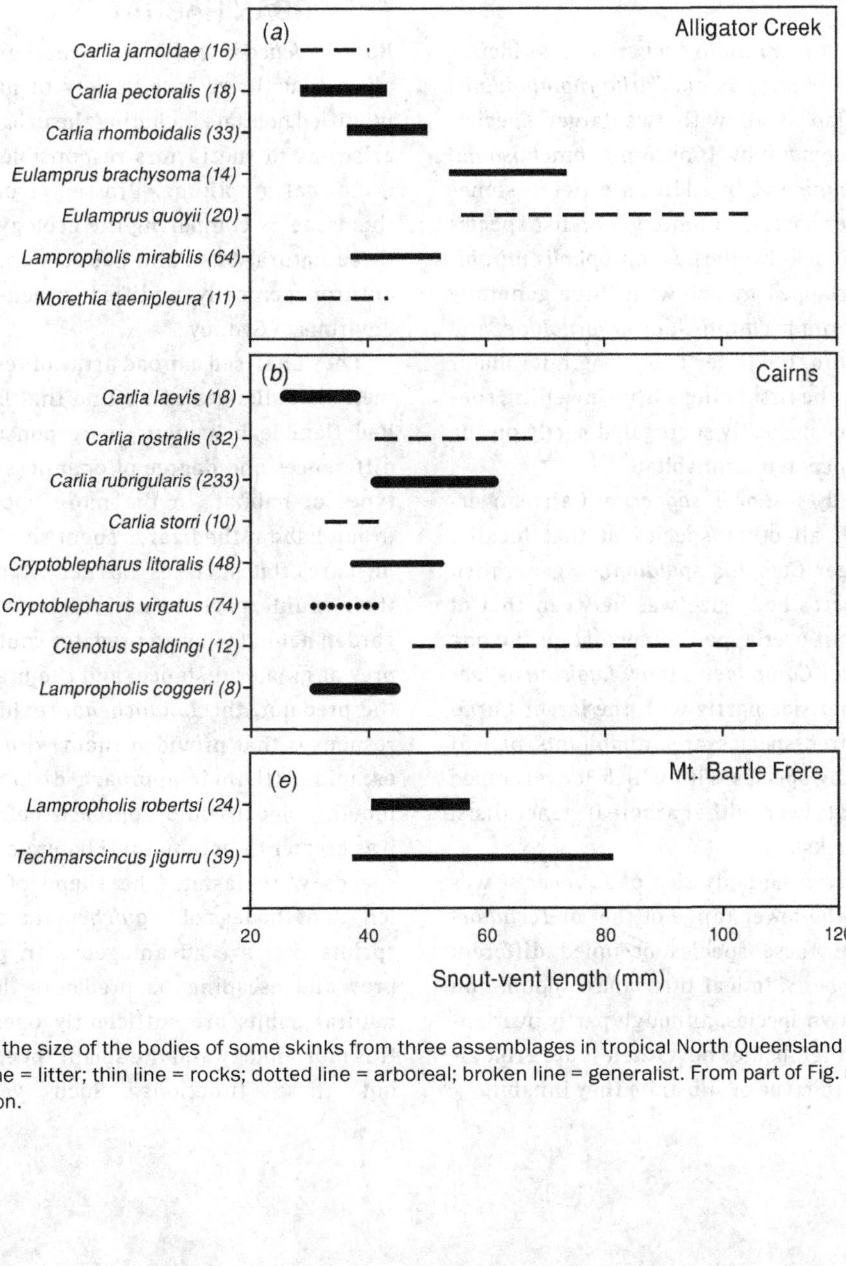

Fig. 11.12. Range in the size of the bodies of some skinks from three assemblages in tropical North Queensland and their microhabitat. Thick line = litter; thin line = rocks; dotted line = arboreal; broken line = generalist. From part of Fig. 5 in Goodman (2007). Printed with permission.

relatively high activity at other times of day. *Carlia laevis* (now *Lygisaurus laevis*), *Carlia rostralis*, and *Ctenotus spaldingi* had bimodal activity with their early peaks coinciding with the peak of *L. coggeri*. By contrast *Carlia rubrigularis, Cryptoblepharus litoralis*, and *Cryptoblepharus virgatus* had afternoon peaks that were out of phase with the earlier one of *L. coggeri*. In summary, there was one species that unimodally shared the early peak with *L. coggeri*, some that shared its peak plus exhibiting a non-shared afternoon one, and some that had a non-shared unimodal peak in the afternoon.

Finally, at Mt Bartle Frere there were only two species of skinks and they had opposite activity periods, *Lampropholis robertsi* with an afternoon peak of activity at 13:00–15:00 hours and *Techmarscincus jigurru* with an early peak at 09:00–11:00 hours. Thus, these two species were partially separated by different times of main activity.

At Alligator Creek *L. mirabilis* had an almost identical range in its size of body as did *Carlia rhomboidalis*; they did not overlap at all with two larger species, *Eulamprus brachysoma* (now *Concinnia brachysoma*) and *Eulamprus quoyii*, and in addition differed somewhat from the latter species, a more generalist species than *L. mirabilis* (a rock-dweller). *Lampropholis mirabilis* only partly overlapped in size with three generally smaller species, *Carlia pectoralis, Carlia jarnoldae*, and *Morethia taeniopleura*, the latter two being microhabitat generalists, and the first being a litter-dwelling species and therefore ecologically segregated partly on the criterion of difference in microhabitat.

The range in body size of *L. coggeri* at Cairns overlapped partly with all other species at that locality except for the larger *Ctenotus spaldingi*, a generalist. *Lampropholis coggeri*'s body size was between that of two other species but overlapped somewhat on the one side with the smaller *Carlia laevis* (now *Lygisaurus laevis*) and on the other side partly with the larger *Carlia rubrigularis*; all three species are inhabitants of leaf litter. The remaining species with which it overlapped had different habitats (were either arboreal, generalists, or inhabitants of rocks).

At Mt Bartle Frere, the body size of *L. robertsi* was completely within the lower range of that of *Techmarscincus jigumu* but these species occupied different microhabitats, *L. robertsi* in leaf litter and *T. jigumu* on rocks. Thus, these two species, although partly overlapping in the sizes of their bodies nevertheless are ecologically segregated by the type of substrate they inhabit.

In general, and including the areas where there were no *Lampropholis*, all species used available microhabitats non-randomly, that is, they were associated with specific features. In assemblages with a greater number of species, each species occupied a narrower, more restricted, niche than was the case in areas where there were fewer species, i.e. there was greater 'species packing'. Also generalists are more likely to share microhabitats with specialists or even other generalists than do two species with the same choice of microhabitat. The author concluded that 'Despite subtle differences in temporal activity and body size, differences in structural microhabitat use appears to be the dominant niche axis allowing the coexistence of species within these assemblages of tropical skinks'.

THE URBAN HABITAT

Both *L. delicata* and *L. guichenoti* are ecological generalists that thrive in a variety of natural and human-modified habitats including the urban one. The question arises as to the factors responsible for their amazing ecological amplitude. Prosser *et al.* (2006) addressed this issue by comparing the ecology of *L. guichenoti* in three natural bushland habitats and three more open, anthropogenically modified garden-like habitats in the environs of Sydney.

They analysed a broad array of responses to environmental challenges and found that *L. guichenoti* exhibited flexible behaviour in response to the structural differences and degree of openness between these two types of habitat. In the more open structure of the urban habitat the lizards spent significantly more time on bare, flat surfaces farther from refuges than did their counterparts in natural areas. In the more open garden habitat where a predator could visually detect its prey at greater distance, and the prey also could detect the predator, the *L. guichenoti* exhibited two adaptive responses that provided them with a better chance of escaping: (1) their approach distance (the distance a moving object could approach before the lizard fled) was greater than in natural habitats, and (2) their sprint speeds were faster. The shape of the body and the length of the legs of L. *guichenoti* are suited to the rapid sprints that are advantageous in pursuing its insect prey and escaping its predators. Both modified and natural habits are sufficiently open to accommodate this morphology and the sprint speeds required to carry out these functions. Such versatility probably contributes to the success of this species as a generalist. Accordingly, there were no significant differences between the size of the legs of *L. guichenoti* in relation to the length of its body between urbanised areas and natural habitat.

Walsh *et al.* (2018) confirmed the high degree of adaptation of *L. delicata* when faced with the demands of urban living and tested its boldness in response to stimuli such as novel scents, noises, and simulated attacks by predators. In contradistinction to some other taxa of lizards they found that neither 'neophobic' behaviour (measured as time spent within a refuge) nor 'bold' behaviour (measured as time spent basking) differed between urban and non-urban populations and hypothesised that 'inherent behavioural traits may have predisposed the delicate skink to successfully inhabit both urban and non-urban environments', i.e. is pre-adapted to urban living. Thus, it appears that either the thermal regimes of its urban and non-urban habitats are similar enough that the same duration of basking required for behavioural thermoregulation is the same in both, or that its thermal preferences are sufficiently wide to accommodate different body-temperatures. It also would be instructive to assess whether predation by domestic animals, such as cats, approximate the danger posed by wild predators. *Lampropholis delicata* is a highly eurytopic species and it would be germane to test whether the species confined to more restricted habitats would perform differently from *L. delicata*.

See Chapter 4 for a treatment of responses to urbanisation as a part of personality syndromes and Chapter 12 for a discussion of urbanisation as a threat to species of *Lampropholis*.

THE BIOTIC COMMUNITY AS A WHOLE: FOOD WEBS AND INTERACTIVE DIAGRAMS

The brief example from the section immediately above indicates how structure of assemblages can emerge from a list of its components and a knowledge of their diets and preferences for microhabitat. This can now be expanded to the community as a means of ascertaining how the assemblage fits into the community at large. For example, a list of the prey that avian predators are known to, or are likely to, include in their diet might be a guide to interpreting their potential effect on a known member of their community, such as *Lampropholis* skinks. Similarly, a list of the prey known, or likely, to be included in the diet of the lizards themselves, may give insight into potential competitive interactions with other ecologically similar lizards, or even with taxonomically distant members of their guild. Seldom have comprehensive lists of this nature been assembled. When species of predator and prey are found to occur in the same community, an interaction between them would be likely.

From Homan's (2008) study we know that the snakes *Parasuta flagellum* and *Pseudonaja textilis* occur in heathy dry forest, and thus likely prey on the species of the lizard assemblage, except for *T. rugosa*, the adults of which may be too large to be ingested. These species occur together in the heathy dry forest and all are diurnal; thus, both snakes probably include both species of lizard in their diet. The diets of the other lizards were noted above so they can be added to the food web.

From this simple list we can make some educated conjectures to test by further, more-detailed research. Both species of snakes feed on small lizards and thus are potential predators. Because detailed lists of the other taxa of vertebrates were not given the same detailed treatment specific for this or other habitats, similar hypotheses cannot be formed for possible interactions for the community beyond the reptilian assemblage. Were this study expanded to include individual components for each habitat, comparison of the resulting structures of assemblages in different habitats would provide another level of testable hypotheses. One way of testing such hypotheses is by ecological comparison of individual pairs or groups of individuals within given assemblages, then comparing how interactions may change in the context of different habitats.

REFERENCES

Abercrombie M, Hickman CJ, Johnson ML (1964) *A Dictionary of Biology*. Aldine Publishing Company, Chicago.

Anderson L, Burgin S (2002) Influence of woodland remnant edges on small skinks (Richmond, New South Wales). *Animal Ecology* **27**(6), 630–637. doi:10.1111/j.1442-9993.2002.tb00212.x

Anderson RO, Goulet CT, Chapple DG (2023) Acclimation of thermal physiology to new basking regimes in a widespread Australian skink. *Journal of Thermal Biology* **113**, 103530. doi:10.1016/j.jtherbio.2023.103530

Baker JK (1979) The rainbow skink, *Lampropholis delicata*, in Hawaii. *Pacific Science* **33**(2), 207–212.

Belmont C (1977) Resource partitioning and habitat selection in ground skinks. BSc (Hons) thesis. Department of Zoology, University of New England, Armidale, NSW.

Bilcke J, Downes S, Büscher I (2006) Combined effect of incubation and ambient temperature on the feeding performance of a small ectotherm. *Austral Ecology* **31**, 937–947. doi:10.1111/j.1442-9993.2006.01663.x

Brown GW (1983) Comparative feeding ecology of South-Eastern Australian scincids. PhD Dissertation. La Trobe University, Melbourne.

Brown GW, Nelson JL (1993) Influence of successional stage of *Eucalyptus regnans* (mountain ash) on habitat use by reptiles in the Central Highlands, Victoria. *Australian Journal of Ecology* **18**(4), 405–417. doi:10.1111/j.1442-9993.1993.tb00468.x

Camaiti M, Evans AR, Hipsley CA, Hutchinson MN, Meiri S, Anderson RO, Slavenko A, Chapple DG (2022) A database of the morphology, ecology and literature of the world's limb-reduced skinks. *Journal of Biogeography* **49**(7), 1397–1406. doi:10.1111/jbi.14392

Camaiti M, Evans AR, Hipsley CA, Hutchinson MN, Meiri S, de Oliveira Anderson R, Slavenko A, Chapple DG (2023) Macroecological and biogeographical patterns of limb reduction in the world's skinks. *Journal of Biogeography* **50**(2), 428–440. doi:10.1111/jbi.14547

Carranza S, Arnold, EN, Geniez P, Roca J, Mateo JA (2008) Radiation, multiple dispersal and parallelism in the skinks, *Chalcides* and *Sphenops* (Squamata: Scincidae), with comments on *Scincus* and *Scincopus* and the age of the Sahara Desert. *Molecular Phylogenetics and Evolution* **46**(3), 1071–1094. doi:10.1016/j.ympev.2007.11.018

Chapple DG, Miller KA, Kraus F, Thompson MB (2013) Divergent introduction histories among invasive populations of the delicate skink (*Lampropholis delicata*): has the importance of genetic admixture in the success of biological invasions been overemphasized? *Diversity and Distributions* **19**, 134–146. doi:10.1111/j.1472-4642.2012.00919.x

Chapple DG, Slavenko A, Tingley R, Farquhar JE, Camaiti M, Roll R, Meiri S (2023) Built for success: distribution, morphology, ecology and life history of the world's skinks. *Ecology and Evolution* **13**(12). doi:10.1002/ece3.10791

Child J, Child P (1971) *Australian Dictionary of Biology*. Periwinkle Books, Lansdowne Press, Melbourne.

Clarke CJ (1965) A comparison between some Australian five-fingered lizards of the genus *Leiolopisma* Duméril and Bibron (Lacertilia: Scincidae). *Australian Journal of Zoology* **13**, 577–592. doi:10.1071/ZO9650577

Cogger HG (2018) *Reptiles & Amphibians of Australia*. 8th edn. CSIRO Publishing, Colllingwood.

Cott HB (1957) *Adaptive Coloration in Animals*. Methuen and Company, London.

Cottam G, Curtis JT (1956) The use of distance measures in phytosociological sampling. *Ecology* **37**(3), 451–460. doi:10.2307/1930167

CRA/RFA Steering Committee (1997) 'Systematic vertebrate fauna survey project. Stage I - Vertebrate fauna survey in the South East Queensland Bioregion'. Forest Assessment Unit, Queensland Department of Environment, Indooroopilly. <https://www.agriculture.gov.au/sites/default/files/sitecollectiondocuments/rfa/regions/qld-south-east/enviroment/vertebrate-fauna-survey/qld_se_eh1b.pdf>

Crome B (1981a) Herpetofaunal Assemblages in three Woodlands of the New England Tableland of Australia. Honours Thesis, Department of Zoology, University of New Englnd, Armidale, NSW.

Crome B (1981b) The diet of some ground-layer lizards in three woodlands of the New England Tableland of Australia. *Herpetofauna* **13**(1), 4–11.

Díaz JA, Cabezas-Díaz S (2004) Seasonal variation in the contribution of different behavioural mechanisms to lizard thermoregulation. *Functional Ecology* **18**(6), 867–875. doi:10.1111/j.0269-8463.2004.00916.x

Farquhar JE, Russellk W, Chapple DG (2023) Identifying the abiotic factors that determine the inland range limits of a mesic-adapted species. *Integrative and Comparative Biology* **64**(1), 55–66. doi:10.1093/icb/icad124

Goodman BA (2007) Microhabitat separation and niche overlap among five assemblages of tropical skinks. *Australian Journal of Zoology* **55**, 15–27. doi:10.1071/ZO06066

Gorissen S, Greenlees M, Shine R (2017a) A skink out of water: impacts of anthropogenic disturbance on an endangered reptile in Australian highland swamps. *Oryx* **51**, 610–618.

Gorissen S, Greenlees M, Shine R (2017b) Habitat and fauna of an endangered swamp eosystem in Australia's eastern highlands. *Wetlands* **37**, 269–276.

Graham KA, Mahony SV, Chapple DG, Farquhar JE (2023) The long unknown: rediscovery of the long sunskink, *Lampropholis elongata* (Squamata: Scincidae) – after almost a decade, and after 50 years of data deficiency. *Austral Ecology* **48**(5), 877–884. doi:10.1111/aec.13314

Hanson HC (1962) *Dictionary of Ecology*. Philosophical Library, New York.

Heatwole H (1966) Factors affecting orientation and habitat selection in some geckos. *Zeitschrift für Tierpsychologie* **23**(3), 301–314. doi:10.1111/j.1439-0310.1966.tb01698.x

Heatwole H (1971) Marine-dependent terrestrial biotic communities on some cays in the Coral Sea. *Ecology* **52**(2), 363–366. doi:10.2307/1934598

Heatwole H (1977) Chapter 3. Habitat selection in reptiles. In *Biology of the Reptilia. Vol. 7: Ecology and Behaviour A*. (Ed. C Gans and DW Tinkle) pp. 137–155. Academic Press, New York.

Heatwole H (1989) The concept of the econe, a fundamental ecological unit. *Tropical Ecology* **30**(1), 13–19.

Heatwole H (1999) *Sea Snakes*. University of New South Wales Press, Sydney.

Heatwole H, Taylor J (1987) *Ecology of Reptiles*. Surrey Beatty & Sons, Chipping Norton, NSW.

Heatwole H, Done T, Cameron E (1981) *Community Ecology of a Coral Cay*. Dr. W. Junk, The Hague.

Heatwole H, Trémont S, Broese E (2013) Point-diversity, a critical tool for assessing dynamics of guilds of scavenging ants (Hymenoptera: Formicidae): an example from a eucalypt woodland. *Systematics and Biodiversity* **11**(2)1-32. doi:10.1080/14772000.2013.788578

Homan P (2008) Surveys of vertebrate fauna in the Grampians National Park, 2003–2007. *The Victorian Naturalist* **125**(2), 47–55.

Hoskin CJ (2025) A new species of *Lampropholis* skink (Lacertilia: Scincidae) from Scawfell Island, mid-east Queensland, Australia. *Australian Journal of Taxonomy* **85**, 1-13. doi:10.54102/ajt.6uihp

Howard R, Williamson I, Mather P (2003) Structural aspects of microhabitat selection by the skink *Lampropholis delicata*. *Journal of Herpetology* **37**(3), 613–617.

Howland B, Slojanovic D, Gordon IJ, Manning AD, Fletcher D, Lindenmayer DB (2014) Eaten out of house and home: impacts of grazing on ground-dwelling reptiles in Australian grasslands and grassy woodlands. *PLoS One* **9**(12), e105966. doi:10.1371/journal.pone.0105966

Huey RB, Bennett AF (1987) Phylogenetic studies of coadaptation: preferred temperatures versus optimal performance temperatures of lizards. *Evolution* **41**(5), 1098–1115. doi:10.1111/j.1558-5646.1987.tb05879.x

Hutchinson MN (1993) Chapter 31. Family Scincidae. In *Fauna of Australia*. (Eds CJ Glasby, GJB Ross, PL Beesley). pp. 261–279. Australian Government Publishing Service, Canberra.

Hutchinson MN, Donnellan SC (1992) Taxonomy and genetic variation in the Australian lizards of the genus *Pseudemoia* (Scincidae: Lygosominae). *Journal of Natural History* **26**(1), 215–264. doi:10.1080/00222939200770091

Lunney D, Ashby E, Grigg J, O'Connell M (1989) Diets of scincid lizards *Lampropholis guichenoti* (Dumeril and Bibron) and *Lampropholis delicata* in Mumbulla State Forest on the South Coast of New South Wales. *Australia Wildlife Research* **16**(3), 307–312. doi:10.1071/WR9890307

Lunney D, Eby P, O'Connell M (1991) Effects of logging, fire and drought on three species of lizards in Mumbulla State Forest on the south coast of New South Wales. *Australian Journal of Ecology* **16**(1), 33–46. doi:10.1111/j.1442-9993.1991.tb01479.x

Metcalfe DC, Annable TJ (2016) *Lampropholis delicata* (delicate skink) mortality. *Herpetological Review* **47**(2), 299.

Michelangeli M, Goulet CT, Kang HS, Wong BM, Chapple DG (2018) Integrating thermal physiology within a syndrome: locomotion, personality and habitat selection in an ectotherm. *Functional Ecology* **32**(4), 970–981. doi:10.1111/1365-2435.13034

Milton D (1980) Some aspects of the population dynamics of *Lampropholis guichenoti* in Toohey Forest near Brisbane. *Herpetofauna* **11**, 19–21.

Mulhall SJ, Di Stefano J, Dorph A, Swan M, Sitters H (2024) Do reptile responses to habitat structure and time since fire depend on landscape structure? *Forest Ecology and Management* **553**, 121564. doi:10.1016/j.foreco.2023.121564

Pengilley R (1972) Systematic relationships and ecology of some lygosomine lizards from southeastern Australia (2 Vols). PhD thesis. Australian National University, Canberra.

Pianka ER (1973) The structure of lizard communities. *Annual Review of Ecology and Systematics* **4**, 53–74. doi:10.1146/annurev.es.04.110173.000413

Prosser C, Hudson S, Thompson MB (2006) Effects of urbanization on behavior, performance, and morphology of the garden skink, *Lampropholis guichenoti*. *Journal of Herpetology* **40**(2), 151–159. doi:10.1670/38-05A.1

Rawlinson PA (1974) Biogeography and ecology of the reptiles of Tasmania and the Bass Strait area. *Monographiae Biologicae* **25**, 291–338.

Ruibal R, Philibosian R (1974) The population ecology of the lizard *Anolis anolis*. *Ecology* **55**(3), 525–537. doi:10.2307/1935144

Schulz M (2009) Notes on the Delicate Skink *Lampropholis delicata* occurring on Lord Howe Island. *Herpetofauna* **39**, 48–52.

Sexton OJ, Heatwole H, Knight D (1964) Correlation of microdistribution of some Panamanian reptiles and amphibians with structural organization of the habitat. *Caribbean Journal of Science* **4**, 261–295.

Shattuck S (1999) *Australian Ants: Their Biology and Identification*. CSIRO Publishing, Collingwood.

Shoo LP, Wilson R, Williams YM, Catterall CP (2014) Putting it back: woody debris in young restoration plantings to stimulate return of reptiles. *Ecological Management & Restoration* **15**(1), 84–88. doi:10.1111/emr.12079

Simbotwe MP (1985) Comparative ecology and demography of two populations of Australian reptiles (Lacertilia, Scincidae). PhD thesis. University of New England, Armidale, NSW.

The Reptile Database (nd) <http://www.reptile-database.org> (accessed 11 December 2023).

Tingley R, Thompson MB, Hartley S, Chapple DG (2016) Patterns of niche filling and expansion across the invaded ranges of an Australian lizard. *Ecography* **39**(3), 270–280. doi:10.1111/ecog.01576

Walsh S, Goulet CT, Wong BBM, Chapple DG (2018) Inherent behavioural traits enable a widespread lizard to cope with urban life. *Journal of Zoology* **306**(3), 189–196. doi:10.1111/jzo.12582

Warner FJ (1995) A comparison of the dietary contents of two sympatric scincid lizards, *Lampropholis delicata* (de Vis, 1888), and *Lampropholis guichenoti* (Duméril and Bibron, 1839) (Lacertilia, Scincidae). BAppSc (Hons) thesis. University of Western Sydney, Hawkesbury. 119 pp.

Webb GA (1983) Diet in a herpetofaunal community on the Hawkesbury Sandstone Formation in the Sydney area. *Herpetofauna* **14**, 87–91.

Webb GA (1985) Habitat use and activity patterns in some southeastern Australian skinks. In *Biology of Australasian Frogs and Reptiles*. (Eds G Grigg, R Shine and H Ehmann) pp. 23–30. Surrey Beatty & Sons, Chipping Norton, NSW.

Webb GA (1991) A survey of the reptiles and amphibians of Bondi State Forest and surrounding areas, near Bombala, New South Wales. *Australian Zoologist* **27**(1–2), 14–19. doi:10.7882/AZ.1991.002

12
Threats and conservation

INTRODUCTION

Biodiversity is a central component of Earth's life supporting systems and directly relevant to human societies. Any attempt to defend a social cause such as conserving biodiversity should rest on the best facts available. With growing demand for credible taxonomic [scientific] information, taxonomists [scientists] must facilitate, not roadblock, in stopping what may be the irreversible loss of biodiversity. The time is ripe for multiple stake-holders, spanning from researchers to land managers, professionals to entrepreneurs, and from policy makers to general public, to recognise the discipline of taxonomy [science] as planetary scale science.

We expanded this quote from Khuroo *et al.* (2007) to relate to science generally rather than to taxonomy alone. It succinctly describes the urgent importance of a comprehensive approach to conservation. Human-induced degradation of the natural environment is largely responsible for the sixth major mass extinction in global history (Geyle *et al.* 2021) and amphibians and reptiles have the highest proportion of threatened species among terrestrial vertebrates (Mi *et al.* 2023). Scientists have been warning for decades of the impending crisis, but with inadequate response from governments, commercial enterprises, and the public at large; those warnings were not heeded, and a recent prediction stated that 'The current and next few generations are going to be critical to ensure that future generations inherit this planet in a healthy state' (Garcia-Gonzalez *et al.* 2024). The outlook has become increasingly acute, and the warnings more urgent; belatedly the public is becoming aware of the crisis by directly experiencing the fruits of past neglect (merely read a daily newspaper or watch the news on television). Some action is being taken that may ameliorate the worst-case scenario, but it is already too late to completely halt the present catastrophic decline in biodiversity. It is happening now. The future is already here.

From a sample of 1,500 species of reptiles worldwide Böhm *et al.* (2013) found that one in five species (20%) was threatened with extinction and that in the Australasian region there were insufficient data for evaluation of 14.6% of the total sample of 219 species of lizards; of the 187 species that could be evaluated the categories of threat were: Critically Endangered 2.8%; Endangered 7.5%; Vulnerable 5.4%; Near Threatened 4.8%; Least Concern 79.7%. Thus, a total of 20.5% of the species that could be evaluated were under some degree of risk.

Gibbons *et al.* (2000) listed six major reasons for the worldwide decline of reptiles: (1) loss and degradation of habitat, (2) interaction with invasive species, (3) environmental pollution, (4) disease, (5) unsustainable use by humans, and (6) global climatic change. Subsequent opinion has not changed. For example, much more recently, others (e.g. Chapple *et al.* 2019; Falaschi *et al.* 2019; Howard *et al.* 2020; Geyle *et al.* 2021) have reinforced the validity of the above six reasons although the sources of threat may vary both within and among species. Hoskin and Shea (2018) identified invasive species and climatic change as threatening 7% of the Australian reptilian fauna with extinction. The effects of parasites and pathogens on *Lampropholis* are covered in Chapter 10 and that of predation in Chapter 9.

Senior *et al.* (2021) used the IUCN Red List and publicly available datasets for 14 intrinsic and extrinsic traits to assess those likely to threaten Australian squamates with extinction. They concluded that life-history traits, ecological characteristics, and the harm caused by humans were all collectively implicated and that no single pathway to extinction could be identified. Clearly this means that producing effective policy for conservation of Australian lizards will be difficult and will require comprehensive research into many aspects of their biology. The relative impact of different factors may vary between different locations and habitats, and between the taxa of lizards. A single policy may not be applicable to all situations.

CONSERVATION STATUS

Australian legislation and protective actions for reptiles are based almost entirely on the threatened-species concept and the assessments of individual species are based on their vulnerability to threats and on estimates of the extent and rate of decline in their numbers. From these data, estimates of the cost of actions required to reverse the declines and reduce species' vulnerability may be calculated and relevant action plans developed for individual species, together with the estimated costs involved (Cogger 2018a). A compilation of such data on Australian skinks is contained within *The Action Plan for Australian Lizards and Snakes* (Chapple *et al.* 2019). Clayton *et al.* (2006) provided a state-by-state account of the conservation status of the Australian species of reptiles and amphibians. No species of *Lampropholis* was listed as among the 20 species of Australian squamates expected to be imperiled with extinction in the wild by 2040 (Geyle *et al.* 2021), nor included on the EPBC Act's List of Threatened Fauna of Australia (Webb *et al.* 2015). In fact, most species are listed by the IUCN Red list as of 'Least Concern' and, except for those whose population-trend was unknown, as 'stable' (Table 12.1). *Lampropholis elongata* is the most endangered species as it was listed as unequivocally 'Critically Endangered' despite a reserve having been set aside especially for it (DCCEEW 2023; Table 12.1).

Lampropholis amicula is classified as Data Deficient in the IUCN Red List; Shea *et al.* (2018a–g) suggested that it likely is more widespread than previously recorded and hence when more data are available, this species may prove to be less vulnerable than would be the case were its distribution more restricted. By contrast, some species may now be more vulnerable than originally thought. For example, *L. bellendenkerensis* is not listed as a species at risk on IUCN Red List of Threatened Species; however, Chapple *et al.* (2019) reported that a draft had been prepared to have this species and *L. robertsi* re-classified as 'Vulnerable'. Still other species, e.g. *L. colossus* and *L. mirabilis*, may have different classifications in different parts of their geographic range or are classified differently by different authorities (Table 12.1). No entry for *L. similis* appears in the IUCN Red List for Threatened Species although Chapple *et al.* (2019) reported that a draft was being developed for the species that would recommend listing this species as 'Critically Endangered'.

In contrast to Australia where all species of *Lampropholis* are considered valued species, in New Zealand where *L. delicata* was unintentionally introduced (see Chapter 3), that species is considered undesirable so instead of seeking ways to conserve it, the means of eradicating it have been a subject of study (Wairepo 2015).

Unlike most threatened species, of which between 17.2% and 21.9% are not represented in any protected area (Tingley *et al.* 2019), all described species of *Lampropholis* are protected in at least one such area. Currently, no specific plans for conservation of any species of *Lampropholis* are in place (or being developed) (Table 12.1). This does not imply that species worthy of protection are necessarily neglected. For example, it was not considered necessary to develop a recovery plan for the listing of *L. elongata* as Critically Endangered because the actions required are already well understood and may be guided appropriately by currently available advice (DCCEEW 2023).

Of the species of *Lampropholis* reported in the IUCN Red List in 2024, at least some substantial segment of the range of each species was within one or more protected area and were classified in the IUCN Red List as of Least Concern. As mentioned above, *L. elongata* recently was classified as Critically Endangered. At the other extreme, *L. guichenoti* is widespread and abundant and second only to *L. delicata* in the breadth of range of any species within the genus *Lampropholis*. Its distribution spans south-eastern Queensland, the full length of eastern New South Wales, most of Victoria, and much of eastern Tasmania (see Chapter 2). There also are disjunct

Table 12.1. Data relevant to the conservation of species of *Lampropholis*.

E = eastern; N = northern; NE = north-eastern; S = southern; SE = south-eastern.

Scientific name	IUCN Red List Status	Population trend	States where present	Status[1] and reference
Lampropholis adonis	Least Concern	Stable	Apparently disjunct populations: mid-E and SE Qld. and Carnarvon National Park	No threats; occurs in many protected areas (Shea *et al.* 2018a).
Lampropholis amicula	Least Concern	Stable	Disjunct populations: Far SE Qld. and N NSW	No major threats; presumed present in some protected areas (Shea *et al.* 2018b).
Lamprophlis bellendenkerensis	Least Concern, Drafted as Vulnerable[6]	Unknown	Mt Bellenden Ker, Mt Bartle Frere, part of southern Atherton Tableland	Occurs in Wooroonooran National Park.
Lampropholis caligula	Least Concern	Unknown	NSW	Occurs in protected areas (Shea *et al.* 2018c; Eldridge 2020).
Lampropholis coggeri	Least Concern	Stable	NE Qld.	Occurs in protected areas (McDonald *et al.* 2018).
Lampropholis colossus	Data Deficient; Near Threatened[7]	Stable	SE Qld.	Occurs in Bunya Mountains National Park (Shea *et al.* 2018d).
Lampropholis couperi	Least Concern[8]	Stable	Qld.	Occurs in multiple protected areas (Shea *et al.* 2018e).
Lampropholis delicata	Least Concern	Stable	Qld., NSW, SA, Vic., Tas.	Occurs in multiple protected areas (Chapple *et al.* 2018; Eldridge *et al.* 2020).
Lampropholis elliotensis[3]	Vulnerable[3]	Unknown	Mt Elliot, Qld.	Occurs in Bowling Green Bay National Park, on Mt Elliot (pers. inf.).
Lampropholis elongata	Critically Endangered	Unknown	Ngulin Nature Reserve and surrounding areas, southern edge of the Walcha Plateau in NE NSW	Reserve created specifically for this species (Shea *et al.* 2018f). Graham *et al.* (2023) indicated that this species and its habitat are of significant concern.
Lampropholis guichenoti	Least Concern	Unknown	S Qld., NSW, Vic., SA, Kangaroo Island	Occurs in protected areas (Shea *et al.* 2018g).
Lampropholis isla	Least Concern	Unknown	Qld: Scawfell Island only	Entire range is within the South Cumberland Islands National Park (Wikipedia 2025). Fire is the only potential threat (Hoskin 2025).
Lampropholis mirabilis[4]	Least Concern; Near Threatened	Unknown	N Qld.	Occurs in two national parks – mainland and offshore island (Hoskin *et al.* 2018).
Lampropholis robertsi[2]	Least Concern; Drafted as Vulnerable	Unknown	High elevation NE Qld.	Disjunct populations occur across multiple protected areas (Hoskin and Shea 2018).
Lampropholis similis[5]	Critically Endangered[5]	Unknown	Near Townsville, Qld.	Occurs in Paluma Range National Park (Burgin pers. inf.).

[1]No species of *Lampropholis* currently has a species-specific Action Plan.
[2]Chapple *et al.* (2019) reported that a draft had been prepared suggesting that this species be re-evaluated as Vulnerable.
[3]Chapple *et al.* (2019) reported that a listing was being drafted with a recommendation that this species be listed as Vulnerable.
[4]Originally listed as 'Unknown–Least Concern' on the IUCN Red List but as state-based 'Near Threatened' in Queensland (Hoskin *et al.* 2018; Chapple *et al.* 2019).
[5]Not listed by IUCN but indicated as Critically Endangered by the Environment Protection and Biodiversity Act (EPBA), effective 15 November 2023.
[6]Status assigned by *Nature Conservation Act 1992* of the Queensland Government in 2021, updated in 2023: Drafted as Vulnerable (Nature Conservation Act [Qld] 2023).
[7]State-based listed as Near Threatened in Queensland (Chapple *et al.* 2019) but as Deficient by IUCN.
[8]Also listed as Least Concern by the Queensland Government's *Nature Conservation Act 1992* (Nature Conservation Act [Qld] 2023).

populations, for example in South Australia (Cogger 2018b). Another species (*L. colossus*) was recorded as having insufficient data for assessment (Table 12.1). Although the populations of this species were considered stable, it was noted in the IUCN Red List (Shea *et al.* 2018d) that if *L. colossus* proves to be a valid species, it should be considered rare. It currently is known only from the edges of dry rainforest adjacent to 'balds' (open grassy areas) and savannah woodland containing rainforest elements at high elevation in the Bunya Mountains National Park in southern Queensland (Smith *et al.* 2015).

Likewise, Greer (1997) reported *L. elongata* to be rare. This suggestion of rarity was supported by a survey of lizards and snakes by Heatwole *et al.* (2003). Across a 34-year study (1966–2000) in the New England area, NSW (including a segment of the range of *L. elongata*), they encountered 106 reptilian species, of which 47 were skinks; *L. elongata* was encountered at only two sites.

As with most species listed in Table 12.1, *L. mirabilis* currently is classified as of 'Least Concern' in the IUCN Red List alongside the comment that it is 'localised but common' with 'no major threats' (Hoskin and Shea 2018). It is, nevertheless, classified as 'Near Threatened' under the Queensland Government's *Nature Conservation Act 1992* (Nature Conservation Act [Qld] 2023). Fire does not seem to be a threat to this species as Burgin (2010) indicated that its habitat probably has been on fire over geological time.

Another species, *L. robertsi*, is reported to be of high vulnerability under the present predicted scenario of climatic change for the Wet Tropics Bioregion (Hoskin *et al.* 2018; Weber *et al.* 2021). This species may, however, be in an even more precarious position than recognised when that update was completed for the IUCN Red List. This is because, in addition to the species of *Lampropholis* that are listed by Cogger (2018b) and in the IUCN Red List (Table 12.1), two additional morphologically cryptic species (*L. similis* and *L. elliotensis*), derivatives of *L. coggeri*, have been formally described by Singhal *et al.* (2018) (see Chapter 14).

CRYPTIC SPECIES AND CONSERVATION

More than 90% of Australian species of reptiles are endemic to the country; however, some cryptic lineages remain unclassified (Donnellan *et al.* 1993; Geyle *et al.* 2021).

Cryptic species may pose special issues for conservation. The extensive geographic ranges of eurytopic species such as *L. delicata* and *L. guichenoti* would suggest that they are unlikely to be threatened with extinction. If, however, neither is a single species but rather each is a congeries of contiguously distributed, genetically distinct species, then the derivative species may be adapted to the local conditions within their limited geographic ranges and hence be more vulnerable to climatic change and more susceptible to extinction. Chapple *et al.* (2011a, b; also see Chapter 2) discovered that *L. delicata* and *L. guichenoti* each consisted of genetically diverse clades and subclades. Retention of this complete range of genetic diversity is essential, as environmental conditions are changing at an accelerating rate and successful conservation of all such variants increases the likelihood that at least some of them will be able to conform to the new conditions. It is for this reason, it is essential to retain the complete range of present biodiversity and effort needs to be expended to conserve each entity, whether a cryptic species, a clade, or subclade. Having such an entity lost would be much worse than merely having the range of a widespread, genetically uniform species reduced in size.

While currently all described species of *Lampropholis*, including cryptic ones, have at least some segment of their range within protected areas, other populations are suspected of being still-undescribed cryptic species; some of these may be restricted to geographic areas not included in any protected area, thus making their long-term persistence more precarious.

THREATS

Loss and degradation of habitat

Invasion by exotic plant species into native ecosystems generally is considered a significant environmental issue (Virkki *et al.* 2012; Lindenmayer *et al.* 2017), and management typically is considered advantageous (Bower *et al.* 2014; Lindenmayer *et al.* 2017). However, control measures often change the physical properties of a habitat's characteristics, including the structure and growth rate of vegetation. Such controls also may alter microhabitat and availability of shelter sites for reptiles. In turn, there are likely to be changes in faunal assemblages (Bower *et al.* 2014). It has often been observed, for example, that the density and species richness of reptiles are not equivalent or improved in rehabilitated habitats compared with undisturbed ones although the condition of animals may be equivalent in the two habitat types (Cristescu *et al.* 2012).

Traditional restoration practices have focused on plants with the assumption that fauna, including reptiles, will automatically recolonise sites (Triska *et al.* 2016). However, reptiles have been reported to be relatively slow to return to, for example, rehabilitated mine sites (Twigg and Fox 1991; Kanowski *et al.* 2006). Taylor and Fox (2001a) investigated a chronosequence (4, 8, 14, 20 years) after sand was mined on the Central Coast of New South Wales. When sand-mined dunes were compared with unmined forest, *L. guichenoti* remained in low abundance at 20 years post-mining and *L. delicata* did not appear on the mined site until 20 years after the completion of mining. These differences in timing are consistent with the observations of Twigg and Fox (1991) who reported that, in general, lizards typically are slower than other animal groups to recolonise after disturbance and that it may take substantially more than 20 years before they return to even a completely restored path through open forest after sand had been mined there.

Bragg *et al.* (2005) reported that across the interface between regenerating sand-mined areas and open forest, in the study area where Twigg and Fox (1991) undertook their survey, the number of *L. delicata* captured in pitfalls increased from the interface into the forest. This observation was consistent with the species being a forest specialist. The species' response was also positively correlated with height of the understorey. However, the composition of the assemblage of reptiles was not solely related to the time since disturbance, whether by fire or mining. When Taylor and Fox (2001a) considered the association between the accumulation of leaf litter and the abundance of *L. delicata*, there was a significant relationship between the density of leaf litter and the abundance of skinks.

In the tropics, areas of ecological restoration and old plantations where the canopy tends to be more open than in native rainforests, 'litter skinks', including *Lampropholis*, were the most common skinks. Kanowski *et al.* (2003, 2006) suggested that this may be due to the preference of these species for edges of small clearings and other openings in the canopy (i.e. sunlit patches). Such areas were observed either adjacent to, or connected to, rainforest via vegetated corridors or a matrix of timber plantations and remnant forests (Kanowski *et al.* 2006). The lack of dispersal otherwise may have been influenced by the small home ranges of many species (Stow *et al.* 2001; Osterwalder *et al.* 2004; Kanowski *et al.* 2006). Regional differences in the occupancy of reforested sites by rainforest reptiles may be due to the lizard faunas in subtropical rainforests being depauperate compared to those in tropical rainforests (Heatwole and Taylor 1987; Kanowski *et al.* 2006).

One of the consequences of loss of habitat is that large, once-contiguous areas of favourable habitat are broken up into smaller isolated patches. Such loss of connectivity of suitable habitat for *L. coggeri* has been shown to have occurred already; its original habitat was an extant area of rainforest in the Australian Wet Tropics that, beginning with European settlement of Australia, became increasingly fragmented into isolated, smaller patches (Macdonald *et al.* 2017). The advent of global warming and the certainty that it will become worse before its progress can be arrested poses challenges for the conservation of various species of *Lampropholis* (see section *Climatic change* below).

Invasive exotic plants

Australia has many introduced species of plants, especially around settled areas, that alter the structure of the vegetation and the microclimate. One such plant is the blue periwinkle (*Vinca major*), a sprawling weed originally from Europe and western Asia. It is prostrate and forms dense mats of broad elliptical leaves and thin wiry stems that root at the tips and nodes. The distribution of *L. delicata* and this plant overlap in coastal New South Wales. Downes and Hoefer (2007) studied the effect of the plant upon the ecology and behaviour of this species. Where they obtained the lizards for their project the patches of *V. major* varied in size from 4 to 60 m^2 and covered 40% of the ground cover; 53% of the 257 lizards used in their experiments were caught in such patches.

Hatchlings from clutches produced by the initial captives were divided equally into three enclosures: (1) high exposure; substrate entirely covered with *V. major*, (2) intermediate exposure; half of the substrate covered by *V. major*, half by grass, and (3) no exposure; substrate entirely covered by grass. Various aspects of the lizards' morphology, thermal biology, activity, performance, behaviour, reproduction, and growth were compared among the three treatments. The effect of the presence of *V. major* was surprisingly complex and comprehensive and some of the effects were, in turn, causative. For example, animals from all three coverages, when placed in a laboratory gradient, exhibited the same preferred temperature (~30°C). However, in the enclosures with high density of *V. major*, shading by the broad leaves kept the substrate under the plant as low as 19.9 ± 0.1°C when the maximum temperatures on the surface of its canopy was 48.5 ± 0.5°C and the temperature throughout the grass was 45.1 ± 0.9°C. Only by climbing up onto the canopy of *V. major* to bask, could the animals in weeds reach their preferred temperature, whereas their siblings that hatched from the same clutches of eggs as they had, but subsequently kept in enclosures of only grass did not climb. The climbers developed significantly longer hind limbs relative to the size of their bodies than did their non-climbing siblings. The front limbs retained the same size relative to the rest of the body in both groups.

The relatively longer hind limbs translated into another difference between the two groups: the speed of sprinting. At the beginning of the experiment, the lizards (8 days old) from all three densities of *Vinca* did not differ significantly in their speed of sprinting, but as they grew and were tested again at 94, 172, 257, and 354 days of age, as the disparity of the lengths of their hind limbs increased so did the differences in their sprinting speed. By age of 94 days the speed of sprinting by the animals reared in high densities of *Vinca* already was significantly higher than in the other two treatments ($P < 0.05$) and thereafter the differences were highly significant ($P < 0.01$). Thus, even among such genetically similar individuals as siblings from the same clutches of eggs, merely the structure of the vegetative cover of the microhabitat caused a change in behaviour (mode of thermoregulatory basking) which in turn caused a change in morphology (relative size of hind limbs), which finally caused an increase in their speed of sprinting. Some such causes seem to be maladaptive, e.g. basking in a more exposed location would seem to increase the risk of predation, whereas others, for example greater speed of sprinting, would seem to enhance the likelihood of the lizard escaping a predator or of it capturing its own prey.

Despite the thermal disadvantage of not being able to bask at ground level in the high-density areas of *V. major*, lizards in those enclosures were more likely to remain hidden during the day than were animals in the areas of intermediate or low density of *V. major*. However, during the time that lizards were on the surface the proportion of time being active did not differ among different coverages of vegetation, nor between the sexes of animals, and these attributes did not change over the entire duration of the study (1 year).

There were still other consequences of the invasion of the weed *V. major*. Growth rates were markedly lower in the lizards reared in high density of the weed than in the other treatments; growth in the enclosure with intermediate density of *V. major* was only slightly lower than in the enclosure lacking this weed altogether. Thus, another cascade of non-genetic factors may have come into play: the less time spent on the surface reduced the time devoted to foraging, which in turn may have had consequences for overall growth rates and hence for the lighter weight of eggs and the resulting hatchlings produced by the denizens of the weed-infested enclosure. These intricacies on interaction are also discussed in other contexts in Chapter 6.

Urbanisation

Expanding urbanisation continues to impact species of reptiles worldwide (Hamer 2011) due to the loss, modification, and fragmentation of native habitats (Jellinek *et al.* 2004). In Australia, more than 70% of the human population resides in the eastern states where most of the major cities are situated (ABS 2021), the area of the country that also encompasses the range of most known species of *Lampropholis*. Some indication of the decline that may occur in reptiles due to urbanisation is provided by Hamer (2011), who reported that since 1850, in Melbourne (a city within the range of *L. delicata* and *L. guichenoti*), 66.7% ($n = 39$) of the reptilian species were recorded in fewer Local Government Areas between 1990 and 2006 than between 1850 and 1989. Within Greater Sydney, White and Burgin (2004) reviewed the loss of reptiles from six urban bushland reserves and observed that skinks tended to fare better than other terrestrial species, and that *L. delicata* and *L. guichenoti* were recorded from all reserves.

While the reserves where White and Burgin (2004) focused their surveys were in bushland, elsewhere in urban areas within the geographic range of *L. delicata* and *L. guichenoti*, these species frequently are abundant in parklands and urban gardens (Burgin 1993). In many such urban open spaces, the landscape is maintained quite differently than in the natural habitat of these species. For example, at ground level, the habitat is much less complex than in adjacent remnant bushland. Often there may be, for example, more open space than appropriate cover for skinks.

The skinks also may behave differently in an urban environment than in natural habitat. For example, Prosser *et al.* (2006) compared the difference in behaviour and performance of *L. guichenoti* in response to a predator in natural and modified habitats. Overall, skinks fled sooner and at greater speed when disturbed in an open area than in natural habitat. By contrast, comparative studies of the learning response between urban and non-urban *L. delicata* showed equivalent responses between these two groups (Kang *et al.* 2018; Walsh *et al.* 2018). For example, Kang *et al.* (2018) did not observe a single significant difference in negotiating a maze that indicated learned behaviour between urban and non-urban *L. delicata*.

Another study of the effects of urbanisation on native reptiles was undertaken by Garden *et al.* (2007) in 59 sites across Brisbane's urban remnant bushland. *Lampropholis delicata* and *Cryptoblepharus pulcher* (formally *C. virgatus*) were the most common species of reptiles encountered throughout the urban matrix, whereas *L. amicula* also was present, although in substantially fewer numbers. These data indicate that some species of *Lampropholis* are pre-adapted to urban conditions and are thus able to persist in such areas despite significant changes to their habitat.

For a discussion of the role of personality syndromes to adaptation to urbanisation see Chapter 4, and for treatment of the performance of *Lampropholis* in urban situations see Chapter 11.

Road-kill

Wolfe *et al.* (2019) suggested that roadways are a significant threat to reptiles. Despite observing that motorists did not intentionally drive over models of reptiles placed on the road, they estimated that the probability of strike rate of slow-moving reptiles was >75%. This is despite most motorists slowing to avoid a reptile. Due to their small size, a motorist would be highly unlikely to notice a *Lampropholis* in their vehicle's path. Not surprisingly, mention of collision of vehicles with small reptiles the size of *Lampropholis* is rare, undoubtedly, in part, due to their size and rapid desiccation after death. The carcass would also be the target of a range of scavenging birds, especially in urban/peri-urban areas where birds cross-forage between urban and local natural habitats (see Anderson and Burgin 2008) and they would thus be removed without their demise being tallied.

The observations of Wotherspoon and Burgin (2011) are a rare exception to the lack of data on the impact of road traffic on *Lampropholis*. Herpetofaunal road-kills collected on early morning walks included both *L. delicata* and *L. guichenoti*, both of which occur naturally at Falconbridge (adjacent to Blue Mountains National Park, west of Sydney). Although seldom observed on roadways, these two species often have been observed to forage and thermoregulate on the verges of roads (Burgin, pers. obs.) and to use overpasses as habitat (McGregor *et al.* 2015).

Pet trade

Gippet and Bertelsmeier (2021) suggested that invasive species, including reptiles, are strongly overrepresented in the pet trade. There is no evidence, however, that any species of *Lampropholis* is currently valued as a pet. This is despite ample evidence that *L. delicata* is sufficiently robust to stowaway in cargo and become established and thrive on the islands of Hawaii, New Zealand, and Lord Howe Island (see Chapter 2). It may be assumed that the lack of appeal is because individuals of all species within the genus are small and innocuous. This explanation may, however, be dispelled in the future because Gippet and Bertelsmeier (2021) reported that there is an emerging trade for pets as small as ants.

Changes in global trends in human populations and in the ownership of pets may result in greater popularity of small reptiles such as *Lampropholis*. For example, 55% of the world's population of humans already live in cities and it is predicted that this figure will rise to ~70% by 2050 (World Bank 2020). This increasing urbanisation has resulted in a demand for smaller dwellings. For example, between 2008 and 2013, the number of apartments rose by 17.3% (4.9% detached, 7.0% semi-detached; Lennard 2014). Within this context, and with reptiles already being the most popular exotic pets worldwide (Valdez 2021) a trend towards smaller animals as pets is expected to continue (Gippet and Bertelsmeier 2021). The ease of maintaining, at least *L. delicata* and *L. guichenoti*, in captivity (Burgin 1993), and the under-regulation of the global pet trade (Marshall *et al.* 2020) alone indicate that these likely will enter the pet trade in the future.

Agriculture and grazing

Brown *et al.* (2011), Kay *et al.* (2013) and Pulsford *et al.* (2017) suggested that the conservation of biodiversity in rural landscapes has become increasingly problematic with the on-going intensification of agriculture. Such landscapes do, however, often encompass areas of remnant vegetation within a matrix of non-native plants that may be detrimental to the maintenance of native species such as reptiles (Driscoll *et al.* 2013; Pulsford *et al.* 2017). Such fragmentation of the agricultural landscape typically results in a reduction of native habitat to remnant patches of vegetation, fence-lines, and open areas of grazed or cropped land.

In an investigation of the negative effects of agriculture on the biodiversity of reptiles, Pulsford *et al.* (2017) observed that the abundance of *L. guichenoti* was positively associated with the cover of trees and of leaf litter. By contrast, *L. delicata* were more likely to be found in association with logs in remnant vegetation than with those in open areas. Subsequently Pulsford *et al.* (2018) reported that most recaptures (surveys undertaken 5 days to 1 year apart) indicated that *Lampropholis* tended not to move far from where they were first captured. It was concluded that grazed paddocks, especially those with key features such as fences, trees, leaf litter, and shrubs may enable reptiles to successfully disperse through larger areas of the landscape than through areas without such habitat.

Even where natural habitat was present at the interface of paddocks and fence-lines, trees, shrubs, and grasses, Anderson and Burgin (2002) found that, overall, there were significantly more *L. delicata* and *L. guichenoti* in core areas than at the edges of small remnants of natural habitat surrounded by an agricultural landscape. Whilst both species were more common within the remnants than they were on the edges, *L. guichenoti* were more abundant in the edge habitat than were *L. delicata*. The latter species tended to be more common in the inner area of the small remnants of farmland abutting open land that was being grazed or cropped. The size-structure of *L. guichenoti*, but not *L. delicata*, also differed between edge and core. Subadults of *L. guichenoti* were found in larger numbers in the edge than within remnants. In addition, significantly fewer adults and juveniles occupied the outer area of remnants although no such differences were observed in the size-structure of *L. delicata*.

Howland *et al.* (2014) investigated the effect of grazing in grassland and grassy woodlands and found that *L. delicata*, the single species of *Lampropholis* present in that study area, was not impacted by the intensity of grazing.

Predation

See Chapter 9 for a detailed coverage of predation upon *Lampropholis*.

Fire and drought

A substantial body of the research on the impacts of fire on *L. delicata* and *L. guichenoti* was undertaken in Mumbulla State Forest, southern New South Wales (Lunney *et al.* 1989, 1991; Penn *et al.* 2003). A different pattern in distribution occurred between the two species due to fire. Although there was no obvious change in numbers of *L. delicata*, there was a reduction in the number of *L. guichenoti*. Drought resulted in a substantial reduction in both species, although, subsequently, the numbers of *L. guichenoti* recovered more quickly than did those of *L. delicata*. Based, on research that occurred at Mumbulla (and nearby), Penn *et al.* (2003) concluded that the impacts on biodiversity due to burning were complex and multidimensional. The diet of *L. delicata* and *L. guichenoti* that occur in sympatry in Mumbulla State Forest were assessed before an intense fire in summer (November 1980) and again post-fire (December 1980, 1981, 1983, 1984) which spanned a period of intense drought (1980–1983). These generalist feeders foraged over a range of substrates. Their diet was similar and consisted predominately of insects and spiders although in drought there were substantial differences between the two species in the proportions of some taxa consumed. Lunney *et al.* (1989) suggested that, at least in part, the skinks' survival through fire and drought was likely due to their flexibility in foraging.

A further investigation of the effects of fire on reptiles in Mumbulla State Forest (and beyond) that also included *L. delicata* and *L. guichenoti* was undertaken by Penn *et al.* (2003) who compared the impacts of a patchy, low-intensity, hazard-reduction burn in logged regrowth and unlogged sites that had previously been burnt in 1980. In contrast to the former burn that resulted in a reduction in the abundance of *L. guichenoti* on the ridges of logged sites (Lunney *et al.* 1991), 20 months after the hazard-reduction burn *L. guichenoti* was more abundant than before in the areas previously burnt. Conversely, *L. delicata* were observed to be more abundant after the 1980 fire and less so after the 1999 fire. The differences were assumed to have been because of increased visibility due to the lack of cover since the survey was undertaken soon after the fire (i.e. 2 weeks) compared with 10 months after the hazard-reduction fire (Penn *et al.* 2003), rather than to any real differences in numbers.

In a more northerly study site (Central Coast, NSW) Taylor and Fox (2001a) observed that *L. guichenoti* were more common in the unmined forested area than in the adjacent sand-mined site. There was an increase in the species' abundance over the 3–4 years post-fire although 14 years later there had not been any further increase in numbers. There were, however, significantly more *L. guichenoti* in forested areas after disturbance than where mining had occurred. *Lampropholis delicata* were only present in mined sites and not in burnt areas. There was, however, no significant increase in the number of *L. delicata* with time since fire.

In a rare study of the effects of fire on *Lampropholis* other than *L. delicata* and *L. guichenoti*, Hannah *et al.* (1998) compared assemblages of reptiles in Bauple State Forest, in southern Queensland, that experienced different prescribed fire regimens (annual and periodic burning) over ~50 years. Abundance, species richness, and species diversity all were higher in the unburnt habitat. The authors recorded that all species in their study were common on the coastal lowlands of the area, including *L. delicata*, *L. amicula*, and *L. couperi*. However, these three species were rarely observed in areas that were burnt annually. Changes in microhabitat due to the frequent burning was assumed to be the reason for the substantial differences observed between the two regimens of burning.

Another approach to determining the impact of hazard-reduction burns on native fauna, including *L. delicata* and *L. guichenoti*, was undertaken by Croft *et al.* (2010). They used charred and uncharred fence posts as artificial refuges to compare their use by fauna in adjacent areas of forest, one burnt and the other unburnt. Vertebrates were slower than invertebrates to use either type of introduced refuge in either burnt or unburnt forest. Overall, however, lizards were less likely to use the burnt posts as retreats in unburnt forest than those that were unburnt in either forest type. For example, only two species (*L. delicata* and *Saiphos equalis*) used this habitat and of all the skinks observed during the study only *L. delicata* were sometimes seen basking in the experimental refuges, and then only on unburnt posts.

In a review of a wide range of organisms, including lizards, Harris *et al.* (2023) concluded that interference with natural fire regimens may shape their natural selection, dispersal, and repopulation.

Logging

The effect of logging on *Lampropholis* varied between different regions and types of forest. Webb (1994) compared the abundance of *L. guichenoti* in logged versus unlogged areas of eucalypt forest at Eden, NSW, and found that by two separate methods, pitfall trapping and counts along transects, this species was more abundant in the unlogged areas than in the logged ones although the difference was significant only for the results from pitfall traps. Lunney *et al.* (1991) focused on the effects of logging on the resident *Lampropholis* at Mumbulla State Forest, NSW, by comparing areas that were unlogged, just logged, one year after logging, and after 10–15 years of regrowth. Similar patterns in the use of habitat were observed in response to the perturbations caused by logging, as occurred during overlapping drought and fire (see above). *Lampropholis delicata* and *L. guichenoti* were in approximately equal numbers in unlogged and recently logged forest although fewer of each species occurred in areas of regrowth 10–15 years post-logging. This change was attributed to the differences in the amount and pattern of sunshine reaching the forest floor. These results were tempered somewhat by aspect; although the direction that slopes faced had no effect otherwise, the abundance of *L. guichenoti* on northern aspects were lower than on eastern ones. Daly and Lemckert (2011) concluded that logging in forests near Tenterfield, NSW, had a positive impact on the density of *L. delicata*.

Pollution

Fisher and Ineich (2012) suggested that the detrimental effects of pesticides and other chemicals have been poorly studied in reptiles. Although most studies have concluded that most chemicals are not lethal, Burgin (1993) noted that local pest controllers or gardeners, keen on pesticides and/or herbicides, may spray small reptiles without consideration of their plight even though the spray was known to be detrimental to *L. delicata* and *L. guichenoti*. However, since this detrimental effect was reported over 25 years ago, it is very likely that alternatives are now used that are not as harmful to small skinks. Alternatively, some animals may have acquired resistance to pesticides (Moreno *et al.* 2011; Goldson *et al.* 2015). The Environment Protection Authority (2017, p. 21), however, warned that many pesticides are hazardous chemicals. An indication that some current pesticides may be detrimental to reptiles, including *L. delicata*, was implied by Eason *et al.* (2013) in a recommendation that *L. delicata* be used as a surrogate for native New Zealand reptiles in testing for toxicity.

In Australia, the use of the herbicide glyphosate in various formulations is widely used for the control of invasive plants (Martin and Murray 2013; Lindenmayer *et al.* 2017); however, the impacts of such control of weeds on native animals have rarely been evaluated (Lindenmayer *et al.* 2017). In some instances, a significant impact has not been observed. For example, the focus of the study of Martin and Murray (2013) was primarily on the impact that the spraying of glyphosate on *Chrysanthemoides monilifera* ssp. *rotundata* (bitou bush) had on the abundance, species richness, and composition of assemblages of reptiles, including *L. delicata* and *L. guichenoti*. Over 7–10 months post-spraying, they did not observe a significant difference between assemblages of reptiles in any of these attributes. More recently, Lindenmayer *et al.* (2017) undertook a similar, but much more comprehensive study (8 years) to investigate the impact on native species of spraying glyphosate on bitou bush, interspersed with controlled burning. They found no impact of glyphosate on native ground-dwelling animals, including *Lampropholis*.

Another pollutant, Fluorine (F^-), typically is present in its ionic form in the environment (Palczewska-Komsa *et al.* 2016). Taylor and Fox (2001b) studied the effects of atmospheric fluoride pollution on reptiles on the coastal dunes in dry sclerophyll forest at Tomago, NSW. The area had been mined for sand and subjected to fire 4 and 8 years before their study commenced. In unmined areas where fluoride levels were low (<0.25 μmol $F.g^{-1}$), values of species richness and of the total abundance of lizards were significantly higher than in mined areas where fluoride levels were higher (1.85–3.4 μmol $F.g^{-1}$ and 8.00–13.2 μmol $F.g^{-1}$).

Research on the succession of reptiles after the mining of sand in the area used by Taylor and Fox (2001b) was undertaken by Letnic and Fox (1997). They found limited evidence that the fallout of fluoride associated with the mining of sand had a direct toxic effect on the local reptilian assemblage, which included *L. delicata* and *L. guichenoti*. They also concluded that the major impediment to the recovery of the reptilian assemblage was habitat structure. For example, in areas with a reduced cover of canopy due to the fallout of fluoride, *L. delicata* were absent. It was concluded that where there was a source of fluoride emissions sufficient to reduce the canopy, the levels of pollution had, at least indirectly, a detrimental impact on reptiles. The reduction in the cover of the canopy that accompanied mining appeared to be the most important factor in the loss of reptiles from the area (Taylor and Fox 2001b).

Exacerbating the effects of fluoride pollution on the canopy and the prevalence of *Lampropholis* in association with the mining of sand in a sclerophyll forest, is the significant impact on arthropod succession of such pollution. For example, Madden and Fox (1997) found that there were direct negative effects on arthropods and some other invertebrates, many of which form a major component of *Lampropholis'* diet (see Chapter 8).

There is an apparent absence of data from New Zealand where *L. delicata* is a feral species. Eason *et al.* (2013) suggested, however, that there was a need to retain and continue to refine and develop alternatives to the use of pest-control chemicals used to target terrestrial vertebrates. While acknowledging that it was necessary to limit environmental persistence of chemicals and improve control mechanisms, the achieving of such outcomes was considered a significant challenge.

Climatic change

Lampropholis coggeri, *L. colossus*, *L. couperi*, and *L. robertsi* have natural distributions mainly in rainforests. Several others (*L. delicata*, *L. guichenoti*, and *L. mirabilis*) have disjunct distributions that include populations on the Australian mainland and small near-shore islands (Cogger 2018a). These distributions are consistent with the result of the isolation of sub-populations of species during past periods of climatic change (Bell *et al.* 2010; Chapple *et al.* 2011a, b; Llewelyn *et al.* 2018). Predications are that within the current episode, the greatest effect is likely to be on ectotherms in the tropics (Llewelyn *et al.* 2016), which is the area that includes at least part of the range of the majority of species of *Lampropholis*.

The climate is once again warming rapidly. For example, in Australia, there has been a rise in temperature since 1910[1] of 1.44 ± 0.24°C. This change has been accompanied by an increase in duration and frequency of extreme-heat events. In addition, both diurnal and nocturnal temperatures have increased in all months to the extent that most years are now warmer than any recorded in the 20th century. Much of this warming has occurred post-1950 with each subsequent decade being warmer than the previous one with an associated increased risk of dangerous bushfires. Within the natural geographic range of the genus *Lampropholis*,

1 Year of commencement of national records.

warming conditions are predicated to continue with more extremely hot days and fewer extremely cool days. In addition, it is predicted that rainfall will decrease in the cooler period of the year and there will be more prolonged annual fire seasons with an increased number of days of 'dangerous fire weather'. There will also potentially be substantial variations in climate among years, with more rainfall events that are heavier, but of shorter duration, and a changed frequency of high-intensity cyclones (CSIRO 2020).

There has already been an erosion of the diversity of lizards worldwide due to changing climate (Sinervo *et al.* 2010). In addition, there is increasing evidence of rapid and on-going changes in phenology, shifts in geographic range, and in declines of populations of some Australian reptiles (Bell *et al.* 2010; Llewelyn *et al.* 2017; Macdonald *et al.* 2017). Under such conditions, isolation of sub-populations may occur with subsequent genetic and phenotypic divergence (Llewelyn *et al.* 2018). For example, Bell *et al.* (2010) compared the responses of two rainforest-dwelling taxa of north-eastern Queensland (*L. robertsi* and *L. coggeri*) subjected to past climatic fluctuations and now occurring in disjunct populations. They found genetic diversity of both species to be consistent with the persistent isolation of rainforest remnants with divergence sufficient to develop new species. One of these taxa, *L. robertsi*, is a montane specialist whereas its congener, *L. coggeri*, is more widely distributed. The modelling by Bell *et al.* (2010) of the potential future changes in the populations of these two species under current predicted climatic change, indicated that sub-populations of *L. robertsi* will once again become increasingly isolated into disjunct populations under warming environmental conditions. These changes are predicated to be consistent with those that persisted throughout the climatic cycling of the Pleistocene. The past changes in climate have resulted in species with narrow ranges within the *L. coggeri* radiation. By contrast, their modelling predicted that the habitat of *L. coggeri* would be climatically more stable and connected than that of *L. robertsi*. Based on their current genetic profile, both species have experienced low dispersal and high persistence across multiple isolated regions during the previous period of climatic warming.

As indicated above, the range of species in the genus *Lampropholis* largely are encompassed within the eastern states of Australia. This area is expected to be vulnerable to climatic change. For example, in 2015, Welbergen *et al.* reported that in the Wet Tropics (i.e. much of north-eastern Queensland) regimens of extreme events, but particularly extremes of temperature, will continue into the foreseeable future due to anthropogenically driven climatic change. Within this context, resilience (ability of a species or population to recover after an environmental insult) and vulnerability (ability to adapt to changing conditions, most likely due to physiological and/or behavioural plasticity) are predicated to be determinates of a species' ability to adapt to climatic extremes (Welbergen *et al.* 2015, p. 8). Under the current predictions of future climates for the montane rainforests of northern Queensland, Bell *et al.* (2010) suggested that of the five species of skinks on the list that were deemed to be among the 20 least resilient skinks to climatic change, one was *L. robertsi*. However, this species has subsequently been reclassified as two different species. It may thus be assumed that individually these species would be even less likely to be resilient to future climatic change than the previous composite one. Of the six species of skinks among the most exposed taxa, *L. mirabilis* was listed second, and *L. robertsi* fifth.

Other more widespread species, particularly *L. delicata* and *L. guichenoti*, are currently abundant. Unlike species already restricted to upland areas, they are more likely to persist within a warming climate, albeit probably within a more fragmented habitable range. For example, in their examination of the impact of past climatic oscillations and biogeographic barriers on the evolutionary history of *L. delicata*, Chapple *et al.* (2011a, b) noted that there were nine geographically structured, non-overlapping clades, some probably originating from populations fragmented by topographic or aridity barriers and others existing as populations isolated on uplands.

A species' vulnerability to climatic change is likely influenced by the magnitude of exposure, susceptibility to a warming environment, and the species' ability to adapt to changing conditions (Ofori *et al.* 2017). For example, Llewelyn *et al.* (2016) found that *L. coggeri* had a negative relationship between critical environmental temperature and thermal maximum (CTMax) and a subsequent study confirmed that thermoregulatory behaviour played a role (Llewelyn *et al.* 2017). *Lampropholis coggeri* from a warm environment preferred lower temperatures than did those from a cooler environment. It was further determined that these differing preferences were driven by shifts in behaviour. The CTMax of

L coggeri is, therefore, 'plastic' and is influenced by acclimation. Within a warmer environment this species will seek cooler microenvironments and their physiology acclimates to the temperatures to which they are exposed.

More generally, Llewelyn *et al.* (2016) discovered that the extent to which a species of *Lampropholis* will be vulnerable to increasing environmental temperatures depends on the individual's interaction between thermal physiology and thermoregulatory behaviour. Since physiological modification to rate of heat exchange is minor in small skinks compared with behavioural adjustments (Bartholomew 1982), thermoregulatory behaviour will predominate. Unlike larger reptiles, small skinks, including *L. delicata* and *L. guichenoti*, heat more slowly than they cool which is the reverse of larger reptiles (Fraser and Grigg 1984) (see Chapter 6). We therefore hypothesise that, as Llewelyn *et al.* (2016) noticed in *L. coggeri*, individuals of other species of the genus will require sufficiently variable habitat for individuals to be able to successfully respond to changing conditions that will occur with current climatic changes and, if they are to survive, to either adjust their geographic range or to adapt to new prevailing conditions within their current one.

Plasticity in thermal response to climatic change is not, however, the only factor that will contribute to resilience among species of *Lampropholis.* There are a range of processes that span factors influencing the capacity for adaptation, and the risk of exposure that likely relate to local adaptation in the face of a rapidly changing climate. Although such adaptation may occur rapidly when an appropriate trait occurs within the population and conditions are favourable (Catullo *et al.* 2019), in the absence of appropriate traits, species diversity may erode due to increasing warming of the climate and/or by the alteration of thermal niches (Sinervo *et al.* 2010).

Macdonald *et al.* (2017) suggested that as climatic change progresses, a possible strategy for conservation might be targeted gene flow by moving pre-adapted individuals in peripheral isolates into declining populations.

Temporal and spatial scope

Llewelyn *et al.* (2016) held temporal scope constant for *L. coggeri* and studied the spatial one separately. The acclimation of four different traits, CTMin, CTMax, performance (thermal optimum for sprinting), and desiccation (rate of evaporative loss of water), were compared among populations from different localities at the same time. All four of these traits showed marked differences among localities. The CTMin varied from 8.2 ± 0.3°C to 16.3 ± 0.3°C and the CTMax from 36.6 ± 0.2°C to 41.3 ± 0.2°C in different populations measured at the same time; optimal temperature for sprinting ranged from 27.1 ± 0.8°C to 37.1 ± 1.1°C, the desiccation rate from 1.8 ± 0.2 mg/hr to 6.9 ± 0.6 mg/hr. In addition to the spatial scope there was a temporal one with values varying over time within the same locality and, finally, there also were other differences; for example, large skinks tested at the higher temperatures lost more weight than did smaller ones. In summary, there were local adjustments to environmental conditions independent of seasonal ones.

The most surprising finding was counter-intuitive: the relationship between CTMax and temperature of the environment was negative, that is, the CTMax was higher in areas with lower prevailing temperatures. In a later paper (Llewelyn *et al.* 2017), however, the reverse was found. The paradox was resolved by analysing the lizards' thermoregulatory behaviour: animals from hot environments chose cooler microenvironments than did those from habitats that were cool and the physiology of the animals from hot environments acclimated to the chosen temperature, rather than to the prevailing one of the habitat in general. Such nuances need to be assessed in predicting the response of lizards to impending hotter environments.

Finally, in their triad of papers Llewelyn *et al.* (2016, 2017, 2018) compared the traits of populations of *L. coggeri* known to differ physiologically from two areas, Paluma (cooler, wetter site) and Hervey Range (hotter, drier site). In addition to the traits used in previous studies, maximum speed of sprinting, breadth of the range of temperatures permitting sprinting, and heat shock (see section *Adaptation to changing thermal conditions* above) were added. Results were compared between animals recently after capture, again after acclimation, and finally a genetic component was added by studying the subjects' offspring.

Initially, there were small, but statistically significant, differences between the two sites in most of the traits. In comparison to Hervey Range, lizards from Paluma had lower CTMinima, higher performances in several measures of the speed of sprinting, and higher desiccation rates. Acclimation also affected most of the traits, bringing them closer to laboratory conditions. An exception was the response to heat shock; it was significantly higher for animals from Paluma than for those from Hervey Range.

Eggs from the subjects of the experiments described above were incubated at two different temperatures and the progeny tested for some of the same attributes as those to which their parents had been subjected. CTMinima did not differ significantly between locations when the second-generation individuals were tested at 1 month of age (Paluma 13.12°C; Hervey Range 13.12°C) or as adults. CTMax also did not differ significantly between populations at 1 month of age (Paluma 38.73°C, Hervey Range 38.51°C). However, animals had higher CTMaxima at Paluma than at Hervey Range when retested as adults.

This array of physiological, behavioural, and genetic responses by so many traits from different localities within a small area attests to a surprising versatility of this species. Even if one combination of traits fails to adapt to changing conditions, others may succeed – to use a familiar notation in a different context, the situation denotes 'the survival of the fittest'. The authors suggest that because of this versatility the species with small geographic ranges may not be as vulnerable to climatic change as originally thought. A critical question remains, however, as to whether the breadth of any of these combinations of responses are sufficient to cope with the extent and rapidity of the impending march of climatic changes.

PROTECTION FROM THREATS

Natural refugia

The threats facing lizards are somewhat ameliorated by legally vetting areas of natural, undisturbed habitat for the purpose of providing biota with a haven from the destructive forces described above. Although many such areas are designed specifically to protect mammals, birds, or a particular type of vegetation, the preservation of natural ecosystems coincidentally protect whatever ancillary biota that resides within them. However, Roll *et al.* (2017) found that existing protected areas represent the needs of birds and mammals better than those of reptiles and additional conservation actions are needed to effectively protect reptiles, especially lizards, and Chapple *et al.* (2021) found that the global network of protected areas currently is insufficient to conserve most species of skink. Michael *et al.* (2021) identified *L. elongata* as a species whose small distributional range had low overlap with any reserves.

Given that ~500 reptilian species are predicted likely to become extinct within the next century because of pending climatic changes (Mi *et al.* 2023), identifying refugia that conserve the greatest number of threatened species is of paramount importance. Mi *et al.* (2023) estimated that the geographic ranges of 7.8% of known herpetofaunal species do not include protected areas within their borders, and that large spatial gaps remain.

Fortunately, all species of *Lampropholis* have at least part of their geographic ranges within a protected area (see Table 12.1). This is not as reassuring as one might think, however, for two reasons. (1) Some of the species are restricted to small areas at high elevations (see Chapter 2) and as global warming proceeds, the elevational zones of vegetation likely will creep upward and perhaps extinguish small areas of suitable alpine habitat for montane species; such changes probably will occur too rapidly for the stenotopic montane species of *Lampropholis* to adapt to the new conditions. (2) The two widespread eurytopic species, *L. delicata* and *L. guichenoti*, are not genetically uniform, but each consists of various clades and subclades (see Chapter 2). Consequently, as habitats occupied by some local clades or subclades are altered or eliminated, along with their populations of *Lampropholis*, the species will lose part of its array of genetic diversity and its capacity to adapt to varying situations. Bell *et al.* (2010) suggested an alternative prediction for point 1 above. They reasoned that the persistence of species in small isolated tropical montane regions is indicative of a resilience to former changes in climate that may auger well for successfully meeting future changes. Which of these two scenarios will be enacted as global warming proceeds will depend on whether these species can adapt quickly enough to keep pace with the speed at which future climatic changes progress.

THE FUTURE

Approximately 10% of the herpetofauna of the world occurs in Australia – the largest proportion for any country. Many of these species are in decline and the 20 most imperiled ones have a restricted range (Geyle *et al.* 2021). Most species of *Lampropholis*, however, are designated as either Data Deficient or Least Concern (see Table 12.1). The exceptions are *L. elongata* (Critically Endangered) and *L. mirabilis* (Near Threatened); in addition, *L. elliotensis* has been recommended for reclassification to Vulnerable and *L. similis* to Critically Endangered. *Lampropholis colossus* is considered to be rare.

Of the *Lampropholis* recorded by Cogger (2018a), ~65% (n = 11) could be regarded as having a restricted range. At least some segment of the range of all described species, however, includes one or more protected area.

As has been previously noted throughout this book, published ecological research associated with most species of this genus is limited for all except *L. delicata* and *L. guichenoti*. It is likely that, as has been recently determined for *L. coggeri* and *L. robertsi*, additional species are likely to be identified and further species may ultimately prove to be at greater long-term risk than currently recorded on the IUCN Red List.

About 25 years after the first comprehensive assessment of the status of Australian squamate reptiles (Cogger *et al.* 1993), a second assessment noted the rapid rate at which species had been described in the intervening years. The current rate is likely to continue (see Fig. 3.1 in Chapple *et al.* 2019). The description of *Lampropholis* species has followed this trajectory. For example, Cogger (1975), listed only two species (*L. delicata* and *L. guichenoti*, then included in the genus *Leiolopisma*). In the most recent edition of *Reptiles and Amphibians of Australia* (Cogger 2018a) there are 11 species listed and, since that time, a further three species have been described (see Chapter 1). Descriptions of additional species are expected in the future because some existing ones may consist of multiple cryptic species (see Chapple *et al.* 2019).

Even without such knowledge, the fact that all described *Lampropholis* species have at least some protection in reserves certainly reduces some potential anthropogenic threats such as broad-scale clearing of land. However, without more specific management of predatory feral species, particularly cats, and with continued degradation of habitat due to the expansion of urbanisation and agriculture, weeds, fire, development of infrastructure in parks (e.g. internal tracks and roadways that may impinge on critical habitat), all exacerbated by the future impacts of climatic change, there are likely to be future major issues facing the conservation of species of *Lampropholis*. To date, there has been limited research of these threatening processes, and even basic ecological requirements for most of the species require further investigation. There is much that remains to be learned about the species of *Lampropholis*!

It is likely that in the future climatic instability, driven by heating of the oceans will impose stresses on many ectotherms and threaten them with extinction because they will be unable to adapt sufficiently quickly. Consequently, not only will physical environments change, but the elimination of some species may change the composition of assemblages and alter the biotic context impinging on the remaining species. Kearney *et al.* (2009) applied a biophysical model to ectotherms that used macroclimatic, topographic, and locational data to construct hourly microclimatic conditions above and below ground (hourly changes in solar and infrared radiation; wind speeds; and temperatures of the air, soil surface, and subsoil). They also used a biotic model that coupled energy and mass in balanced equations to calculate suitable core temperatures of ectotherms within the available microclimates as a function of the known morphology, physiology, and behaviour of the animals. By coupling these two types of models they were able to map the animals' niches to particular landscapes. The validity of these models was verified by empirical tests in which the results were within an average of 1°C of predictions. Application of this technique by matching species of *Lampropholis* with the range of conditions in their present habitats and with predicted future landscapes arising from global warming in different parts of their geographic range could facilitate management of future landscapes in ways enhancing survival of these lizards. For example, because the efficacy of behavioural thermoregulation is tied strongly to the availability of shade, altering the cover of vegetation to provide more shade might improve the habitats' sustainability for *Lampropholis*. Such efforts may also be applicable to the conservation of other species of similar size and ecologies and enhance the value of *Lampropholis* as model organisms (see Chapter 15). Kearney *et al.* (2009) suggested several ways of improving these models, such as adding the physical variables of monthly changes in cloud cover, vegetation cover, and albedo.

Plasticity in life history may be important in the survival of such widespread, eurytopic species as *L. delicata* and *L. guichenoti*. The fact that *L. guichenoti* can modify the number of ovarian cycles per year in response to annual environmental variation (see Chapter 5) may increase its likelihood of survival.

TURNABOUT IS FAIR PLAY

Where species like *L. delicata* have been introduced concentration no longer is centred on protecting it from competition from local species, but rather how to protect local species from it as an invasive species. One of the main features of this species in terms of its competitive ability is the wide range of taxa of prey it consumes (Smith 2016).

REFERENCES

ABS (2021) *National, State and Territory Population*. Australian Bureau of Statistics. <https://www.abs.gov.au/statistics/people/population/national-state-and-territory-population>

Anderson L, Burgin S (2002) Influence of woodland remnant edges on small skinks (Richmond, New South Wales). *Austral Ecology* **27**(6), 630–637. doi:10.1046/j.1442-9993.2002.x

Anderson L, Burgin S (2008) Pattern of bird predation on reptiles in small woodland remnant edges in peri-urban north-western Sydney, Australia. *Landscape Ecology* **23**(9), 1039–1047. doi:10.1007/s10980-008-9252-5

Bartholomew GA (1982) Physiological control of body temperature. In *Biology of the Reptilia*, Vol. 12. (Eds C Gans and FH Pough) pp. 167–211. Academic Press, New York.

Bell RC, Parra JL, Tonione M, Hoskin CJ, Mackenzie JB, Williams SE, Moritz C (2010) Patterns of persistence and isolation indicates resilience to climate change in montane rainforest lizards. *Molecular Ecology* **19**(12), 2531–2544. doi:10.1111/j.1365-294X.2010.04676.x

Böhm M, Collen B, Ballie JEM, Bowles P, Chanson J, Cox N, Hammerson G, Hoffmann N, Livingstone SR, Ram M *et al.* (2013) The conservation of the world's reptiles. *Biological Conservation* **157**, 372–385. doi:10.1016/j.bicon.2012.07.015

Bower DS, Valentine LE, Grice AC, Hodgson LE, Grice A, Hodgson L, Schwarzkopf L (2014) A trade-off in conservation: weed management decreases the abundance of common reptile and frog species while restoring and invaded floodplain. *Biological Conservation* **179**, 123–128. doi:10.1016/j.bicon.2014.09.003

Bragg JG, Taylor JE, Fox BJ (2005) Distributions of lizard species across edges delimiting open-forest and sand-mined areas. *Austral Ecology* **30**(2), 188–200. doi:10.1111/j.1442-9993.2005.01436.x

Brown GW, Dorrough JW, Ramsey DSL (2011) Landscape and local influences on patterns of reptile occurrence in grazed temperate woodlands of southern Australia. *Landscape and Urban Planning*, **103**(3–4), 277–288. doi:10.1016/j.landurbplan.2022.08.002

Burgin S (1993) *Lampropholis* the 'new' laboratory animal. In *Herpetology in Australia: A Diverse Discipline*. (Eds D Lunney and D Ayers) pp. 279–282. Transactions of the Royal Zoological Society of New South Wales. Surrey Beatty & Sons, Chipping Norton, NSW.

Burgin S (2010) A natural history of the Ham Common, the lands sandwiched between Richmond and Windsor – two of Governor Macquarie's five towns. In *A Natural History of Sydney*. (Eds D Lunney, P Hutchings and D Hochuli) pp. 263–276. Royal Zoological Society of New South Wales, Sydney.

Catullo RA, Llewelyn J, Phillips BL, Moritz CC (2019) The potential for rapid evolution under anthropogenic climate change. *Current Biology* **29**(19), R996–R1007. doi:10.1016/j.cub.2019.08.028

Chapple DG, Hoskin CJ, Chapple SNJ, Thompson MB (2011a) Phylogeographic divergence in the widespread delicate skink (*Lampropholis delicata*) corresponds to dry habitat barriers in eastern Australia. *BMC Evolutional Biology* **11**(1), 191. doi:10.1186/1471-2148-11-191

Chapple DG, Chapple SNJ, Thompson MB (2011b) Biogeographic barriers in south-eastern Australia drive phylogeographic divergence in the garden skink, *Lampropholis guichenoti*. *Journal of Biogeography* **38**(9), 1761–1775. doi:10.11861471-2148-11-191

Chapple DG, Shea G, Dickman C, Wilson S, Hobson R, Sanderson C (2018) Rainbow skink *Lampropholis delicata*. *The IUCN Red List of Threatened Species* 2018: e. T109473102A109473141. <https://www.iucnredlist.org/species/109473102/109473141>

Chapple DG, Tingley R, Mitchell NJ, Macdonald SL, Keogh S, Shea GM, Bowles P, Cox NA, Woinarski JCZ (2019) *The Action Plan for Australian Lizards and Snakes 2017.* CSIRO Publishing, Clayton South, Vic. doi:10.1071/9781486309474

Chapple DG, Roll U, Böhm M, Aguilar R, Amey AP, Austin CC, Baling M, Barley AJ, Bates MF, Bauer AM, Blackburn DG, Bowles P, Brown RM, Candramouli SR, Chirio L, Cogger H, Colli GR, Conradie W, Couper PJ, Cowan MA, Craig MD, Das I, Datta-Roy A, Dickman CR, Ellis RJ, Fenner AL, Ford S, Ganesh SR, Gardner MG, Geissler P, Gillespie GR, Glaw F, Greenlees MJ, Griffith OW, Grismer LL, Haines ML, Harris DJ, Hedges SB, Hitchmough RA, Hoskin CJ, Hutchinson MN, Ineich I, Janssen J, Johnston GR, Karin BR, Keogh JS, Kraus F, LeBreton M, Lymberakis P, Masroor R, McDonald PJ, Mecke S, Melville J, Melzer S, Michael DR, Miralles A, Mitchell NJ, Nelson NJ, Nguyen TQ, Nogueira CDC, Ota H, Pafilis P, Pauwels OSG, Perera A, Pincheira-Donoso D, Reed RN, Ribeiro-Junior MA, Riley JL, Rocha S, Rutherford PL, Sadlier RA, Shacham B, Shea GM, Shine R, Slavenko A, Stow A, Sumner J, Tallowin OJS, Teale R, Torres-Carvajal O, Trape JF, Uetz P, Ukuwela KDB, Valentine L, Van Dyke JU, van Winkel D, Vasconcelos R, Vences M, Wagner P, Wapstra E, While GM, Whiting MJ, Whittington CM, Wilson S, Ziegler T, Tingley R, Meiri S (2021) Conservation status of the world's skinks (Scincidae): taxonomic and geographic patterns in extinction risk. *Biological Conservation* **257**, 109101.

Clayton M, Wombey JC, Mason IJ, Chester RT, Wells A (2006) *CSIRO Checklist of Australian Vertebrates: A Reference with Conservation Status*. 2nd edn. CSIRO Publishing, Collingwood, Vic.

Cogger H (1975) *Reptiles and Amphibians of Australia*. A.H. & A.W. Reed, Sydney.

Cogger H (2018a) *Reptiles and Amphibians of Australia*. Updated 7th edn. CSIRO Publishing, Clayton South, Vic.

Cogger H (2018b) A brief demographic overview of Australia's native amphibians. In *Status of Conservation and Decline of Amphibians: Australia, New Zealand, and Pacific Islands*. (Eds H Heatwole and JL Rowley) pp. 5–14. CSIRO Publishing, Clayton South.

Cogger HG, Cameron EE, Sadlier RA, Eggler P (1993) *The Action Plan for Australian Reptiles*. Australian Nature Conservation Agency, Canberra.

Cristescu RH, Frère C, Banks P (2012) A review of fauna in mine rehabilitation in Australia: current state and future directions. *Biological Conservation* **149**(1), 60–72. doi:10.1016/j.biocon.2012.02.003

Croft P, Reid N, Hunter JT (2010) Experimental burning changes the quality of fallen timber as habitat for vertebrate and invertebrate fauna: implications for fire management. *Wildlife Research* **37**(7), 574–581. doi:10.1071/WR10053

CSIRO (2020) *State of the Climate 2020*. CSIRO (Commonwealth Science Industry Research Organisation) and Bureau of Meteorology. <https://www.csiro.au/en/research/environmental-impacts/climate-change/state-of-the-climate/previous/state-of-the-climate-2020>.

Daly G, Lemckert F (2011) Survey of the reptiles and amphibians of the montane forests near Tenterfield on the north coast of New South Wales. *Australian Zoologist* **35**(4), 957–972. doi:10.7882/AZ.2011.048

DCCEEW (2023) Conservation Advice for *Lampropholis elongata* (long sunskink). Department of Climate Change, Energy, the Environment and Water. <https://www.environment.gov.au/biodiversity/threatened/species/pubs/64728-conservation-advice-15112023.pdf>.

Donnellan SC, Adams M, Hutchinson M, Baverstock PR (1993) The identification of cryptic species in the Australian herpetofauna: a high research priority. In *Herpetology in Australia: A Diverse Discipline*. (Eds D Lunney and D Ayres) pp. 121–126. Transactions of the Royal Zoological Society of Australia, Surrey Beatty & Sons, Chipping Norton, NSW.

Downes S, Hoefer A-M (2007) An experimental study of the effects of weed invasion on lizard phenotypes. *Oecologia* **153**(3), 775–785. doi:10.1007/s00442-007-0775-2

Driscoll DA, Banks SC, Barton PS, Lindenmayer DB, Smith AL (2013) Conceptual domain of the matrix in fragmented landscapes. *Trends in Ecology and Evolution* **28**(10), 605–613. doi:10.1016/j.tree.2013.06.010

Eason CT, Fairweather A, Ogilvie S, Blackie H, Miller A (2013) A review of recent non-target toxicity testing of vertebrate pesticides: establishing generic guidelines. *New Zealand Journal of Zoology* **40**(3), 226–235. doi:10.1080/03014223.2013.772067

Eldridge MDB, Ingleby S, King AG, Mahony SV, Parnaby HE, Beatson CA, Divljan A, Frankham GJ, Hay AC, Major RE, Reader SE, Sadlier RA, Tsang LR (2020) Australian Museum surveys of the vertebrate fauna of Coolah Tops National Park, NSW. *Technical Reports of the Australian Museum Online* **30**, 1–20. doi:10.3853/j.1835-4211.30.2020.1757

Environment Protection Authority (2017) *EPA Guidelines for Responsible Pesticide Use* (Updated April 2017). WaterCare, EPA South Australia. <https://www.epa.sa.gov.au/files/477372_guide_pesticides.pdf>

Falaschi M, Manenti R, Thuiller W, Ficetola GF (2019) Continental-scale determinants of population trends in European amphibians and reptiles. *Global Change Biology* **25**(10), 3504–3515. doi:10.1111/gcb.14739

Fisher R, Ineich I (2012) Cryptic extinction of a common Pacific lizard *Emoia impar* (Squamata, Scincidae) from the Hawaiian Islands. *Oryx* **46**(2), 187–195. doi:10.1017/S003060531001778

Fraser S, Grigg GC (1984) Control of thermal conductance is insignificant to thermoregulation in small skinks. *Physiological Zoology* **57**(4), 392–400. doi:10.1086/physzool.57.4.30163341

Garcia-Gonzalez F, Ripple WJ, Malo AF (2024) Scientists' warnings to humanity for long-term planetary thinking on biodiversity and humankind preservation, a cosmic perspective. *BioScience* **74**(2), 82–85. doi:10.1093/biosci/biad108

Garden JG, McApline CA, Possingham HP, Jones DN (2007) Habitat structure is more important than vegetation composition for local-level management of native terrestrial reptile and small mammal species living in urban remnants: a case study from Brisbane, Australia. *Austral Ecology* **32**(6), 669–685. doi:10.1111/j.1442-9993.2007.01750.x

Geyle HM, Tingley R, Amey AP, Cogger H, Couper PJ, Cowan M, Craig MD, Doughty P, Driscoll DA, Ellis R, Emery J-P, Fenner A, Gardner MG, Garnett ST, Gillespie GR, Greenlees MJ, Hoskin CJ, Keogh JS, Lloyd R, Melville J, McDonald PJ, Michael DR, Mitchell NJ, Sanderson C, Shea GM, Sumner J, Wapstra E, Woinarski JCZ, Chapple DG (2021) Reptiles on the brink: identifying the Australian terrestrial snake and lizard species most at risk of extinction. *Pacific Conservation Biology* **27**, 3–12. doi:10.1071/PC20033

Gibbons JW, Scott DE, Ryan TJ, Buhlmann KA, Tuberville TD, Metts BS, Greene JL, Mills T, Leiden Y, Poppy S, Winne CT (2000) The Global decline of reptiles, Dèjà Vu amphibians. *BioScience* **50**, 653–666. doi:10.1641/0006-3568(2000)050[0653:TGDORD]2.0.CO;2

Gippet JMW, Bertelsmeier C (2021) Invasiveness is linked to greater commercial success in the global pet trade. *Proceedings of the National Academy of Sciences of the United States of America* **118**(14), e2016337118. doi:10.1073/pnas.2016337118

Goldson SL, Bourdôt GW, Brockerhoff EG, Byrom AE, Clout MN, McGlone MS, Nelson WA, Popay AJ, Suckling DM, Templeton MD (2015) New Zealand pest management: current and future challenges. *Journal of the Royal Society of New Zealand* **45**(1), 31–58. doi:10.1080/03036758.2014.1000343

Graham KA, Mahony SV, Chapple DG, Farquhar JE (2023) The long unknown: rediscovery of the long sunskink, *Lampropholis elongata* (Squamata: Scincidae) – after almost a decade, and after 50 years of data deficiency. *Austral Ecology* **48**(5), 877–884. doi:10.1111/aec.13314

Greer A (1997) A new species of *Lampropholis* (Squamata: Scincidae) with a restricted, high altitude distribution in eastern Australia. *Australian Zoologist* **30**(3), 360–368. doi:10.7882/AZ.1997.014

Hamer AJ (2011) The herpetofauna of Melbourne: using past and present distributions to assess impacts of urbanisation. *The Victorian Naturalist* **128**(5), 162–174. doi:10.3316.informit.69102199390524

Hannah DS, Smith GC, Agnew G (1998) Reptile and amphibian composition in prescribed burnt dry sclerophyll forest, southern Queensland. *Australian Forestry* **61**(1), 34–39. doi:10.1080/00049158.1998.10674714

Harris J, Chapple DG, MacLean CA, Melville J (2023) Genetic footprints of fire: understanding the genetic implications of fire regimes for fauna and effective conservation strategies. *Biological Conservation* **284**, 110169. doi:10.1016/j.biocon.2023.110169

Heatwole H, Taylor J (1987) *Ecology of Reptiles*. Surrey Beatty & Sons, Chipping Norton, NSW.

Heatwole H, de Bavay J, Webber P (2003) Faunal survey of New England. V. The lizards and snakes. *Memoirs of the Queensland Museum* **49**(1), 299–325.

Hoskin CJ (2025) A new species of *Lampropholis* skink (Lacertilia: Scincidae) from Scawfell Island, mid-east Queensland, Australia. *Australian Journal of Taxonomy* **85**, 1–13. doi:10.54102/ajt.6uihp

Hoskin C, Shea G (2018) Grey-bellied sunskink. *Lampropholis robertsi. The IUCN Red List of Threatened Species* 2018: e. T109473231A109473236. <https://www.iucnredlist.org/species/109473231/109473236>

Hoskin C, Amey A, Couper P (2018) Saxicoline sunskink. *Lampropholis mirabilis. The IUCN Red List of Threatened Species* 2018: e.T109473218A109473223. <https://www.iucnredlist.org/species/109473218/109473223>

Howard C, Flather CH, Stephens PA (2020) A global assessment of the drivers of threatened terrestrial species richness. *Nature Communications* **11**(1), 993. doi:10.1038/s41467-020-144771-6

Howland B, Stojanovic D, Gordon IJ, Manning AD, Fletcher D, Lindenmayer DB (2014) Eaten out of house and home: impacts of grazing on ground-dwelling reptiles in Australian grasslands and grassy woodlands. *PLoS One* **9**(12), e105966. doi:10.1371/journal.pone.015966

Jellinek S, Driscoll DA, Kirkpatrick JB (2004) Environmental and vegetation variables have a greater influence than habitat fragmentation in structuring lizard communities in remnant urban bushland. *Austral Ecology* **29**(3), 294–304. doi:10.1111/j.1442-9993.2004.01366.x

Kang F, Goulet CT, Chapple DG (2018) The impact of urbanization on learning ability in an invasive species. *Biological Journal of the Linnean Society* **123**(1), 55–62. doi:10.1093/biolinnean/blx131

Kanowski JJ, Catterall CP, Wardell-Johnson GW, Proctor H, Reis T (2003) Development of forest structure on cleared rainforest land in eastern Australia under different styles of reforestation. *Forest Ecology and Management* **183**(1–3), 265–280. doi:10.1016/S0378-1127(03)00109-9

Kanowski JJ, Reis TM, Catterall CP, Piper SD (2006) Factors affecting the use of reforested sites by reptiles in cleared rainforest landscapes in tropical and subtropical Australia. *Restoration Ecology* **14**(1), 67–76. doi:10.1111/j.1526-100X.2006.00106.x

Kay GM, Michael DR, Crane M, Okada S, MacGregor C, Florance D, Trengove D, McBurney L, Blair D, Lindenmayer DB (2013) A list of reptiles and amphibians from Box Gum Grassy Woodlands in south-eastern Australia. *Check List* **9**, 476–481. doi:10.15560/9.3.476

Kearney M, Shine R, Porter WP (2009) The potential for behavioral thermoregulation to buffer 'cold-blooded' animals against climate warming. *Proceedings of the National Academy of Sciences of the United States of America* **106**(10), 3835–3840. doi:10.1073/pnas.0808913106

Khuroo AA, Dar GH, Khan ZS, Malik AH (2007) Exploring an inherent interface between taxonomy and biodiversity: current problems and future challenges. *Journal for Nature Conservation* **15**(4), 256–261. doi:10.1016/jnc.2007.07.003

Lennard C (2014) *Global Urbanisation Trend Boosting Apartment Dwelling*. Blog post. Euromonitor International, <euromonitor.com>

Letnic MI, Fox BJ (1997) The impact of industrial fluoride fallout on faunal succession following sand mining of a dry sclerophyll forest at Tomago, NSW.—I. Lizard recolonisation. *Biological Conservation* **80**(1), 63–81. doi:10.1016/S0006-3207(96)00064-X

Lindenmayer DB, Wood J, MacGregor C, Hobbs RJ, Catford JA (2017) Non-target impacts of weed control on birds, mammals, and reptiles. *Ecosphere* **8**(5), e01804. doi:10.1002/ecs2.1804

Llewelyn J, Macdonald SL, Hatcher A, Moritz C, Phillips BL (2016) Intraspecific variation in climate-relevant traits in a tropical rainforest lizard. *Diversity and Distributions* **22**(10), 1000–1012. doi:10.1111/ddi.12466

Llewelyn J, Macdonald SL, Hatcher A, Moritz C, Phillips BL (2017) Thermoregulatory behaviour explains countergradient variation in the upper thermal limit of a rainforest skink. *Oikos* **126**(5), 748–757. doi:10.1111/oik.03933

Llewelyn J, Macdonald SL, Moritz C, Martins F, Hatcer A, Phillips BL (2018) Adjusting to climate change: acclimation, adaptation and developmental plasticity in physiological traits of a tropical rainforest lizard. *Integrative Zoology* **13**(4), 411–427. doi:10.1111/1749-4877.12309

Lunney D, Ashby E, Grigg J, O'Connell M (1989) Diets of scincid lizards *Lampropholis guichenoti* (Dumeril and Bibron) and *Lampropholis delicata* (De vis) in Mumbulla State Forest on the south coast of New South Wales. *Australian Wildlife Research* **16**(3), 307–312. doi:10.1071/WR9890307

Lunney D, Ebby P, O'Connell M (1991) Effects of logging, fire and drought on three species of lizard in Mumbulla State Forest on the south coast of New South Wales. *Australian Journal of Ecology* **16**(1), 33–46. doi:10.1111/j.9993.1991.tb01479

Macdonald SL, Llewelyn J, Moritz C, Phillips BL (2017) Peripheral isolates as sources of adaptive diversity under climate change. *Frontiers in Ecology and Evolution* **5**, 88. doi:10.3389/fevo.2017.00088

Madden KE, Fox BJ (1997) Arthropods as indicators of the effects of fluoride pollution on the succession following sand mining. *Journal of Applied Ecology* **34**(5), 1239–1256. doi:10.2307/2405235

Marshall BM, Strine C, Hughes AC (2020) Thousands of reptile species threatened by under-regulated global trade. *Nature Communications* **11**(1), 4728. doi:10.1038/s41467-020-18523-4

Martin LJ, Murray BR (2013) A preliminary assessment of the response of a native reptile assemblage to spot-spraying invasive bitou bush with glyphosate herbicide. *Ecological*

Management and Restoration **14**, 59–62. doi:10.1111/emr.12019

McDonald P, Hoskin C, Shea G (2018) *Lampropholis coggeri*. *The IUCN Red List of Threatened Species* 2018: e.T109472898A473056. <https://www.iucnredlist.org/species/109473045/109473056>

McGregor ME, Wilson SK, Jones DN (2015) Vegetated fauna overpass enhances habitat connectivity for forest dwelling herpetofauna. *Global Ecology and Conservation* **4**, 221–231. doi:10.1016/j.gecco.2015.07.002

Mi C, Ma L, Yang M, Li X, Meiri S, Roll U, Oskyrko O, Pincheira-Donoso D, Harvey LP, Jablonski D, Safei-Mahroo B, Ghaffari H, Smid J, Jarvie S, Kimani M, Masroor R, Kazemi SM, Nneji LM, Fokoua AMT, Taboue GCT, Bauer A, Nogueira C, Meirte D, Chapple DG, Das I, Grismer L, Avila LJ, Júnior MAR, Tallowin, OJS, Torres-Carvajal O, Wagner P, Ron SR, Wang Y, Itescue Y, Nagy ZT, Wilcove DS, Liu X, Du W (2023) Global protected areas as refuges for amphibians and reptiles under climate change. *Nature Communications* **14**(1), 1389. doi:10.1038/s41467-023-36987-y

Michael DR, Moore H, Wassens S, Craig MD, Tingley R, Chapple DG, O'Sullivan J, Hobbs RJ, Nimmo DG (2021) Rock removal associated with agricultural intensification will exacerbate loss of reptile diversity. *Journal of Applied Ecology* **58**, 1557–1565.

Moreno V, Aguayo CA, Brunton DH (2011) A survey for the amphibian chytrid fungus *Batrachochytrium dendrobatidis* in New Zealand's endemic Hochstetter's frog (*Leiopelma hochstetteri*). *New Zealand Journal of Zoology* **38**(2), 181–184. doi:10.1080/03014223.2010.548073

Nature Conservation Act [Qld] (2023) *Nature Conservation Act 1992*. Last updated 2023. Queensland Government. <legislation.qld.gov.au>.

Ofori BY, Stow AJ, Baumgartner JB, Beaumont LJ (2017) Influence of adaptive capacity on the outcome of climate change vulnerability assessment. *Scientific Reports* **7**, 12979, doi:10.1038/s41598-017-13245-y

Osterwalder K, Klingenböck A, Shine R (2004) Field studies on a social lizard: home range and social organization in an Australian skink, *Egernia major. Austral Ecology* **29**(3), 241–249. doi:10.1111/j.1442-9993.2004.01339.x

Palczewska-Komsa M, Wilk A, Stogiera A, Chlubek D, Buczkowska-Radlińska J, Wiszniewska B (2016) Animals in biomonitoring studies of environmental fluoride pollution. *Fluoride* **49**(3), 279–292.

Penn AM, Sherwin WB, Lunney D, Banks P (2003) The effects of a low-intensity fire on small mammals and lizards in a logged, burnt forest. *Wildlife Research* **30**(5), 477–486. doi:10.1071/WR02080

Prosser C, Hudson S, Thompson MB (2006) Effects of urbanization on behavior, performance, and morphology of the garden skink, *Lampropholis guichenoti*. *Journal of Herpetology* **40**(2), 151–159. doi:10.1670/38-05A.1

Pulsford SA, Driscoll DA, Barton PS, Lindenmayer DB (2017) Remnant vegetation, planting and fences are beneficial for reptiles in agricultural landscapes. *Journal of Applied Ecology* **54**(6), 1710–1719. doi:10.1111/1365-2664.12923

Pulsford SA, Barton PS, Driscoll DA, Kay GM, Lindenmayer DB (2018) Reptiles and frogs use most land cover types as habitat in a fine-grained agricultural landscape. *Austral Ecology* **43**(5), 502–513.

Roll U, Feldman A, Novosolov M, Allison A, Bauer AM *et al.* (2017) The global distribution of tetrapods reveals a need for targeted reptile conservation. *Nature Ecology and Evolution* **1**, 1677–1682.

Senior AF, Bóhm M, Johnstone CP, McGee MD, Meri S, Chapple DG, Tingley R (2021) Correlates of extinction risk in Australian squamate reptiles. *Journal of Biogeography* **48**(9), 2144–2152. doi.org/10.1111/jbi.14140

Shea G, Wilson S, Venz M, Hobson R (2018a) Diamond Shielded Sunskink. *Lampropholis adonis. The IUCN Red List of Threatened Species* 2018: e.T109472898A109472952. <https://www.iucnredlist.org/species/109472898/109472952>

Shea G, Wilson S, Sanderson C, Hobson R (2018b) Friendly Sunskink. *Lampropholis amicula. The IUCN Red List of Threatened Species* 2018: e.T109472962A109472985. <https://www.iucnredlist.org/species/109472962/109472985>

Shea G, Cogger H, Greenlees M (2018c) Montane Sunskink. *Lampropholis caligula. The IUCN Red List of Threatened Species* 2018: e.T109473000A109473024. <https://www.iucnredlist.org/species/109473000/109473024>

Shea G, Hobson R, Wilson S (2018d) Bunya Sunskink. *Lampropholis colossus. The IUCN Red List of Threatened Species* 2018: e.T109473064A109473067.en. <https://www.iucnredlist.org/species/109473064/109473067>

Shea G, Venz M, Sanderson C, Hobson R, Wilson S (2018e) Plain-backed Sunskink. *Lampropholis couperi. The IUCN Red List of Threatened Species* 2018: e.T109473083A109473088. <https://www.iucnredlist.org/species/109473083/109473088>

Shea G, Cogger H, Greenlees M (2018f) Elongate Sunskink *Lampropholis elongata. The IUCN Red List of Threatened Species* 2018: e.T109473149A109473158. <https://www.iucnredlist.org/species/109473149/109473158>

Shea G, Cogger H, Greenlees M (2018g) Pale-flecked garden Sunskink. *Lampropholis guichenoti. The IUCN Red List of Threatened Species* 2018: e.T109473177A109473189. <https://www.iucnredlist.org/species/109473177/109473189>

Sinervo B, Mendez-de-la-Cruz F, Miles DB, Heulin B, Bastiaans E, Villagran-Santa Cruz M, Lara-Resendiz R, Martínez-Méndez N, Calderón-Espinosa ML, Meza-Lázaro RN, Gadsden H, Avila LJ, Morando M, De la Riva IJ, Sepulveda PV, Duarte Rocha CF, Ibargüengoytía N, Puntriano CA, Massot M, Lepetz V, Oksanen TA, Chapple DG, Bauer AM, Branch WR, Clobert J, Sites JW (2010) Erosion of lizard diversity by climate change and altered thermal niches. *Science* **328**(5980), 894–899. doi:10.1126/science/1184695

Singhal S, Hoskin CJ, Couper P, Potter S, Moritz C (2018) A framework for resolving cryptic species: a case study from the lizards of the Australian Wet Tropics. *Systematic Biology* **67**(6), 1061–1075. doi:10.1093/sysbio/syy026

Smith AH (2016) Invasion ecology of the plague skink (*Lampropholis delicata*) in Hawaii. MSc thesis, University of Hawaii, Manoa.

Smith GC, Fensham RJ, Ferguson D, Hogal L, Mathieson M (2015) Fauna of the grassland-forest landscape mosaics of the Bunya Mountains, eastern Australia. *Australian Zoologist* **37**(3), 302–310. doi:10.7882/AZ.2014.035

Stow AJ, Sunnucks P, Briscoe DA, Gardner MG (2001) The impact of habitat fragmentation on dispersal of Cunningham's skink (*Egernia cunninghami*): evidence from allelic and genotypic analyses of microsatellites. *Molecular Ecology* **10**(4), 867–878. doi:10.1046/j.1365-294X.2001.01253.x

Taylor JE, Fox BJ (2001a) Disturbance effects from fire and mining produce different lizard communities in eastern Australian forests. *Austral Ecology* **26**(2), 193–204. doi:10.1046/j.1442-9993.2001.01105.x

Taylor JE, Fox BJ (2001b) Assessing the disturbance impact on vegetation and lizard communities of fluoride pollution interacting with fire and mining in eastern Australia. *Austral Ecology* **26**(4), 321–337. doi:10.1046/j.1442-9993.2001.01117.x

Tingley R, Macdonald SL, Mitchell NJ, Woinarski JCZ, Meiri S, Bowles P, Cox NA, Shea GM, Böhm M, Chanson J, Tognelli MF, Harris J, Walke C, Harrison N, Victor S, Woods C, Amey AP, Bamford M, Catt G, Clemann N, Chapple DG (2019) Geographic and taxonomic patterns of extinction risk in Australian squamates. *Biological Conservation* **238**, 108203.

Triska MD, Craig MD, Stokes VL, Pech RP, Hobbs RJ (2016) The relative influence of in situ and neighborhood factors on reptile recolonization in post-mining restoration sites. *Restoration Ecology* **24**(4), 517–527. doi:10.1111/rec.1234

Twigg LE, Fox BJ (1991) Recolonization of regenerating open forest by terrestrial lizards following sand mining. *Australian Journal of Ecology* **16**(2), 137–148. doi:10.1111/j.1442-9993.1991.tb01041.x

Valdez JW (2021) Using Google Trends to determine current, past, and future trends in the reptile pet trade. *Animals* **2021**, 11, 676. doi:10.3390/ani11030676

Virkki DA, Tran C, Castley JG (2012) Reptile responses to lantana management in a wet sclerophyll forest, Australia. *Journal of Herpetology* **46**(2), 177–285. doi:10.1670/11-225

Wairepo J (2015) Developing biosecurity strategies for an invasive reptile, the plague skink (*Lampropholis delicata*) on Great Barrier Island. Master of Science in Conservation Biology thesis, Masssey University, Albany, New Zealand.

Walsh S, Goulet CT, Wong BBM, Chapple DG (2018) Inherent behaviour traits enable a widespread lizard to cope with urban life. *Journal of Zoology* **306**(3), 189–196. doi:10.1111/zo.12582

Webb GA (1994) Effects of logging on lizards in eucalypt forest at Eden, New South Wales. *Australian Forestry* **58**(4), 155–159. doi:10.1080/00049158.1995.10674660

Webb JK, Harlow PS, Pike DA (2015) Australian reptiles and their conservation. Chapter 17 in *Austral Ark: The State of Wildlife in Australia and New Zealand*. (Eds A Stow, N Maclean and I Holwell) pp. 354–381. Cambridge University Press, Cambridge.

Weber ET, Catterall CP, Locke J, Ota LS, Prideaux B, Shirreffs L, Talbot L, Gordon IJ (2021) Managing a World Heritage site in the face of climate change: a case study of the Wet Tropics in northern Queensland. *Earth* **2**(2), 248–271. doi:10.3390/earth2020015

Welbergen JA, Meade J, Storlie C, Van Der Wal C, Dalziell J, Hodgson AH, Larson L, Krockenberger A, Williams SE (2015) *Climate Change and the Impacts of Extreme Events on Australia's Wet Tropics Biodiversity. Report to the National Environmental Research Program*. Reef and Rainforest Research Centre, Cairns.

White AW, Burgin S (2004) Current status and future prospects of reptiles and frogs in Sydney's urban-impacted bushland reserves. In *Urban Wildlife: More than Meets the Eye*. (Eds D Lunney and S Burgin) pp. 109–123. Royal Zoological Society of New South Wales, Sydney.

Wikipedia (2025) South Cumberland Islands National Park. <https://en.wikipedia.org/wiki/South_Cumberland_Islands_National_Park>. (Accessed 21 May 2025).

Wolfe AK, Fleming PA, Bateman PW (2019) Surveying attitudes toward reptiles on roads: questionnaire responses do not directly relate to behavioural action. *Anthrozoös* **32**(3), 333–346. doi:10.1080/08927936.2019.1598652

World Bank (2020) *Urban Development*. <http://www.worldbank.org>

Wotherspoon D, Burgin S (2011) The impact on native herpetofauna due to traffic collision at the interface between a suburban area and the Greater Blue Mountains World Heritage Areas: an ecological disaster? In *Wildlife Response to Disasters*. (Eds C Dickman, D Lunney and P Banks) pp. 1040–1046. Royal Zoological Society of New South Wales, Sydney.

13 Genetics

KARYOLOGY

Donnellan (1991) investigated the karyotypes of all the then recognised species of *Lampropholis*. Of the original nine species, however, only five are still recognised as valid members of this genus. Using standard and C-banded karyotypes, he found that the diploid chromosome number differed between *L. delicata* ($2N = 28$) and the other species of *Lampropholis* that were investigated (*L. amicula, L. caligula, L. guichenoti, L. mirabilis*; $2N = 30$). All species had nine macrochromsome pairs, and either five (*L. delicata*) or six microchromosome pairs (all other species). Sampling from 56 individuals across the range of *L. delicata* revealed an exception to the above number of chromosomes: a single individual had $2N = 29$; this mutation was due to a Robertsonian arrangement (i.e. the breakage of two chromosomes at, or near, their centromeres and the subsequent fusion of the two long arms and the loss of the short segments of the original acrocentric chromosomes).

Lampropholis delicata had a variable chromosomal format at the infraspecific level, with individuals from the core of its range heterozygous for five different chromosomal variants. These variants are distinct from each other and from those of other species of *Lampropholis*. Although Donnellan did not make further comment, he may have been dealing with multiple species since at that time there still were some undescribed species in the genus (e.g. Greer 1974; Mather 1990).

GENETIC AND MOLECULAR ANALYSES

A main pillar of evolution is the change in the genetic constitution of organisms under the aegis of natural selection. Individuals that are genetically best suited to prevailing local conditions have enhanced survival there and contribute their favourable genes to the next generation proportionately more than do the lesser-adapted segment of the population. Over time this leads to populations increasingly becoming better adjusted to local conditions as individuals with less adapted genomes become selectively eliminated. As environmental conditions change, or as populations expand their geographic range into areas with different physical environments, or a different suite of competitors, prey, and predators, a genetic constitution that was advantageous previously may no longer be the most adaptive one and the direction of selection may change. The operative conditions may vary geographically or temporally; they may be naturally occurring or they may be brought on by human modification of the environment, even over relatively short intervals. For example, Harris *et al.* (2023) concluded that anthropogenic changes in the deployment of controlled burning, whether differing temporally, spatially, or in intensity, can have genetic consequences at the individual, populational, and meta-populational scale for endemic animals, including reptiles.

Apart from characteristics that are determined by genes, variation in other attributes may be modified by the environment (see Banks *et al.* 2007) and, of course, there may be interaction between the relative effects of genetics and ecology on the final constitution of the individual. Especially with the advent of climatic change and the need to offset its effect on biodiversity by instituting appropriate protocols for management and conservation, the need to assess the heritability of the key traits of threatened species is paramount.

The study of mitochondrial DNA (mtDNA) has revolutionised the study of the evolutionary history of animals.

It provides a means for constructing hypotheses at the intraspecific level. The rapid substitution of mtDNA nucleotide, combined with maternal non-recombining of its inheritance, offers advantages for phylogenetic analysis at the microevolutionary level. Avise *et al.* (1987) pointed out a linkage between taxonomy and population genetics and further suggested that mtDNA constitutes a basis for thinking about the phylogeography of genes and noted that intraspecific phylogeography should assume a place in evolutionary biology at least commensurate with that of ecogeography, with mutual benefit resulting to both disciplines and that theories of speciation and macroevolution must now recognise and accommodate the reality of phylogeographic differentiation at the intraspecific level.

Avise *et al.* (1987) presented three phylogeographic hypotheses motivated by the mtDNA data: (1) Most species are composed of geographic populations whose members occupy different branches of an intraspecific, phylogenetic tree; (2) Species with limited phylogeographic population structure have life histories conducive to dispersal and have occupied ranges free of firm impediments to gene flow; and (3) Monophyletic groups distinguished by large phylogenetic gaps usually arise from long-term extrinsic (zoogeographic) barriers to gene flow.

Genetic and molecular analysis of *Lampropholis*

Harris and Johnston (1977) undertook to define the species *L. delicata* and *L. guichenoti* and to find ways to distinguish between the two. They discovered that there was overlap in most morphological features and details of the colour pattern but of those the colour pattern was somewhat better to differentiate between species but was not infallible. They then subjected five enzymes from livers of both species to electrophoresis and two of them, phosphoglucomutase and esterase, showed different electrophoretic patterns that clearly indicated that these two entities are maintaining separate gene pools and that they are reproductively isolated – hence they are different species.

Martins *et al.* (2019) investigated the heritabilities and genetic correlations of four morphological and six physiological traits in *L. coggeri*. They found high heritability of the morphological traits of mass, total length, and snout-to-vent length of hatchlings and of growth rates (but see Chapter 9 for an account of the effect on these characteristics by the temperatures under which embryos were incubated). Growth rate also was highly heritable. By contrast the physiological effects related to temperature (CTMin, CTMax, thermal optimum for sprinting, the maximum speed of sprinting at the thermal optimum, and the range of temperatures over which the lizards speed of sprinting was more than 80% of its maximum speed) had low to moderate heritability ($h^2 < 0.31$). This result is not surprising as many physiological processes adjust via acclimation to a range of rising and lowering temperatures, thereby attaining a degree of adaptive flexibility in coping with environmental change (see Chapter 6). The resistance to desiccation, however, had a higher, and significant, value ($h^2 > 0.51$). The authors concluded that their results indicated 'little potential [for this species] to respond to selection on thermal traits in this population and provide new insights into the capacity of tropical ectotherms to adapt *in situ* to rapid climate change'.

An early molecular analysis of the relationship within *Lampropholis* was conducted by Baverstock and Donnellan (1990). They used microcomplement fixation of albumen to assess the relationships of Australian agamids and skinks. Four species that, at that time were considered to be in the genus *Lampropholis*, as well as those of several other genera all radiated from each other ~10 million years ago (phylogenetic tree in their Fig. 2). This group exhibited a disparity between greater molecular evolution than morphological evolution. Two of the putative *Lampropholis* in their study, *L. basiliscus* and *L. challengeri*, later were designated as members of the closely related genus *Saproscincus* (see Chapter 14).

Genetic analysis also has been used to define the relationships between *Lampropholis* and its closely related genera (Smith *et al.* 2007; Brandley *et al.* 2015; Chapple *et al.* 2023; see Chapter 14) as well as confirm infraspecific lineages and cryptic species within this genus (Chapple *et al.* 2011a, b; Singhal *et al.* 2018). The former study used mtDNA (ND2) and nuclear (RAG-1, c-mos) DNA to define three lineages of Australian Eugongylini and to construct a phylogenetic tree of the Australian genera; their study also confirmed that the origin of this tribe was Asia with dispersal from there to Africa, Australasia, and Oceania.

Because the mutation of DNA, either by alteration of sequences of nucleotides or by damages or errors in its replication, varies across species depending on internal factors such as metabolic rate and external ones such as environmental stresses like temperature, Ivan *et al.* (2021) assembled data on these causative agents impinging on species of *Lampropholis* and other taxa of eugongylines and used that information to construct a phylogenetic tree of that taxon (for the subset of *Lampropholis* see Fig. 14.3 in Chapter 14). They used a large set of exon-sequence data spanning a large latitudinal gradient. The association between temperature (the strongest positively correlated predictor of variation in substitution rate) and other traits was assessed on the basis of 44 pairs of species using analyses of regression. Forty-five per cent of variation in substitution rate was explained, with 11% being left after controlling for all other factors. Synonymous rate of substitution was negatively associated with body size, with 6% of the variation being explained by this factor after controlling for the effects of temperature. Other factors were not associated with synonymous substitution rate after controlling for temperature.

For a discussion of genetically coded polymorphism in the colour pattern that bestows protection of *Lampropholis* against predation by snakes, see Chapter 9.

Lampropholis delicata

Chapple and Thompson (2009) assessed the genetic variation in 12 polymorphic microsatellite loci for *L. delicata* from Sydney, eastern Victoria, Tasmania, and Lord Howe Island and found that the number of alleles per locus ranged from 2 to 23. The heterozygosity was high (Sydney: 0.579–0.895, Victoria–Tasmania: 0.080–0.875), and the level of genetic variation was generally greater in the Sydney population than in the Victoria–Tasmania region.

Chapple *et al.* (2011a) collected tissue samples from 238 specimens of *L. delicata* from which they extracted total genomic DNA, sequenced portions of four mitochondrial genes, and estimated the genetic diversity (number of haplotypes, haplotypic diversity, number of polymorphic sites, and nucleotide diversity). Their samples comprised 2,426 characters. They identified nine clades with 1–4 subclades per clade, for a total of 18 subclades (see Table 13.1A and Chapters 2 and 14) and suggested that others may need to be investigated in more detail. Isolation by the identified barriers accounted for 66.5% of the total variation within the species, a value that was elevated to 97.7% by inclusion of the diversity of all individual populations within those clades.

The molecular diversity was high among the clades (Table 13.1) and comprised 165 unique haplotypes. The substantial genetic distances among the clades were statistically significant except for a few cases that were based on small sample sizes. Divergences relate to past and present ecological conditions and the time and location of the imposition of barriers to gene flow between populations, such as corridors of dry habitat between what are now subclades and clades (see Fig. 14.1 in Chapter 14); for example, the Kroombit Tops and Blackdown Tableland, inland from the main distributional range of this species. The genetic divergences relate to past and present ecological conditions and the time and location of barriers to gene flow between what are now subclades and clades (see Fig. 14.1 in Chapter 14).

The accidental introduction of *L. delicata* into New Zealand from different localities in Australia brought populations into contact that had been separated in their native range for more than 2 million years. Chaplin *et al.* (2022) analysed data on mitochondrial DNA and microsatellites to assess the extent of hybridisation among these recently united populations. They found that hybridisation occurred among all the populations originating from the four different sources and concluded that divergence among them had not reached the level of speciation.

Often the established population in a new country is founded by very few individuals and hence the full range of the genetic diversity present in the donor population in the native country is reduced (founder effect) and consequently it would be expected that the immigrant population would be less versatile and able to compete with the native species in the new land. If, however, the introduced population subsequently is supplemented by the introduction of additional migrants from the country of origin, especially if from populations other than that of the original immigrants (admixture), then the competitive ability of the immigrant population should be enhanced. Chapple *et al.* (2013) tested this hypothesis by using sequences of mitochondrial DNA to compare the genetic distances and molecular variances of *L. delicata* across 120 populations across its native range in Australia with 371 samples from its introduced populations in Hawaii, New Zealand, and Lord Howe Island. There were five identified Australian sources of the introduced populations (Brisbane, Tenterfield, Border Ranges, Yamba–Coffs Harbour, and Sydney) representing four distinct Australian genetic lineages. A single introduction from Brisbane founded the Hawaiian populations, one or more introduced populations in New Zealand originated from Tenterfield, and multiple introductions from all five donor regions founded populations on Lord Howe Island.

Chaplin *et al.* (2022) found 14 haplotypes in the introduced range (one in Hawaii, seven in New Zealand, and

Table 13.1. Estimated genetic diversity of the various clades and subclades of *Lampropholis delicata* (Species A) and clades and lineages of *Lampropholis guichenoti* (Species B).

Data from Chapple *et al.* (2011a, b).

Clade Subclade	Sample size	Number of haplotypes	Haplotypic diversity	Mutations	Segregating sites	Nucleotide diversity
			A. ***Lampropholis delicata***			
1	**14**	**13**	**0.987**	**191**	**187**	**0.026**
a	2	2	1.0	38	38	0.016
b	3	3	1.0	7	7	0.002
c	9	8	0.972	19	19	0.003
2	**8**	**7**	**0.893**	**12**	**12**	**0.001**
3	**19**	**15**	**0.965**	**170**	**164**	**0.022**
a	4	4	I	4	4	0.001
b	3	1	0	0	0	0
c	10	8	0.933	54	53	0.008
d	2	2	1.0	4	4	0.002
4	**48**	**43**	**0.996**	**213**	**206**	**0.017**
a	29	24	0.998	119	119	0.009
b	14	14	1.0	30	30	0.003
c	5	5	1.0	47	47	0.011
5	**18**	**13**	**0.948**	**87**	**87**	**0.013**
a	11	7	0.873	18	18	0.002
b	7	6	0.952	35	35	0.006
6	**4**	**2**	**0.5400**	**1**	**1**	**<0.001**
7	**55**	**29**	**0.937**	**127**	**123**	**0.005**
a	3	3	1.0	5	5	0.001
b	52	26	0.930	68	68	0.002
8	**2**	**1**	**0**	**0**	**0**	**0**
9	**70**	**42**	**0.943**	**205**	**201**	**0.023**
a	24	16	0.913	48	48	0.005
b	1	1	Not applicable	Not applicable	Not applicable	Not
c	9	8	0.972	17	17	applicable
d	36	17	0.821	24	24	0.003
						0.002
Overall	**238**	**16 5**	**0.990**	**715**	**625**	**0.044**
			B. ***Lampropholis guichenoti***			
Northern lineage (N)						
N1	**3**	**2**	**0.667**	**11**	**11**	**0.011**
N2	**18**	**10**	**0.922**	**27**	**26**	**0.010**
N3	**3**	**2**	**0.667**	**3**	**3**	**0.003**
N4	**2**	**2**	**1.000**	**4**	**4**	**0.006**
N5	**2**	**2**	**1.000**	**3**	**3**	**0.004**
Overall	**28**	**18**	**0.963**	**110**	**104**	**0.035**
Southern lineage (S), Sublineage A						
S1	**34**	**23**	**0.964**	**51**	**48**	**0.014**
S2	**18**	**14**	**0.954**	**31**	**31**	**0.010**
S3	**6**	**5**	**0.933**	**17**	**17**	**0.011**
Overall	**58**	**42**	**0.983**	**94**	**88**	**0.028**

Clade Subclade	Sample size	Number of haplotypes	Haplotypic diversity	Mutations	Segregating sites	Nucleotide diversity
Southern lineage (S), Sublineage B						
S4	7	7	1.000	27	27	0.013
S5	16	5	0.683	4	4	0.002
S6	10	8	0.933	33	33	0.014
S7	4	3	0.833	3	3	0.002
Total (S) B	37	23	0.937	85	75	0.027
S overall	95	65	0.984	154	133	0.047
Lampropholis guichenoti overall	123	83	0.989	232	193	0.074

seven on Lord Howe Island). Contrary to expectations, *L. delicata* was successful in sustaining established populations either with admixture or without it. These authors concluded that the expected advantage of admixture probably has been overemphasised.

Lampropholis guichenoti

Chapple *et al.* (2011b) carried out an investigation on *L. guichenoti* similar to their study of *L. delicata*. They collected tissue samples from 123 specimens representing 64 populations. Two separate genetic lineages were found, a northern one with five clades in south-eastern Queensland and north-eastern New South Wales (NSW) and a southern one with seven clades from south-eastern NSW, the Australian Capital Territory, Victoria, and south-eastern South Australia (Table 13.1B; see Fig. 14.2 in Chapter 14). Paired comparisons between lineages, clades, and subclades were substantial and except when samples sizes were too small for reliable testing, were statistically significant. The genetic divergence between the northern and southern lineages was 12.9–16.5%, with the number of mutations high. Despite the relatively restricted distribution of the northern lineage (see Chapter 14) the mean genetic distance between *L. guichenoti* clades was equivalent to that of the more widespread southern lineage (again except when samples were too small for adequate testing). The mean divergence between the two sublineages of the southern lineage was 7.1%.

Lampropholis coggeri and *Lampropholis robertsi*

These two species are morphologically almost identical, genetically closely related, and both are endemics in montane tropical rainforests in north-eastern Queensland (Bell *et al.* 2010; and see Chapters 2 and 14). Such species are especially sensitive to climatic fluctuations because their narrow thermal tolerances and elevational distributions minimise their ability to either occupy, or disperse across, other types of habitats. Consequently, both *L. coggeri* and *L. robertsi* occur in multiple isolated centres of genetic diversification in uplands during the current inclemently warm conditions, from which refuges they can expand into lowland areas during more favourable, cooler conditions such as those occurring during the glacial periods of the Pleistocene (reviewed by Bell *et al.* 2010). Their local endemism is likely due to their long-term persistence and the divergence between isolated populations on the tops of mountains.

Both of these species exhibit pronounced phylogeographic structuring of mitochondrial and nuclear genes, thereby indicating low capacity for dispersal and high persistence across multiple isolated regions. These characteristics are more prominent in *L. robertsi*, however, than in *L. coggeri*. Spatial modelling of *L. robertsi* predicted strong isolation among disjunct montane refuges during warm climates but with potential for localised genetic exchange during the most recent glacial period. In contrast *L. coggeri* is more widespread and its populations are more connected (Bell *et al.* 2010; and see Chapter 2).

Much of the research on genetics was directed toward its evolutionary consequences for taxonomy or adaptation to the physical or biological context (see Chapters 11 and 14 for a further exposition of this topic.

Lampropholis coggeri consists of two lineages that are phenotypically the same, yet genetically distinct (Singhal and Moritz 2012). They formed when the rainforest that they inhabited contracted into two refugia during the glaciation of the Late Quaternary but ~8,000 to 2,000 years ago the rainforest expanded again and joined the two lineages. At the junction they now meet in a narrow zone (mean width only 403 m) in which they hybridise. Singhal and Moritz (2012) compared the morphological phenotypes of these two lineages and assessed hybridisation and introgression in this zone. They found no morphological differences between the two lineages. Their analyses suggested that individuals in the zone of contact consisted of two parental lineages, each as a genetic cluster, and the hybrids between them as an admixture between them. Clines across the contact zone had coincident centres within 100 m of each other. At the central locations 55% of the individuals were hybrids, of which 65% had mitochondrial types of the northern lineage and 35% had those of the southern lineage.

After reviewing several possible hypotheses for the results they obtained Singhal and Moritz (2012) suggested selection against hybrids to be paramount in maintaining such narrow clines in the zone of contact.

Lampropholis similis

This species is reliably distinguished from both *L. coggeri* and *L. elliotensis* by 17 nucleotide differences in the mitodhondrial gene NADH dehydrogenase 4, that, in turn, results in 15 differences in amino acids (Singhal *et al.* 2018).

Lampropholis bellendenkerensis and *Lampropholis robertsi*

These two species, although sharing a similar morphology, differ genetically by 13 nucleotides in their mitochondrial gene NADH dehydrogenase 4 that results chemically in nine differences in amino acids between them (Singhal *et al.* 2018).

Lampropholis isla and some other members of the genus

Sequencing of DNA by Hoskin (2025) allowed assessment of the phylogenetic relationships of *L. isla* with other species of *Lampropholis* as well as the genetic relationships of various other species in the genus whose genetic affinities had not been examined previously: *L. elliotensis, L. similis, L. bellendenkerensis*, and *L. adonis* (compare Fig. 13.1 below with the right-hand diagram in Fig. 14.3 in Chapter 14).

GENETICS VERSUS PHENOTYPIC PLASTICITY

Animals adapt to their environments in two ways, either through: (1) evolutionary adaptation to specific local conditions under the influence of natural selection, or (2) as a result of physiological flexibility involving either broad tolerances to environmental conditions, temporal (e.g. seasonal) shifts of more narrow tolerances, or responses that track changing environmental conditions through acclimation (see Chapter 6). Both types of responses are, of course, ultimately determined by the genetic constitution of the animal (Macdonald *et al.* 2017). Once a species' geographic range is fragmented into isolated patches,

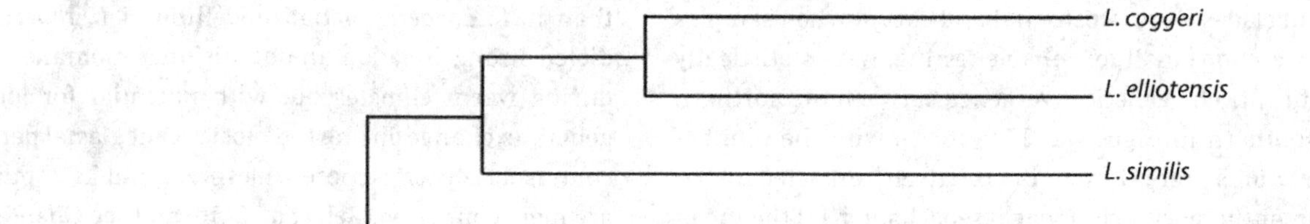

Fig. 13.1. Phylogeny of rainforest *Lampropholis* species based on approximately 850 bp ND4 mtDNA, with *L. delicata* used as the outgroup. Simplified by Susan King based on Fig. 2 in Hoskin (2025).

each patch can diverge evolutionarily in conformity to the nuances of environmental differences between patches. The extent of such phenotypic divergence will depend not only on the degree of difference between the environments of the patches (connectivity) but also on the extent of gene flow and exchange of individuals between patches. Macdonald *et al.* (2017) studied the climatic variables of temperature and precipitation in patches of habitat of *L. coggeri* and used connectivity as a surrogate for gene flow. They concluded that physiological traits showed stronger evidence of local adaptation than did morphological traits, e.g. both annual mean precipitation and precipitation of the driest quarter of the year correlated with connectivity (see Chapter 6 for results on CTMax and CTMin) and that gene flow impedes adaptation.

REFERENCES

Avise JC, Arnold J, Ball RM, Bermingham E, Lamb T, Neigel JE, Reeb CA, Saunders NC (1987) Intraspecific phylogeography: the mitochondrial DNA bridge between population genetics and systematics. *Annual Review of Ecology and Systematics* **18**, 489–522. doi:10.1146/annurev.es.18.110187.002421

Banks AC, Piggott MP, Stow AJ, Taylor AC (2007) Sex and sociability in a disconnected world: a review of the impacts of habitat fragmentation on animal social interactions. *Canadian Journal of Zoology* **85**, 1065–1079. doi:113910/Z07-094

Baverstock PR, Donnellan SC (1990) Molecular evolution in Australian dragons and skinks: a progress report. *Memoirs of the Queensland Museum* **29**(2), 323–331.

Bell RC, Parra JL, Tonione M, Hoskin CJ, Mackenzie JB, Williams SE, Moritz C (2010) Patterns of persistence and isolation indicate resilience to climate change in montane rainforest lizards. *Molecular Ecology* **19**(12), 2531–2544. doi:10.1111/j.1365-294X.2010.04676.x

Brandley MC, Bragg JG, Singhal S, Chapple DG, Jennings CK, Lemmon AR, Lemmon EM, Thompson MB, Moritz C (2015) Evaluating the performance of anchored hybrid enrichment at the tips of the tree of life: a phylogenetic analysis of Australian *Eugongylus* group scincid lizards. *BMC Ecology and Evolution* **15**(1), 62. doi:10.1186/s12862-015-0318-0

Chaplin K, Smith Date K, Bray RD, Miller KA, Lutz ML, Razeng E, Thompson MB, Chapple DG (2022) Intraspecific hybridisation of an invasive lizard on Lord Howe Island. *Australian Journal of Zoology* **69**(5), 184–196. doi:10.1071/ZO21045

Chapple DG, Thompson MB (2009) Isolation and characterization of microsatellite loci from the invasive delicate skink (*Lampropholis delicata*), with cross-amplification in other Australian *Eugongylus* group species. *Conservation Genetics Resources* **1**, 55–58. doi:10.1007/s12686-009-9013-z

Chapple DG, Hoskin CJ, Chapple SNJ, Thompson MB (2011a) Phylogenetic divergence in the widespread delicate skink (*Lampropholis delicata*) corresponds to dry habitat barriers in eastern Australia. *BMC Evolutionary Biology* **11**(1), 191. doi:10.1186/1471-2148-11-191

Chapple DG, Chapple SNJ, Thompson MB (2011b) Biogeographic barriers in south-eastern Australia drive phylogeographic divergence in the garden skink, *Lampropholis guichenoti*. *Journal of Biogeography* **38**(9), 1761–1775. doi:10.1111/j.1365-2699.2011.02531.x

Chapple DG, Miller KA, Kraus F, Thompson MB (2013) Divergent introduction histories among invasive populations of the delicate skink (*Lampropholis delicata*): has the importance of genetic admixture in the success of biological invasions been overemphasized? *Diversity and Distributions* **19**(2), 134–146. doi:10.1111/j.1472-4642.2012.00919.x

Chapple DG, Chapple SNJ, Smith SA, Shea GM, Brennan IG, Sadlier RA (2023) Phylogenetic relationships in the Eugongylini (Squamata: Scincidae): generic limits and biogeography. *Australian Journal of Zoology* **70**(6), 165–203. doi:10.1071/ZO23007

Donnellan SC (1991) Chromosomes of Australian lygosomine skinks (Lacertilia: Scincidae), II: The genus *Lampropholis*. *Genetica* **83**, 223–234. doi:10.1007/BF00126228

Greer AE (1974) The generic relationships of the scincid lizard genus *Leiolopisma* and its relatives. *Australian Journal of Zoology, Supplementary Series* **22**(13), 1–67.

Harris BF, Johnston PG (1977) Electrophoretic evidence for the specific status of the lizards *Lampropholis guichenoti* (Dumeril & Bibron, 1839) and *L. delicata* (de Vis, 1887). *Australian Zoologist* **19**(2), 149–154.

Harris J, Chapple DG, McLean CA, Melville J (2023) Genetic footprints of fire: understanding the genetic implications of fire regimes for fauna and effective conservation strategies. *Biological Conservation* **284**, 110169. doi:10.1016/j.biocon.2023.110169

Hoskin CJ (2025) A new species of *Lampropholis* skink (Lacertilia: Scincidae) from Scawfell Island, mid-east Queensland, Australia. *Australian Journal of Taxonomy* **85**, 1–13. doi:10.54102/ajt.6uihp

Ivan J, Moritz C, Potter S, Bragg J, Turakulov R, Hua X (2021) Temperature predicts the rate of molecular evolution in Australian Eugongylinae skinks. *Evolution* **76**(2), 252–261. doi:10.1111/evo.14342

Macdonald SL, Llewelyn J, Moritz C, Philips BL (2017) Peripheral isolates as sources of adaptive diversity under climate change. *Frontiers in Ecology and Evolution* **5**, doi:10.3389/fevo.2017.00088

Martins F, Kruuk L, Llewelyn J, Moritz C, Phillips B (2019) Heritability of climate-relevant traits in a rainforest skink. *Heredity* **122**, 41–52. doi:10.1038/s41437-018-0085-y

Mather P (1990) Electrophoretic and morphological comparisons of *Lampropholis delicata* (Lacertilia: Scincidae) populations from Eastern Australia, and a resolution of the taxonomic status of this species.

Australian Journal of Zoology **37**(5), 561–574. doi:10.1071/ZO9890561

Singhal S, Moritz C (2012) Strong selection against hybrids maintains a narrow contact zone between morphologically cryptic lineages in a rainforest lizard. *Evolution* **66**(5), 1474–1489. doi:10.1111/j.1558-5646.2011.01539.x

Singhal S, Hoskin CJ, Couper P, Potter S, Moritz C (2018) A framework for resolving cryptic species: a case study from the lizards of the Australian Wet Tropics. *Systematic Biology* **67**(6), 1061–1075. doi:10.1093/sysbio/syy026

Smith SA, Sadlier RA, Bauer AM, Austin CC, Jackman T (2007) Molecular phylogeny of the scincid lizards of New Caledonia and adjacent areas: evidence for a single origin of the endemic skinks of Tasmantis. *Molecular Phylogenetics and Evolution* **43**, 1151–1166. doi:10.1016/j.ympev.2007.02.007

14 Taxonomic and evolutionary history

INTRODUCTION

Phylogeny is the evolutionary history of a group of organisms (Toothill 1988) and it subsumes a number of other disciplines such as palaeontology (the study of extinct species including their fossil remains and impressions left by them in rocks), systematics (the study of the diversity of organisms and their relationships with each other), taxonomy (the naming and classification of organisms into groups that reflect their genetic affinities), morphology (the study of the anatomical structure of organisms), and genetics (the study of inheritance). These topics are treated in the present book only in relation to their immediate relationship to the genus *Lampropholis*. For example, there are no fossils known of this genus, so palaeontology is not included herein despite it being very important in the history of reptiles generally. Some topics are relevant to chapters in addition to the present one and consequently cross-referencing occurs (e.g. genetic analyses of the various species of *Lampropholis* as it relates to their taxonomy is covered in Chapter 13).

DEFINING THE GENUS *LAMPROPHOLIS*

Encompassing ~1,745 described species and ~460 taxa spanning 46 genera (Chapple *et al.* 2023), Family Scincidae constitutes the largest and most widely distributed taxon of extant lizards (Greer 1989; Uetz *et al.* 2019; Shea 2021). It is not surprising that the history of such an exceptionally rich and diverse family has been tortuous and complicated by many uncertainties about homologies and relationships. Species have been allocated to various species groups, subgroups, complexes, subdivisions, and subfamilies (e.g. see Smith 1937; Mittleman 1952; Greer 1979, 1989; Greer and Kluge 1980; Ingram and Rawlinson 1981; Couper *et al.* 2006; Shea 2021).

The present book treats one taxon of these skinks, the genus *Lampropholis*. This genus is now recognised as comprising a monophyletic core of closely related species. Although the genus previously shared in the taxonomic confusion at the generic and specific level, its familial relationship (Scincidae) has never been disputed. Three major tribes within this family are recognised as occurring within Australia: Sphenomorphini, Eugongylini, and Tiliquini; *Lampropholis* belongs in the Eugongylini (Chapple *et al.* 2023).

Within the Subfamily Lygosominae (*sensu lato*), a group of genera are now recognised as belonging to the Tribe Eugongylini (see Chapple *et al.* 2023); some authors even raise that tribe to the level of a subfamily (Eugongylinae) on a par with Lygosominae (e.g. see Ivan *et al.* 2021).

In summary, the hierarchical classification of the genus *Lampropholis* is: Family Scincidae; Subfamily Lygosominae (*sensu lato*); Tribe Eugongylini; genus *Lampropholis*. Some consider the Tribe Eugongylini to merit even higher status as the Subfamily Eugongylinae on a par with Lygosominae (*sensu stricto*), e.g. Ivan *et al.* (2021), with the alternative classification of Family Scincidae, Subfamily Eugongylinae, genus *Lampropholis*.

At the generic level, Fitzinger (1843) erected the genus *Lampropholis* and designated *L. guichenoti* as the type species. This species was originally described as *Tiliqua trilineata* and *Tiliqua duperreyi* from Australia and New Zealand by Gray (1838) and as *Lygosoma guichenoti* in Australia as the lectotype by Duméril and Bibron (1839). Later Gray (1845) designated *Lygosoma guichenoti* as the lectotypic species of his genus *Mocoa*. These early descriptions of species either in part (*duperreyi*;

presumably separating the Australian species from those of New Zealand), or in full (the rest of the species) were synonymised under the name *Lampropholis guichenoti* (see Cogger *et al.* 1983). *Lampropholis delicata* was described originally as *Mocoa delicata* by De Vis (1888). All other species now included in the genus *Lampropholis* were described in 1981 or more recently, and all were directly allocated to the genus *Lampropholis* by their original describers.

Boulenger (1885–1887) followed by Loveridge (1934), Smith (1935), and McCann (1955) all commented on the significant overlap in the quantitative features used to separate the genera *Lampropholis* and *Leiolopisma*. Cogger *et al.* (1993) reported that P. A. Rawlinson planned to designate the lectotype of the species *L. guichenoti* as originating from Kangaroo Island, off the South Australian coast, rather than the more generic term 'Australia'. It appears that this intention was not carried out.

In reviewing the genus *Lygosoma*, Smith (1937) separated the Australian *Leiolopisma* into two complexes, one group based on fused frontoparietal scales and another based on two distinct frontoparietal scales. Mittleman (1952) formalised this division by designating the first complex to *Lampropholis* (*L. ocellata*, *L. metallica*, *L. pretiosa*, *L. trilineata*, and *L. guichenoti*) and the latter group to *Leiolopisma* (*L. entrecasteauxi*, *L. mustelina*, *L. challengeri*). Most subsequent workers ignored these divisions and regarded all these species as *Leiolopisma* (e.g. see Clarke 1965; Greer 1970; Pengilley 1972). Subsequently, the genus was recognised as being based on plesiomorphic characters and, as such, were poorly defined (Greer 1974).

To rectify this situation, Greer (1974) reinvestigated the genus *Leiolopisma* using osteological and morphological characters. He resurrected and redefined the genus *Lampropholis* as follows: 'small (snout-vent length up to 62 mm) Australian beta skinks with five toes on the front foot, prefrontals and a moveable lower eyelid with a clear transparent window; clutch size variable, up to 10. Frontoparietals distinct (2 species) or fused (2 species)' (Greer 1974, p. 25).

The species that originally were assigned to the genus *Lampropholis* were *Lampropholis challengeri* (Boulenger 1885–1887), *Lampropholis delicata* (De Vis 1888), *Lampropholis guichenoti* (Duméril and Bibron 1839), and *Lampropholis mustelina* (O'Shaughnessy 1874). Greer (1974) reported that he also knew of 11 or 12 additional undescribed species; however, it is not clear what overlap there was within and between the undescribed species identified by numerous other authors (e.g. Schuster 1981; Mather 1986, 1989a, b; Burgin 1989; Donnellan 1985, 1991; Mather and Hughes 1992), some of which have been described as distinct species subsequently. In addition, Greer and Kluge (1980) described *L. tetradactyla*; almost certainly it was among those later described by Ingram and Rawlinson (1981) who proposed that the genus consisted of two complexes, the *Challengeri* Complex and the *Delicata* Complex.

In 1980, Greer and Kluge suggested that the two complexes within *Lampropholis* were worthy of generic rank and subsequently Wells and Wellington (1983, 1985) elevated the two complexes to that level (Table 14.1). The *Challengeri* Complex thus disappeared, leaving it unnecessary to retain the status of 'Complex' for the remaining species in the genus *Lampropholis*.

Hutchinson (1983), using immunoelectrophoretic comparisons, showed that *Lampropholis* (*sensu* Wells and Wellington 1985) were allied with the genera *Carlia* and *Menetia* (i.e. Greer's *Pseudemoia* Subgroup) while *Saproscincus* was more closely allied with the *Leiolopisma–Morethia* radiation (i.e. Greer's *Eugongylus* Subgroup). Hutchinson's molecular-level data therefore supported the separation of the two groups at the generic level.

The work of Joss and Minard (1985) also supported the supposition that at least one of the species relegated to the *Challengeri* Complex was not closely related to other members of the genus (*sensu* Greer 1974). They studied ovarian steroid production in the genus and observed that there was a similar pattern of synthesis of steroid hormones in *L. guichenoti* and *L. delicata*, both from the former *Delicata* Complex (*sensu* Greer 1974), but that *Saproscincus mustelinus* (formerly *L. mustelina*) and *Saproscincus challengeri* (formerly *L. challengeri*) from the former *Challengeri* Complex (*sensu* Greer 1974) showed a very different pattern. The former pair of species synthesised steroids in thecal glands prior to ovulation; however, *S. challengeri* had higher concentrations of 17β-estradiol and greater abundance of thecal gland tissue than either of the other species. It also had a very different pattern of steroid-hormone synthesis. The species, *S. mustelinus*, was not observed to have thecal gland tissue, and steroid levels were low and confined to vitellogenic follicles. Joss (pers. comm.) suggested that these differences were of such magnitude that she suspected *S. challengeri* to be a quite distant relative of the other species she investigated. Burgin (1989) found that there was substantial molecular differentiation within and

Table 14.1. Species once allocated to the genus *Lampropholis* but later found to belong to a different genus.

Compiled from The Reptile Database (nd).

Original name	Authority	Present name
Lampropholis basiliscus Ingram and Rawlinson 1981	Ingram and Rawlinson (1981)	*Saproscincus basiliscus* Wells and Wellington 1985
Lampropholis challengeri (Boulenger 1885–1887)	Cogger *et al.* (1983)	*Saproscincus challengeri* Sadlier, Colgan, and Shea 1993
Lampropholis czechurai Ingram and Rawlinson 1981	Ingram and Rawlinson (1981)	*Saproscincus czechurai* 1985
Lampropholis mustelina (O'Shaughnessy 1874)	Cogger *et al.* (1983)	*Saproscincus mustelinus* Wells and Wellington 1985
Lampropholis tetradactyla Greer and Kluge 1980	Greer and Kluge (1980)	*Saproscincus tetradactylus* Wells and Wellington 1985

between some species in the former complexes of *Lampropholis* that supported the observation that *S. mustelinus* was a distant relative of the other two species.

Burgin (1989) investigated the evolutionary relationships of *Lampropholis* (*sensu* Greer and Kluge 1980) using morphological and meristic characterisation, and molecular techniques current at the time. Predictably when using a range of analyses, there was some variation; however, the results provided very similar topologies. Although she selected *L. caligula* and *L. amicula* as the 'bench-mark' for separating species, she acknowledged that, alone, the 10% fixed difference in allozymes between the two species was too low to use for separation at the species level; however, other available data, such as current taxonomic placement in different complexes within the genus based on morphological data (Greer and Kluge 1980), supported that separation. Chromosomal variation (Donnellan 1985); molecular data (Hutchinson 1983); and data on range and habitat (Cogger 1983) subsequently indicated that *L. caligula* and *L. amicula* were valid species (see Cogger 2018). Burgin (1989) also provided an alternative, much more conservative, topology. Although the former provided greater fragmentation of the genus, even the most conservative outcome indicated that the '*Lampropholis* Complex' was a distinct genus with several undescribed species. Later the two complexes within the genus *Lampropholis* (*sensu* Greer and Kluge 1980) were raised to generic status (*Lampropholis* and *Saproscincus*) by Wells and Wellington (1985) (Table 14.1). On the basis of the three-locus mito-clear dataset (ND2, Rag-1, cmos) estimated using CMCree, Chapple *et al.* (2023) constructed a tree showing *Saproscincus* as distinct from *Lampropholis* but as the genus most closely related to it.

In addition to the species of *Lampropholis* listed by Cogger (2018) and more recently in the IUCN Red List, three additional morphologically cryptic species were formally described by Singhal *et al.* (2018) (see section *Cryptic species* below) and there are now 15 species within the genus *Lampropholis* that are recognised as valid by The Reptile Database (nd; and including the newly described species *L. isla* (Hoskin 2025)). Of these species, *L. elongata* was listed under the *Environment Protection and Biodiversity Conservation Act 1999* (EPBC Act) as Critically Endangered in November 2023. The factors listed as underpinning the listing included the species' very restricted geographic distribution in a location that is prone to frequent and severe fire, together with a continuing decline in abundance due to the loss of the species' declining niche (DCCEEW 2023). Coincidently, Graham *et al.* (2023) rediscovered the species after what had been 50 years of 'Data Deficiency'. Their findings suggested that *L. elongata* were relatively easily detected using active survey methods. They stated that they did not consider the species to be endangered although they commented that, in the future, 'forestry activity' may be an issue.

Lampropholis guichenoti and *L. delicata* are very similar in size, morphology, and general appearance and usually are not recognised as separate species by the public even when both occur together syntopically. The traditional common name (penny lizard) was shared indiscriminately by both species. Cogger's key (see Box 14.1 below) separates them based on differences in colour pattern and they generally are distinct (see Burgin 1989). The other species of the genus may be recognised by their geographic location, a combination of differences in scutellation and colour pattern, and/or genetic constitution.

WHAT IS A SPECIES?

Operationally, a species is a 'population of organisms, all the members of which are able to breed amongst themselves and produce fertile offspring' (Toothill 1988).[1] By implication, populations that are genetically, physiologically, or behaviourally incapable of interbreeding belong to different species. As clear-cut as this definition seems, reality is much more complicated and there are variant populations below the species level that can be identified either morphologically and/or genetically. Hanson (1962) defined these taxa hierarchically as: (1) '*subspecies*, a taxon of distinct, geographically separated complexes of genes immediately below species and above variety ... [but] sometimes considered as synonymous with *variety*, or as an incipient species', followed by (2) race as 'a population within a species that differs in one or more inherited characteristics from other populations but not sufficiently distinct to rate as [a separate] taxon'.

Except for monozygotic siblings, no two individuals generated by sexual reproduction are completely identical genetically. In general, those of the same race or subspecies are genetically more similar than those from different subspecific categories; those of the same species share greater genetic similarity with each other than they do with other species in their genus, and so on up the taxonomic hierarchy. The degree of genetic difference between individuals of one taxon and another, however, may not be distinct. Certainly, there are no abrupt transformations over geological time as one species diverges into another type of organism (Dawkins 2004), i.e. there is not an immediate formation of a new species between one generation and its immediate offspring, but rather a gradual accumulation of genetic differences over many generations, eventually leading to reproductive incompatibility.

Historically, taxonomists almost never tested whether populations of a putative new species were reproductively compatible with previously described populations but rather merely based their classification on the degree of morphological differences they subjectively deemed sufficient to warrant designating two groups as separate species. Accordingly, there were different interpretations between 'lumpers' (requiring greater divergence) and 'splitters' (more prone to accept fewer differences to validate specific status). Seldom was it feasible to check whether the various populations actually interbred as most species were described and named solely from preserved specimens.

Because of the variation within a species there was need to incorporate some standard way of defining its limits for any given species and to provide an official name for it so that it could be recognised universally despite a given species having diverse common names in countries with different languages, and even in different localities within a single monolingual country (see Appendix 1).

Accordingly, Karl von Linné (also known a Carolus Linnaeus and Caroli Linnaei) devised a nomenclatural system of a specific name consisting of a specific epithet (such as *delicata*) preceded by a generic name (e.g. *Lampropholis*) that comprised a group of very similar species, in turn grouped hierarchically into Families, Orders, Classes, Phyla, and various intermediate categories designated as 'sub' or 'supra', as in 'Subfamily' or 'Supraorder'. Linnaeus' classification became the universal system of naming species and the tenth edition of his *Systema Naturae* (Linnaei 1758) became the standard for naming species in an orderly way by following the rules of Latin grammar. It has served science well ever since.

In descriptions of new species, one individual, known as the holotype, is nominated to represent the species and against which all future specimens can be compared; descriptions of other specimens available at the time of description, known as paratypes, often are designated as indicating variations occurring within the species.

Once the role of evolution in the diversity of life was understood (Darwin, undated) it became recognised that this nomenclature also denoted the degree of genetic relationship of species, and the degree from which they had diverged from each other over geologic time. The pattern of variation within the confines of a given species sometimes showed geographic correlates and infraspecific taxonomic categories such as 'subspecies', 'races', and 'varieties' came into use.

Today there is an International Code of Zoological Nomenclature (ICZN) (2023) that consists of rules for the formal naming of animals, determining whether a

1 There are gradations, however, in the ability of clearly different species of animals to interbreed; in some cases, the hybrids usually are sterile and hence are evolutionary dead-ends (e.g. mules or hinnies resulting from crosses between horses and donkeys). In others, hybridisation may have evolutionary consequences (see review by Adavoudi and Pilot 2022).

proposed name is available (not already used), ruling on the validity of applied names, and deciding how scientific names should be cited. Decisions regarding zoological nomenclature are published by the International Commission on Zoological Nomenclature (also abbreviated ICZN). The code is used by the scientific community worldwide. Any changes in nomenclature are governed by guidelines in the code (see Preface).

The differences between lumpers and splitters have been evident in interpretations of the classification of species in the genus *Lampropholis* (see Chapter 16 and the section *Historical development of the concept of species in the genus* Lampropholis below).

Cryptic species

Modern techniques of genetic analysis have introduced greater refinement of taxonomic comparisons than merely morphology alone and has revealed that sometimes there are populations that are practically identical morphologically but sufficiently diverse genetically as to warrant description as different species (see section *Defining the genus* Lampropholis above and Chapter 13). Singhal *et al.* (2018) recognised that *L. coggeri* (*sensu lato*) in fact consisted of three cryptic species. They retained *L. coggeri* (*sensu stricto*) and described two new cryptic ones: *L. similis* and *L. elliotensis*. Similarly *L. robertsi* (*sensu lato*) also consists of two cryptic species: itself, *L. robertsi* (*sensu stricto*) and a new cryptic one (*L. bellendenkerensis*). Singhal *et al.* (2018) presented a flow-chart outlining procedural steps to follow in validating putative cryptic lineages and evaluated the benefits and limitations of each approach.

Convergent species

Convergent species are those that are distantly related genetically but, nevertheless, have independently evolved remarkably similar morphological, physiological, or ecological attributes. Familiar examples are: (1) sharks (cartilaginous fish) and porpoises (bony mammals) resembling each other in shape as a consequence of both taxa evolving a streamlined shape that adapts them to locomotion through water, and the (2) marsupials in Australia, such as taxa that resemble placental counterparts elsewhere in the world, for example, *Notoryctes* (marsupial 'moles'), quolls (marsupial 'cats'), *Antechinus* (marsupial 'mice'), and the thylacine (Tasmanian 'tiger', now extinct). Such convergence occurs as a consequence of genetically distant taxa evolving comparable, but independent, adaptations to similar environments. Convergence is obvious in the extreme examples given above, but when occurring within taxa that are more closely related phylogenetically, it may be difficult to detect. For example, two species in the same family or genus may adapt to a particular environment (e.g. rainforest or desert) in similar ways not because both inherited those characteristics from a common ancestor, but because a particular kind of environment selected for similar adaptative traits. Their overt similarity may thus mislead the investigator to consider them more closely related genetically, say, in the same genus rather than in separate genera, merely because of their independent (convergent) adaptation to similar environments. For this reason genetic analysis has become important in modern taxonomic assessments. It provides insight into subtle differences and permits refinement of taxonomic decisions to a more precise scale.

Clades

Populations that on genetic grounds, perhaps supplemented by slight morphological differences (but even if morphologically practically identical), can be distinguished as distinct entities that are not sufficiently divergent to warrant designation as separate species are called clades, or if less distinct, as subclades. Where one draws the line may be subjective (see Chapter 16).

Clines

There also is a spatial, as well as a temporal component to selective pressures leading to genetic change. A widespread species encounters different sets of environmental conditions (e.g. temperature and/or rainfall) from north to south or from lowlands to higher elevations and local adaptation to such environmental nuances may occur in which gradual changes in physiological, behavioural, and genetic characteristics may be manifest along the environmental continuum without identifiable intervening boundaries that could separate population into clades or subclades. The special case of a series of gradual changes in populations within a species along a geographic or elevational gradient is called a cline.

In extreme cases populations at the two ends of the cline may be sufficiently divergent as to be reproductively incompatible, and hence technically are different species. However, by being connected sequentially by a series of populations, each of which may be reproductively compatible with the populations on either side of it, but not with more distant ones, where does one draw the boundary between species? In the case of clines there is no sharp one. That is their distinguishing feature. The situation can be even more complex. If a geographic barrier interposes for sufficient time for clades to differentiate, but then contact later re-joined, a narrow zone of genetic introgression may occur between adjacent clades previously on opposite sides of the barrier rather than the more gradual transition between adjacent population in a cline.

HISTORICAL DEVELOPMENT OF THE CONCEPT OF SPECIES IN THE GENUS *LAMPROPHOLIS*

The first scientific descriptions of species in the *Lampropholis* genus were of the two widespread, eurytopic species, *L. guichenoti* in 1839 and *L. delicata* in 1888 (Table 14.2); the majority of the other species that currently are accepted as valid were described in the latter part of the 20th century (1981–1990) with various authors suggesting that additional species remained yet to be described. As true for most taxa of that time, the features used for distinguishing between species were primarily morphological despite the fact the criterion for defining them was inability to interbreed (see section *What is a species?*). Some of the more recently described species were populations previously considered to be *L. delicata* (Table 14.2), and both *L. delicata* and *L. guichenoti* previously were considered members of different genera (see section *Defining the genus* Lampropholis).

Traditionally crosses between populations seldom were attempted so taxonomists merely based their classification on the degree of morphological difference they deemed sufficient to warrant designation as separate species. Now modern technology has made more precise decisions possible, and one can analyse genomes and estimate the degree of genetic difference between populations and even estimate the particular span of geologic time over which various changes have occurred (Dawkins 2004; Chapple *et al.* 2011a). In many cases it is now known that morphologically indistinguishable populations (or at least so similar that traditionally they would have been considered conspecific on morphological grounds) are found on genetic evidence to be divergent in their haplotypes.

Smith *et al.* (2007) and Chapple *et al.* (2023) used comparison of mitochondrial and nuclear DNA to assess the relationships of the skinks of Tribe Eugongylini. They corroborated the conclusion of many previous authors that the Eugongylini constitutes a monophlyletic group within the Subfamily Lygosominae with diversification within Australia, New Guinea, and some Pacific Islands. From Australia, they dispersed naturally overwater initially to New Caledonia, then to Lord Howe Island, and finally to New Zealand. The genus *Lampropholis* occurred natively only within Australia (but see Chapter 3 for an account of its inadvertent introduction by humans into Lord Howe Island, New Zealand, and Hawaii).

Chaplin *et al.* (2022) studied populations of *L. delicata* that originated from four different areas in Australia before being introduced onto Lord Howe Island. In their native land, these populations had been separated for more than 2 million years. All four populations were successfully interbreeding on Lord Howe Island, thereby confirming that original divergence of the source populations consisted of relatively minor differences in morphology, life history, and phenotype that did not reach the level of speciation.

In contrast to the outcomes of research carried out by Chaplin *et al.* (2022) on *L. delicata*, Singhal and Moritz (2012) found phenotypically cryptic lineages in *L. coggeri* between which was an extremely narrow hybrid zone. Based on knowledge of the region's biogeography, the natural history of the species, the results of the authors' simulations, and the relatively narrow range of introgression, there was no apparent strong selection against hybrid individuals. Since all clines had a relatively narrow range of introgression, it seems that the hybrid zone had not yet attained equilibrium. Nevertheless, the results clearly showed that phylogeographic lineages had evolved that resulted in substantial reproductive isolation despite a lack of morphological diversification. Subsequently, two of these lineages were described as new species, *L. elliotensis* and *L. similis*, on the basis of genetic characteristics (Singhal *et al.* 2018).

Morphology and colouration as taxonomic characters

Clarke (1965) conducted a survey of the morphological characteristics of *L. delicata* and *L. guichenoti* and summarised her results and those of previous investigators for *L. guichenoti* as: 'snout-forelimb length contained from once to twice the distance between axilla and groin, usually 6 superciliaries (scales immediately above the eyes), adpressed limbs meet, overlap, or fail to meet depending on axilla-groin length, usually 26-30 mid-body scale

Table 14.2. The synonymy of species of the genus *Lampropholis*.

The synonymy contains names that are no longer deemed to be valid, including: (1) later new names erroneously applied to a species that had previously been named, (2) names of species that were found to belong to a different genus than the one to which they had been allocated when originally described, thereby necessitating that the older generic name be replaced by the correct generic designation, or (3) names of species assigned to a new genus that was not deemed to be different from the one to which the species was originally assigned. Synonymy based on Cogger *et al.* (1983), Shea and Sadlier (1999), Cogger (2018), and The Reptile Database (last accessed 10 May 2024).

Scientific name, describer and date of publication	Type locality	Synonymy
Lampropholis adonis Ingram 1991	18 km N of Dalrymple Heights, Qld. 21° 04′ S, 148° 36′ S	*Lampropholis delicata* Form B (Mather and Hughes 1992), *Ndurascincus adonis* (Wells 2002)
Lampropholis amicula Ingram and Rawlinson 1981	3 km E of Lake Cooloomera, Qld. 26° 02′ S, 153° 05′ E	*Adrasteia amicula* (Wells 2002), *Adrasteiascincus amicula* (Wells 2010)
Lampropholis bellendenkerensis Singhal, Hoskin, Couper, Potter, and Moritz 2018	Bellenden Ker Range, Qld. 17° 20′ S, 145° 52′ E	Holotype had also been a paratype of *L. robertsi* Ingram (1991)
Lampropholis caligula Ingram and Rawlinson 1981	Polblue, Barrington Tops, NSW. 31° 59′ S, 151° 30′ E	*Adrasteia caligula* (Wells 2002), *Adrasteiascincus caligula* (Wells 2010)
Lampropholis coggeri Ingram 1991	Shiptons Flat, Cape York Peninsula, Qld. 15° 48′ S, 145° 16′ E	*Lampropholis delicata* Form D (Mather and Hughes 1992), *Ndurascincus coggeri* (Wells 2002)
Lampropholis colossus Ingram 1991[1]	Bunya Mountains National Park, Qld. 26° 53′ S, 151° 37′ E	*Allengreerus colossus* (Hoser 2009)
Lampropholis couperi Ingram 1991	Kondalilla National Park, Qld. 26° 41′ S, 152° 52′ E	*Lampropholis delicata* Form C (Mather and Hughes 1992), *Ndurascincus couperi* (Wells 2002), *Allengreerus couperi* (Hoser 2009)[1]
Lampropholis delicata (De Vis 1888)	Warro Station, Port Curtis, Qld. 28° 38′ S, 151° 25′ E	*Mocoa delicata* (De Vis 1888), *Leiolopisma hawaiiensis* (Loveridge 1939), *Lygosoma hawaiiensis* (Fisher 1948), *Lampropholis longleyi* (Wells and Wellington 1985), *Allengreerus delicata* (Hoser 2009)[2], *Allengreerus ronhoseri* (Hoser 2009)[2], *Lampropholis ronhoseri* (Wells 2010)[1], *Allengreerus delicata jackyhoserae* (Hoser 2012)[2]
Lampropholis elliotensis Singhal, Hoskin, Couper, Potter, and Moritz 2018	Mt Elliot, Qld. 19° 28′ S, 146° 59′ E	None
Lampropholis elongata Greer 1997	Riamukka State Forest, NSW. 31° 19′ S, 151° 39′ E	*Adrasteia elongata* (Wells 2002), *Adrasteiascincus elongata* (Wells 2010)
Lampropholis guichenoti (Duméril and Bibron 1839)	Australia in original; lectotype from Kangaroo Island (see Cogger *et al.* 1983)	*Lygosoma guichenoti* (Duméril and Bibron 1839), *Lygosoma duperreyii* [part] (Duméril and Bibron 1839), *Lygosoma guttulatum* (Peters 1887), *Lygosoma guichenoti* (Lucas and Frost 1894), *Lampropholis lunneyi* (Wells and Wellington 1984), *Lampropholis swani* Wells and Wellington (1985)
Lampropholis mirabilis Ingram and Rawlinson 1981	Magnetic Island, Qld. 19° 08′ S, 146° 50′ E	*Helioscincus mirabilis* (Wells 2002)
Lampropholis robertsi Ingram 1990	Thornton Peak via Daintree, N. Qld. 16° 10′ S, 145° 23′ E	*Ndurascincus robertsi* (Wells 2002)

continued

Table 14.2. continued

Scientific name, describer and date of publication	**Type locality**	**Synonymy**
Lampropholis similis Singhal, Hoskin, Couper, Potter, and Moritz 2018	The Pinnacles, south-west of Townsville, Qld. 19° 23′ 42″ S, 146° 39′ 07″ E	None

[1] According to Torkkola *et al.* (2024) there is some doubt whether this species is distinct from *L. delicata*.
[2] Following is an excerpt from The Reptile Database (accessed 10 May 2024) as its justification for relegating some names to synonymies: 'Hoser (2009) established a new genus *Allengreerus* Hoser 2009 for *L. delicata*. He diagnosed the genus as follows: "Separated from *Lampropholis* (type species *guichenoti*) to which it/they would otherwise be identified as, by the general lack of a distinct midvertebral stripe as seen in adult specimens." The validity of *Allengreerus* remains unclear, given that "there is clinal variation within *Lampropholis delicata*, which Hoser has failed to take into account. Although he claims *ronhoseri* lacks the pale mid-lateral stripe of *delicata*, there is a gradual decrease in the frequency of the pale mid-lateral stripe with increasing latitude in New South Wales." [Glen Shea, pers. comm]. Without cladistic analysis it is impossible to decide whether Hoser's form is different from *Lampropholis* and whether *A. ronhoseri* is different from *L. delicata*'.
The following 14 names have neither been listed as names of valid species nor as synonyms by The Reptile Database (accessed 10 May 2024); however, the genus *Allengreerus* was considered to be a synonym of *Lampropholis* (see immediately above). These 14 entities were described by Hoser (2022) as species derived from *L. delicata*: *Allengreerus alengreeri* [the spelling *alengreeri* was used instead of *allengreeri* in the original description and therefore this misspelling is the official name and must be used despite the fact that the correct form of the name was used later in the manuscript]; *Allengreerus dorsei*; *Alllengreerus merceicai*; *Allengreerus robertwatsoni*; *Allengreerus angelikadennesae*; *Allengreerus scottgranti*; *Allengreerus rosswellingtoni*; *Allengreerus kaputarensis*; *Allengreerus trevorhawkeswoodi*; *Allengreerus richardwellsei*; *Allengreerus adelynhoseerae*; *Allengreerus matheri*; *Allengreerus marulnanensis*; *Allengreersi oxyi*. For further discussion of these entities see Chapter 16.

rows'. The equivalent for *L. delicata* was: 'Very similar to *L. guichenoti* but differing in colour and in having seven superciliaries and 24-28 mid-body scale rows'. Some of the morphometric and meristic characteristics were variable (Plate 8) with broadly overlapping ranges between species (also see Chapter 5), or with the proportions changing with age, and consequently were useful for distinguishing between these two species only if multiple specimens of both were available. She concluded that the only morphological character that could serve as a diagnostic character for a single individual was the number of superciliary scales. There were also some differences between these two species in colour pattern, the one that she considered the most definitive being a conspicuous narrow, lateral, white stripe ventrally bordering the wide, black, lateral stripe in *L. guichenoti* but lacking in *L. delicata*. However, it has since been learned that both *L. delicata* and *L. guichenoti* are dimorphic for this character (Forman and Shine 1995).

Pengilley (1972) carried out a detailed morphological comparison that not only confirmed the great similarity between these two species but also between them and eight species of other genera that are so closely related that all were formerly grouped with *Lampropholis* in the genus *Leiolopisma* (see above). For example, the number of premaxillary teeth were mostly 5–6 in all species, the prefrontals met the nasals, and they all have a pterygoid process on the palatine bone; the palatine bones and various other bones were similarly shaped; in addition to osteological features other characteristics were similar such as details of scalation. For example, the most interesting finding of Pengilley's study was that although there was broad overlap among all these species with each other and with both widespread species of *Lampropholis*, in the numbers of supraciliary scales (Fig. 14.1A) and of scale rows at mid-body (Fig. 14.1B), *L. guichenoti* and *L. delicata* were more divergent from each other than each were from some of their relatives in other Lygosomine genera. *Lampropholis delicata* resembled one subset of the related genera to which they were compared and *L. guichenoti* to a different subset; for supraciliaries there was also a third subset that did not closely resemble either of these species of *Lampropholis* (Fig. 14.1A).

Of even greater surprise was the comparisons of the number of lamellae on the toes of the forefeet and hind feet (Fig. 14.1C). In this characteristic, there were regional differences between populations of *L. delicata*, with those from Queensland differing from the ones from New South Wales and each of these resembling different subsets of the other lygosomine species. This unexpected pattern perhaps can be explained by *L. delicata* adapting to different structures within the habitat. There would be no selection for the ability to climb, and hence for a greater number of lamellae on the toes in a grassland, than in a forest where tree trunks added a habitat for scansorial species. An alternative possibility is that the greater number of lamellae might be an adaptation reducing competition; in a forested habitat without any scansorial species, competition would be reduced by adapting to such an unoccupied niche (see Chapter 11) such as occupancy of tree trunks or other microhabitats that require greater climbing ability than does dwelling on the ground. In the first example the structure of the habitat would be the determining element, in the second one it would be the absence of competitors in the local assemblage of lizards for the given physical structure of the habitat.

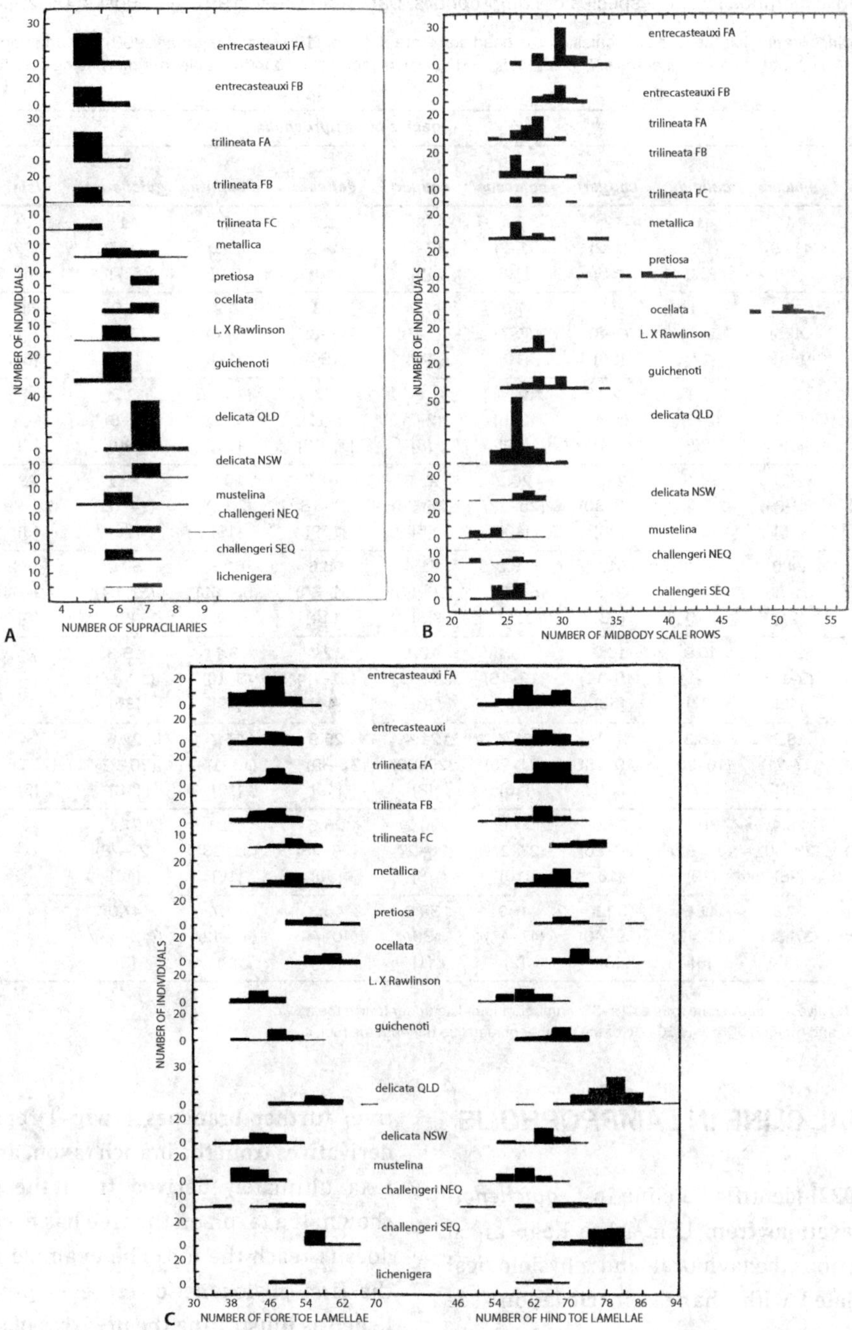

Fig. 14.1. Frequency distributions of supraciliary scales (A), number of scale rows at mid-body (B), and lamellae on the toes (C), and in populations of various species of closely related genera of lygosomine skinks. Modified from Pengilley (1972). Species FA, FB, FC refer to designated 'forms' of various species.

Morphological criteria have been difficult to apply successfully for defining the species of *Lampropholis*. Many of the anatomical characteristics that were conveniently measured or counted were so variable, even within a local population, and overlap of ranges so great between populations, that it was necessary to define a species, not on some outstanding distinctive feature, but on a combination of characters that individually were insufficient (Table 14.3). Also note that in Table 14.3 the extent of overlap probably was underestimated as some sample sizes were so low that they unlikely included the complete range of variation.

Table 14.3. Comparative morphology of the species of *Lampropholis*. Data from Greer (1997) and Hoskin (2025).

N = sample size. For drawings identifying the types of scales on the head see Plate 8, Greer (1997), and Ingram (1990). For additional data on morphological features, see Table 14.4. Note that *L. bellendenkerensis*, *L. elliotensis*, and *L. similis* (not included in the table) are morphologically indistinguishable from *L. coggeri*.

Trait Number (Range) [N]	Species of *Lampropholis*											
	adonis	***amicula***	***caligula***	***coggeri***	***colossus***	***couperi***	***delicata***	***elongata***	***guichenoti***	***isla***	***mirabilis*****	***robertsi***
Preocular scales	2 (2–2) [18]	1 (1–2) [49]	1 (1–1) [27]	2 (1–2) [55]	1.4 (1–2) [10]	2 (2–2) [8]	1.2 (1–2) [40]	1 (1–1) [19]	1 (1–1) [27]	2 (2–2) [6]	1 (1–1) [25]	2 (2–2) [16]
Supraciliary scales	7 (7–8) [21]	5 (4–5) [56]	5.1 (5–6) [27]	7.0 (7–8) [28]	7.0 (7–7) [10]	7.0 (7–7) [8]	6.9 (6–8) [150]	5.2 (5–6) [19]	5.9 (5–7) [19]	7.0 (7–7) [6]	6.9 (5–8) [25]	7.1 [*sic*] (7–7) [16]
Nuchal scales	2.1 (2–4) [40]	3.6 (2–4) [56]	2.2 (2–3) [22]	2.1 (1–4) [94]	1.2 [*sic*] (2–5) [10]	2 (2–2) [8]	2.7 (1–11) [1,102]	2 (2–2) [19]	2.2 (2–5) [30]	4 (4–4) [6]	2.2 (2–4) [25]	2.1 (2–3) [11]
Scale rows at mid-body*	27.4 (26–30) [16]	22.1 (22–24) [30]	21.6 (20–23) [27]	28.3 (26–30) [26]	26.2 (25–28) [10]	25.8 (25–26) [5]	25.4 22–28 [159]	20.2 (20–2) [19]	28.1 (24–31) [20]	28 (28–29) [6]	30.6 (30–32) [19]	27.7 (26–30) [12]
Paravertebral scales	51.6 (49–55) [16]	54.0 (51-56) [27]	61.6 (9–64) [27]	51.9 (49–57) [23]	55.0 (52–57) [10]	51.4 (50–53) [8]	55.6 (51–60) [114]	62.8 (55–65) [19]	59.0 (57–61) [9]	51 (49–54) [6]	57.2 (54–60) [19]	52.6 (49–56) [16]
Number of scales beneath toe 4	10.6 (10–13) [43]	10.0 (9–11) [42]	10.9 (10–11) [27]	10.2 (10–11) [53]	13.1 (12–16) [10]	10.1 (10–11) [8]	12.7 (11–14) [44]	9.1 (8–10) [19]	13.3 (12–14) [35]	-	16.6 (15–19) [25]	10.6 (9–12) [12]
Number of lamellae beneath toe 4	23.6 (20–26) [28]	18.2 (17–21) [19]	18.3 (16–20) [27]	22.5 (19–25) [29]	23.4 (21–25) [10]	21.4 (20–23) [8]	25.9 (21–30)** [49]	15.7 (13–17) [19]	22.6 (20–27) [18]	24 (22-26) [6]	30.9 (29–33) [17]	22.4 (21–24) [11]
Presacral vertebrae	26.0 (26–27) [30]	28.4 (28–30) [28]	30.8 (30–32) [19]	26.0 (26–26) [16	27.0 (27–27) [10]	26.2 (26–27) [8]	26.5 (26–28) [96]	31.9 (31–33) [16]	27.1 (27–28) [16]	-	26.0 (26–26) [4]	26.1 (26–27) [11]
Postsacral vertebrae	40.6 (38–42) [9]	37.8 (37–38) [3]	41.6 (41–43) [3]	39.3 (39–40) [3]	47.0 (47–47) [2]	39.0 39 [1]	45.3 (40–49) [16]	41.7 (41–42) [3]	47.0 (45–49) [3]	-	48.0 (48) [1]	-

*Ingram and Rawlinson (1981) give mid-body scale rows as 29–33; number of lamellae under fourth toe as 27.

**De Vis (1888) indicates 20 and Green (1965) lists 16–25 for the number of lamellae beneath toe four.

AN ELEVATIONAL CLINE IN *LAMPROPHOLIS GUICHENOTI*

Anderson *et al.* (2022) identified a cline in *L. guichenoti* over a range of elevations from 11 m asl to 1,546 m asl and examined various behavioural and physiological adaptations associated with changes in elevation (see Chapters 6 and 7).

DIAGRAMMATIC PORTRAYAL OF TAXONOMIC RELATIONSHIPS

There are several ways of visually portraying the relationships of various lineages to each other within an evolutionary framework. The simplest form is a phylogenetic tree in which the most ancestral taxon is considered the stem from which taxa derived from it are shown as 'branches' leading off from it, much like those of a tree; further branches ('twigs') represent evolutionary derivatives from the branch taxon, and so on until all the taxa ultimately derived from the ancestral stem are shown. If a taxon on the tree has become extinct its twig doesn't reach the top. The example in Fig. 14.2 is likely the first phylogenetic tree ever produced. It was published as illustrating the progress of speciation under the influence of natural selection in Darwin's monumental work *The Origin of Species by Means of Natural Selection or the Preservation of Favored Races in the Struggle for Life*. The final result in this hypothetical example was that only eight species reached the top line (the present time); they represent the fittest that eventually survived. Note that at least 80 species became extinct along the way (plus those becoming extinct during the gap separating the lower and upper parts of the diagram) and which now would only be known as fossils.

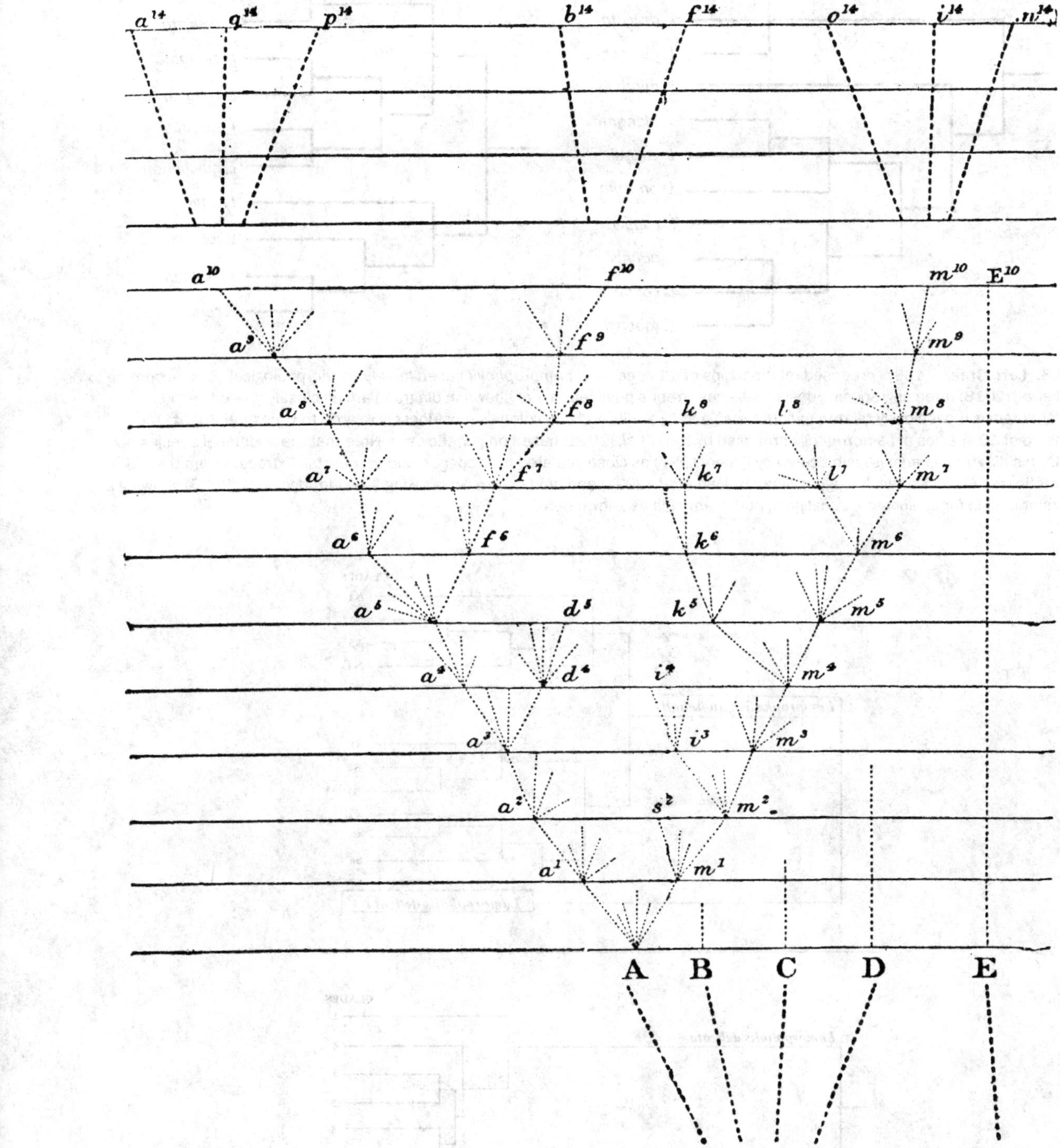

Fig. 14.2. Part of a hypothetical phylogenetic tree prepared by Charles Darwin. The horizontal lines represent successive geological strata containing fossils. Upper-case letters from A to E designate ancestral lineages. Note that lineages B to E became extinct at different geological times (they didn't reach the top of the diagram) and didn't diverge into any new taxa. Lower-case letters indicate points of divergence into different branches or twigs. Lineage A diverged into two major stems with the left one diverging into three branches, each of which generated multiple twigs (species); the right stem divided into four branches, also speciating into multiple species. Note that the three upper layers are not connected to the lower one, signifying an indeterminate gap in time between the two parts of the diagram. This kind of tree does not necessarily need to have the geological times specified or extinct forms shown, but only the radiating lines portraying relationships of extant forms. Simplified from Darwin's (undated) sixth edition.

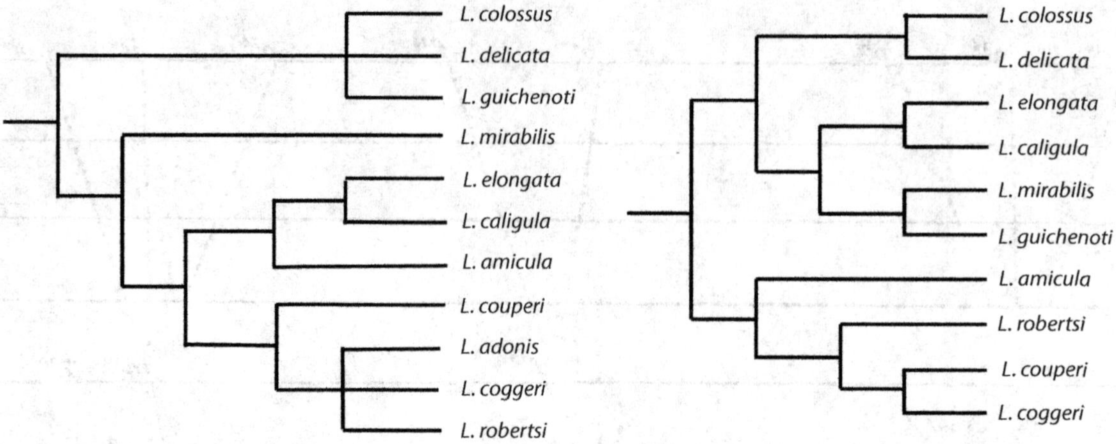

Fig. 14.3. Left: Greer's (1997) proposed relationships of 11 species of *Lampropholis* based largely on morphological data. According to Singhal *et al.* (2018) three new species described since Greer's publication (not shown in diagram) are all closely related, with *L. bellendenkerensis* most closely related to *L. robertsi,* and *L. similis* and *L. elliotensis* most loosely related to *L. coggeri.* **Right**: The relationship of 10 species of *Lampropholis* proposed by Ivan *et al.* (2021) based on genetic data. Note that the additional species not shown in this diagram, *L. adonis*, is indicated by Greer (1997) as closely related to *L. coggeri* and *L. robertsi.* Extracted from the phylogenetic tree of the Subfamily Eugongylinae by Ivan *et al.* (2021) and modified by Susan King. See Eldridge *et al.* (2020) showing a similar relationship for *L. elongata, L. caligula*, and *L. amicula* as shown here.

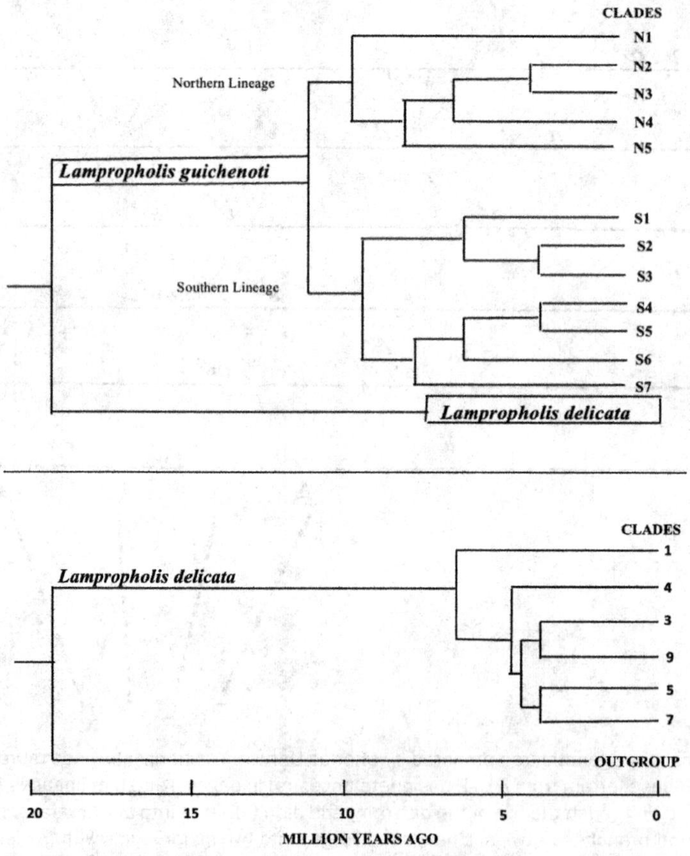

Fig. 14.4. Cladigram showing the degree of relationship between the two lineages and the 12 clades of *Lamprophois guichenoti* (**Top**) and the seven clades of *Lampropholis delicata* (**Bottom**) Based on data from Chapple *et al.* (2011a, b). Note that the scale of time on the *x*-axis relates only to the bottom cladogram.

Table 14.4. Features considered by the original describers of the species of *Lampropholis* to be diagnostic of their new species and the institutions housing the holotype and paratype specimens.

Note that some describers did not identify which characters they considered diagnostic and in those cases, we have included much of their broader description of morphology. Data are from De Vis (1888), Cogger *et al.* (1983); Ingram (1990), Greer (1997), Shea and Sadlier (1999), Singhal *et al.* (2018), Hoskin (2025), and The Reptile Database (accessed 31 December 2023). For coloured photographs of *L. amicula, L. caligula, L. coggeri, L. delicata, L. guichenoti, L. robertsi, L. guichenoti* and *L. delicata* see Cogger (2014) and Plate 1. Black and white photographs of *L. adonis, L. coggeri, L. colossus, L. couperi,* and *L. robertsi* are available from Ingram (1990), and of *L. elongata* from Greer (1997). Photographs of some of these species also can be found on the website of The Reptile Database (www.reptile-database.org), Wikipedia ('Lampropholis' en.wikipedia.org), iNaturalist, and the covers and introductory pages of this book. Some diagnoses in the table were slightly edited by the present authors.

Species	Diagnostic characters	Location of holotyipe (H) and paratypes (P), and type locality
Lampropholis adonis Ingram 1990	A large (maximum snout-to-vent length [SVL] 51 mm), robust (maximum width of head 18 mm) species with fused frontoparietals, a fused interparietal, seven supralabials, seven supraciliaries, short hind limbs and darker colouration of the upper lateral surface grading evenly into lighter lower lateral colouration. A mid-lateral light brown to white line or series of spots is never present.	H: Queensland Museum 335097 P: Queensland Museum Type Locality: 18 km N of Dalrymple Heights, NE Qld; 21° 04′ S, 148° 36′ S
Lampropholis amicula Ingram and Rawlinson 1981	A little, small-limbed *Lampropholis* distinguished from all other members of this genus (except *L. caligula)* by its 5 supraciliaries. Further distinguishable from *L. caligula* by its higher number of supraoculars (4 versus 3) and upper labials (7 versus 6).	H: Australia Museum R96587 (formerly Queensland Museum J37287) P: Australian Museum Type Locality: 3 km E of Lake Cooloomera, SE Qld; 26° 02′ S, 153° 05′ E
Lampropholis bellendenkerensis Singhal, Hoskin, Couper, Potter, and Moritz 2018	A large species with dark flanks and prominent spotting on the posterior ventral surfaces, a row of dark-edged pale spots on the underside of the tail. Although *L. bellendenkerensis* is morphologically similar to its closest congener, *L. robertsi*, it can be reliably distinguished from it by genetic and chemical differences (see Chapter 13)	H: Queensland Museum 339855 P: Queensland Museum Type Locality: Bellenden Ker Range, Qld; 17° 20′ S, 145° 52′ E
Lampropholis caligula Ingram and Rawlinson 1981	A large, drab *Lampropholis* distinguished from all other members of this genus by its 5 supraciliaries and 3 supraoculars. Further distinguished from *L. amicula* by its low number of upper labials (5 versus 7).	H: Australian Museum R13313 P: Queensland Museum Type Locality: Polblue, Barrington Tops, mid-eastern NSW; 31° 59′ S, 151° 30′ E
Lampropholis coggeri Ingram 1990	A small (maximum SVL 41 mm), robust (maximum width of head 18 mm species with fused frontoparietals, a free interparietal, 7 supralabials, 7 supraciliaries, short hind limbs (<40 mm) and with darker upper lateral surface colouration grading evenly into lighter lower lateral colouration. A mid-lateral light brown to white line or series of spots is never present. Further distinguished from *L. couperi* by colour pattern: dorsal colour reddish to olive-brown usually flecked with pale spots and black longitudinal dashes; the dashes can coalesce to form up to six dorsal lines; the outer two lines can delimit a paler brown dorsolateral stripe; lateral ground colour dark brown to reddish-brown flecked with paler spots and, usually, black dashes versus dorsal colour uniform olive-brown; upper lateral surface evenly black; dorsolateral area pale brown, edged strongly in black below.	H: Queensland Museum 127133 P: Queensland Museum Type Locality: Shiptons Flat, Cape York Peninsula, Qld; 15° 48′ S, 145° 16′ E

continued

Table 14.4. continued

Species	Diagnostic characters	Location of holotyipe (H) and paratypes (P), and type locality
Lampropholis colossus Ingram 1990	A very large (maximum SVL 56 mm), delicate (maximum width of head 14 mm) *Lampropholis* species with fused frontoparietals, a free interparietal, seven supralabials, seven supraciliaries, short hind limbs (<40 mm) and with darker upper lateral surface distinctly demarcated mid-laterally from lighter lower lateral colouration. The demarcation can be due to a sharp transition of dark to light pigments or be delineated by a white mid-lateral line or series of spots. Further distinguished from *L. delicata* by large size (maximum SVL 56 mm versus 45 mm) and ventral colouration (bright yellow to cream versus white). Further distinguished from *L. robertisi* by ventral colour pattern.	H: Queensland Museum 149687 P: Queensland Museum Type Locality: Bunya Mountains National Park, in rainforest near Tim Shea Falls, Qld; 26° 53′ S, 151° 37′ E
Lampropholis couperi Ingram 1990	A large (maximum SVL 49 mm), delicate (maximum width of head 16 mm) species with fused frontoparietals, a free interparietal, 7 supralabials, 7 supraciliaries, short hind limbs (<40 mm) and with darker upper lateral surface colour grading evenly into lighter lower lateral colour. A mid lateral light brown to white line or series of spots is never present. Further distinguished from *L. coggeri* by colour pattern (see above).	H: Queensland Museum 149575 P: Queensland Museum Type Locality: Kondalilla National Park, southern Qld; 26° 41′ S, 152° 52′E
Lampropholis delicata (De Vis 1888)	'But for its enlarged preanals, would be comparable with *L. guichenoti*, but on direct comparison its slenderer form, feebler limbs, and entirely different physiognomy show that its affinity to *guichenoti* is by no means close' (see De Vis 1888).	H: Queensland Museum (presumed lost) Neotype: Queensland Museum J45765 H: Museum of Comparative Zoology, Harvard 44234 (called *hawaiiensis*) P: None recorded Type Locality: Warro Station, Port Curtis, Central Qld; 28° 38′ S, 151° 25′ E
Lampropholis elliotensis Singhal, Hoskin, Couper, Potter, and Moritz 2018	*Lampropholis elliotensis* is a small, dark-sided rainforest skink with pentadactyl limbs (usually separated by several scales rows when adpressed) and a movable lower eyelid containing a transparent disc. It is reliably distinguished from its sibling species (*L. similis* and *L. coggeri*) by 17 nucleotide differences in the mitochondrial gene NADH dehydrogenase 4 that result in 15 amino acid differences among the species.	H: Queensland Museum J91116 P: Queensland Museum Type Locality: Mt Elliot, Qld; 19° 28′ 55″ S, 146° 59′ E
Lampropholis elongata Greer 1997	Differs from all other members of the genus in two characters: (1) three supraocular scales, the first two in contact with the frontal scale, and (2) a phalangeal formula [number of phalanges in the digits of the hands/feet] of 2.3.4.4.3/2.3.4.4.3 instead of 2.3.4.5.3/2.3.4.5.3	H: Australian Museum R148161 P: Australian Museum Type Locality: just south-east of Grundy firetower, Riamukka State Forest, NSW; 31° 19.5′ S, 151° 39.4′ E
Lampropholis guichenoti (Duméril and Bibron 1839)	Nasal scales small, entirely lateral; internasales very enlarged, articulated with the rostral by a large rectilinear edge; fronto-nasales almost contiguous; one large diamond-shaped front-parietal, truncated behind; interparietal small, triangular; two parietals; ears somewhat large, sub-circular, with a simple edge, uncovered; two of the four scales of the last preanal row much larger than the others. Back a greenish-bronze; a blackish stripe or band on each side of the body. [translated from the French]	H: Muséum national d'Histoire naturelle, Paris P: None recorded Type Locality: 'Australia' [Kangaroo Island]

Species	Diagnostic characters	Location of holotyipe (H) and paratypes (P), and type locality
Lampropholis isla Hoskin 2025	A free interparietal scale; bright orange flanks on breeding males; 7 supraciliary and 4 supraocular scales; no pale mid-lateral stripes or pale dashes on flanks. This species is shorter in body, has longer forelimbs and different markings on the tip of the rostral scale than does *L. couperi*; is distinguished from *L. coggeri*, *L. similis*, and *L. elliotensis* by dorsal and lateral surfaces generally not marked with pale or dark flecks or spots; differs from *L. robertsi*, *L. bellendenkerensis*, *L. delicata*, *L. colossus,* and *L. guichenoti* by lacking sharp demarcation between dark upper and lighter lower flanks; distinguished from *L. mirabilis* by having shorter limbs and lack of pattern on the body (versus dark and white spots and flecks in *L. mirabilis*); contrasts with *L. amicula* by being larger and having 7 supraciliary scales (versus 5 in *L. amicula*); differs from *L. adonis* by having a fused interparietal scale and by difference in the colouration of males; differs from *L. caligula* and *L. elongata* by a shorter length of the body and by having 4 supraocular scales (versus 3 in those two species).	H: Queensland Museum J98079 P: Queensland Museum J98077-J98011 Type Locality: Scawfell Island, mid-east Qld; 20° 52′ S; 149° 37′ E
Lampropholis mirabilis Ingram and Rawlinson 1981	Has longer legs, a higher number of toe lamellae and mid-body scale rows, and dark colouring with white flecking, when compared with to its congeners.	H: Queensland Museum J24439 P: Australian Museum Type Locality: Magnetic Island, Qld; 19° 08′ S, 146° 50′ E
Lampropholis robertsi Ingram 1990	A large (maximum SVL 49 mm), robust (maximum width of head 17 mm) species with fused fronto-parietals, a free interparietal, 7 supralabials, 7 supraciliaries, short hind limbs and with darker upper lateral surface distinctly demarcated mid-laterally from lighter lower lateral colouration. Demarcation can be due to a sharp transition of dark to light pigments or be delineated by a pale brown to golden mid-lateral stripe or series of spots. Further distinguished from *L. colossus* and *L. delicata* by ventral colouration: grey with black-edged scales flecked with black: throat, chin, lower part of body and tail usually heavily black-spotted; under the tail, the black markings often coalesce to outline a longitudinal series of white chevrons versus bright yellow to cream (*L. colossus*) or white (*L. delicata*) with or without black flecking on throat and tail, which can coalesce into lines.	H: Queensland Museum 143911 P: Queensland Museum Type Locality: Thornton Peak, via Daintree, NE Qld; 16° 10′ S, 145° 23′ E
Lampropholis similis Singhal, Hoskin, Couper, Potter, and Moritz 2018	A small, dark-sided rainforest skink with pentadactyl limbs (overlapping or very narrowly separated when adpressed) and a movable lower eyelid containing a transparent disc. It is reliably distinguished from its sibling species (*L. coggeri* and *L. elliotensis*) by 17 nucleotide differences in the mitochondrial gene NADH dehydrogenase 4 that result in 15 differences in amino acids.	H: Queensland Museum Q91380 P: Queensland Museum Type Locality: N Pinnacles, SW of Townsville, North Qld; 19° 23′ S, 146° 39′ 07″ E

Charles Darwin illustrated what probably was the first phylogenetic tree ever produced. It portrayed how he envisioned taxa to evolve over geological time with only the fittest still surviving (Fig. 14.2).

A more formalised kind of tree is oriented horizontally instead of vertically and the shape is formatted into a box-like, right-angled structure rather than in radiating lines (Fig. 14.3). Greer (1997) constructed such a diagram, based on morphological characters, for the 11 species of *Lampropholis* that were known at that time. Three new species have been described and named since then: *L. bellendenkerensis* is deemed to be most closely related to *L. robertsi* and both *L. similis* and *L. elliotensis* are closely related to *L. coggeri* (Singhal *et al.* 2018).

Such portrayals are enhanced if a timeframe can be added to show the chronology of divergences of taxa in real time (Fig. 14.4). Analysis of DNA in such populations can reveal the extent of genetic divergence between them (see Chapter 13). *Lampropholis* have been subjected to such investigations and various cryptic species have been identified (Singhal *et al.* 2018) or relationships between first-order infraspecific populations (clades) and second-order ones (subclades) assessed (Chapple *et al.* 2011a, b) (see sections *Cryptic species* and *Clades* above).

Not all characteristics included in the description of a new species are considered diagnostic of its differentiation from other species. Those considered diagnostic for the species of *Lampropholis* are shown in Table 14.4.

BOX 14.1 KEY TO THE SPECIES OF *LAMPROPHOLIS*[1]

Harold Cogger

Some species have been defined primarily by their genetic distinction and may be almost indistinguishable in appearance and morphology. For example, the original describers of *Lampropholis similis* pointed out that '*Lampropholis similis* and *L. coggeri* are not morphologically distinct. But *L. similis* can be distinguished from both its sibling species (*L. coggeri* and *L. elliotensis*) by 17 nucleotide differences in the mitochondrial gene NADH dehydrogenase subunit 4 that result in 15 differences in amino acids' (Singhal *et al.* 2018). Further identification may also be assisted by comparing the locality of a specimen with the known distributions of two or more similar species in the Atlas of Living Australia (https://www.ala.org.au/tag/wildlife/).

Often, the distribution maps of two similar species will indicate whether a specimen to be identified is outside the known range of similar species and therefore likely to be the species within whose range it has been recorded.

1. Distribution restricted to Scawfell Island, Queensland; mature males in breeding condition with red flanks .... ***L. isla***
   Occurs elsewhere but not on Scawfell Island; mature males in breeding condition lacking red colour on flanks 2
2. No dark vertebral stripe; dark lateral band not bordered below by well-defined narrow light-coloured band ... 3
   An irregular dark vertebral stripe; dark lateral band bordered below by a well-defined, narrower band, lighter in colour than the dorsal surface .... ***L. guichenoti***
3. Fewer than 29 mid-body scale rows; fewer than 27 lamellae under the fourth toe 4
   29 or more mid-body scale rows; 27 or more lamellae under the fourth toe .... ***L. mirabilis***
4. 3 supraoculars (B); usually 6 supralabials 6
   6 supraoculars (A); usually 7 supralabials 5
5. First two supraoculars contacting the frontal .... ***L. elongata***
   Only the first supraoculars contacting the frontal .... ***L. caligula***
6. Usually only 5 (C) supraciliaries .... ***L. amicula***
   Usually 7 (D) supraciliaries 7
7. Parietal scales fused to form a single shield but distinct from the interparietal (E) 8
   Parietal and interparietal scales fused to form a single shield (F) .... ***L. adonis***
8. Dark mid-lateral zone sharply contrasting and demarcated below from the paler lower lateral zone 9
   Dark mid-lateral zone not sharply demarcated (contrasting) with lower lateral zone but gradually merging into it 12
9. Lower surfaces whitish or yellow 11
   Lower surfaces usually grey, heavily spotted with black, with the spots under the tail coalescing to create a series of white chevrons 10
10. North-east Queensland usually below 900 m .... ***L. robertsi***
    North-east Queensland mountains and uplands above 900 m .... ***L. bellendenkerensis***
11. Lower surfaces yellow; species confined to rainforest .... ***L. colossus***
    Lower surfaces whitish; species found in open forest, woodland, heath, and urban gardens .... ***L. delicata***
12. Dark upper lateral zone medium- to dark-brown, peppered with whitish and dark brown dots and flecks 13
    Dark upper lateral zone usually intense dark brown or black, without included pale dots or flecks .... ***L. couperi***
13. Adpressed limbs usually separated by several scales .... ***L. elliotensis***
    Adpressed limbs usually touch or overlap (morphologically similar) .... ***L. similis & L. coggeri***

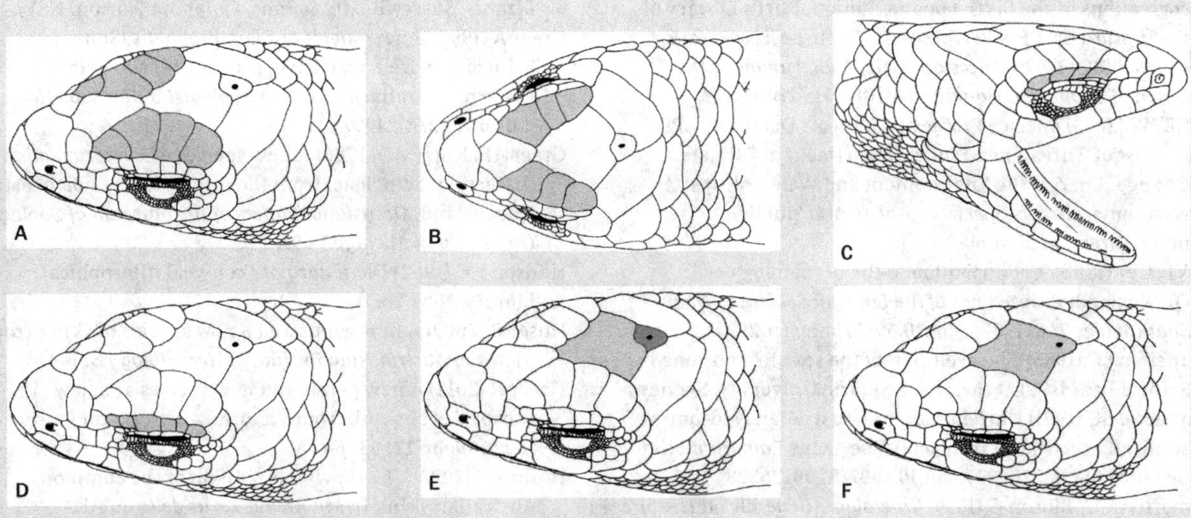

Patterns of scales on the heads of *Lampropholis*: (A) and (B) Supraoculars; (C) and (D) Supraciliaries; (E) and (F) Parietal scales.

[1]Modified and updated from Cogger (2018) by Harold Cogger 20 May 2025.

REFERENCES

Adavoudi R, Pilot M (2022) Consequences of hybridization in mammals: a systematic review. *Genes* **13**(1), 50. doi:10.3390/genes13010050

Anderson RO, Alton LA, White CR, Chapple DG (2022) Ecophysiology of a small ectotherm tracks environmental variation along an elevational cline. *Journal of Biogeography* **49**(2), 405–415. doi:10.1111/jbi.14311

Boulenger GA (1885–1887) *Catalogue of Lizards in the British Museum (Natural History)*. 2nd edn. Taylor & Francis, London.

Boulenger GA (1890) First report on additions in the lizard collection in the British Museum (Natural History). *Proceedings of the Zoological Society of London* **1890**, 77–86.

Burgin S (1989) The taxonomic and phylogenetic relationships of Australian scincid lizards (Scincidae: Lygosominae). PhD thesis. Macquarie University, Sydney.

Chaplin K, Smith Date K, Bray RD, Miller KA, Lutz ML, Razeng E, Thompson MB, Chapple DG (2022) Intraspecific hybridisation of an invasive lizard on Lord Howe Island. *Australian Journal of Zoology* **69**(5), 184–196. doi:10.1071/ZO21045

Chapple DG, Hoskin NJ, Chapple SNJ, Thompson MB (2011a) Phylogeographic divergence in the widespread delicate skink (*Lampropholis delicata*) corresponds to dry habitat barriers. *BMC Evolutionary Biology* **11**, 191. doi:10.1186/1471-2148/11/191

Chapple DG, Chapple SNJ, Thompson MB (2011b) Biographic barriers in south-eastern Australia drive phylogeographic divergence in the garden skink, *Lampropholis guichenoti*. *Journal of Biogeography* **38**(9), 1761–1775. doi:10.1111/j.1365-2699.2011.02531.x

Chapple DG, Chapple SNJ, Smith SA, Shea GM, Brennan IG, Sadlier RA (2023) Phylogenetic relationships in the Eugongylini (Squamata: Scincidae): generic limits and biogeography. *Australian Journal of Zoology* **70**(6), 165–203. doi:10.1071/ZO23007

Clarke CJ (1965) A comparison between some Australian five-fingered lizards of the genus *Leiolopisma* Duméril and Bibron (Lacertilia: Scincidae). *Australian Journal of Zoology* **13**(4), 577–592. doi:10.1071/ZO9650577

Cogger HG (1983) *Reptiles and Amphibians of Australia*. Revised edn. AH and AW Reed, Sydney.

Cogger HG (2018) *Reptiles and Amphibians of Australia*. Updated 7th edn. CSIRO Publishing, Clayton South, Vic.

Cogger HG, Cameron EE, Cogger HE (1983) *Zoological Catalogue of Australia. Vol. 1 – Amphibia and Reptilia*. Bureau of Flora and Fauna, Canberra.

Cogger HG, Cameron EE, Sadlier RA, Eggler P (1993) *The Action Plan for Australian Reptiles*. Australian Nature Conservation Agency, Sydney.

Couper PJ, Covacevich JA, Amey AP, Baker AM (2006) The genera of skinks (Family Scincidae) of Australia and its island territories: diversity, distribution and identification. Chapter 20. In *Evolution and Biography of Australian Vertebrates*. (Eds JR Merrick, M Archer, GM Hickey and MSY Lee) pp. 367–383. Auscipub, Oatlands, NSW.

Darwin C (undated) *The Origin of Species by Means of Natural Selection or the Preservation of Favored Races in the*

Struggle for Life. Undated reprint with all additions and corrections of the Sixth London Edition. Burt's Library of the World's Best Books. A.L. Burt, Publisher, New York.

Dawkins R (2004) *The Ancestor's Tale, A Pilgrimage to the Dawn of Evolution*. Houghton Mifflin Harcourt, NY.

DCCEEW (2023) Species Profile and Threats Database: EPBC Act List of Threatened Fauna. Department of Climate Change, Energy, the Environment and Water. <https://www.environment.gov.au/cgi-bin/sprat/public/publicthreatenedlist.pl>

De Vis CW (1888) A contribution to the herpetology of Queensland. *Proceedings of the Linnean Society of New South Wales* **2**, 811–826. doi:10.5962/bhl.part.29213

Donnellan SC (1985) The evolution of the sex chromosome in Scincid lizards. PhD thesis. Macquarie University, Sydney.

Donnellan SC (1991) Chromosomes of Australian lygosomine skinks (Lacertilia: Scincidae) II The genus *Lampropholis*. *Genetica* **83**(3), 223–234. doi:10.1007/BF00126228

Duméril AMC, Bibron G (1839) *Erpétologie Générale ou Histoire Naturelle Complete des Reptiles*. Roret, Paris.

Eldridge MDB, Ingleby S, King AG, Mahony SV, Parnaby HE, Beatson CA, Divljan A, Frankham GJ, Hay AC, Major RE, Reader SE, Sadlier RA, Tsang LR (2020) Australian Museum surveys of the vertebrate fauna of Coolah Tops National Park NSW. *Technical Reports of the Australian Museum Online* **30**, 1–20. doi:10.3853/j.1835-4211.30.2020.1757

Fisher HI (1948) Locality records of Pacific Island reptiles and amphibians. *Copeia* **1948**, 69.

Fitzinger LJFJ (1843) *Systema Reptilium. Fasciculus primus, Amblyglossae*. Braumüller et Seidel, Vienna.

Forsman A, Shine R (1995) The adaptive significance of colour pattern polymorphism in the Australian scincid lizard *Lampropholis delicata*. *Biological Journal of the Linnean Society* **55**(4), 273–291. doi:10.1111/j.1095-8312.1995.tb01066.x

Graham KA, Mahony SV, Chapple DG, Farquhar JE (2023) The long unknown: rediscovery of the sunskink, *Lampropholis elongata* (Squamata: Scincidae) – after almost a decade, and after 50 years of data deficiency. *Austral Ecology* **48**(5), 877–884. doi:10.1111/aec.13314

Gray JE (1838) Catalogue of the slender-tongued saurian, with descriptions of many new genera and species. *Annals of Natural History Museum* **2**, 287–293. doi:10.1080/00222933809496676

Gray JE (1845) *Catalogue of the Specimens of Lizards in the Collection of the British Museum*. British Museum, London.

Green RH (1965) Two skink lizards newly recorded from Tasmania. *Records of the Queen Victoria Museum* **19**, 1–4.

Greer AE (1970) A subfamilial classification of scincid lizards. *Bulletin of the Museum of Comparative Zoology, Harvard* **139**, 151–184.

Greer AE (1974) The generic relationships of the scincid lizard genus *Leiolopisma* and its relatives. *Australian Journal of Zoology, Supplementary Series* **22**(31), 1–67. doi:10.1071/AJZS031

Greer AE (1979) A phylogenetic subdivision of Australian skinks. *Records of the Australian Museum* **32**(8), 339–371. doi:10.3853/j.0067-1975.32.1979.459

Greer AE (1989) *The Biology and Evolution of Australian Lizards*. Surrey Beatty & Sons, Chipping Norton, NSW.

Greer A (1997) A new species of *Lampropholis* (Squamata: Scincidae) with a restricted, high altitude distribution in eastern Australia. *Australian Zoologist* **30**(3), 360–368. doi:10.7882/AZ.1997.014

Greer AE, Kluge AG (1980) A new species of *Lampropholis* (Lacertilia: Scincidae) from the rainforests of northeastern Queensland. *Occasional Papers of the Museum of Zoology University of Michigan* **691**, 1–11.

Hanson HC (1962) *Dictionary of Ecology*. Philosophical Library, New York.

Hoser R (2009) A new genus and a new species of skink from Victoria. *Australasian Journal of Herpetology* **3**, 1–6.

Hoser R (2012) A new genus and new species and new subspecies of skink from Victoria. *Australasian Journal of Herpetology* **12**, 63–64.

Hoser RT (2022) The inevitable split up of the common Australian skink lizard *Allengreerus delicata* AKA *Lampropholis delicata* into resurrected and new species (Reptilia: Squamata: Scincidae). *Australasian Journal of Herpetology* **57**, 28–52.

Hoskin CJ (2025) A new species of *Lampropholis* skink (Lacertilia: Scincidae) from Scawfell Island, mid-east Queensland, Australia. *Australian Journal of Taxonomy* **85**, 1–13. doi:10.54102/ajt.6uihp

Hutchinson M (1983) The generic relationships of Australian lizards of the family, Scincidae. A review and immunological reassessment. PhD thesis. La Trobe University, Melbourne.

Ingram G (1990) Five new species from Queensland rainforests. *Memoirs of the Queensland Museum* **30**, 443–453.

Ingram G, Rawlinson P (1981) Five new species of skinks (genus *Lampropholis*) from Queensland and New South Wales. *Memoirs of the Queensland Museum* **20**(3), 311–317.

International Code of Zoological Nomenclature (2023) Wikipedia. <https://en.wikipedia.org/wiki/International_Code_of_Zoological_Nomenclature> (accessed 27 November 2023).

Ivan J, Moritz C, Potter S, Bragg J, Turakulov R, Hua X (2021) Temperature predicts the rate of molecular evolution in Australian Eugongylinae skinks. *Evolution* **76**(2), 252–261. doi:10.1111/evo.14342

Joss JMP, Minard JA (1985) On the reproductive cycles of *Lampropholis guichenoti* and *L. delicata* (Squamata: Scincidae) in the Sydney Region. *Australian Journal of Zoology* **33**(5), 699–704. doi:10.1071/ZO9850699

Linnaei C (1758) *Systema Naturae per Regna Tria Naturae Secundum Classes, Ordines, Genera, Species, cum Characteribus Differentiis. Synonymis, Locis*. Vol. 1, 10th edn, Revised. Holmiae Impensis, Laurentii Salvii.

Loveridge A (1934) *Australian reptiles in the Museum of Comparative Zoology Bulletin*. Museum of Comparative Zoology, Cambridge.

Lucas AHS, Frost C (1894) The Lizards indigenous to Victoria. *Proceedings of the Royal Society of Victoria (new series)* **6**, 24–92 plus 11 Plates.

Mather PB (1986) The taxonomic status of *Lampropholis delicata* in southeast Queensland. PhD thesis. La Trobe University, Melbourne.

Mather PB (1989a) A comparison of the normal habitats of skinks of three electrophoretically distinguishable forms of *Lampropholis delicata* (Lacertilia, Scincidae) in South-Eastern-Queensland. *Australian Wildlife Research* **16**(2), 159–165. doi:10.1071/WR9890159

Mather PB (1989b) Electrophoretic and morphological comparisons of *Lampropholis delicata* (Lacertilia, Scincidae) populations from Eastern Australia, and a resolution the taxonomic status of this species. *Australian Journal of Zoology* **37**(5), 561–574. doi:10.1071/ZO9890561

Mather PB, Hughes JM (1992) Genetic variation in three species in the *Lampropholis delicata* (Lacertilia: Scincidae) complex. *Biological Journal of the Linnean Society* **47**(2), 135–146. doi:10.1111/j.1095-8312.1992.tb00660.x

McCann C (1955) *The Lizards of New Zealand: Gekkonidae and Scincidae*. No 17, Dominion Museum, Wellington.

Mittleman MB (1952) A generic synopsis of the lizards of the subfamily Lygosominae. *Smithsonian Miscellaneous Publication* **117**, 1–35.

O'Shaughnessy AWE (1874) A description of a new species of Scincidae in the collection of the British Museum. *The Annals and Magazine of Natural History* **13**, 298–301.

Pengilley R (1972) Systematic relationships and ecology of some lygosomine lizards from southeastern Australia (2 Vols). PhD thesis. Australian National University, Canberra.

Peters AJ (1887) Catalogue of the Lizards in the British Museum (Natural History). 2nd edn. *Vol. III: Lacertidae, Gerrhosauridae, Scincidae, Anelytropidae, Donamidae, Chamaeleontidae*. (Ed. GA Boulenger) pp. 281–282. Taylor & Francis, London.

Sadlier RA, Colgan DJ, Shea GM (1993) Taxonomy and distribution of the Australian scincid lizard *Saproscincus challengeri* and related species in southeastern Australia. *Memoirs of the Queensland Museum* **33**, 139–158.

Schuster MN (1981) Origins and historical biogeography of Eastern Australian rainforest Scincidae. In *Proceedings of the Melbourne Herpetological Symposium*. (Eds CB Banks and AA Martin) pp. 222–229. Zoological Board of Victoria, Melbourne.

Shea GM (2021) Nomenclature of supra-generic units within the Family Scincidae (Squamata). *Zootaxa* **5067**(3), 301–351. doi:10.11646/zootaxa.5067.3.1

Shea GM, Sadlier RA (1999) A catalogue of the non-fossil amphibian and reptile type specimens in the collection of the Australian Museum: types currently, previously and purportedly present. *Technical Reports of the Australian Museum* **15**, 1–91. doi:10.3853/j.1031-8062.15.1999.1290

Singhal S, Moritz C (2012) Strong selection against hybrids maintains a narrow contact zone between morphological cryptic lineages in a rainforest lizard. *Evolution* **66**(5), 1474–1489. doi:10.1111/j.1558-5646.2011.01539.x

Singhal S, Hoskin CJ, Couper P, Potter S, Moritz C (2018) A framework for resolving cryptic species: a case study from the lizards of the Australian Wet Tropics. *Systematic Biology* **67**(6), 1061–1075. doi:10.1093/sysbio/syy026

Smith MA (1935) *The Fauna of British India, including Ceylon and Burma. Reptilia and Amphibia. Vol. II Sauria*. Taylor & Francis, London.

Smith MA (1937) A review of the genus *Lygosoma* (Scincidae: Reptilia) and its allies. *Records of the Indian Museum* **39**(3), 213–234. doi:10.26515/rzsi/v39/i3/1937/162283

Smith SA, Sadlier RA, Bauer AM, Austin CC, Jackman T (2007) Molecular phylogeny of the scincid lizards of New Caledonia and adjacent areas: evidence for a single origin of the endemic skinks of Tasmania. *Molecular Phylogenetics and Evolution* **43**, 1151–1166. doi:10.1016/j.ympev.2007.02.007

The Reptile Database (nd) <http://www.reptile-database.org> (accessed May 2024).

Toothill E (1988) *The Facts on File Dictionary of Biology*. Facts on File Publications, New York.

Torkkola J, Hines H, Chauvenet A, Oliver P (2024) Are species richness and endemism hotspots correlated within a biome? A test case in the firer-impacted subtropical rainforests of Australia. *Research Square*. doi:10.21203/rs-4249551/v1

Uetz P, Cherikh S, Shea G, Ineich I, Campbell PD, Doronin IV, Rosado J, Wynn A, Tighe KA, McDiarmid R, Lee JL, Köhler G, Ellis R, Doughty P, Raxworthy CJ, Scheinberg L, Resetar A, Sabaj M, Schneider G, Franzen M, Glaw F, Böhme W, Schweiger S, Gemel R, Couper P, Amey A, Dondorp E, Offer G, Meiri S, Wallach V (2019) A global catalog of primary reptile type specimens. *Zootaxa* **4695**, 438–450.

Wells RW (2002) Some taxonomic changes to the genus *Lampropholis* (Reptilia: Scincidae) from Australia. *Australian Biodiversity* **8**, 1–24.

Wells RW (2010) Some taxonomic and nomenclatural considerations on the Class Reptilia in Australia. Comments on the genus *Lampropholis* and related genera in the Family Scincidae. *Australian Biodiversity Record* **1**, 1–22.

Wells RW, Wellington CR (1983) A synopsis of the class Reptilia in Australia. *Australian Journal of Herpetology* **1**, 63–129.

Wells RW, Wellington CR (1984) A synopsis of the class Reptilia. *Australian Journal of Herpetology* **1**, 73–129.

Wells RW, Wellington CR (1985) A classification of the Amphibia and Reptilia of Australia. *Australian Journal of Herpetology*, Supplementary Series **1**, 1–61.

15
Lampropholis: An emerging model taxon

THE CONCEPT OF MODEL TAXA

Some taxa have been singled out for intense scientific scrutiny and are known as 'model organisms' because of their fundamental importance in biological research. Matthews and Vosshall (2020) recognised only six species, a bacterium (*Escherichia coli*), two yeasts (*Saccharomyces cerevisiae* and *Schizosaccharomyces pombe*), a nematode (*Cenorhabditis elegans*), a fruit fly (*Drosophila melanogaster*) and a rodent (the house mouse *Mus musculus*) as qualifying for this category although a number of other candidates come to mind, such as two fungi (*Aspergillus nidulans* and *Neurospora crassa*); a plant (*Arabidopsis thaliana*); *a* protist (*Tetrahymena*); fish (zebrafish, *Danio rerio*; guppies, *Poecilia reticulata*; stickleback, *Gasterosteus aculeatus*); a frog (*Xenopus laevis*); some birds (zebra finch, *Taeniopygia guttata*; budgerigar, *Melopsittacus undulatus*); and two more rodents (laboratory rat, *Rattus norvegicus*; hamster, *Mesocricetus auratus*).

The more that is known about a species, the more valuable it becomes as a target of research because any new findings are easier to incorporate into a more complete biological context, and into a better understanding of the species' ecology and role in the assemblages and communities of which it is a component. Although some species that are especially suited for investigation of particular topics (e.g. flatworms for studying regeneration) and thus could qualify as candidates for model organisms, others could not. Imagine, for example, how inappropriate that would be for such late-maturing, long-lived species as elephants, giant tortoises (Milinkovitch *et al.* 2012; Loire *et al.* 2013), and the tuatara (*Sphenodon punctatus*) or a species such as the jaguar (*Panthera onca*) with a home range of many square kilometres (Crawshaw and Quigley 1991), or one like the red panda (*Ailurus fulgens*) with a highly specialised diet that includes >90% 'carefully chosen' bamboo leaves and shoots which represents >1.5 kg fresh leaves or 4 kg of fresh shoots' daily (Eriksson *et al.* 2010), thereby making it difficult to rear in captivity and thus to be used as a model species.

The species best suited to be model organisms are those that lend themselves to experimental investigations in a wide variety of biological disciplines. Important attributes are small size, widespread distribution, short mean-generation time, inexpensive, commensal of humans, and breeding well in captivity (Matthews and Vosshall 2020). Although unusual spectacular species serve to illustrate the diversity of adaptation and the limits to which natural selection can operate, the common, small, and easily managed species are more amenable for unravelling the basic concepts of functional biology. The quintessential such taxon is *Drosophila*. At one time, the early work on *Drosophila* was viewed as trivial and the animals themselves as unimportant; yet, other than the laboratory rat, it is difficult to find a species that has contributed so much to the present-day understanding of important biological principles (Rosenblueth and Wiener 1945; O'Kane 2003; Rose 2008; Kogan *et al.* 2015; Shenoi *et al.* 2022).

Lizards have been considered potential model organisms (Van Damme *et al.* 2008; Baird 2013; Mezzasalma *et al.* 2021; Lailvaux *et al.* 2022) and at least some species of *Lampropholis* are endowed with features needed to qualify as a candidate for such status. Some are abundant and widespread in eastern Australia and one, *L. delicata*, has been introduced into Hawaii, New Zealand, and Lord Howe Island (see Chapter 3). Almost everyone living within their broad geographic range has them in the gardens around their homes and they also are common in local parklands and adjacent bushland. They adapt well to living in the laboratory and quickly adjust to the presence of humans (Burgin 1993); they also have a broad diet of almost any insect or other small invertebrate within an appropriate size range (see Chapter 8). They thus fulfil the criteria for status as a model organism (Shine 2017; Catullo *et al.* 2019; Goulet *et al.* 2021). Van Dyke *et al.* (2021) opined that Australian skinks are ideal models for testing adaptive hypotheses given their diversity in reproductive mode, especially taxa that overlap in distribution and ecology, and he signaled *Lampropholis* as one of the four most important genera in this regard. *Lampropholis guichenoti* and *L. delicata* are remarkably alike in their life cycles, ecologies, and morphologies, yet are broadly sympatric and even syntopic. However at some localities one species is more abundant than the other. They both show flexibility in their basic attributes in ways that would seem to be adaptive. Some of the more important questions requiring resolution are: How can two such similar species coexist? Why don't they compete, or do they? How do they fit into the varied assemblage of lizards they encounter over their geographic range? Is it because they are flexible enough to modify their ecological niche to conform to whatever axes may be left vacant by other components of their immediate assemblage? Does their flexibility extend to ways of coping with different physical environments? Are there subtle differences in their ecologies that explain the extensions of geographic range of one species beyond that of the other? The information summarised in this book has posed these questions and hints at possible answers (see especially Chapters 5 and 10) but the information in the scientific literature is patchy and incomplete. What is needed now is comprehensive exposition of the community ecology of the assemblages in which the species of *Lampropholis* are a part both in the centre and at the edges of their common geographic ranges, as well as in areas where *L. guichenoti* or *L. delicata* do not overlap.

SUITABILITY OF *LAMPROPHOLIS* AS A MODEL TAXON

The above discussion refers mainly to the two eurytopic, widespread, abundant species, *L. delicata* and *L. guichenoti*, that collectively constitute excellent canditates for a model taxon. They have a wide distribution over which they encounter divergent physical environments, both elevationally and latitudinally, such as temperature and moisture. They also occur in a multiplicity of assemblages of species with which they interact ecologically and behaviourally against a partially understood background of genetic and environmentally induced characteristics that have both direct and interactive influences. Unravelling these sources of influence in the genus *Lampropholis* would lead to it playing an even more important role as a model taxon for investigating the processes of evolution and adaptation (see Chapter 15). Indeed, Goulet *et al.* (2021) recognised the importance of this genus by indicating that 'Skinks in the genus *Lampropholis* have become a model system for behavioral syndrome research'. However, unlike some model taxa, this genus also includes several other species that are closely related to the two eurytopic forms but possess some contrasting attributes (see Chapters 2 and 3); for example, they have restricted distributions that suggest ecological divergences. Comparing these species with the euytropic ones enhances the value of the genus as a model taxon and may provide unique insights into the adaptive scopes of related genomes.

Lampropholis already has contributed significantly in ways that are characteristic of a model taxon (see Burgin 1993; DeMeester and Baeckens 2021; Van Dyke *et al.* 2021). The preceding chapters indicate the contribution of this genus to a deeper understanding of the transition from oviparity to viviparity in reptiles (Chapter 5), and of ecology (Chapter 11), physiology (Chapter 7) genetics (Chapter 13), and phylogeny (Chapter 14). Qualls and Shine (1998) suggested that lizards 'may be ideal model systems with which to further investigate the contributions of developmental plasticity in embryogenesis to geographic variation in phenotypic traits; *L. guichenoti* was their species of choice in their study of this topic. More recently, collaborative research by David Chapple and his colleagues have propelled this genus to the forefront of heuristic research through a myriad of papers on many aspects of its biology. The present book assembles the widely scattered data on these fascinating animals and we hope that our summary of the accumulated information into a more comprehensive and integrated background will contribute to this genus becoming an even more effective reptilian model taxon.

Intensive research on a taxon has significant benefit to that taxon itself. The more that is known about a species, the better one can provide an appropriate environment for it and manage its conservation and health.

Comparison of the populations both within *L. delicata* and *L. guichenoti* throughout their respective distributions, for example, has demonstrated an amazing geographic diversity of attributes (Chapple *et al.* 2011a, b) and underscores the necessity of preserving the totality of the diversity over the entire geographic ranges of these species, and that conservation measures may need to be tailored to specific localities rather than uncritically applied broadly. Thus, the sensitive and well-meaning, but naïve, suggestion that research upon these animals should be discouraged or restricted (Stiles 1993) is counter to the best interest of these species. We hope that the information in the present book will alert the general public to the need to conserve the complete range of diversity in the genus *Lampropholis* throughout its entire distribution and to conduct research upon it not only to expand understanding of its biology in general but for the benefit of all its species, clades, and subclades.

SUMMARY

Two species of *Lampropholis* (*L. delicata* and *L. guichenoti*) have precisely the attributes that are requisite for a successful model taxon and a group of satellite congeners with narrow distributions contrast markedly with the two eurytopic ones. The combination provides a valuable contrast and expands our understanding of the adaptive repertory of species with a common genetic heritage – in effect being the saurian counterpart of the insect *Drosophila*. The genus *Lampropholis* unwittingly is rapidly becoming a reptilian model taxon; in fact it has already reached that status.

REFERENCES

Baird TA (2013) Lizards and other reptiles as model systems for the study of contest behaviour. In *Animal Contests*. (Eds TA Baird, ICW Hardy and M Briffa) pp. 258–286, Cambridge University Press, Cambridge.

Burgin S (1993) *Lampropholis*: the new 'laboratory' animals. In *Herpetology in Australia: A Diverse Discipline*. (Eds D Lunney and D Ayers) pp. 279–281. Royal Zoological Society of New South Wales and Surrey Beatty & Sons, Chipping Norton, NSW. doi:10.7882/RZSNSW.1993.041

Catullo RA, Llewelyn J, Phillips BL, Moritz CC (2019) The potential for rapid evolution under anthropogenic climate change. *Current Biology* **29**(19), R996–R1007. doi:10.1016/j.cub.2019.08.028

Chapple DG, Hoskin CJ, Chapple SNJ, Thompson MB (2011a) Phylogenetic divergence in the widespread delicate skink (*Lampropholis delicata*) corresponds to dry habitat barriers in eastern Australia. *BMC Evolutionary Biology* **11**(1), 191. doi:10.1186/1471-2148-11-191

Chapple DG, Chapple SNJ, Thompson MB (2011b) Biogeographic barriers in south-eastern Australia drive phylogeographic divergence in the garden skink, *Lampropholis guichenoti*. *Journal of Biogeography* **38**(9), 1761–1775. doi:10.1111/j.1365-2699.2011.02531.x

Crawshaw Jr, PG, Quigley HB (1991) Jaguar spacing, activity and habitat use in a seasonally flooded environment in Brazil. *Journal of Zoology* **223**, 357–370. doi:10.1111/j.1469-7998.1991.tb04770.x

DeMeester G, Baeckens S (2021) Reinstating reptiles: from clueless creatures to esteemed models of cognitive biology. *Behaviour* **158**(12–13), 1057–1076. doi:10.1163/1568539X-00003718

Eriksson P, Zidar J, White D, Westander J, Andersson M (2010) Current husbandry of red pandas (*Ailurus fulgens*) in zoos. *Zoo Biology* **29**(6), 732–740. doi:10.1002/zoo.20323

Goulet CT, Hart W, Phillips BL, Llewelyn J, Wong BBM, Chapple DG (2021) No behavioral syndromes or sex-specific personality differences in the southern rainforest sunskink (*Lampropholis similis*). *Ethology* **127**(2), 102–108. doi:10.1111/eth.13103

Kogan V, Molodtsov I, Menshikov LI, Shmookler R, Fedichev P (2015) Stability analysis of a model gene network links aging, stress resistance and negligible senescence. *Scientific Reports* **5**, 13589. doi:10.1038/srep13589

Lailvaux SP, Mishra A, Pun P, Ul Kabir MW, Wilson RS, Herrel A, Hoque MT (2022) Machine learning accurately predicts the multivariate performance phenotype from morphology in lizards. *PLoS One* **17**(1), e0261613. doi:10.1371/joounal.pone.0261613

Loire E, Chiari Y, Bernard A, Cahais V, Romiguier J, Nabholz B, Lourenço JM, Galtier N (2013) Population genomics of the endangered giant Galápagos tortoise. *Genome Biology* **14**(12), R136. doi:10.1186/gb-2013-14-12-r136

Matthews BJ, Vosshall LB (2020) How to turn an organism into a model organism in 10 'easy' steps. *Journal of Experimental Biology* **223**(Supplement 1), jeb218198. doi:10.1242/jeb.218198

Mezzasalma M, Guarino FM, Odiema G (2021) Lizards as model organisms of sex chromosome evolution: what we really known from a systematic distribution of available data? *Genes* **12**(9), 1341. doi:10.3390/genes12091341

Milinkovitch MC, Kanitz R, Tiedemann R, Tapia W, Caccone A, Gibbs JP, Powell JR (2012) Recovery of a nearly extinct Galápagos tortoise despite minimal genetic variation. *Evolutionary Applications* **6**(2), 377–383. doi:10.1111/eva.12014

O'Kane CJ (2003) Modelling human diseases in *Drosophila* and *Caenorhabditis*. *Seminars in Cell and Developmental Biology* **14**(1), 3–10. doi:10.1016/S1084-9521(02)00162-3

Qualls FJ, Shine R (1998) Geographic variation in lizard phenotypes: importance of the incubation environment. *Biological Journal of the Linnean Society* **64**(4), 477–491. doi:10.1006/bijl.1998.0236

Rose MR (2008) Making SENSE: strategies for engineering negligible senescence evolutionary. *Rejuvenation Research* **11**(2), 527–534. doi:10.1089/rej.2008.0688

Rosenblueth A, Wiener N (1945) The role of models in science. *Philosophy of Science* **12**(4), 316–321. doi:10.1086/286874

Shenoi VN, Brengdahl MI, Grace JL, Eriksson B (2022) A genome-wide test for paternal indirect genetic effects on lifespan in *Drosophila melanogaster. Proceedings of the Royal Society B: Biological Sciences* **289**(1974), 20212707. doi:10.1098/rspb.2021.2707

Shine R (2017) Biological invasions and animal behaviour. *Copiea* **105**, 601–649. doi:10.1643/OT-17-649

Stiles K (1993) Response: *Lampropholis*: the new laboratory animals. In *Herpetology in Australia: A Diverse Discipline*. (Eds D Lunney and D Ayers) p. 282. Royal Zoological Society of New South Wales and Surrey Beatty & Sons, Chipping Norton, NSW. doi:10.7882/RZSNsw.1993.041

Van Damme R, Entin P, Vanhooydonck B, Herrel A (2008) Causes of sexual dimorphism in performance traits: a comparative approach. *Evolutionary Ecology Research* **10**(2), 229–250.

Van Dyke JU, Thompson MB, Burridge CP, Castelli MA, Clulow S, Dissanayake DSB, Dong CM, Doody JA, Edwards DL, Ezaz T, Griesen CR, Gardner MG, Georges A, Higgie M, Hill PL, Holleley CE, Hoops D, Hoskin CJ, Merry DL, Riley JL, Wapstra E, While GM, Whiteley SL, Whiting MJ, Zozaya SM, Whittington CM (2021) Australian lizards are outstanding models for reproductive biology research. *Australian Journal of Zoology* **68**(4), 168–199. doi:10.1071/ZO21017

16
Where are we now and where do we go from here?

WHERE ARE WE NOW?

In this book we summarised what is known of the biology of the genus *Lampropholis* but probably have raised as many questions as we have answered. Perhaps it is time to briefly take stock of where we are and make recommendations of how to proceed into the future. The picture that has unfolded is of two species *L. delicata* and *L. guichenoti*, similar in behaviour, appearance, size, distribution, diet, physiology, and ecology, as discussed in Chapters 2, 4, 5, 7, and 8. In addition there are a group of species that have small geographic ranges (some even restricted to individual mountain tops). Some of these species are morphologically so similar to other species as to be indistinguishable from them except by analysis of their DNA. Indeed, some populations of one of the widespread species (*L. delicata*) are morphologically more different from each other, than are some of the genetically defined satellite species from each other (see Chapter 14).

Complicating this already complex situation are the facts that: (1) some authors disagree as to whether or not *L. delicata* and *L. guichenoti* differ in respect to a particular ecological predilection (see Chapter 11), and that (2) a number of morphological and behavioural features of these eurytopic species have been shown to be strongly influenced by environmental factors operating on the developing embryo or even on individuals after hatching and living free of the egg (see Chapter 5). Use of such characteristics as the basis for distinguishing and describing new species is, of course, invalid as they merely represent the effect of local environments upon a common genome, not genetic differences between populations. Even more perplexing is the interaction of ecological and genetic factors in influencing a given characteristic. Table 16.1 shows the relative influence of some ecological and genetic factors that interact in affecting particular attributes of *L. guichenoti*. Note that for some characteristics the genetic influence is stronger; for others the ecological one is. The result of such interactions means that if one relies solely on morphology in defining a species mistakes can be made because the particular structural characteristics employed may merely be environmentally induced variants. Conversely, except for parthenogenetic offspring, or monozygotic siblings, such as identical twins, triplets etc., no two individuals on earth are exact genetic replicates. Just where does one draw the line in deciding how great that genetic difference must be to make two similar populations distinct species? The official criterion for separation into distinct species is whether populations can interbreed and have fertile offspring; if so, they are the same species, if not they are different species. Yet, we have seen that in some cases two groups that clearly are different species by all other criteria can interbreed under some circumstances (e.g. male lion × female tiger = liger and male tiger × female lion = tigon) (Yadav *et al.* 2019) and the familiar hybrids between horses and donkeys to produce mules (donkey stallion and female horse) and hinnies (horse stallion and female donkey); however, the hybrids usually are sterile. We have seen previously that in *Lampropholis* supposedly conspecific populations fail to interbreed simply because males will not mate with females from the opposite population (Forman and Shine 1995). Is this behavioural trait genetically based and therefore legitimate grounds for considering these populations as separate species, or is it because, say, members of one population have a different odour from those from another population merely because they are

Table 16.1. Division of the phenotypic variation in labile traits between hatchlings of *Lampropholis guichenoti* from Sydney (incubated at higher temperature) and the Blue Mountains (incubated at lower temperature) into that which is attributable to the source population (genetic influence) and that attributable to differences in temperatures under which incubation took place (environmental influence).

Slightly modified from Qualls and Shine (1998).

Phenotypic trait	Difference attributable to temperature of incubation (environmental) (%)	Difference attributable to source population (genetic) (%)
Mean initial sprint speed	66	34
Mean speed over 1 m	46	54
Inter-limb length	46	54
Length of tail	53	47

feeding on different species of prey, rub against different species of plants in their separate localities, or some other environmentally induced difference acting upon genetically compatible populations?

Attention PhD students who are looking for a subject for your dissertation! Why not artificially cross-fertilise eggs from one population with sperm from another population as a means of ascertaining which populations are physiologically and genetically capable of producing viable offspring and which are not? Then for those that are compatible in these ways, but behaviourally avoid mating, seek the olfactory, or other, cues responsible for such repellence. For example, an investigator could rub cutaneous secretions from individuals from one population onto females from another population to see if the latter are then rejected by males that are artificially rendered as 'alien', then isolate the chemical components of the skink's cutaneous secretions and test them individually as an entrée into the specific ones that are operative, and whether they are produced by the lizards themselves (and hence based on different genetically controlled pathways of synthesis), or occur only in plants or other environmental substances encountered in some localities but not in others.

This example emphasises the value of model organisms. Clearly, the above hypotheses and guidelines for research could not have been developed without prior knowledge about the environmental induction of characteristics in this model genus, or of the behavioural isolation of populations summarised in the foregoing chapters. *Lampropholis delicata* and *L. guichenoti* are especially suitable because of their attributes described in the preceding chapters.

Taxonomists and geneticists have followed parallel paths in the study of the evolutionary diversification of animals and are now beginning to merge. Taxonomists traditionally used morphology as the basis for classification and nomenclature whereas more recently geneticists have complemented our understanding of the process of evolution and have uncovered cryptic species and other subtleties that so far have not been integrated entirely into the classical system of nomenclature. Blom *et al.* (2016) suggested that 'analogous habitats have provided parallel ecological opportunity and have repeatedly promoted adaptive diversification. By contrast, speciation within the same habitat has resulted in distinct lineages with relatively limited morphological variation' and Burbrink *et al.* (2020) pointed out that genomic data often show little support for the traditional interpretation of phylogenetic relationships based on morphology. The genus *Lampropholis* exemplifies these dichotomies and at the same time points the way toward a reconciliation of these two approaches.

Thanks to two seminal papers by Chapple *et al.* (2011a, b) we now have detailed knowledge of the geographic distribution of the genetic variants within the two widespread, eurytopic species, *L. guichenoti* and *L. delicata*, as well as the type, location, climate, and period of existence of barriers that reproductively isolated populations and led to their divergence; in short, when and where they became separated from each other and the degree of genetic divergence that has occurred during isolation. Definition of the infraspecific variation into precisely analysed clades and subclades gives a much clearer picture of taxonomic structure as it identifies divergences between populations that are morphologically similar but genetically different. This procedure, however, does not prevent the naming of satellite cryptic species, particularly if it can be shown that they are incapable of interbreeding with their parent species (see Chapter 14). We feel that denoting clades and subclades based on a firmer basis of understanding of the relationships of populations is preferable to classifying variants into traditional categories such as subspecies or races and that caution should be exercised in naming variants as new species solely on slight morphological differences unless there is clear evidence of reproductive incompatibility with other populations, that is neither based on clinal relationships nor resulting from environmentally induced characteristics or such cues as odours obtained from the immediate environment. It is more realistic to define relationships in terms of clades and subclades that emphasise the close relationships of the populations involved on the basis of known genetic affinity rather than prematurely proliferating a number of named putative species.

Chapple *et al.* (2011a) identified the Burdekin Gap, St Lawrence Gap, Macpherson Main Range, Hunter Valley, Southern NSW, Gippsland Basin, Bass Strait, the Murray Basin, and the Mt Lofty Ranges as barriers separating populations of *Lampropholis* from maintaining genetic continuity and thus leading to evolutionary divergence between populations (Figs 2.1 and 2.2 in Chapter 2). These gaps were of different dominant features, predominately dry barriers, topographical barriers, and either past or present marine barriers, although some barriers combine several of these features. There are three drier areas separating two types of moister vegetation: (1) the Burdekin Gap is a broad tongue of dry woodland savannah extending from inland to the coast and separating rainforests on its north from mid-eastern Queensland forest on its south; (2) the St Lawrence Gap is another coastal dry barrier separating two more mesic types of forest, the mid-eastern Queensland forest on its north and the south-eastern Queensland forest on its south; and (3) the Hunter Valley is a dry, open valley delineating the southern limit of the eastern biogeographic region from the south-eastern forests. This barrier combines topography as well as moisture. The topographic barriers include Kroombit Tops, an insular-like moist region perched above the surrounding drier lowland of eucalypt woodlands.

Now, in the Anthropocene, the alteration of habitats has increased markedly due to anthropogenic agencies and climatic change, both indirectly through the effect on climate, or more directly through alteration of the habitat through such techniques of management as prescribed burning (Harris *et al.* 2023).

Heuristic summary

Charles Darwin (undated) exhibited great prescience in a preface to the 6th edition of his classic treatise on the origin of species by means of natural selection:

> *Who can explain why one species ranges widely and is very numerous, and why another allied species has a narrow range and is rare? Yet these relations are of the highest importance, for they determine the present welfare and, as I believe, the future success and modification of every inhabitant in the world.* (pp. 4–5)

One would be hard pressed to find a better description characterising the genus *Lampropholis*. The present book has summarised how far herpetologists have gone in answering Darwin's query.

It is tempting to hypothesise on the basis of current information that the northward extension of the range of *L. delicata* may have been because: (1) that species has been more successful than *L. guichenoti* in adapting to the warmer conditions toward the equator, but (2) less successful in adapting to the drier conditions of inland New South Wales, and that (3) the former's appearance on the Bass Strait islands and Tasmania was because of the vagaries of overseas dispersal or its greater propensity for overseas transport by humans at times when there was no terrestrial connection between those areas and the mainland (see Chapter 2), or maybe (4) their slightly off-skewed differences in distribution reflects the nuances of their overlap with the geographically varying constitution of assemblages of contingent predators (see Chapter 9) or competitors (see Chapter 11) in which these species were embedded. Any of these hypotheses alone is simplistic; multiples of factors may interact synergistically in producing the complex array of physical and biotic determinants of the geographic variation in adaptive responses. After the various facets of the biology of this fascinating pair of species have been examined, we will revisit this topic and suggest ways of addressing the issue (see below). Doing so is important within the context of the projected consequences of the instability of climatic conditions we have inflicted upon wildlife with whom we share our planet (see Chapter 12).

The fact that *L. delicata* and *L. guichenoti* occupy a wide range of habitats and are broadly sympatric over much of their collective geographic ranges, or even often are syntopic, would seem to rule out competition between them as influencing their distributions. Also, the fact that both species eat practically any kind of prey (Chapter 8) may mean that they can opportunistically feed on whatever is available, thereby reducing competition. Given their ecological similarity, it seems unlikely that competition with an entirely different third species would eliminate either of these two species over part of their potential range without also affecting the other. There simply is not enough information available to ascertain what factors limit the distributions of these two widespread species of *Lampropholis*. Some such influences may be very subtle and not obvious without careful experimentation and may differ on different edges of their ranges (see Chapter 10 and the section immediately below).

Recent research, especially that conducted by Richard Shine and his students and that carried out currently and in progress by David Chapple and his collaborators have pioneered investigation into these complex interactions and have gone far to establish the genus *Lampropholis* as the truly model taxon that it now is (see Chapter 15).

WHERE DO WE GO FROM HERE?

Wheeler *et al.* (2012) concluded that by building on the existing data, collections, and expertise; using recent advances in technology and the guidance of planning by professionals from diverse fields of expertise, it would be possible to 'vastly enrich and expand the environmental and evolutionary sciences, and to return unlimited dividends to society in innovative solutions to countless challenges' and at reasonable cost. This view may be naïvely optimistic given the present state of the globe and its inhabitants, brought about by ignoring the warnings of scientists and by the burgeoning of avarice, ignorance, and wastage in the conduction of public policies and in the managing of resources. Nevertheless, it is a goal worth striving to attain. To retain a diverse natural world of wonder and beauty we must understand how natural ecosystems work and how they can be maintained as the human population outstrips the resources upon which it depends. Lizards may seem a small and insignificant part of that world, but all organisms can contribute to an understanding of the principles underlying the stability, or lack thereof, of ecosystems. We need to understand the intricacies of adaptation to the nuances of variation in the physical and biotic world.

Emanating from what is known so far, we can recommend some practical approaches for future research: (1) One should not use any morphological characteristic as diagnostic in describing a new species if that feature is known to be influenced directly by environmental (non-genetic) determinants (see Chapter 14). (2) If possible, research into the genetic constitution of populations should be carried out in determining the taxonomic status of any species. (3) Clades should not be elevated to the status of species unless there is compelling evidence of reproductive incompatibility. (4) Research needs to be conducted that assesses the relative roles of the physical environment, phylogenetic heritage, and composition of the assemblages of which the subject is a member. This approach necessitates comparing the competitive interaction of the subject species with the varying composition of its syntopic assemblages in different geographic regions and elevations. For example, how does the presence or absence of a third species in the assemblage affect the interaction of two other coexisting species? Do the two species change the size of their home ranges and tend to forage equally in a wider range of microhabitats? Or do they keep the same size of home range but include more than one habitat within a fixed area? What happens in an area where there is a fourth, or more, species in addition to the first three? Does one, or more, species alter the kind of area in which they forage; how do the dynamics of this more complex assemblage differ?

Regardless of how systematic and nomenclatural dilemmas are resolved, there are ways of analysing different entities (whether designated as species, subspecies, or clades) that are compartmentalised into different habitats and biological assemblages in compatible but varying ways. This approach involves an integrated combination of: (1) ascertaining the degree of genetic divergence versus environmentally induced differences, (2) detailed mapping of microhabitats, such as illustrated by Plate 3, (3) determining the home ranges of individual lizards and the degree of their overlap with other entities, as shown to be possible by the study of Purani (1994) and superimposing those on maps, and finally (4) subjecting the data to an analysis of point-diversity (see Heatwole *et al.* 2013) to ascertain the extent to which individual species either contact each other, or co-occur at the same point, either sequentially or simultaneously, within their overlapping home ranges. Heatwole *et al.* (2013), in a study of ants, developed the concept of point-diversity by ascertaining the extent of co-occurrence of the various species in the assemblage and the extent to which they used the same space. Similar studies of lizards but using comprehensive coverage of pitfall traps, carried out at strategically selected habitats at different latitudes and elevations, and involving different assemblages of species of *Lampropholis* and their close relatives of equivalent size, would be instructive. Note that it may be necessary to distinguish between different econes for some species (Heatwole 1989).

We recommend the following hierarchical assessment of the multiple factors involved in the fitting of species into the Rubik's Cube of their assemblage, all of which already have been demonstrated to be feasible for achieving the goals listed for approach number 4 above.

(1) *The level of the habitat*. Determine the complete range of habitats used by a species in some broad region. Rawlinson (1974) exemplified this approach by listing all the major habitats of lizards in Tasmania and the islands of Bass Strait and the species inhabiting each of them, thereby indicating not only the range of habitats each species occupies but also the species with which it can coexist in each of them (for a fuller discussion of his study, see Chapter 11). Hints for selecting an appropriate array of areas for such an analysis can be gained from Figs 2.1 and 2.2.
(2) *The level of the microhabitat and niche*. Assess the degree of overlap in the daily level of activity, diet, body-size, microhabitat, and degree of specialisation of component species in assemblages of species coexisting in the same habitat. Again there is a precedent. Goodman (2007) described the degree of overlap in some of these axes in assemblages involving species of *Lampropholis* with restricted geographic distributions and found that species broadly overlapping in one axis of the niche, say body size, tended to be separated along a different axis, say, microhabitat (e.g. arboreal versus terrestrial) (see Chapter 11). Ideally, in order to completely resolve the dilemma of *L. delicata* and *L. guichenoti* (see Chapter 11), each subclade of both these species should be analysed in this way, although such an extensive scope probably is logistically challenging.
(3) *The level of the home range*. The fact that two species occupy the same kind of microhabitat in the same general region does not necessarily mean that every patch of that microhabitat contains both species. In the case of lizards that defend territories, an individual male may exclude other conspecific males from its territory by threat or actual combat, and this perhaps could apply to other species of similarly sized lizards. However, even though combat between individuals is known in *Lampropholis* and that they sometimes have agonistic interactions with individuals of other species, they do not seem to be territorial even intraspecifically (see Chapter 4); it seems unlikely that they would exhibit interspecific territoriality. Nevertheless, one should be alive to the possibility that via some mechanism or other syntopic individuals of different species of *Lampropholis* might exclusively occupy individual patches of a given type of microhabitat. Studies of *L. delicata and L. guichenoti* by Purani (1994) indicate broad overlap of individuals' home ranges, not only between the sexes, but also between these two species (see Chapter 4) and certainly *L. delicata* and *L. guichenoti* both share parts of their home ranges for nesting and brumating communally with various species, even from different genera, as well as communally with members of their own species (see Chapter 4). Nevertheless, it might be instructive to set up a series of enclosures, each featuring a different type of microhabitat (e.g. some with leaf litter, others with grass, and still others with mixed patches of two or more types of microhabitats) and introduce equal numbers of individually marked *L. delicata* and *L. guichenoti* so as to observe any differences in the partitioning of space. We provide these suggestions because it is what we would be doing were we not retired and spending our productive time in writing up our backlog of previous research rather than pursuing new projects.
(4) *The level of point-diversity*. Even if the home ranges of two individuals overlap for certain activities such as nesting, brumation, or basking, as they do for species of *Lampropholis*, individuals still could have different home ranges for such activities as foraging, or even if not, they might use different patches or parts of patches preferentially, or at least some more often than others, with infrequent overlap between individuals' use of specific locations, i.e. point-diversity would be low (see Chapter 11). The experiment described in level number three would reveal whether such subtle intraspecific ecological segregation exists.
(5) *The level of the econe*. Econes in which males, females, and juveniles are ecologically segregated by foraging in different microhabitats have been demonstrated in a non-scincid lizard (see Chapter 11) and various attributes have been shown to be related to body size in *Lampropholis* in the preceding chapters. Whether such differences translate into econes in *Lampropholis* could be detected by including juveniles as well as both sexes in appropriately designed enclosures and conducting experiments similar to those described for level number 3 above. Recall that some species of *Lampropholis* have slight overlaps in size of the body

both with generally larger and generally smaller species. If there are econes in either *Lampropholis* or the opposing species, it could mean that the home ranges of foraging adults of two species could be separate, with interspecific overlap being limited to that between the juveniles of one species with the adults of its smaller counterpart.

Superimposed on this already complex situation is the effect of additional numbers of species in the assemblage. For example if you have species A, B, and C each individually capable of occupying a particular habitat either alone or in combinations of either A + B, A + C, or B + C but not A + B + C, the complexity of the possible integration of species within their assemblage becomes obvious. For example, the combination of A + B collectively could dominate the range of crucial microhabitats used by C and competitively exclude C, but that a fourth species D differed sufficiently in its use of habitat that it could coexist with A + B but not with C with which it overlapped in microhabitat and other ecological attributes; thus A + B + D would be a compatible assemblage, but not D + C. Now assume that species D could forage in either of two different microhabitats (say grass or leaf litter) because these were not used by A or B, and the only one of which was used by C. D could narrow its foraging to the available vacant microhabitat, thereby leaving the alternate one for C. In this case the combination A + B + C + D would be possible. Now add species E, F, G, and so on with the occurrences of the other species to increase the diversity of the assemblage, or move to a different geographic area beyond the distributional ranges of C to G, but instead the assemblage now includes H and I instead, which occupy slightly different niches than did any of C to G, meaning that A and B now need to adjust their niches to accommodate a different variety of competitors in a geographic area where not only does the mix of species differ, but so does the seasonal march of temperature and moisture, structure of microhabitats, and composition of vegetation. Add to all of that the fact that some of these environmental influences, such as the structure and composition of the vegetation, may induce non-genetic changes in the morphology, life history, behaviour, growth, and reproductive biology that may mimic what one normally would perceive as genetically based, rather than environmentally induced characteristics (see Qualls and Shine 1998, 2000; Downes and Hoefer 2007; and Chapter 12) and one can begin to understand how individual populations of such widespread species as *L. delicata* and *L. guichenoti* can adapt to local conditions and diverge into clines. Finally, the imposition of physical barriers at different times and places that isolate populations of widespread species and result in describable genetic differences between them reinforces local adaptation that eventually can lead to reproductive incompatibility (i.e. speciation). Just when that threshold is reached is the basis of the perpetual debate between lumpers and splitters (see Chapter 14). Just where are the boundaries between morphs (individual variants within a reproductively compatible population, such as blue eyes versus brown eyes in humans), introgression, hybridisation, and speciation? Certainly in the case of gradual clines in which there are no sharp transitions that clearly identify where the boundary between putative species occurs even though populations at the extreme ends may be reproductively incompatible, should be simply identified as clines, not broken into arbitrary species. We favour the classification scheme of Chapple *et al.* (2011a, b) in which the populations of widespread taxa such as *L. guichenoti* and *L. delicata* (*sensu lato*) are genetically (and possibly morphologically) defined as clades and subclades within these two species. This classification clearly indicates the degree of relationship between populations without prematurely committing to whether those differences have reached the level of reproductively isolated species. It may be found that some clades will be found to be full species and when sufficient information is available to do so, some of the names that are now relegated to the synonymy of one of the two widespread species could be resurrected. Conversely some of the cryptic species may be found to be sufficiently compatible reproductively with nominate *L. delicata* or *L. guichenoti* that there will be consensus that cross-breeding denotes true compatibility, rather than hybridisation. In such a case, the present name would be demoted to synonymy. Again, exactly where to draw the line may be subjective and universal agreement may not ever be achieved between lumpers and splitters.

A HEURISTIC THEORETICAL MODEL

Tracy and Christian (1986) proposed that home ranges be evaluated to reflect their functional relationship with other aspects of a species' biology. In doing so, we find that many previous portrayals of home ranges are misleading.

Consider three species that tend to preferentially carry out their daily activities, each in a different microhabitat, species A in leaf litter, species B in grass, and species C on bare ground, but if any one of them were the sole species of the trio that were represented in the assemblage occurring in an area, it would forage over all three substrates and probably in greater population density than when co-occurring with either of the two remaining species, or with a combination of both. Now suppose that instead of any species being the sole member of the trio to occur in the assemblage that species A + B constitute the assemblage. In this case instead of either occurring throughout the region, A would withdraw selectively into areas of leaf litter and B into the grassy ones and which would continue to forage in the bare areas would be determined by the interaction of these two species in that microhabitat. If A and C constituted the assemblage then A would still dominate in leaf litter and C on bare ground and which would dominate in grassy areas depending on the interaction of A and C on that substrate. An assemblage of B and C would result in B foraging in grass, and C on bare ground, with the forager on litter to be disputed. Now include all three in the assemblage and we get each species retreating into their preferred areas for foraging as described by the home range of an individual of each the three species in a specific theoretical plot (Plate 9).

This map shows a hypothetical configuration of patches of these three types of preferred microhabitat. There are three patches of leaf litter and an individual of species A forages intermittently in all three but must traverse patches of other substrates to do so. An individual of species B forages in grass but this microhabitat is in larger patches than those used by species A, and although B could reach all parts of its structural patch without crossing any other types, it is not constrained to do so. Finally, an individual of species C occupies the matrix in which the other two types of microhabitat are embedded and it, like species B, could reach any part of its microhabitat without crossing any other type of substrate but would also have the option of sometimes taking shortcuts across other kinds of patches. One could discover which option species take by doing what old-fashioned naturalists did so effectively – watch the animals. By spending hours painstakingly watching conspicuously marked lizards through binoculars and/or relying on data from a saturation of pitfall traps, one could ascertain whether an individual spent time foraging over the entire area of a polygon enclosing locations at which it was recorded (small maps in Plate 9) or whether it preferentially foraged only in a specific microhabitat and moved precipitously between patches of that type (large map in Plate 9).

There is one component of our hypothetical model that has not been mentioned yet. So far, we have emphasised only the microhabitats in which animals forage, but they also engage in other activities. Note that there is a log located in individual A's microhabitat of leaf litter, but there is no such cover in any of the other patches. Logs, or equivalent cover, are an important ecological requirement for some *Lampropholis* and a given microhabitat may not suffice for all activities. We have seen that cover is needed for periods of nocturnal or seasonal inactivity and that various species successfully share sites for communal laying and incubation of eggs (see Chapter 4). We have also seen (Chapter 4) that females of *L. delicata* may leave the part of the home range that they usually use for foraging in search of suitable places to lay their eggs. Perhaps it is time to compartmentalise the concept of the home range into the various functions it serves and think of a home range used for foraging, home range for nesting, home range for shelter, and so on. Our theoretical model has incorporated these diversity of functions.

Clearly the use of polygons without sufficient detail as to the amount and location of differential use of space (as seen in the smaller maps in Plate 9, the kind usually featured in studies of home range) can overestimate the size of the home range used for foraging (compare the small maps with the composite map in Plate 9). That method also suggests greater overlap between individuals' home ranges for foraging than actually occurs; again compare the size of each individuals' estimated home range in the individual (small) maps compared to the large map.

The scale of polygons to describe the geographic ranges of the species of *Lampropholis*, and probably of other taxa as well, also is misleading. Take Fig. 2.2 in Chapter 2 for example. The darker line circumscribes a polygon usually designated as the geographic range of *L. guichenoti* (including in the present book) but that is not its true geographic range. On a different scale its geographic range is better described by the series of clades and subclades delimited by lighter lines. The pattern of those differs substantially between *L. delicata* and *L. guichenoti*. There are places where one species occurs but not the other and there are places where the two species are truly sympatric and even syntopic. This disparity is also made clear by Figs 2.1 and 2.2 which show that some faunal surveys found both species in the same areas, whereas elsewhere only one species of the pair occurs. Furthermore, there are a few places where subclades of a given species overlap, for example S3 with S7 and N2 with N4 in *L. guichenoti* (Fig. 2.2 in Chapter 2). The gaps between clades also occur at different places for these two widely distributed species. For example subclade 7b of *L. delicata* overlaps parts of subclades S1 and S2 and their intervening gap (compare Figs 2.1 and 2.2 in Chapter 2). What events in the evolutionary history of these two species generated such different patterns? These anomalies need to be investigated in more detail.

In addition to the varying relationship between *L. delicata* and *L. guichenoti*, there is another species that has been shown to have similar preferences for microhabitat: *Pseudemoia entrecasteauxi* (see Chapter 11). These three species would constitute an ideal triad for exemplifying the theoretical species A, B, and C in the discussion above. Choosing sites where these three co-occurred in different combinations and studying the resulting alterations in their use of microhabitats might provide insight into the ecological dynamics of structuring assemblages.

CONCLUSION

The genus *Lampropholis* has contributed significantly to the understanding of the complex interaction of physical, biotic, and genetic characteristics in determining the ecology, and evolution of a genus of lizards. It is now known that: (1) some morphological characteristics of species in this genus are environmentally affected in addition to those determined genetically; that (2) there are cryptic species within the genus; that (3) some species contain infraspecific clades and subclades as a result of being separated by topographic and climatic barriers; and that (4) the relative abundances and occurrences of particular species may differ under different physical environments and compositions of assemblages.

The challenge now is to provide further documentation and to integrate this mélange of interacting factors into a comprehensive understanding of the biology of this fascinating genus and use that information to preserve its genetic diversity, distributional continuity, and persistence in the face of the accelerating environmental deterioration imposed by humans. We are the cause of the decline of our rapidly disappearing fauna and we need also to be the remedy.

REFERENCES

Blom MPK, Horner P, Moritz C (2016) Convergence across a continent: adaptive diversification in a recent radiation of Australian lizards. *Proceedings of the Royal Society B: Biological Sciences* **283**. doi:10.1098/rspb.2016.0181

Burbrink FT, Grazziotin FG, Pyro RA, Cundall D, Donnellan S, Irish F, Keogh JS, Kraus F, Murphy RW, Noonan B, Raxworthy CJ, Ruane S, Lemmon AR, Lemmon EM, Zaher H (2020) Interrogating genomic-scale data for Squamata (lizards, snakes, and amphisbaenians) shows no support for key traditional morphological relationships. *Systematic Biology* **69**, 502–520.

Chapple DG, Hoskin CJ, Chapple SNJ, Thompson MB (2011a) Phylogeographic divergence in the widespread delicate skink (*Lampropholis delicata*) corresponds to dry habitat barriers in eastern Australia. *Systematics and Biodiversity* **10**, 1–20. doi:10.1186/1471-2148-11-191

Chapple DG, Chapple SNJ, Thompson MB (2011b) Biogeographic barriers in south-eastern Australia drive phylogeographic divergence in the garden skink, *Lampropholis guichenoti*. *Journal of Biogeography* **38**(9), 1761–1775. doi:10.1111/j.1365-2699.2011.02531.x

Darwin C (undated) Introduction. In *The Origin of Species by Means of Natural Selection or the Preservation of Favored Races in the Struggle for Life*. Undated reprint with all additions and corrections of the Sixth London Edition. Burt's Library of the World's Best Books. pp. 1–5. A.L. Burt, Publisher, New York.

Downes S, Hoefer A-M (2007) An experimental study of the effects of weed invasion on lizard phenotypes. *Oecologia* **153**(3), 775–785. doi:10.1007/s00442-007-0775-2

Forsman A, Shine R (1995) Parallel geographic variation in body shape and reproductive life history within the Australian scincid lizard *Lampropholis delicata*. *Functional Ecology* **9**(6), 818–828. doi:10.2307/2389979

Goodman A (2007) Microhabitat separation and niche overlap among five assemblages of tropical skinks. *Australian Journal of Zoology* **1**, 15–27. doi:10.1071/ZO06066

Harris J, Chapple DG, MacLean CA, Melville J (2023) Genetic footprints of fire: Understanding the genetic implications of fire regimes for fauna and effective conservation strategies. *Biological Conservation* **284**, 110169. doi:10.1016/j.biocon.2023.110169

Heatwole H (1989) The concept of the econe, a fundamental ecological unit. *Tropical Ecology* **30**(1), 13–19.

Heatwole H, Trémont S, Broese E (2013) Point-diversity, a critical tool for assessing dynamics of guilds of scavenging ants (Hymenoptera: Formicidae): an example from a eucalypt woodland. *Systematics and Biodiversity* **11**(2), 1–32. doi:10.1080/14772000.2013.788578

Purani K (1994) Population dynamics of *Lampropholis guichenoti* (Dumeril and Bibron) and *L. delicata* (De Vis) in an Open Sclerophyll Forest. MSc (Environmental Science), University of Western Sydney – Hawkesbury. Richmond.

Qualls FJ, Shine R (1998) Geographic variation in lizard phenotypes: importance of the incubation environment. *Biological Journal of the Linnean Society* **64**(4), 477–491. doi:10.1111/j.1095-8312.1998.tb00345.x

Qualls FJ, Shine R (2000) Post-hatching environment contributes greatly to phenotypic variation between two populations of the Australian garden skink, *Lampropholis guichenoti*. *Biological Journal of the Linnean Society* **71**(2), 315–341. doi:10.1111/j.1095-8312.2000.tb01260.x

Rawlinson PA (1974) Biogeography and ecology of Tasmania and the Bass Strait area. *Monographiae Biologicae* **25**, 291–338. doi:1007/978-94-010-2337-5_12

Tracy CR, Christian KA (1986) Ecological relations among space, time, and thermal niche axes. *Ecology* **67**, 609–615.

Wheeler QD, Knapp S, Stevenson DW, Stevenson J, Blum SD, Boom BM, Borisy GG, Buizer JL, De Carvalho MR, Cibrian A, Donoghue MJ, Doyle V, Gerson EM, Graham CH, Graves P, Graves SJ, Guralnick RP, Hamilton AL, Hanken J, Law W, Lipscom DL, Lovejoy TE, Miller H, Miller JS, Naeem S, Novacek MJ, Page LM, Platnick NI, Porter-Morgan H, Raven PH, Solis MA, Valdecasas AG, Van Der Leeuw S, Vasco A, Vermeulen N, Vogel J, Walls RL, Wilson EO, Woolley JB (2012) Mapping the biosphere: exploring species to understand the origin, organization and sustainability of biodiversity. *Systematics and Biodiversity* **10**(1), 1–20. doi:10.1080/ 14772000.2012.665095

Yadav A, Jain A, Sahu J, Dubey A, Gadpayle R, Barwa DK, Verma U, Kashyap K (2019) An overview on species hybridization in animals. *International Journal of Fauna and Biological Studies* **6**(5), 36–42.

Plate 1. Images of the two widespread eurytopic species: **A**. *Lampropholis delicata* from Coolah Tops National Park, NSW; and **B**. *Lampropholis guichenoti* eating a march fly at Noosa River, Qld. Images of some of the stenotopic species with smaller distributions: **C**. *Lampropholis adonis* from Eungella, Qld; **D**. *Lampropholis amicula* from Mt Coot-tha Forest, Brisbane, Qld; **E**. *Lampropholis bellendenkerensis* from Mt Bellenden Ker, Qld; **F**. *Lampropholis caligula* from Polblue, Barrington Tops Plateau, NSW; **G**. *Lampropholis coggeri* from Thornton Peak, Qld; **H**. *Lampropholis colossus* from Bunya Mountains, Qld; **I**. *Lampropholis couperi* from Mt Nebo, Qld; **J**. *Lampropholis elliotensis* from Mt Elliot, Qld; **K**. *Lampropholis elongata* from Ngulin Nature Reserve, NSW; **L**. *Lampropholis mirabilis* from Magnetic Island, Qld; **M**. *Lampropholis robertsi* from Mt Lewis, NSW; **N**. *Lampropholis similis* from Palm Island, Qld. **O**. *Lampropholis isla* from Scawfell Island, mid-east Qld. Photographs **A** by Jules Farquhar; **B**, **G**, **H** by Steve K. Wilson; **C**, **D**, **F**, **I**, **K**, **L**, **M** by Stephen Mahony; and **E**, **J**, **N**, **O** by Conrad Hoskin.

Plate 1. continued

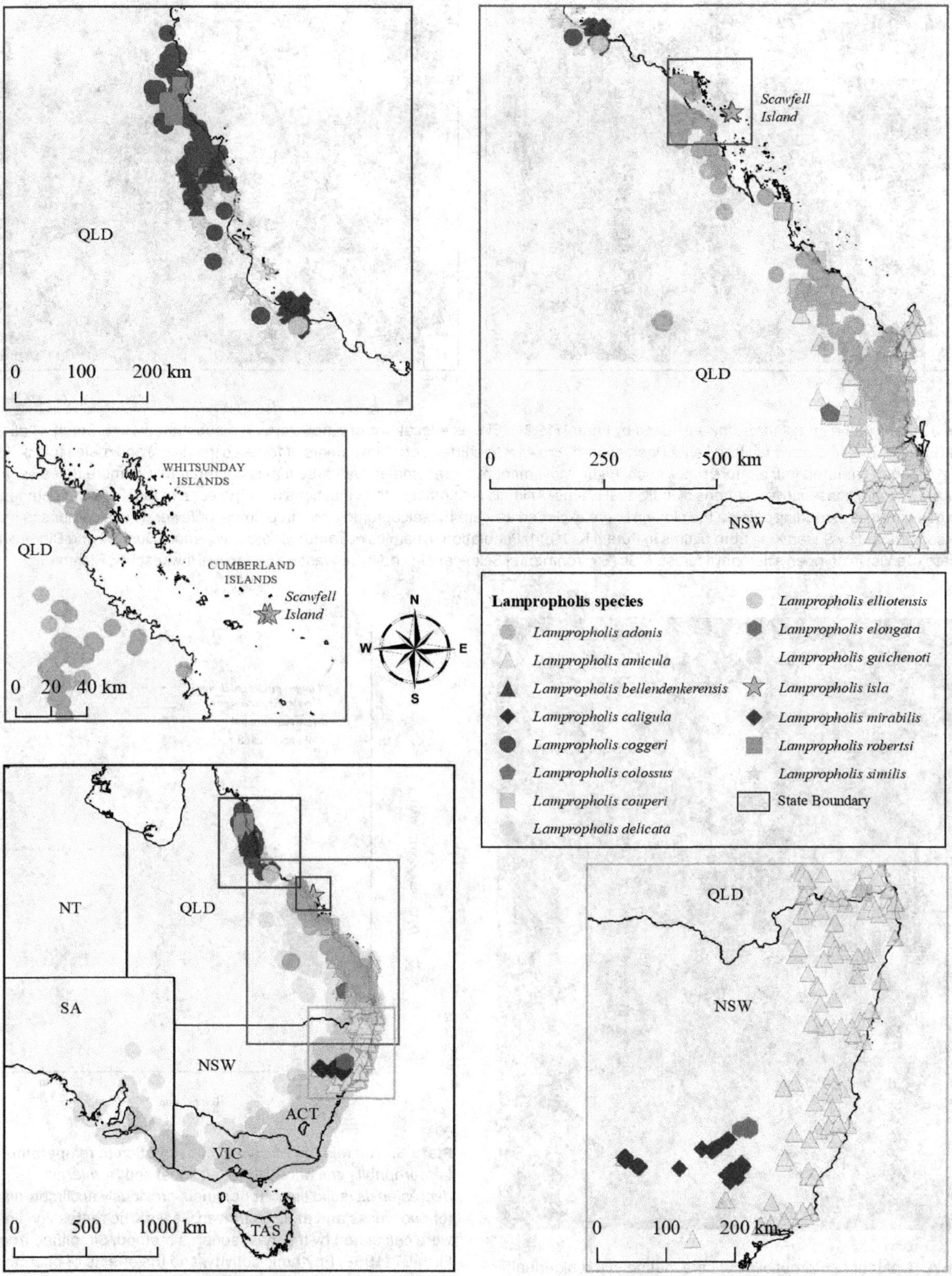

Plate 2. Maps of the distribution of the species of *Lampropholis* within Australia.

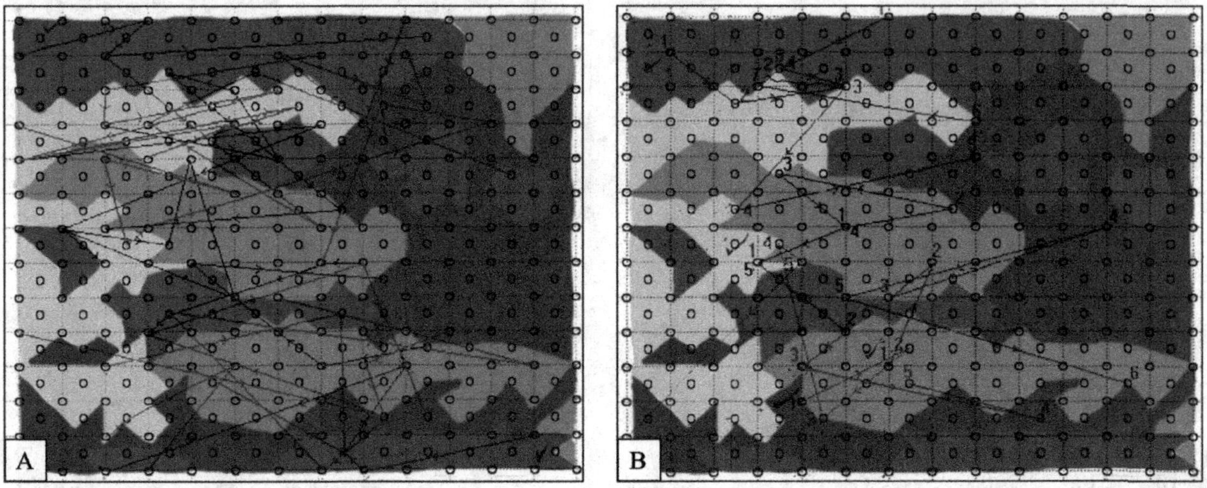

Plate 3. Maps of the grid of the study area used by Purani (1994). Circles = locations of pitfall traps. The substrates: Dark Green = Tall grass; Light Green = Sparse grass and litter; Tan = Open ground. Brown = leaf litter. Left: Movements of females (red lines) and males (black lines) of *L. guichenoti* captured four or more times each. Right: Movement of three females and three males captured four or more times each. The numbers indicate consecutive locations of individual lizards. Animals with most of their locations near the edge of the plot were deemed to have home ranges extending outside the plot and were excluded as were those captured only three times, or fewer. These composite maps were constructed by Susan King from figures in Purani K (1994) Population dynamics of *Lampropholis guichenoti* (Duméril and Bibron) and *L. delicata* (De Vis) in an open sclerophyll forest. MSc (Environmental Science). University of Western Sydney - Hawkesbury, Richmond.

Plate 4. A pair of *Lampropholis delicata* mating in a residential yard in Sydney. Photograph by Terri Shine, used with permission.

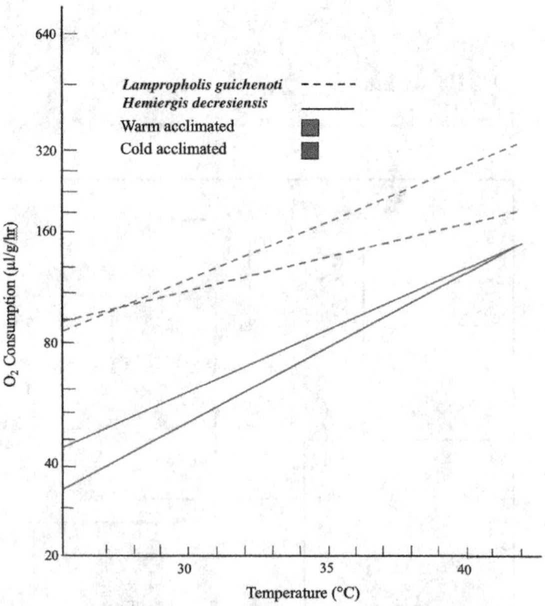

Plate 5. The relation of oxygen consumption to temperature in *Lampropholis guichenoti* (dashed lines) and *Hemiergis decresiensis* (solid lines) for animals previously acclimated at 3°C for two weeks and at 32°C for two weeks. The regression lines were calculated by the least squares method. Simplified from Firth BT (1968) Rhythmic activity, and the effects of photothermal acclimation and temperature on oxygen consumption in the skinks *Lampropholis guichenoti* and *Hemiergis decresiensis*. (Hons) thesis. University of New England, Armidale, NSW.

Plate 6. An adult *Lampropholis delicata* captured and encased in a web by a female red-backed spider (*Latrodectus hasselti*) in a suburban backyard in Sydney. Photograph by B. Tamayo.

Plate 7. A sub-adult *Lampropholis delicata* in the clutches of a wolf spider (*Tasmanicosa* sp.) in a residential area in a wet tropical area, Silkwood, North Queensland. Photograph by G. S. Turner.

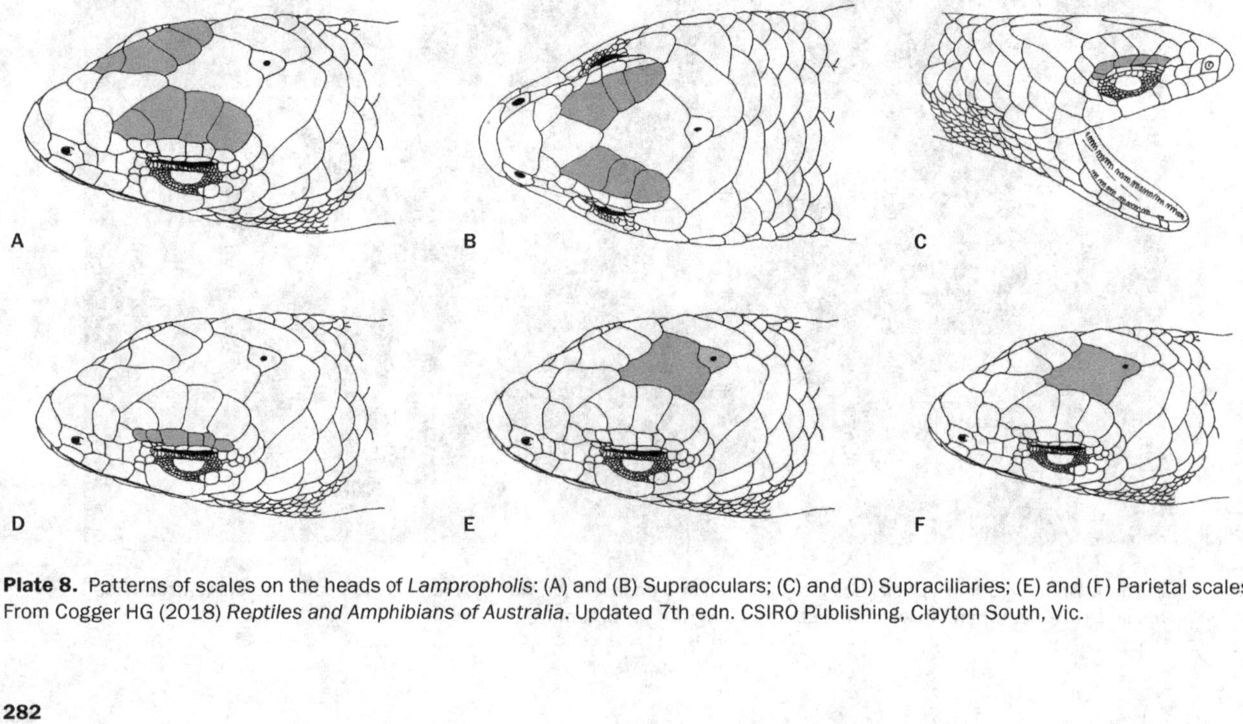

Plate 8. Patterns of scales on the heads of *Lampropholis*: (A) and (B) Supraoculars; (C) and (D) Supraciliaries; (E) and (F) Parietal scales. From Cogger HG (2018) *Reptiles and Amphibians of Australia*. Updated 7th edn. CSIRO Publishing, Clayton South, Vic.

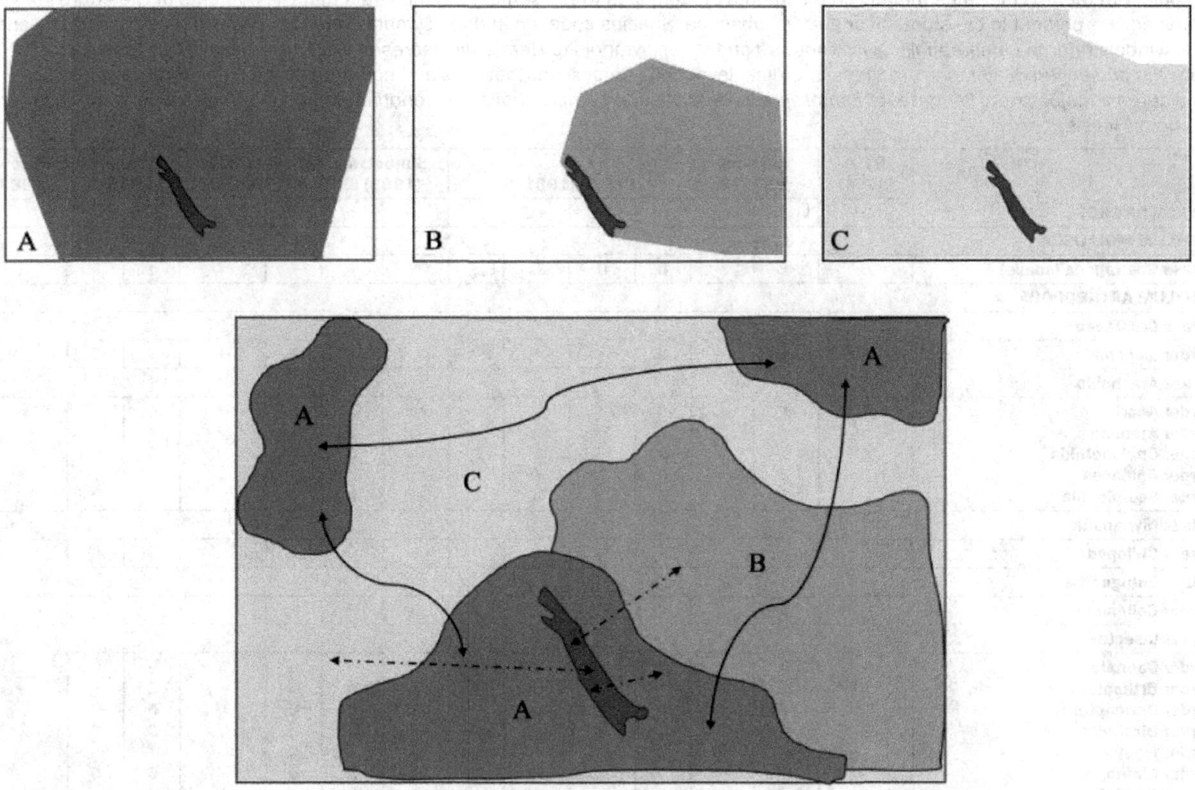

Plate 9. Theoretical map showing the differences in estimating the size and degree of overlap of the home ranges of individuals A, B, and C, each from a different species. Individual A forages on leaf litter (light brown), B on grass (green), and C in the open (grey). The large map shows the home ranges of these individuals based on data from many pitfall traps that saturate an area and are frequently checked. In addition, individuals are watched to ascertain the routes individual A uses in moving between the patches of leaf litter in which it forages. Dark brown represents a log in which all three individuals communally nest, shelter during inactivity, oviposit, and/or brumate and hence it is within the general home range of all three individuals, but outside of the part of the home range used for foraging by individuals B and C. <———> indicates the routes followed by Individual A when moving between patches it uses for foraging. <- - - - > indicates the routes from the area of the general home range to and from the parts used for oviposition or shelter. The small maps separately indicate the less precise polygons that usually are used to indicate the general home ranges for individuals A, B, and C without identifying the enclosed areas that are not used at all or between parts of the home range used for foraging as opposed to other purposes, such as interspecific communal activities or merely for travel between microhabitats used for foraging.

Plate 10. Comparison of the diets of the *Lampropholis delicata, Lampropholis guichenoti*, and some other common skinks that are syntopic with them. S indicates that only the contents of stomachs were examined. G indicates that the contents of the entire gut was examined. + = present in the stomach or gut. 0 = absence. Species code: *Ld* and red symbols refer to *Lampropholis delicata*, *Lg* and blue symbols refer to *Lampropholis guichenoti*, *Hd* and green symbols to *Hemiergis decresiensis*, and *Mb* and yellow entries to *Morethia boulengeri*. Major taxa are those at ordinal level or higher, except that ants and non-ants of the Hymenoptera were each considered a major taxon. For data for *Lampropholis delicata* and *Lampropholis guichenoti* identification of prey was taken to lower taxonomic levels.

	Rose (1974)	Belmont (1977) S		Crome (1981) S				Simbotwe (1985) S		Lunney *et al.* (1989)[1] G		Warner (1995) S		Brown (1983) G
TAXON OF PREY	*Ld*	*Ld*	*Lg*	*Ld*	*Lg*	*Hd*	*Mb*	*Lg*	*Hd*	*Ld*	*Lg*	*Ld*	*Lg*	*Lg*
PHYLUM MOLLUSCA														
Class Gastropoda (snails)	0	+	+	0	0	0	0	+	+	+	+	+	+	+
PHYLUM ARTHROPODA														
Class Crustacea														
Order Isopoda	+[2]	+	+	0	+	+	0	0	0	+	+	+	0	+
Class Arachnida														
Order Acari	0	+	+	0	+	+	0	+[3]	0	0	0	0	+	0
Order Araneae	0	+	+	+	+	+	+	+	+	+	+	+	+	+
Order Chelonethida	0	0	0	0	0	0	0	0	0	0	0	0	0	+
Order Opiliones	0	0	0	0	0	0	0	0	0	0	0	0	0	+
Order Scorpionida	0	0	0	0	0	0	0	0	0	0	0	0	0	+
Class Myriapoda														
Order Chilopoda	0	0	+	+	0	0	0	0	0	+	+	0	0	+
Class Entognatha														
Order Collembola	0	+	+	0	0	0	0	0	0	0	0	+	0	+
Class Insecta														
Order Odonata	0	0	0	0	0	0	0	+	0	0	0	0	0	0
Order Orthoptera	0	+	+	0	+	0	+	+	+	+	+	0	+	+
Order Dermaptera	+	+	+	0	0	0	0	0	0	0	0	0	0	0
Order Blattodea	0	0	+	0	+	+	0	+	+	0	0	0	0	+
Order Isoptera	0	0	+	0	+	+	0	0	0	0	0	+	+	+
Order Mantodea	0	0	+	0	0	0	0	+	0	0	0	0	0	0
Order Hemiptera	0	+	+	+	0	+	0	+	+	+	+	+	+	+
Order Thysanoptera	0	0	0	0	0	0	0	0	0	+	+	0	0	0
Order Psocoptera	0	0	0	0	0	0	0	0	0	0	0	0	0	0
Order Coleoptera	0	+	+	+	+	+	0	+	+	+	+	+	+	+
Order Neuroptera[3]	0	0	0	0	0	0	0	0	0	+	+	0	0	+
Order Hymenoptera														
Non-ants	0	+	+	0	+	0	0	+	+	0	0	+	+	+
Ants	+	+	+	+	+	+	0	+	0	+	+	+	+	+
Order Trichoptera	0	+	+	0	0	0	0	+	+	0	0	0	0	0
Order Lepidoptera	+	+	+	0	0	0	0	+	0	+	+	+	+	+
Order Mecoptera	0	0	0	0	0	0	0	+	0	0	0	0	0	0
Order Diptera	0	+	+	0	+	0	0	+	0	+	+	+	+	+
TOTAL MAJOR TAXA	4	14	18	5	10	8	2	15	8	12	12	11	11	16

[1]Data only from 1980 before the area was disturbed by a prolonged drought and a fire.

[2]Listed as 'Amphipoda' but identification is considered doubtful; if correct, it would represent an aquatic habitat that is not present at the other sites.

[3]Listed as 'Arachnidae larvae' but probably mites.

REFERENCES

Belmont CM (1977) Resource partitioning and habitat selection in the ground skinks. BSc (Hons) thesis. University of New England, Armidale, NSW.

Brown GW (1983) Comparative feeding ecology of South-Eastern Australian Scincids. PhD Dissertation. La Trobe University, Melbourne. 382 pp.

Crome B (1981) The diet of some ground-layer lizards in three woodlands of the New England Tableland of Australia. *Herpetofauna* **13**, 4–11.

Lunney D, Ashby E, Grigg J, O'Connell M (1989) Diets of scincid lizards *Lampropholis guichenoti* (Dumèril & Bibron) and *L. delicata* (de Vis) in Mumbulla State Forest on the South Coast of New South Wales. *Australian Wildlife Research* **16**(3), 307–312. doi:10.1071/WR9890307

Rose AB (1974) Gut contents of some amphibians and reptiles. *Herpetofauna* **7**, 4–8.

Simbotwe MP (1985) Comparative ecology and demography of two populations of Australian reptiles (Lacertilia, Scincidae). PhD Dissertation. University of New England, Armidale, NSW.

Warner FJ (1995) A comparison of the dietary contents of two sympatric scincid lizards, *Lampropholis delicata* (de Vis, 1888), and *Lampropholis guichenoti* (Duméril and Bibron, 1839) (Lacertilia, Scincidae). BAppSc (Hons) thesis. University of Western Sydney – Hawkesbury. Richmond.

Appendix 1 (to Preface). Scientific names and contrived 'common' names of lizards of the genus *Lampropholis*

CSIRO = Commonwealth Scientific and Industrial Research Organisation; For changes in scientific names see the synonymy in Table 14.2, Chapter 14. – indicates that no common name was listed. This table is not a comprehensive list of the contrived names used in various websites and scientific papers; for example, see Preface for discussion of this appendix.

	Contrived 'common' names					
Scientific name	**CSIRO names (Clayton *et al.* 2006)**	**Australian Society of Herpetologists names (2023)**	**The Reptile Database (accessed 31 December 2023)**	**Atlas of Living Australia (accessed January 2024)**	**iNaturalist (accessed January and June 2024)**	**The Action Plan for Australian Lizards and Snakes 2017 (Chapple *et al.* 2019)**
Lampropholis adonis	Diamond Shielded Sunskink	Diamond Shielded Sunskink	Diamond-shielded Sunskink	Diamond-shielded Sunskink; Ingram's Litter Skink	Diamond-shielded Sunskink; Ingram's Litter Skink	Diamond Shielded Sunskink
Lampropholis amicula	Friendly Sunskink	Friendly Sunskink	Friendly Sunskink	Friendly Sunskink	Friendly Sunskink	Friendly Sunskink
Lampropholis bellendenkerensis	–	Southern Montane Sunskink[1]	Bellenden Ker Sunskink	–	–	–
Lampropholis caligula	Montane Sunskink	Mountain Sunskink; Barrington Sunskink	Montane Sunskink	Montane Sunskink	Montane Sunskink	Montane Sunskink; Barrington Sunskink
Lampropholis coggeri	Rainforest Sunskink	Plain-backed Sunskink	Rainforest Sunskink	Northern Sunskink; Northern Rainforest Sunskink; Rainforest Sunskink	Northern Rainforest Sunskink; Rainforest Sunskink	Northern Rainforest Sunskink
Lampropholis colossus	Bunya Sunskink	Bunya Sunskink	Bunya Sunskink	Bunya Sunskink	Bunya Sunskink	Bunya Sunskink
Lampropholis couperi	Plain-backed Sunskink	Plain-backed Sunskink; Couper's Sunskink	Plain-backed Sunskink	Plain-backed Sunskink	Plain-backed Sunskink	Plain-backed Sunskink; Couper's Sunskink
Lampropholis delicata	Dark-flecked Garden Sunskink	Delicate Skink; Delicate Grass Sunskink; Dark-flecked Garden Sunskink; Plague Skink[2]; Rainbow Skink[2]	Dark-flecked Garden Sunskink; Plague Skink[2]; Rainbow Skink; Grass Skink	Dark-Flecked Garden Sunskink; Delicate Skink; Rainbow Skink; Garden Skink; Delicate Grass Skink	Delicate Skink; Rainbow Skink; Garden Skink; Delicate Grass Skink; Dark-flecked Garden Sunskink	Delicate Skink

continued

Appendix 1. continued

	Contrived 'common' names					
Scientific name	**CSIRO names (Clayton *et al.* 2006)**	**Australian Society of Herpetologists names (2023)**	**The Reptile Database (accessed 31 December 2023)**	**Atlas of Living Australia (accessed January 2024)**	**iNaturalist (accessed January and June 2024)**	**The Action Plan for Australian Lizards and Snakes 2017 (Chapple *et al.* 2019)**
Lampropholis elliotensis	–	Mt Elliot Rainforest Sunskink	Mt Elliot Sunskink	–	–	Southern Sunskink
Lampropholis elongata	–	Elongate Sunskink	Elongate Sunskink; Long Sunskink	Elongate Sunskink; Long Sunskink	Elongate Sunskink; Long Sunskink	Elongate Sunskink
Lampropholis guichenoti	Pale-flecked Garden Sunskink	Pale-flecked Garden Sunskink; Garden Skink	Pale-flecked Garden Skink; Garden Skink	Common Garden Skink; Pale-Flecked Garden Sunskink; Grass Skink; Guichenot's Grass Skink	Common Garden Skink; Pale-Flecked Garden Sunskink; Grass Skink; Guichenot's Grass Skink	Garden Skink; Grass Skink
***Lampropholis isla*[3]**	–	–	–	–	–	–
Lampropholis mirabilis	Saxicoline Sunskink	Saxicoline Sunskink	Saxicoline Skink	Saxicoline Skink	Saxicoline Skink	Saxicoline Sunskink
Lampropholis robertsi	Grey-bellied Sunskink	Grey-bellied Sunskink; Northern Montane Sunskink	Grey-bellied Sunskink	Grey-bellied Sunskink	Grey-bellied Sunskink	Grey-bellied Sunskink; Roberts' Sunskink; Northern Montane Sunskink
Lampropholis similis	–	Southern Rainforest Sunskink	–	Southern Rainforest Sunskink	Southern Rainforest Sunskink	–

[1]Singhal *et al.* (2018), in their original description and naming of this species, gave it the contrived name of Bellenden Ker Sunskink and the wildnet database managed by the Queensland Department of Environment and Science lists it as the 'grey-bellied sunskink'.
[2]Note that Plague Skink and Rainbow Skink may not be contrived names, but rather a true common name for *L. delicata* in New Zealand and Hawaii respectively, where it was inadvertently introduced from Australia (see Chapter 14).
[3]The original description of the species (Hoskin 2025) supplied a contrived 'common' name of Scawfell Island Sunskink.

REFERENCES

Chapple DG, Tingley R, Mitchell NJ, Mcdonald SL, Keogh JS, Shea GM, Bowles P, Cox NA, Woinarski JCZ (2019) *The Action Plan for Australian Lizards and Snakes 2017.* CSIRO Publishing, Clayton South, Vic.

Clayton M, Wombey JC, Mason IJ, Chesser RT, Wells A (2006) *CSIRO List of Australian Vertebrates: A Reference with Conservation Status.* 2nd edn. CSIRO Publishing, Collingwood, Vic.

Hoskin CJ (2025) A new species of *Lampropholis* skink (Lacertilia: Scincidae) from Scawfell Island, mid-east Queensland, Australia. *Australian Journal of Taxonomy* **85**, 1–13. doi:10.54102/ajt.6uihp

Singhal S, Hoskin CJ, Couper P, Potter S, Moritz C (2018) A framework for resolving cryptic species: a case study from the lizards of the Australian Wet Tropics. *Systematic Biology* **67**, 1061–1075. doi:10.1093/sysbio/syy026

Appendix 2 (to Preface). Results of a survey in early to mid-2022 of the perceived common names in Sydney and Armidale, NSW, for species of *Lampropholis*

Name	Professional herpetologists (%)	Lay persons (%)	Total (%)
Penny lizard	4 (17.4)	16 (61.5)	20 (40.8)
Garden skink	12 (52.2)	0	12 (24.5)
Grass skink	3 (13.0)	0	3 (6.1)
Sun skink	2 (8.7)	0	2 (4.1)
Delicate skink	1 (4.4)	0	1 (2.0)
Brown skink	1 (4.4)	0	1 (2.0)
Small lizard	0	4 (15.4)	4 (8.2)
Don't know*	0	6 (23.1)	6 (12.3)
TOTAL	**23 (100.1)**	**26 (100)**	**49 (100)**

*Two of these respondents did not list any common name but merely gave the scientific name of a different genus of skinks.

Appendix 3 (for Chapter 2). Occurrence of *Lampropholis delicata* and *Lampropholis guichenoti* in faunal surveys conducted within the general geographic range of both species

Elevations obtained from cited references and various websites. When the two species were known to be represented at different densities, the most prominent one is shown in **boldface**.

Location	Habitat	Elevation (metres)	Species: *Lampropholis delicata*	Species: *Lampropholis guichenoti*	Authority
Beerwah, SE Qld Scientific Purposes Area	Heath, herblands, eucalypt woodland	32	**Present**	Absent	Hannah and Smith (1996)
Brisbane, Qld	Stunted eucalyptus scrub, swamp, cleared grassland	45	Absent	**Present**	Lyon (1972)
Tenterfield Management Area, NE coast of NSW	Mesic to dry tablelands, various types of forest		Individuals in transects and caught in (pitfall traps)	Individuals in transects and caught in (pitfall traps)	Daly and Lemckert (2011)
Bookookoorara	Mesic tableland	875	**52 (1)**	0 (0)	
Boonoo	Tall open forest	870–1,105	**75 (1)**	0 (0)	
Boorook	Tall open forest	905–980	**150 (6)**	2 (0)	
Forest Land	Closed forest	975–1,070	**365 (9)**	0 (0)	
Gilgurry	Dry tableland	680–690	**40 (0)**	0 (0)	
Girard[1]	Closed forest	570–965	**46 (1)**	0 (0)	
Girard extension	Coastal forest	430–560	**9 (0)**	0 (0)	
Jenner	Coastal forest	975	**19 (0)**	0 (0)	
Malara	Tall open forest	950–1,035	**53 (0)**	5 (0)	
Spiraba	Tall open forest	985–1,140	**121 (11)**	0 (0)	
Armidale, NSW					Belmont (1977)
Pine Forest	Pine forest	980	Present	Present	
Puddledock	Eucalypt forest (closed canopy)	980	Present	Present	
Blue Hole	Eucalypt forest (sparse canopy)	980	Absent	**Present**	
Mumbulla State Forest SE coast of NSW	Eucalypt forest	113	Present	Present	Lunney *et al.* (1991)
Bendalong South coast of NSW	Various woodlands and other coastal habitats	Near sea level	**Abundant in all areas**	Only one specimen from 1974 to 1997[2]	Scanlon (2000)

Location	Habitat	Elevation (metres)	Species: *Lampropholis delicata*	Species: *Lampropholis guichenoti*	Authority
Morton National Park Southern coast of NSW	Heath and eucalypt woodlands	0–600	**271 specimens**	64 specimens	Daly (2006)
Wadbilliga National Park Southern coast of NSW	Eucalypt forest	5–1,210	**26% of the individuals in the survey of reptiles**	15% of the individuals in the survey of reptiles	Daly (2007)
Reserves on the south-western slopes of NSW	Eucalypt woodlands	200–890			Daly (2004)
Overall			27 individuals	**152 individuals**	His Table 3
Microhabitats listed			2 individuals associated with logs	**82 individuals, 66 associated with litter, 15 with logs, and 1 with rocks**	His Table 6
Budderoo National Park Southern coast of NSW	Heath and woodland; open forest; tall open forest	150–650	**31 animals at 11 sites**	17 animals at 11 sites	Daly and Jackett (2009)
Bombaderry Bushland Southern coast of NSW	Eucalypt woodland and open forest	Near sea level	Absent	Absent	Murphy (1996)
Jervis Bay, Southern coast of NSW		0–170			Daly (2000)
Beecroft Peninsula	Heath, shrubland, woodland, and forest		Present	Present	
Booderee National Park			Present	Present	
Bowen Island	Heath, woodland, and sedges		**Present**	Absent	
Jervis Bay National Park			Present	Present	
Vincentia					
Long Bow Point	Open forest,		**Present**	Absent	
Currambene State Forest	woodland, and heath		**Present**	Absent	
Escarpment and Riverine forests, near Nowra, NSW	Eucalypt woodlands and forests, subtemperate rainforests; heathland, cleared grassland, sandstone cliffs	0–68	**Common in 5 habitats: woodland, heathland, open forests, rainforest, and grassland**	Present but uncommon in 3 habitats: heathland, open forest, grassland	Murphy and Daly (1998)
Kurnell Peninsula, near Sydney	Native bushland plus areas developed for pharmaceutical plants, chemical industries, and oil refineries	0–60	Common and widespread	Common and widespread	White (2010)
Sydney Urban Reserves		0–449			White and Burgin (2004)
Rockland Corridor	Coastal dune forest		Present	Present	
Wolli Creek Valley	Gully forest		Present	Present	
Middle Harbour	Gully forest and Sandstone complex		Present	Present	
Burnt Bridge Creek	Sandstone complex		Present	Present	
University Western Sydney, Richmond	Eucalypt woodland		Present	Present	

continued

Appendix 3. continued

Location	Habitat	Elevation (metres)	Species: *Lampropholis delicata*	Species: *Lampropholis guichenoti*	Authority
Western Sydney University, Hawkesbury	Remnant eucalypt woodland		Present	Present	Mo (2022)
Suburbs of Sydney	Urban habitats		Absent	**Common**	Green (1973)
Sydney, NSW	Urban habitats				Shea (2010)
Eastern Suburbs			Present	Present	
Inner West			Present	Present	
Liverpool			Present	Present	
North Shore			Present	Present	
Manly Peninsula			**Present**	Absent	
Northern Beaches			Present	Present	
Hornsby			Present	Present	
Ryde Peninsula			Present	Present	
Southern Beaches			Present	Present	
North West			Present	Present	
South West			Present	Present	
			Number of museum specimens	**Number of museum specimens**	
Eastern Suburbs			**53**	46	
Inner West			36	**68**	
Manly Peninsula			**92**	0	
North Shore			**38**	8	
Ryde Peninsula			**13**	8	
Former Maroota State Forest, Now Deerubin Aboriginal Land, NW of Sydney	Eucalypt forest with low shrubs on Hawkesbury Sandstone	~200 m	**Present**	Absent	Webb (1983)
Coolangubra State Forest	Eucalypt forest	800–850	Absent	**Present**	Webb (1985)
Seven Mile Beach National Park, southern coast of NSW	Various: see columns 4 and 5	Coastal	**Commonest reptile in the park; abundant in teatree scrub, blackbutt/bangalay forest, littoral rainforest, and swamp mahogany woodland**	Abundant on the foredune; absent or uncommon elsewhere	Murphy (1994)
Gore Creek Reserve and Lane Cove Bushland Park, Sydney	Closed eucalypt forest and open eucalypt forest	71	**Common**	Uncommon	Murphy (1996, 2010)
Bushland near Sydney	Remnant urban bushland	80–180	Present at all 5 sites	Present at all 5 sites	
Near Goulburn NSW	Native and improved pasture with scattered eucalypts and shrubs	640–685	Present	Present	Daly *et al.* (2008)
Upper Billabong Creek Catchment Area, near Holbrook, Southern NSW	Dry eucalypt forest; grassy box woodland; riverine forest and woodland	220–889	Present	Present	Sass *et al.* (2008)

Location	Habitat	Elevation (metres)	Species: *Lampropholis delicata*	Species: *Lampropholis guichenoti*	Authority
Malabar Headland, near Maroubra, NSW	Bushland	Coastal	**123**	53	White (2007)
Wagga Wagga, NSW		220	**Present**		Sass (2003)
Mates Gully	Low eucalypt forest		Absent	**Present**	
Mt Moorong	Eucalypt forest		Absent	Absent	
Plum Pudding	Eucalypt forest		Absent	**Present**	
Mt Kiora	Eucalypt forest		Absent	**Present**	
Ballandry	Grazing land		Absent	Absent	
Bald Hill	Grazing land		Absent	Absent	
East Central NSW	All with eucalypt understorey and ground cover				Daly *et al.* (2001)
Razorback Nature Reserve		540–840	Present	Present	
Keverstone State Forest		680–1,000	**Present**	Absent	
Abercromie Caves		580–700	Absent	Absent	
Topographic comparison					
Ridge 1			**Present**	Absent	
Ridge 2			Present	Present	
Ridge 3			**Present**	Absent	
Mid-ridge 1			**Present**	Absent	
Mid-ridge 2			**Present**	Absent	
Gully 1			Present	Present	
Gully 2			Absent	**Present**	
Goobang National Park, NSW	Grassy eucalypt woodland	–	**Present**	Absent	LeBreton *et al.* (2002)
Coolah Tops National Park, Western Liverpool Range, NSW	Tall, moist montane forest	>1,000	**Present**	Absent	Eldridge *et al.* (2020)[3]
Southwest Slope Restoration Area, SE NSW	Restored plantings of woodlands on agricultural sites	Up to 600	Present	Present	Lindenmayer *et al.* (2023)
South-eastern Australia	Box gum grassy woodlands	Various			Kay *et al.* (2013)
Murrumbidgee			Rare/Uncommon	Sparse/Absent	
Lachlan			Rare/Uncommon	**Sparse**	
Central West			Absent	**Sparse/Absent**	
Namoi			Absent	Absent/Common	
Gwydir			**Common**	Absent	
Maranoa-Balonne			Rare/Absent	Rare/Absent	
Condamine			Absent	Absent	
Total specimens			**69**	12	
Lower western slopes of Mt Difficult Range, Vic.	Woodland	377	1 animal Uncommon	**11 animals** Uncommon	Homan (2008); Homan and Shultz (2012)
Condowie Reservoir, Almurta, Vic.		60–255			Lloyd (2006)
	Eucalypt woodland		Absent	**Common**	
	Grass/sedgeland		Absent	**Common**	
Wonthaggi Heathland and Coastal Reserve, Vic.	Coastal heathland, woodland, and sedgeland	20	Absent	**Uncommon**	Homan (2003)

[1]In addition *Lampropholis amicula* was found at this site.

[2]Near a trailer and suspected of having been introduced into the site.

[3]Only noted that it had been reported from the area in general but not recorded from any of the surveyed sites.

REFERENCES

Belmont CM (1977) Resource partitioning and habitat selection in the ground skinks. BSc (Hons) thesis. University of New England, Armidale, NSW.

Daly G (2000) Island populations: reptiles and amphibians of the Jervis Bay region on the South Coast of New South Wales. *Herpetofauna* **30**, 11–17.

Daly G (2004) Surveys of reptiles and amphibians on the south-western slopes of New South Wales. *Herpetofauna* **34**, 2–16.

Daly G (2006) Reptiles and amphibians of Morton National Park and environs on the south coast of New South Wales. *Herpetofauna* **36**(1), 5–22.

Daly G (2007) Reptiles and amphibians of Wadbilliga National Park and environs on the South Coast of New South Wales. *Herpetofauna* **37**(1), 45–62.

Daly G, Jackett N (2009) Surveys of reptiles at Budderoo National Park and environs on the south coast of New South Wales. *Herpetofauna* **39**(1), 80–90.

Daly G, Lemckert F (2011) Survey of the reptiles and amphibians of the montane forests near Tenterfield on the north coast of New South Wales. *Australian Zoologist* **35**(4), 957–972. doi:10.7882/AZ.2011.048

Daly G, Pennay M, Gosper C (2001) Survey of reptiles and amphibians at Razorback Nature Reserve, Keverstone State Forest and the Abercombie Caves Region of New South Wales. *Herpetofauna* **31**, 89–91.

Daly G, Virtue B, Stone G (2008) Results of a survey for the striped legless lizard *Delma impar* near Goulburn, New South Wales. *Herpetofauna* **38**, 51–58.

Eldridge MDB, Ingleby S, King AG, Mahony SV, Parnaby HE, Beatson CA, Divljan A, Frankham GJ, Hay AC, Major RE, Reader SE, Sadlier RA, Tsang LR (2020) Australian Museum surveys of the vertebrate fauna of Coolah Tops National Park NSW. *Technical Reports of the Australian Museum Online* **30**, 1–20. doi:10.3853/j.1835-4211.30.2020.1757

Green D (1973) The reptiles of the outer north-western suburbs of Sydney. *Herpetofauna* **6**(2), 2–5.

Hannah DS, Smith DC (1996) The baiting of pitfall traps with vegetable oils: its effect on the captures of herptiles in a coastal lowland community. *Herpetofauna* **26**, 16–18.

Homan P (2003) A reptile and amphibian survey of the Wonthaggi Heathland and Coastal Reserve. *The Victorian Naturalist* **120**, 147–152.

Homan P (2008) Surveys of vertebrate fauna in the Grampians National Park, 2003–2007. *The Victorian Naturalist* **125**(2), 47–55.

Homan P, Schultz N (2012) Further records of mammal and reptile fauna from the Black Range, near Stalwell and the Grampians National Park, Western Victoria. *The Victorian Naturalist* **129**(2), 36–45.

Kay GM, Michael DR, Crane M, Okada S, MacGregor C, Florance D, Trengove D, McBurney L, Blair D, Lindenmayer DB (2013) A list of reptiles and amphibians from Box Gum Grassy Woodlands in south-eastern Australia. *Check List* **9**(3), 476–481. doi:10.15560/9.3.476

LeBreton M, Faulkner B, Ellis M (2002) Frogs and reptiles of Goobang and Nangar National Parks, Central Western NSW. *Herpetofauna* **32**(1), 13–19.

Lindenmayer DB, Florance D, Smith D, Crane C, Siegrist A, Lang E, Crane M, Michael DR, Scheele BC, Evans MJ (2023) Temporal trends in reptile occurrence among temperate old-growth, regrowth and replanted woodland. *PLoS One* **18**(9), e0291641. doi:10.1371/journal.pone.0291641

Lloyd R (2006) A herpetofauna survey of the Candowie Reservoir Section of the Tennent Creek Catchment, Almurta, Victoria. *Herpetofauna* **36**(2), 75–79.

Lunney D, Ebby P, O'Connell M (1991) Effects of logging, fire and drought on three species of lizard in Mumbulla State Forest on the south coast of New South Wales. *Australian Journal of Ecology* **16**(1), 33–46. doi:10.1111/j.1442-9993.1991.tb01479.x

Lyon B (1972) Area survey of reptiles in the outer north-eastern Brisbane suburbs – Geebung, Wavell Heights, Virginia, part Zillmere and Boondall. *Herpetofauna* **5**(3), 2–4.

Mo M (2022) Reptiles and amphibians on a university campus in a peri-urban area of Sydney, New South Wales, Australia. *Reptiles & Amphibians* **29**, 122–133. doi:10.17161/randa.v29i1.16317

Murphy MJ (1994) Reptiles and amphibians of Seven Mile Beach National Park, NSW. *Herpetofauna* **24**, 24–30.

Murphy MJ (1996) Relict herpetofauna of a small bushland remnant in Sydney, NSW. *Herpetofauna* **26**, 45.

Murphy MJ (2010) Additions to the herpetofauna of Gore Creek Reserve and Land Cove Bushland Park in inner suburban Sydney, NSW. *Herpetofauna* **40**, 103–110.

Murphy MJ, Daly G (1998) Survey of the reptiles and amphibians of the escarpment and riverine forests north west of Nowra, NSW. *Herpetofauna* **28**, 16–21.

Sass S (2003) Survey of the reptiles of Wagga Wagga, New South Wales. *Herpetofauna* **33**, 18–23.

Sass S, Watson DM, Wilson A (2008) The reptile fauna of the Upper Billabong Creek Catchment Area, southern New South Wales. *Herpetofauna* **38**, 41–50.

Scanlon JD (2000) Notes on herpetofauna of Bendalong (south coast of NSW), with reproductive data on elapid snakes and a range extension for *Hemiaspis signata*. *Herpetofauna* **30**, 36–41.

Shea G (2010) The suburban terrestrial reptile fauna of Sydney – winners and losers. In *The Natural History of Sydney*. (Eds D Lunney, P Hutchings and D Hochuli) pp. 154–197. Royal Zoological Society of NSW, Sydney. doi:10.7882/FS.2010.015

Webb GA (1983) Diet in a herpetofaunal community on the Hawkesbury Sandstone Formation in the Sydney area. *Herpetofauna* **14**(2), 87–91.

Webb GA (1985) Habitat use and activity patterns in some southeastern Australian skinks. In *Biology of Australian Frogs and Reptiles*. (Eds G Grigg, R Shine and H Ehmann) pp. 23–30. Surrey Beatty & Sons, Chipping Norton, NSW.

White AW (2007) Herpetofauna of the Malabar Headland. *Herpetofauna* **37**, 36–41.

White AW (2010) Herpetofauna of the Kurnell Peninsula. *Herpetofauna* **40**, 56–75.

White AW, Burgin S (2004) Current status and future prospects of reptiles and frogs in Sydney's urban-impacted bushland reserves. In *Urban Wildlife: More than Meets the Eye*. (Eds D Lunney and S Burgin) pp. 109–123. Royal Zoological Society of New South Wales, Sydney. doi:10.7882/9780958608572

Appendix 4 (for Chapter 8). Taxonomic classification of the known prey of *Lampropholis delicata* and *Lampropholis guichenoti*

Higher-level classification has been updated to reflect recent phylogenetic revisions. Data are from Rose (1974), Belmont (1977), Crome (1981), Brown (1983, 1991), Webb (1983), Simbotwe (1986), Lunney *et al.* (1989), and Warner (1995). X means that the entry is known to be consumed by that species. Most literature on the diets of *Lampropholis* do not identify prey further than Order. Some taxa, however, were identified as low as the familial level. When the category 'Unidentified' appears under a taxon it means that the prey was identified only to that level and could represent species from lower taxa, such as families, otherwise not listed, but also could be unidentified members of listed families. See Chapter 8 for discussion of this appendix. For data for species of other genera see Plate 10.

		Lampropholis	
PHYLUM/CLASS/Order/Family	**Common name**	***delicata***	***guichenoti***
MOLLUSCA	Molluscs		X
GASTROPODA			
Unidentified	Snails	X	X
Megagastropoda	Snails	X	X
Stylommatophora	Snails		X
Charopidae	Snails		X
Punctidae	Snails		X
ARTHROPODA	Arthropods		
Unidentified	–		X
SUBPHYLUM CHELICERATA	Chelicerates		
ARACHNIDA	Arachnids		
Acari	Mites		X
Araneae	Spiders		
Unidentified	–	X	X
Araneidae	Orb-weaving spiders	X	X
Dictynidae	Mesh-web weavers		X
Linyphiide	Money spiders	X	
Lycosidae	Wolf spiders		X
Zodariidae	Ant spiders		X
Salticidae	Jumping spiders	X	X
Clubionidae	Sac spiders	X	X
Sparassidae	Huntsman spiders		
Chelonethida	Pseudoscorpions		X

Appendix 4 (for Chapter 8) – Taxonomic classification of the known prey of *Lampropholis delicata* and *L. guichenoti*

		Lampropholis	
PHYLUM/CLASS/Order/Family	**Common name**	***delicata***	***guichenoti***
Opiliones	Harvestmen		X
Scorpionida	Scorpions		X
SUBPHYLUM MYRIAPODA	Myriapods		
CHILOPODA	Centipedes	X	X
SUBPHYLUM CRUSTACEA	Crustaceans		
Unidentified	–	X	X[1]
MALACOSTRACA			X
Amphipoda (*Talitris* sp.[1])		X	
Isopoda			X
HEXAPODA	Hexapods		
ENTOGNATHA	Entognathans		
Collembola	Springtails		
Unidentified	–		X
Sminthuridae	Springtails		X
Entomobryidae	Springtails	X	
INSECTA			
Unidentified adults	Insects		X
Unidentified larvae	–		X
Odonata			
Adults	Dragonflies/Damselflies		X
Nymphs	Naiads		X
Orthoptera	Crickets and grasshoppers		
Unidentified			X
Suborder Ensifera	Crickets		
Superfamily Grylloidea	Crickets		
Unidentified			X
Gryllidae	Crickets	X	X
Superfamily Gryllacrocoidea	Camel crickets		X
Unidentified	–		
Suborder Caelifera	Grasshoppers		X
Isoptera	Termites		
Family Termitidae	Termites	X	X
Blattodea	Roaches		
Unidentified			X
Blaberidae			X
Blattidae		X	X
Blatellidae			X
Polyphagidae		X	
Mantodea	Praying mantises		X
Dermaptera		X	X
Thysanoptera		X	X
Hemiptera			
Unidentified	–	X	X

continued

Appendix 4. continued

PHYLUM/CLASS/Order/Family	**Common name**	***Lampropholis***	
		delicata	***guichenoti***
Suborder Homoptera			
Unidentified			X
Suborder Auchenorrhyncha	Autorrhynchids		
Unidentified	–	X	
Cicadellidae	Leafhoppers		X
Superfamily Fulgoroidea	Planthoppers		
Fulgoridae	Planthoppers		X
Dictyopharidae	Planthoppers		X
Delphacidae	Leafhoppers		X
Peloridiidae	Moss bugs		X
Superfamily Aphidoidea			
Aphididae	Aphids		X
Superfamily Aleyrodoidea			
Aleyrodidae	Whiteflies		X
Superfamily Phylloidea	Jumping plant lice		
Unidentified			X
Superfamily Coccoidea			
Pseudococcidae	Mealybugs	X	
Suborder Heteroptera	True bugs		
Unidentified	–	X	X
Superfamily Lygaeoidea			
Lygaeidae	Seed bugs		X
Superfamily Pyrrhocoroidea			
Pyrrhocoridae	Red bugs		X
Superfamily Reduvioidea			
Reduviidae	Assassin bugs		
Superfamily Tingoidea			X
Psocoptera	Bark lice		
Suborder Psocomorpha	Bark lice	X	
Coleoptera	Beetles		
Unidentified adults	–	X	X
Unidentified larvae	–	X	X
Unidentified pupae	–		X
Suborder Polyphaga	Polyphagans		
Unidentified	–	X	
Dascilolidae	Plant beetles	X	X
Chrysomelidae	Leaf beetles		X
Curculionidae	Weevils		X
Buprestidae	Jewel beetles		X
Byrrhidae	Pill beetles		
Lycidae	Net-winged beetles		X
Nosodendridae	Wound-tree beetles		X
Staphylinidae	Rove beetles		

PHYLUM/CLASS/Order/Family	Common name	*Lampropholis*	
		delicata	***guichenoti***
Suborder Adephaga	Adephagans		
Carabidae	Ground beetles		X
Trichoptera	Caddisflies		X
Hydroptilidae	Purse-case caddisflies		X
Mecoptera	Scorpionflies		X
Lepidoptera	Lepidopterans		
Unidentified adults	Moths	X	X
Unidentified larvae	Caterpillars	X	X
Hepialidae	Ghost moths	X	X
Geometridae	Geometer moths	X	X
Diptera	Flies		
Unidentified	–	X	X
Suborder Nematocera	Threadhorns		
Culicidae	Mosquitoes		X
Sciaridae	Dark-winged fungus-gnats	X	X
Bibionidae	March flies		X
Cecidomylidae	Gall midges		X
Ceratopogonidae	Biting midges		X
Mycetophilidae	Fungus-gnats		X
Simuliidae	Blackflies		X
Tipulidae	Craneflies		X
Suborder Brachycera	Brachycerans		
Superfamily Astineoinea			
Xenasteidae	Xenasteids		X
Superfamily Drosophiloidea			X
Chloropidae	Grass flies		X
Superfamily Empidoidea			
Dolichopodidae	Long-legged flies		X
Empididae	Dance flies		X
Superfamily Muscoidea			
Unidentified			X
Calliphoridae	Blowflies	X	X
Superfamily Phoroidea			
Phoridae	Hump-backed flies	X	X
Superfamily Sciomyzoidea			
Sciomyzidae			X
Chamaemylidae	Silver flies		X
Lauxaniidae	Lauxaniids		X
Superfamily Tabanoidea			
Unidentified			X
Stratiomyidae			X
Superfamily Tephritoidea			

continued

Appendix 4. continued

PHYLUM/CLASS/Order/Family	Common name	*Lampropholis* *delicata*	*Lampropholis* *guichenoti*
Tephritidae			X
Neuroptera	Lacewings		
Unidentified adults	Lacewings	X	X
Unidentified larvae	Ant lions		X
Hymenoptera	Hymenopterans		
Unidentified adults	–	X	X
Unidentified larvae	–		X
Suborder Symphyta	Symphytans		
Unidentified	–		X
Superfamily Tenthredinoidea	Tenthredinoidians		X
Unidentified	–		
Tenthredinidae	Sawflies		X
Suborder Apocrita	Apocritans		
Unidentified	–	X	X
Superfamily Proctorupoidea	Proctorupoideans		
Unidentified			X
Diapriidae	Diapriids	X	X
Scelionidae			X
Superfamily Platygastroidea	Platygastroideans		
Platygastridae		X	
Superfamily Chalcidoidea	Chalcid wasps		
Unidentified			X
Eurytomidae	Chalcid wasps	X	
Encyrtidae			
Superfamily Ichneumonoidea			X
Braconidae	Braconids		X
Ichneumonidae	Ichneumon wasps		X
Superfamily Cynipoidea	Cynipoideans		
Cynipidae	Gall wasps		X
Superfamily Megalyroidea	Megalyroideans		X
Megalyridae	Megalyrids		X
Superfamily Vespoidea	Wasps and ants		
Unidentified		X	X
Superfamily Formicoidea	Ants		
Formicidae	Ants	X	X
TOTAL TAXA		**40**	**115**

[1]Listed by Crome (1981) as a 'freshwater crustacean' but later questionably identified as the landhopper, *Talitris* sp. It possibly was misidentified and may have been an isopod.

REFERENCES

Belmont CM (1977) Resource partitioning and habitat selection in the ground skinks. BSc (Hons) thesis. University of New England, Armidale, NSW.

Brown GW (1983) Comparative feeding ecology of South-Eastern Australian scincids. PhD Dissertation. La Trobe University, Melbourne.

Brown GW (1991) Ecological feeding analysis of south-eastern Australian scincids (Reptilia, Lacertilia). *Australian Journal of Zoology* **39**(1), 9–29. doi:10.1071/ZO9910009

Crome B (1981) The diet of some ground-layer lizards in three woodlands of the New England Tableland of Australia. *Herpetofauna* **13**(1), 4–11.

Lunney D, Ashby E, Grigg J, O'Connell M (1989) Diets of scincid lizards *Lampropholis guichenoti* (Dumèril & Bibron) and *L. delicata* (De Vis) in Mumbulla State Forest on the South Coast of New South Wales. *Australian Wildlife Research* **16**(3), 307–312. doi:10.1071/WR9890307

Rose AB (1974) Gut contents of some amphibians and reptiles. *Herpetofauna* **7**, 4–8.

Simbotwe MP (1985) Comparative ecology and demography of two populations of Australian reptiles (Lacertilia, Scincidae). PhD Dissertation. University of New England, Armidale, NSW.

Warner FJ (1995) A comparison of the dietary contents of two sympatric Scincid lizards, *Lampropholis delicata* (de Vis, 1888), and *Lampropholis guichenoti* (Duméril and Bibron, 1839) (Lacertilia, Scincidae). BAppSc (Hons) thesis. University of Western Sydney – Hawkesbury. Richmond.

Webb GA (1983) Diet in a herpetofaunal community on the Hawkesbury Sandstone Formation in the Sydney area. *Herpetofauna* **14**, 87–91.

Index

Note: **Bold** page numbers refer to illustrations.

9781041014454